£10.99

PENGUIN REFERENCE

The Penguin D

David Nelson, born in 1938, was educated at Calday Grange Grammar School, Cheshire, and Christ's College, Cambridge. After postgraduate studies in mathematical logic at Cambridge and Bristol Universities, he entered the teaching profession. He has published papers in mathematical journals and is a Fellow of the Institute of Mathematics and its Applications.

He was a lecturer at the University of Manchester from 1981 to 2001, specializing in mathematical cognition and the history of mathematics. He is now an Honorary Fellow. His publications include *Extensions of Calculus*, *Multicultural Mathematics* and *The Penguin Dictionary of Statistics*. He has lectured in Poland, Italy, France and the USA, and in 1995, at the invitation of the State Education Committee of the People's Republic of China, organized and led a three-week National Seminar on Mathematics Education to initiate reform of the Chinese national mathematics curriculum.

A former member of the National Youth Orchestra, his main recreations are playing the flute and the piano. He also enjoys gardening and golf. His wife, Gillian, is a writer and they have three children.

FOURTH EDITION

The Penguin Dictionary of
MATHEMATICS

Edited by David Nelson

PENGUIN BOOKS

PENGUIN BOOKS

Published by the Penguin Group
Penguin Books Ltd, 80 Strand, London WC2R 0RL, England
Penguin Group (USA) Inc., 375 Hudson Street, New York, New York 10014, USA
Penguin Group (Canada), 90 Eglinton Avenue East, Suite 700, Toronto, Ontario, Canada M4P 2Y3
(a division of Pearson Penguin Canada Inc.)
Penguin Ireland, 25 St Stephen's Green, Dublin 2, Ireland
(a division of Penguin Books Ltd)
Penguin Group (Australia), 250 Camberwell Road, Camberwell,
Victoria 3124, Australia (a division of Pearson Australia Group Pty Ltd)
Penguin Books India Pvt Ltd, 11 Community Centre,
Panchsheel Park, New Delhi – 110 017, India
Penguin Group (NZ), 67 Apollo Drive, Rosedale, North Shore 0632,
New Zealand (a division of Pearson New Zealand Ltd)
Penguin Books (South Africa) (Pty) Ltd, 24 Sturdee Avenue,
Rosebank, Johannesburg 2196, South Africa

Penguin Books Ltd, Registered Offices: 80 Strand, London WC2R 0RL, England

www.penguin.com

First published 1989
Second edition 1998
Third edition 2003
Fourth edition 2008
1

Copyright © Penguin Books Ltd, 1989, 1998, 2003, 2008
All rights reserved

Set in 8/9.75pt Monotype Times Roman
Typeset by Data Standards Ltd, Frome, Somerset
Printed in England by Clays Ltd, St Ives plc

ISBN: 978-0-141-03023-4

www.greenpenguin.co.uk

Mixed Sources
Product group from well-managed
forests and other controlled sources
www.fsc.org Cert no. SA-COC-1592
© 1996 Forest Stewardship Council

Penguin Books is committed to a sustainable future
for our business, our readers and our planet.
The book in your hands is made from paper
certified by the Forest Stewardship Council.

SPECIALIST CONTRIBUTORS TO THE SECOND, THIRD, AND FOURTH EDITIONS

DEREK GJERTSEN *University of Liverpool*

NICK HIGHAM *University of Manchester*

TERENCE JACKSON *Open University*

MARK POLLICOTT *University of Warwick*

ELMER REES *University of Bristol*

PETER SPRENT *University of Dundee*

Derek Gjertsen was a Lecturer in Philosophy at the University of Ghana and then an Honorary Research Fellow at the University of Liverpool. He died in 2008.

Nick Higham is Richardson Professor of Applied Mathematics at the University of Manchester and a Fellow of the Royal Society.

Terence Jackson was a Senior Lecturer in Mathematics at the University of York and is now an Associate Lecturer at the Open University.

Mark Pollicott is a Professor of Mathematics at the University of Warwick.

Elmer Rees is Director of the Heilbronn Institute for Mathematical Research at the University of Bristol and Emeritus Professor of Mathematics, University of Edinburgh.

Peter Sprent is Emeritus Professor of Mathematics (Statistics), University of Dundee.

Preface to the Fourth Edition

The Penguin Dictionary of Mathematics aims to provide school and university students with concise explanations of mathematical terms. It tries to cover all the branches of mathematics, both pure and applied, and to include entries and examples that will be helpful to scientists and others who use mathematics in their work. It is hoped that it will also be a useful reference source for non-specialists. Terms used in computer science are not included unless they are of particular mathematical interest.

The dictionary now contains over 3750 headwords, including over 200 biographies of important mathematicians. Chinese names are given in both the modern Pinyin and the older Wade-Giles form (e.g. Beijing/Peking). Diagrams are provided where they help with the understanding of a term.

There is a network of cross-references. Some entries simply refer the reader to another entry. This may indicate that the terms are synonyms, as with '**prime pair** *See* twin primes'. Alternatively it may indicate that the first term is more conveniently discussed or defined within the entry for the second term, in which case the first term is printed in italics within the entry for the second term, as with '**bisector** *See* bisect'. An asterisk placed before a term, as in '*prime', indicates that this term has its own entry in the dictionary which will provide additional information. Reference is also made to the tables in the Appendix. One of these, Table 7, is designed to help the reader find entries where common signs or symbols such as $\neq$ or Σ are explained.

The first edition of the dictionary had ten specialist contributors: Jane Farrill Southern, George Galfalvi, Derek Gjertsen, Valerie Illingworth, Alan Isaacs, Terence Jackson, Richard Maunder, Margaret Preece, Peter Sprent and Ian Stewart; and it was co-edited by John Daintith and the present editor. Each subsequent edition has had the team of six specialist contributors listed on the previous page, and been edited by the present editor. The preparation of each of these editions has provided an opportunity to revise, update, and expand the work in accordance with its overall aims, taking into account the many helpful comments and suggestions made by reviewers and correspondents. In particular, for this fourth edition, increased coverage is given to the area of coding theory.

The editor is deeply grateful to the present team of specialist contributors, three of whom have worked on every edition, for their cooperative dedication to the project. As well as cheerfully drafting and redrafting new or revised entries they have also commented helpfully on the editor's own contributions. He is also thankful for help, advice, and comments from Francis Coghlan, Philip Davis, Raymond Lickorish, James McKee, Peter Milligan, Peter Neumann and Jeff Paris. Thanks are also due to two former editors at Penguin, Donald McFarlan and Ravi Mirchandani, for their support of the first two editions, and to David Duguid, Caroline Pretty, Ellie Smith and Ruth Stimson for seeing the four books through the press. Finally, we are greatly indebted to our copy-editor, John Woodruff, whose expertise and unwavering tenacity have improved every edition.

David Nelson, 2008

Contents

A_n Symbol for the *alternating group for a set of n elements.

Abel, Niels Henrik (1802–29) Norwegian mathematician noted for his proof (1824) that the general quintic equation is unsolvable algebraically. Also important were his work in the field of elliptic and transcendental functions, and on the convergence of infinite series, and his publication of the first rigorous proof of the binomial theorem.

Abelian group A *set associated with a *binary operation (usually denoted by $+$) that forms a *group and also satisfies the commutative law $x + y = y + x$. Examples of such groups are the set of integers with the operation of addition, and the set of integers modulo n with the operation of addition modulo n, where n ($\geqslant 1$) is itself an integer.

Abelian groups are of central importance in abstract algebra and other branches of modern mathematics, notably algebraic topology, where they provide a starting point for homology and cohomology theory. They are named after Niels Abel, although he did not make explicit use of the concept.

Abel Prize A major prize awarded annually by the Norwegian Government for achievement in mathematical research, and regarded as having the same status as a Nobel Prize. It was first awarded in 2003 to the French mathematician J.-P. Serre for his fundamental contributions to several branches of mathematics, especially number theory and algebraic geometry.

Abel's test A test for *convergence of a *series. Let Σa_n be a convergent series. If the numbers b_n constitute a positive decreasing sequence (i.e. $b_1 \geqslant b_2 \geqslant \cdots > 0$) then the infinite series

$$a_1 b_1 + a_2 b_2 + \cdots + a_n b_n + \cdots$$

converges. This test can also be used to determine whether a functional series has *uniform convergence. *See also* Dirichlet's test.

abridged multiplication Multiplication to give a product of a required accuracy, in which digits that do not affect the accuracy are dropped in each part of the multiplication. For example, if 5.6982 is to be multiplied by 23, the full multiplication would be $(5.6982 \times 3) + (5.6982 \times 20) = 17.0946 + 113.9640 = 131.0586$. To two decimal places the result is 131.06. Under abridged multiplication in which the result is required to two decimal places, only the third decimal place is needed in each part. So (5.6982×3) is abridged to 17.094 and (5.6982×20) to 113.964. The product is 131.058, which to two decimal places is 131.06.

abscissa (*plural* **abscissae**) The x-coordinate, measured parallel to the x-axis in a *Cartesian coordinate system. *Compare* ordinate.

absolute error The magnitude (ignoring sign) of the deviation of an observation from its true or predicted value. *See* error; relative error.

absolute frequency *See* frequency.

absolutely convergent series *See* convergent series.

absolutely normal number *See* normal number.

absolute maximum *or* **minimum** *See* turning point.

absolute number A number that has a single value; a number represented by figures – for example, 2, $\sqrt{5}$, $\frac{2}{3}$, 1.976 – as distinguished from a number represented by a letter or other symbol, which might take more than one value.

absolute term A constant term in an expression; a term that does not contain a variable.

absolute value 1. A positive number that has the same magnitude as a given number. Thus, the absolute value of 6 is 6, and the absolute value of −6 is 6. The absolute value of a number a is written using the notation $|a|$.
2. (of a complex number) *See* modulus.
3. The length of a *vector. For a vector $x\mathbf{i} + y\mathbf{j} + z\mathbf{k}$, the absolute value is given by

$$\sqrt{(x^2 + y^2 + z^2)}$$

The absolute value is written as

$$|x\mathbf{i} + y\mathbf{j} + z\mathbf{k}|$$

absorbing barrier *See* random walk.

absorbing state *See* Markov chain.

absorption laws The two laws

$$x \cap (x \cup y) = x \quad \text{and} \quad x \cup (x \cap y) = x$$

See Boolean algebra.

abstract algebra The theory of algebraic structures seen from the modern viewpoint as sets equipped with various operations, assumed to satisfy some specified system of axiomatic laws. In abstract algebra it is the consequences of these laws, rather than the specific objects that make up the set, that are emphasized. The commonest types of structure involved are the *group, the *ring, and the *field, but there are many others. *See* axiom.

abstraction 1. The process of considering certain features of objects while discounting other features that are not relevant. Abstraction is the basis of classification. It is a procedure that results in the formation of a set whose members have a certain property. Such a set is often denoted by $\{x: P(x)\}$, where P is the property that members of the set must satisfy. For example, $\{x: x \text{ is a man}\}$ denotes the set that includes all men and only men.
2. *See* axiom of abstraction.

abstract space A set of entities, together with a set of *axioms for operations on and relationships between these entities. Examples are *metric spaces, *topological spaces, and *vector spaces.

abundant number *See* perfect number.

acceleration Symbol: **a**. The rate of change of *velocity with respect to time, expressed in metres per second per second (m s^{-2}) or similar units. The rate of decrease of velocity with time, i.e. 'negative acceleration', is called *deceleration*. The *average* acceleration during some interval is equal to the change of velocity during this interval divided by the elapsed time. If this time interval is made to approach zero, then the average acceleration approaches the *instantaneous* acceleration.

Thus, when a point or particle moves in space its acceleration is the first derivative of the velocity **v** (and the second derivative of the position vector **r**) with respect to time:

$$\mathbf{a} = \frac{d\mathbf{v}}{dt} = \frac{d^2\mathbf{r}}{dt^2}$$

$$= \frac{d^2x}{dt^2}\mathbf{i} + \frac{d^2y}{dt^2}\mathbf{j} + \frac{d^2z}{dt^2}\mathbf{k}$$

where $\mathbf{i}$, $\mathbf{j}$, and $\mathbf{k}$ are unit vectors.

Since velocity is a *vector quantity, so too is acceleration. The velocity and acceleration of a point or particle moving along a straight line will both be directed along the line, so that $\mathbf{v} = (ds/dt)\mathbf{i}$ and $\mathbf{a} = (d^2s/dt^2)\mathbf{i}$, where s is the distance from some origin.

The acceleration of a point which moves in a plane curved path is conveniently described by two components. One component is directed along the tangent to the curve and is equal in magnitude to the rate of change of speed at that point, dv/dt; this *tangential component* is zero for uniform circular motion. The second component is normal to the tangent, directed inwards towards the centre of *curvature; this *centripetal* (or *normal*) *component* has magnitude v^2/ρ, where ρ is the radius of curvature. The resultant is given by the vector sum of the components.

Alternatively, if the moving point has *polar coordinates (r, θ), the acceleration can be described by two perpendicular components: a *radial component* of $\ddot{r} - r\dot{\theta}^2$ (directed away from the origin) and a *transverse component* of $r\ddot{\theta} + 2\dot{r}\dot{\theta}$ (anticlockwise), where $\dot{r}$, $\dot{\theta}$, $\ddot{r}$, and $\ddot{\theta}$ represent first and second derivatives with respect to time.

acceleration of free fall (acceleration due to gravity) Symbol: g. The acceleration with which an object falls freely to earth (or to another specified celestial body), unimpeded by air resistance or other disturbing forces. It is mainly a result of the gravitational attraction of the body. If the earth is assumed to be an isotropic solid sphere, then the acceleration is directed towards the earth's centre, and its value can be obtained from Newton's law of *gravitation. In practice, the acceleration is to the earth's surface; the standard value is 9.806 65 metres per second per second, but the magnitude varies slightly with locality, owing mainly to the non-spherical shape of the earth and also to geological variations. At the poles and the equator it is $9.8321\,\mathrm{m\,s^{-2}}$ and $9.7799\,\mathrm{m\,s^{-2}}$, respectively; in the UK it varies from 9.81 to $9.82\,\mathrm{m\,s^{-2}}$.

accent *See* prime symbol.

acceptance region *See* hypothesis testing.

acceptance sampling A *quality control procedure in which a sample is taken from a batch of items and a decision to accept or reject the batch is based on the number of defectives in that sample.

In practice, a producer will not want a batch with a small number of defectives to be rejected, and a consumer (or buyer) will not want to accept a batch with a large number of defectives. Thus the choice of d_{max}, the maximum number of defectives, d, in the sample for which a batch is acceptable, is made first to ensure that there is a small probability, α, of a buyer rejecting a batch having a very small number of defectives which would be acceptable to that buyer or consumer. This probability is called the *producer's risk*. Second, the choice of d_{max} is made also so that there is a small probability, β, of a buyer accepting a batch with a number of defectives which that buyer would regard as unacceptable. This probability is called the *consumer's risk*.

By appropriate statistical arguments it can be shown that if the sample size is 200, and the rule is to accept batches for which $d \leqslant 3$, then the probability of the buyer rejecting batches with 0.5 percent or less defective is $\alpha = 0.019$ (producer's risk), and the probability of the buyer accepting batches with 5 percent or more defective is $\beta = 0.009$ (consumer's risk). Economic factors such as production costs, the cost of rectifying (where possible) faulty items, or scrapping or selling rejected batches at a lower price are taken into account in determining sampling schemes with acceptable characteristics. The probabilities α and β are equivalent to the probabilities of *errors of the first* and *second kind* in *hypothesis testing.

For a given sampling procedure, a graph showing the probability of accepting a batch as a function of the proportion of defectives in the batch is called the *operating characteristic curve*.

accumulation factor *See* interest.

accumulation point *See* limit point.

acnode *See* isolated point.

acoustical property The *focal property of a conic. *See* ellipse; hyperbola; parabola.

action 1. A quantity in *dynamics, defined by the line integral

$$\sum_i \int_A^B p_i \, dq_i$$

where q_i are the *generalized coordinates of the system and p_i are the corresponding generalized *momenta for a given segment, from point A to point B, on the trajectory of the system. This is equivalent to twice the mean kinetic energy of the system over a given time interval multiplied by the time interval. *See also* least action, principle of. **2.** The force applied to a body, producing an equal but opposite *reaction. *See* Newton's laws of motion; least action, principle of. **3.** (of a group on a set) If G is a *group with identity e and X is a set, then the action of G on X is a *map (often written as juxtaposition) whose domain is the Cartesian product $G \times X$ and whose range is X, i.e. given $g \in G$ and $x \in X$, the map produces an element of X denoted by gx. The map satisfies

$$ex = x \quad \text{for every } x \in X$$

and

$$g_1(g_2 x) = (g_1 g_2)x \quad \text{for every } x \in X$$
$$\text{and } g_1, g_2 \in G$$

The group is said to *act on* the set. For example, the n-element *cyclic group whose elements are $e, a, a^2, \ldots, a^{n-1}$ acts on the vertices of a *regular polygon by the map for which ex is x for each vertex x, and $a^k x$ is the vertex obtained when x is rotated through $2\pi k/n$ radians about the centre of the polygon.

action at a distance The concept of action being initiated or transmitted without direct contact of the interacting entities. An early explanation involved the existence of aethers, which were thought to be weightless fluids pervading matter and allowing optical, electromagnetic, or heat disturbances to be propagated. Since the late 18th century the idea has been developed of a *field of force surrounding and under the influence of some physical agency, such as charge or mass. Thus a mass affects the space around it, producing a gravitational field. Another mass placed in this field of force interacts with the field and experiences a force. The remote effect of one body on another is thereby explained by a local interaction. Another model for such interactions is that of exchange (absorption and emission) of virtual particles.

acute angle An angle between $0°$ and $90°$.

acute triangle A triangle that has all three interior angles less than $90°$.

acyclic Not cyclic; having no cycles. For example, an acyclic *graph is one in which there are no *paths (or directed paths in a directed graph) that start and end at the same vertex.

adaptive quadrature A *numerical integration algorithm, designed for use on a computer, in which the points at which the integrand f is evaluated are chosen adaptively, based on the behaviour of the function. If f changes rapidly around a point x in the interval of integration, then the algorithm will choose small subintervals near x, but if f changes slowly, larger subintervals will be chosen. There are many adaptive quadrature algorithms, varying in the basic numerical integration rule used and the strategy for determining the subintervals.

addend One of the numbers combined in forming a sum. *See* addition.

addition A mathematical operation performed on two numbers (*addends*) to give a third (the *sum*). It can also be regarded as the process of increasing one number (the *addend*) by another (the *augend*). Addition of integers is equivalent to the process of accumulating sets of objects. Addition of fractions is performed by putting each in terms of a common denominator, and adding the numerators. To define addition of irrational numbers, a more formal definition is required (*see* Dedekind cut).

Addition of numbers is both commutative and associative:

$$a + b = b + a$$

$$a + (b + c) = (a + b) + c$$

Complex numbers are added by adding the real and imaginary parts separately:

$$(a + ib) + (c + id) = (a + c) + i(b + d)$$

Similarly, polynomials are added by accumulating terms of the same degree, for example

$$(x^2 + 2x + 3) + (2x^2 + x + 5)$$

$$= (x^2 + 2x^2) + (2x + x) + (3 + 5)$$

$$= 3x^2 + 3x + 8$$

The concept of addition can also be applied to other entities, such as *vectors, *matrices, and *sets.

addition formulae Formulae in plane trigonometry that express a trigonometric function of a sum in terms of trigonometric functions, as follows:

$$\sin(x + y) = \sin x \cos y + \cos x \sin y$$

$$\cos(x + y) = \cos x \cos y - \sin x \sin y$$

$$\tan(x + y) = \frac{\tan x + \tan y}{1 - \tan x \tan y}$$

The *subtraction formulae* are similar, but with the signs reversed:

$$\sin(x - y) = \sin x \cos y - \cos x \sin y$$

$$\cos(x - y) = \cos x \cos y + \sin x \sin y$$

$$\tan(x - y) = \frac{\tan x - \tan y}{1 + \tan x \tan y}$$

The addition and subtraction formulae were originally derived from *Ptolemy's theorem on cyclic quadrilaterals and are sometimes known as *Ptolemy's formulae*. *See also* double-angle formulae; multiple-angle formulae.

additive function A *function f such that $f(x + y) = f(x) + f(y)$, for all x and y in its domain. If

$$f(x + y) \leqslant f(x) + f(y)$$

the function is *subadditive*; if

$$f(x + y) \geqslant f(x) + f(y)$$

it is *superadditive*. *Compare* multiplicative function.

additive group A *group for which the result of combining a and b is written as $a + b$, and the group identity is denoted by 0. In a *ring the term is often used to refer to the group obtained by considering all the elements and just the ring's addition operation.

additive inverse *See* inverse; ring.

ad infinitum Continuing without end. [Latin]

adj *See* adjoint.

adjacency matrix (vertex matrix) A *matrix representing a *graph. If the vertices of the graph are v_1, v_2, ..., the matrix has elements a_{ij}, where a_{ij} equals the number of edges joining vertex v_i to vertex v_j, and $a_{ij} = 0$ if there is no edge. For a directed graph, a_{ij} equals the number of edges directed from vertex v_i to vertex v_j, and $a_{ij} = 0$ if there are no edges.

adjacency matrix of a graph.

adjacent 1. Describing an angle, side, or plane lying next to another angle, side, or plane.
2. Describing a pair of vertices of a *graph joined by an edge.
3. Describing a pair of edges of a *graph with a vertex in common.

adjoined number A number that does not lie in a given *field F but which together with F generates a larger field G, said to be obtained by *adjoining* the new number to F. For example, adjoining the number $\sqrt{2}$ to the field F of rational numbers generates the field G consisting of all numbers $p + q\sqrt{2}$ where p and q are rational.

adjoint (of a matrix) The *transpose of the *matrix formed by taking the *cofactors of the given matrix. It is defined only for square matrices. When divided by the *determinant it yields the *inverse matrix. The adjoint of a square matrix A is denoted by $\text{adj}\,A$. The adjoint is sometimes called the *adjugate*. In quantum mechanics the term *adjoint* is sometimes used for the *Hermitian conjugate.

adjugate *See* adjoint.

AE *Abbreviation for* *almost everywhere.

AES *Abbreviation for* Advanced Encryption Standard. A *cipher developed in 1998 by two Belgian cryptographers, Joan Daemen and Vincent Rijmen, and adopted by the US Government in 2002 as an encryption standard.

affine geometry The study of those properties and types of geometric figure that are unchanged by an *affine transformation. Examples are parallel lines, mid-points of line segments, and ellipses. However, some familiar geometric concepts such as the circle are not relevant to affine geometry, because a circle can be transformed to an ellipse by certain affine transformations.

affine transformation A mapping of a *vector space that is the composition of a *translation and a *linear transformation. An affine transformation of Euclidean space may not conserve angle or length, but it will preserve many geometrical properties.

A map $f: V \to W$ between vector spaces V and W is an affine transformation if $f(sv_1 + tv_2) = sf(v_1) + tf(v_2)$ for all $v_1, v_2 \in V$ and real numbers s and t satisfying $s + t = 1$. If f is affine, then the map L_f defined by $L_f(v) = f(v) - f(0)$ is a linear transformation; f is the composition of the linear transformation L_f and translation by the vector $f(0)$.

affirmation of the consequent The *fallacy of inferring from $A \supset B$ and B that A, or an argument of this form. It is so called since the second premise B is the *consequent of the *conditional statement forming the first premise. *See also* denial of the antecedent.

aggregation The process of collecting terms together in an expression and treating them as a single term. Thus, in $3(6 - 4)$, $6 - 4$ is 'aggregated', as indicated by the brackets, before multiplying by 3. Various forms of brackets are commonly used to indicate this. In addition, a long bar (called a *vinculum*) over the aggregated terms is sometimes employed. The most frequent use of this is in writing square roots. $\sqrt{25 - 9}$ is the square root of the whole expression $25 - 9$ and is equal to $\sqrt{16} = 4$. This can also be written as $\sqrt{(25 - 9)}$. Note that this differs from $\sqrt{25} - 9 \; (= -4)$.

Agnesi, Maria Gaetana (1718–99) Italian mathematician who worked on differential calculus. Her book *Instituzioni analitiche* (1748; *Analytical Institutions*, 1801) contains a discussion of the curve known as the *witch of Agnesi.

Airy functions Special functions first used to describe features of the appearance of a star in a telescope and named after the English astronomer George Biddell Airy (1801–92). They are solutions of the differential equation

$$\mathrm{d}^2 y / \mathrm{d} x^2 = xy$$

The Airy function of the first kind is

$$\mathrm{Ai}(x) = \frac{1}{\pi} \int\limits_0^\infty \cos\left(\frac{t^3}{3} + xt\right) \mathrm{d}t$$

aleph-null Symbol: $\aleph_0$. The *cardinal number of any *set that may be put into *one-to-one correspondence with the natural numbers 1, 2, 3, 4, … . It is the smallest infinite cardinal number and forms the basis of *Cantor's theory of sets. Any set having cardinality aleph-null is said to be *countably infinite* or *countable.

Alexander–Conway polynomial *See* knot polynomial.

Alexander polynomial *See* knot polynomial.

algebra **1.** The branch of mathematics that deals with the general properties of numbers, and generalizations arising therefrom. The name comes from the Arabic *al-jabr w'al-muqabala*, meaning 'restoration and reduction', which first occurs in the works of al-Khwarizmi (*c*.780–*c*.850). In algebra, letters are used to denote arbitrary numbers and to state generally valid properties: for example, the relation

$$(x + y)^2 = x^2 + 2xy + y^2$$

holds for any two numbers x and y. In modern times, the scope has been widened enormously with the development of *abstract algebra and *linear algebra.
2. A *vector space V over a *field F, which has a *binary operation $\circ$ on V (so that $x \circ y \in V$ for all $x, y \in V$) that is *distributive over vector addition, i.e. $x \circ (y + z) = x \circ y + x \circ z$ for vectors x, y, and z; and also $(ax) \circ y = a(x \circ y) = x \circ (ay)$ for all $a \in F$ and $x, y \in V$. A simple example of an algebra is the space of geometrical vectors in three-dimensional space with the *vector product as binary operation.

An *associative algebra* is one in which the binary operation is *associative. A *division algebra* is one in which the binary operation has an *identity element, and every nonzero vector has an *inverse. *See* linear algebra; Frobenius's theorem; Cayley algebra.

algebraic curve A curve defined by a set of *polynomial equations and an example of an *algebraic variety. For example, in the plane with coordinates x, y,

$ax + by + c = 0$ is a line,

$x^2 + y^2 = 4$ is a circle,

quadratic polynomials in x, y, define
    conics, and

$y^2 = x^3 + ax + b$ defines an elliptic curve.

The degree of a plane algebraic curve is the highest degree of the terms in the polynomial that defines it. For some purposes, curves are best considered in the projective plane. *See* Bézout's theorem.

algebraic expression Any algebraic formula obtained by combining letters or other symbols together with the arithmetic operations $+$, $-$, $\times$, $\div$ (and possibly square or higher roots, $\sqrt{}$, $\sqrt[n]{}$). For example, $a^3 + b^3 + (xy + pq)^2$ is an algebraic expression.

algebraic function *See* function.

algebraic geometry *See* algebraic variety; geometry.

algebraic number A number, real or complex, that is the *root of a *polynomial equation with *integer coefficients. Examples of algebraic numbers are -7, $\frac{5}{2}$, $\frac{1}{2}(1+\sqrt{5})$, $3-i$, $\sqrt[3]{6}$, and $\frac{1}{3}(1+i\sqrt{2})$, since they are zeroes of the polynomials $x+7$, $2x-5$, x^2-x-1, $x^2-6x+10$, x^3-6, and $3x^2-2x+1$, respectively. An algebraic number is an *algebraic integer* when the term of highest *degree in its polynomial equation has coefficient 1. Of the above examples, only $\frac{5}{2}$ and $\frac{1}{3}(1+i\sqrt{2})$ are not algebraic integers.

A number that is not an algebraic number is a *transcendental number*. Examples of transcendental numbers are π and e. *See* irrational number.

algebraic operation A rule assigning to elements $x_1, \ldots, x_n$ of a given *set another element $M(x_1, \ldots, x_n)$ of the set. Usually M is chosen to satisfy certain desired properties. In most cases the number of elements n is either 1 or 2. When $n = 2$, M is said to be a *binary operation. For example, the arithmetical operation of addition assigns to any two numbers x and y their sum $x + y$.

algebraic structure A *set equipped with one or more *algebraic operations, usually required to satisfy a system of *axioms. Examples are *groups, *rings, and *fields.

algebraic topology The study of problems in *topology by algebraic methods.

The usual line of attack is to construct a *group $G(X)$ corresponding to each *topological space X, and a *homomorphism $G(f): G(X) \rightarrow G(Y)$ corresponding to each continuous map $f: X \rightarrow Y$ between topological spaces, with the following properties:

(1) If $f: X \rightarrow X$ is the identity map (that is, $f(x) = x$ for all $x \in X$), then $G(f): G(X) \rightarrow G(X)$ is the identity isomorphism.

(2) Given topological spaces X, Y, and Z, and continuous maps $f: X \rightarrow Y$ and $g: Y \rightarrow Z$, then we have that $G(gf) = G(g)G(f): G(X) \rightarrow G(Z)$.

It then follows that, if $f: X \rightarrow Y$ is a *homeomorphism, $G(f): G(X) \rightarrow G(Y)$ is an *isomorphism; it is possible to distinguish between nonhomeomorphic spaces by showing that their corresponding groups are not isomorphic. (The converse must not be assumed to be true: it is perfectly possible for $G(X)$ and $G(Y)$ to be isomorphic even though X and Y are not homeomorphic.)

Various ways of constructing such groups $G(X)$ have been developed, notably the *homology groups and the *homotopy groups. They have proved very fruitful, for example in the topological classification of *manifolds and *knots, and in establishing important *fixed-point theorems.

algebraic variety The *set of all solutions $(x_1, \ldots, x_n)$ of a system of simultaneous *polynomial equations

$$P_1(x_1, \ldots, x_n) = 0$$
$$\vdots$$
$$P_k(x_1, \ldots, x_n) = 0$$

For example, a circle is the set of solutions (x_1, x_2) of the single equation

$$x_1^2 + x_2^2 - r^2 = 0$$

where r is the radius. More generally, an algebraic variety is any set that can be formed by patching together sets of the above type in a particular manner. The study of algebraic varieties, *algebraic geometry*, plays a central role in modern mathematics.

algorithm A mechanical procedure for solving a problem in a finite number of steps (a mechanical procedure is one that requires no ingenuity). An example is the *Euclidean algorithm for finding the highest common factor of two numbers. The term derives from the name of the Arab mathematician *al-Khwarizmi.

al-Haytham, Abu 'Ali al-Hasan ibn (Westernized form **Alhazen**) (*c*.965–1038) Arab scientist, born in present-day Iraq, who is best known for his work in optics. This included detailed measurements of angles of incidence and refraction, and a careful geometrical analysis of the formation of images in spherical and parabolic mirrors. His major work was first published in the West as *Opticae thesaurus* in 1572.

aliasing The phenomenon that sometimes arises when samples taken at regular intervals are out of phase, and so give the wrong impression. A familiar example is that of film clips that appear to show wheels turning backwards.

Alice The name conventionally used for the sender of an encrypted message.

alignment chart *See* nomogram.

aliquot part *See* proper divisor.

al-Khwarizmi, Muhammad ibn Musa (*c*.780–*c*.850) Arab mathematician from Khiva, in present-day Uzbekistan. In his *Al-jam' w'al-tafriq ib hisab al-hind* (Addition and Subtraction in Indian Arithmetic), al-Khwarizmi introduced the Indian system of numerals to the West. He also wrote a treatise on algebra, *Hisab al-jabr w'al-muqabala* (Calculation by Restoration and Reduction); from *al-jabr* comes the word 'algebra'. From al-Khwarizmi's name was derived the term 'algorism' (referring originally to the Hindu–Arabic decimal number system, but later to computation in a wider sense), from which in turn comes 'algorithm'. His arithmetic survives only in a mediaeval Latin translation with the title *Algorithmi de numero indorum* (Calculation with Indian Numbers).

almost everywhere (AE) A property is said to hold almost everywhere if it holds except on a set of zero *measure.

alphabet A set of symbols used to construct *words* (also called *strings*). For example, if the alphabet is $A = \{a, b, x, y\}$ then *xbbay* is a word or string from A. An alphabet can consist of any symbols, e.g. $\{0, 1, 2\}$.

alternant A *determinant in which the element in the ith row and jth column is $f_i(r_j)$, where the f_i are n functions $f_1, f_2, \ldots, f_n$, and the r_j are n quantities $r_1, r_2, \ldots, r_n$. The order of the determinant is n. The *transpose of such a determinant is also called an alternant. An example of a third-order alternant is

$$\begin{vmatrix} 1 & 1 & 1 \\ x & y & z \\ x^2 & y^2 & z^2 \end{vmatrix}$$

alternate angles *See* transversal.

alternate segment theorem *See* circle.

alternating group The *permutation group consisting of all *even permutations of the elements of a given *set. It forms a subgroup of the *symmetric group on the set. If the set has five or more elements, the alternating group on it is a *simple group*.

The alternating group for a set of n elements is denoted by A_n.

alternating series A *series whose terms (a_n) are alternately positive and negative. If each term of an infinite alternating series is numerically less than the one preceding it, and if $a_n \to 0$ as $a_n \to \infty$, then the series is convergent. The alternating series

$$1 - \frac{1}{2} + \frac{1}{3} - \frac{1}{4} + \cdots$$

is therefore convergent.

alternation *See* disjunction.

alternative hypothesis *See* hypothesis testing.

altitude 1. A line segment, or the length of a line segment, giving the height of a polygon, polyhedron, cone, cylinder, or other geometric figure. It is the distance between the bases of the figure (e.g. in a prism) or the distance from the base to the vertex (e.g. in a pyramid).

2. Symbol: h. The angular distance of a point on the *celestial sphere from the horizon taken along a great circle passing through the zenith, the point, and the nadir. Altitude is measured from $0°$ to $90°$ north (taken as positive) or south (taken as negative) of the ecliptic. Sometimes its complement, the *zenith distance, is used. *See* horizontal coordinate system.

ambiguous case A case in the *solution of triangles in which the known values can give two possible solutions. In plane trigonometry the ambiguous case may occur when two sides and a non-included acute angle are known. There may be two possible triangles satisfying these conditions: one acute triangle and one obtuse triangle. If a, b, and A are the given sides and the non-included angle, then the case is ambiguous when $b > a > b \sin A$. The same ambiguity occurs in solving spherical triangles. In spherical trigonometry the case in which two angles and a side opposite one of them are known is also ambiguous.

ambiguous case If a, b, and A are known, then ABC and AB'C are both solutions.

amicable numbers Two numbers such that each is equal to the sum of the *proper divisors of the other. Amicable numbers were first studied by the Pythagoreans.

The smallest pair of such numbers is 220 and 284: 220 has proper divisors 1, 2, 4, 5, 10, 11, 20, 22, 44, 55, and 110, which have a sum of 284; 284 has proper divisors 1, 2, 4, 71, and 142, which have a sum of 220. *See also* perfect number.

Ampère, André-Marie (1775–1836) French mathematician and physicist who in 1827 published the first comprehensive mathematical treatment of the newly discovered interactions between electricity and magnetism, a subject to which he gave the name 'électrodynamique'. In the course of his work he formulated *Ampère's rule* on the relation between the direction of a current and its associated magnetic field, and *Ampère's law* on the strength of the magnetic field induced by an electric current. The unit of electric current is named after him.

ampere Symbol: A. The *SI unit of electric current, equal to the constant current that, when maintained in two parallel rectilinear conductors of infinite length and negligible circular section and placed 1 metre apart in a vacuum, will produce a force of 2×10^{-7} newton per metre between the conductors. [After A.-M. Ampère]

amplitude 1. *See* argument.
2. (of a periodic function or a system undergoing periodic motion) The maximum displacement from a reference level in either a positive or a negative direction. The periodic function can represent a vibration or a wave, or can describe the motion of a point on a pendulum or on a spring balance.
3. The azimuth in a *polar coordinate system.

analysis The branch of mathematics concerned with the use of *limits; for example, in the treatment of infinite series or in *calculus.

analysis of covariance An extension of the *analysis of variance allowing adjustment for *concomitant variables*, which are not influenced by treatments. By using *regression techniques the sensitivity of the analysis is improved by a reduction in the *error mean square. For example, in an experiment to test the effect of several insecticides in reducing a pest on fruit trees, an appropriate concomitant variable might be a measure of the level of infestation on each experimental unit immediately prior to the application of the insecticides. The technique adjusts for the effect of differing initial infestation on the response to the insecticides.

analysis of variance (ANOVA)   (R.A. Fisher, 1921) The partitioning of *variance into two or more components, each associated with a particular source of variation such as treatments, design groupings, or error.

The simplest ANOVA is that for the *one-way classification* in which treatments are allocated at random to experimental units. For example, 3 treatments A, B, and C may be allocated to 13 available units, 5 chosen at random receiving treatment A, 4 treatment B, and 4 treatment C. The sum of squares of deviations from the mean is partitioned into (1) a *between-treatments sum of squares* and (2) an *error* (or *residual*) *sum of squares*. The former is what the sum of squares of deviations from the mean would have been if the response for each unit had equalled the mean response for all units receiving that treatment, and the latter is the sum of squares of deviations of individual responses from the mean for all units receiving that treatment. The between-treatments mean square and *error mean square are obtained by dividing the respective sums of squares by their *degrees of freedom. The ratio of the between-treatments mean square to the error mean square is called the *variance ratio. If the errors are independently normally distributed with mean zero and equal variances, then under the *null hypothesis of no difference between treatment means, the variance ratio has an *F-

distribution. High values of the ratio indicate rejection of the null hypothesis.

For a *randomized block design there is an additional additive component sum of squares corresponding to blocks. The associated *degrees of freedom are also additive. For Latin squares there are additive component sums of squares corresponding to treatments (Latin letters), rows, columns, and error. The method extends to more complicated designs such as *balanced incomplete blocks with greater computational and interpretational complexity.

In the analysis of *factorial experiments the treatment sum of squares may be further partitioned into additive components, often those representing main effects and interactions.

analysis situs   An obsolete term for the branch of mathematics now known as *topology. [Latin: analysis of position]

analytic continuation   A method of extending the domain of definition of an *analytic function. It is based on the fact that an analytic function is completely determined by its values on any open set of its *domain. The method can often lead to a *multiple-valued function, as, for example, for the (complex) logarithmic function.

analytic function (holomorphic function, regular function)   A (single-valued) *function $f(z)$, with a *domain D that is a subset of the *complex plane and a *codomain that is the complex plane, which is defined and differentiable at a point z_0, is said to be analytic (or regular) at $z = z_0$. If the function is analytic at all points of the domain D, it is said to be analytic (or holomorphic or regular) on D. If it is analytic in some circle $|z - z_0| < r$, then in this region the function can be expanded as the Taylor series

$$f(z) = \sum_{n=0}^{\infty} \frac{1}{n!} f^{(n)}(z_0)(z - z_0)^n$$

See also entire function; singular point.

analytic geometry *See* coordinate geometry.

anchor ring *See* torus.

and A truth-functional connective (*see* truth function), often symbolized in a *formal system as &, ∧, or . . Its meaning is given by the *truth table

A	B	A & B
T	T	T
T	F	F
F	T	F
F	F	F

The connective is both *commutative and *associative, and thus obeys the following laws:

$$A \& B \leftrightarrow B \& A$$
$$A \& (B \& C) \leftrightarrow (A \& B) \& C$$

It can also be expressed in terms of *disjunction (∨) and *negation (∼), in accordance with *De Morgan's laws, by the following equivalence:

$$A \& B \leftrightarrow \sim(\sim A \vee \sim B)$$

The use of '&' between two statements indicates that the statements are both true. Thus, in everyday English the sentence 'London is a city and Manchester is a city' asserts that both statements are true. Note, however, that the use of 'and' in everyday English can be somewhat anomalous. It is not, for example, always taken as commutative. The statement 'He fell out of bed and broke his leg' is not equivalent to the statement 'He broke his leg and fell out of bed.' In these examples, 'and' stands for 'and consequently'; in 'He got out of bed and got dressed,' 'and' stands for the weaker connective 'and then'. *See also* conjunction; Boolean algebra.

angle A configuration of two lines (the *sides* or *arms*) meeting at a point (the *vertex*). Often an angle is regarded as the measure of the rotation involved in moving from one initial axis to coincide with another final axis (termed a *direction angle*). If the amount and sense of the rotation are specified, the angle is a *rotation angle*, and is *positive* if measured in an anticlockwise sense and *negative* if in a clockwise sense.

Angles are classified according to their measure (*see* diagram):

Null (or *zero*) *angle*: zero rotation (0°)
Right angle: a quarter of a complete turn (90°)
Flat (or *straight*) *angle*: half a complete turn (180°)
Round angle (or *perigon*): one complete turn (360°)
Acute angle: between 0° and 90°

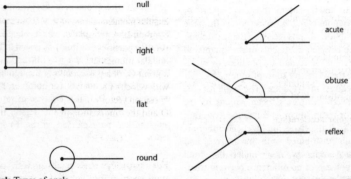

angle Types of angle.

Obtuse angle: between $90°$ and $180°$
Reflex angle: between $180°$ and $360°$

The *angle of elevation* of a point A from another point B is the angle between the line AB and the horizontal plane through B, with A lying above the plane. The *angle of depression* is similarly defined with A lying below the plane.

The angle at point B made by lines AB and CB is denoted by $\angle$ ABC, or $\widehat{ABC}$.

angle bisector theorem The theorem that each *bisector of the angles of a triangle meets the opposite side at a point which divides the side in the ratio of the other two sides (*see* division in a given ratio). Thus, in the diagram, BD : DC = AB : AC.

The bisectors of the *exterior angles of the triangle have a similar property. Each bisector of the exterior angles of a triangle meets the opposite side produced at a point which divides the side externally in the ratio of the other two sides. So in the diagram, BE : EC = AB : AC.

angle bisector theorem

angle of contingence *See* contingence, angle of.

angstrom Symbol: Å. A unit of length equal to 10^{-10} metre. It was formerly used in measurements of wavelength and intramolecular distances, but it is now more usual to use the *SI unit, the nanometre. 1 nanometre = 10 Å. [After A.J. Ångström (1814–74)]

angular acceleration The rate of increase with time of *angular velocity.

angular data *See* directional data.

angular distance *See* distance.

angular frequency Symbol: ω. A measure of the rate of *oscillation of a physical quantity or phenomenon that varies sinusoidally. It is equal to $2\pi v$, where v is the *frequency of the motion, and is usually expressed in radians per second. Trigonometric functions with argument ωt (cos ωt, sin ωt, etc.) occur in equations describing *harmonic motion. Angular frequency is sometimes also called *pulsatance*.

angular measure Measurement of angles. The most common system is *degree* (or *sexagesimal*) *measure* in which one complete turn is divided into 360 degrees, the degree is divided into 60 minutes, and the minute is divided into 60 seconds.

Radian (or *circular*) *measure* of angles is based on an arc of a circle with centre at the vertex of the angle. If r is the radius of the circle and l the length of arc subtending the angle, then the angle is l/r radians. The size of a complete turn ($360°$) is thus the circumference of a circle ($2\pi r$) divided by the radius r, i.e. 2π radians. $180° = \pi$ radians and $90° = \pi/2$ radians. To convert degrees into radians, multiply by $\pi/180$.

A much less common system of angular measure is *centesimal measure*, in which the right angle is divided into 100 degrees, the degree into 100 minutes, and the minute into 100 seconds. In this system, the degree is also called the *grade* (or *grad*), the minute being the *centigrade*.

angular momentum (moment of momentum)
Symbol: **L**. A property of any revolving or rotating particle or system of particles. The angular momentum, relative to (or about) a point O, of a particle of mass m moving with velocity **v** is the *vector product of the *position vector **r** of the particle relative to O and the *momentum **p**:

$$\mathbf{L} = \mathbf{r} \times \mathbf{p} = m(\mathbf{r} \times \mathbf{v})$$

For a particle of mass m moving with constant speed v in a circle of radius r, the angular momentum about the centre is *mvr* in magnitude.

The angular momentum of a system of particles is the sum of the angular momenta of the individual particles. For a *rigid body rotating about an axis with angular velocity **ω**, the angular momentum about a point on the axis is $I\omega$, where I is the *moment of inertia of the body about the axis. The rate of change of this angular momentum is equal to the sum of the *moments of the external forces about the axis. In the absence of external forces, the angular momentum of a system remains constant – no change in configuration can alter the system's angular momentum; there is thus *conservation of angular momentum*. For example, if a rotating cloud of gas in space is contracting under its own gravitation, then it must rotate more rapidly as it contracts so that its angular momentum is conserved.

Angular momentum is important not only in classical mechanics but also in quantum mechanics: an elementary particle, such as an electron, is considered to have *intrinsic angular momentum*, or *spin*, in addition to its *orbital angular momentum* arising from translational motion.

angular speed *See* angular velocity.

angular velocity Symbol: **ω**. A property that is usually associated with *rotational motion: it is a *vector **ω** whose magnitude ω is equal to the number of radians or degrees swept out in unit time; this is known as the *angular speed*. The direction of the vector is that along which a right-handed screw would advance if turned in the same direction as the rotational motion. For a rigid body rotating about an axis, the velocity **v** of any point P relative to any point O on the axis as origin is the *vector product

$$\mathbf{v} = \boldsymbol{\omega} \times \mathbf{r}$$

where **ω** is the instantaneous angular velocity and **r** is the *position vector of P with

respect to the origin O. A particle moving with constant speed v in a circle of radius r has angular speed v/r.

angular velocity: $\mathbf{v} = \boldsymbol{\omega} \times \mathbf{r}$.

anharmonic Neither undergoing nor involving simple *harmonic motion, yet still periodic.

annulus A plane figure bounded by two concentric circles, i.e. that part of a plane lying between two concentric circles. In topology, any *topological space *homeomorphic to such a plane figure is referred to loosely as an annulus.

The area of an annulus is the area of the larger circle (πR^2) minus that of the smaller circle (πr^2), i.e. the area is

$$\pi(R^2 - r^2)$$

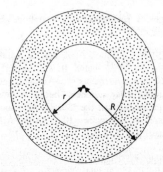

annulus

anomaly The azimuth in a *polar coordinate system.

ANOVA *Acronym for* *analysis of variance. The table setting out the results of an analysis in a standard form is often called an *ANOVA table*.

Ansari–Bradley test *See* homogeneity of variance.

antecedent **1.** The first term in a ratio. Thus, in the ratio $5:7$, 5 is the antecedent (7 is the *consequent*).
2. The part of a *conditional statement that expresses the condition. For example, in the conditional 'if p then q', p stands for the antecedent (q is the *consequent*).

antiderivative An integral. *See* integration.

antidiagonal *See* matrix.

antidifferentiation *See* integration.

anti-Hermitian matrix *See* Hermitian conjugate.

antilogarithm (antilog) A number that has a *logarithm equal to a given number. If $\log_x y = z$, then

$$\text{antilog}_x z = y$$

antinomy A *paradox or *contradiction.

antiparallel vectors *Vectors that have the same or parallel lines of action but point in opposite directions. Two antiparallel vectors have a vector product of zero and a negative scalar product.

antipodal points Two points at opposite ends of a diameter of a *sphere or *ellipsoid.

antiprism A *prismatoid that has two identical bases, one rotated with respect to the other such that the lateral faces are triangles. For example, consider a prism with two square bases – one with its centre directly above that of the other and rotated through 45° with respect to the other. There are eight edges joining the corners of these bases, and the solid has eight lateral triangular faces.

antiprism A symmetrical square antiprism, viewed from above.

antisymmetric matrix *See* symmetric matrix.

antisymmetric relation *See* symmetric relation.

antitrigonometric functions *See* inverse trigonometric functions.

aperiodic Nonperiodic. *See* aperiodic tiling.

aperiodic tiling A set of tiles which tile the plane but can never form a *periodic tiling is said to be *aperiodic*. A tiling with such a set is an *aperiodic tiling*. *See* Penrose tiles.

apex (*plural* apices) The highest point of a geometric figure with respect to some side or plane taken as a base, for example the vertex of a cone or pyramid, or the vertex of a triangle, opposite the base.

Apollonius of Perga (*c.*260–*c.*190 BC) Greek mathematician noted for his *Conics*, of which seven of the original eight books are extant. In about 400 propositions he defined the parabola, hyperbola, and ellipse, and went on to explore some of their more important properties. *See* conic; *see also* problem of Apollonius.

Apollonius' circle (circle of Apollonius) The locus of a point which moves so that its distances from two fixed coplanar points are in a constant ratio not equal to unity is a circle. The locus is sometimes named after Apollonius since the result first appears in his *Plane Loci*.

a posteriori Describing a proposition for which the truth or falsity can be known only through experience. *Compare: a priori*. [Latin: from what comes after]

apothecaries' system 1. A system of units of mass based on the troy ounce (*see* troy system) and formerly used in pharmacy. In this system

20 grains = 1 scruple

3 scruples = 1 drachm (US: dram)

8 drachms = 1 troy ounce (of 480 grains)

12 ounces = 1 troy pound

2. A system of units of fluid volume formerly used in pharmacy. In this system

60 minims = 1 fluid drachm

8 fluid drachms = 1 fluid ounce (of 480 minims)

20 fluid ounces = 1 pint

apothem *See* polygon.

applied mathematics *See* mathematics.

approach *See* limit.

approximation An estimate of a quantity, usually one for which something is known about the accuracy of the estimate. The notation $a \approx b$ (or $a \cong b$) denotes that b is an approximation to a. For example, $\pi \approx 22/7$ gives an approximation to π. Similarly, $\sin x \approx x$ is an approximation to $\sin x$ for small values of x in radians.

approximation theory The branch of *numerical analysis concerned with approximation of a *function $f(x)$ over an interval (a, b) by a simpler function $A(x)$, often a *polynomial. The difference $f(x) - A(x)$ is the *approximation error*; in broad terms the aim of an approximation is to keep this error small over the interval (a, b).

Approximations are often based on known values $y_i = f(x_i)$ at points x_0, x_1, x_2, ..., x_n (*see* interpolation), and $A(x)$ is chosen so as to give zero error at these points. This does not guarantee small errors at intermediate x values, and considerable improvement may be achieved by fitting polynomials P_j (often cubics) piecewise to subintervals of (a, b) so that there is matching not only between $P_j(x_i)$ and $f(x_i)$ at selected points called *nodes*, but also between the first derivatives of these functions (if that for $f(x)$ is known) at the nodes. Polynomials fitted piecewise that pass through the nodes and whose first derivatives agree at the nodes (though not necessarily with $f'(x)$) are called *splines*.

a priori Describing a proposition for which the truth or falsity can be known independently of experience. Logical truths are often held to be *a priori*. *Compare: a posteriori*. [Latin: from what comes before]

apse *See* apsis.

apsidal distance *See* apsis.

apsidal point *See* apsis.

apsis (*plural* **apsides**) A point in an *orbit at which the value r of the radius vector **r** is stationary and at which the orbiting body moves perpendicular to the radius vector. It is sometimes called an *apsidal point* or *apse*. The direction of **r** at such a point is called a *line of apsides* or *apse line*, and r is the *apsidal distance*.

In an elliptical orbit there are two apsides, one nearest and the other farthest from the centre of gravitational attraction. The prefixes *peri-* and *apo-* (or *ap-*) are used to distinguish the two points, as in perihelion (nearest point) and aphelion (farthest point) for a solar orbit.

Arabic numerals (Hindu–Arabic numerals) The numerals 0, 1, 2, 3, etc. *See* number system.

arbitrary origin *See* data coding.

arc 1. A part of a curve (i.e. a segment). The term is also used to mean a complete *open curve. If the circumference of a circle or other closed curve is divided into two unequal parts, the longer of the arcs is the *major arc* (or *long arc*) and the shorter is the *minor arc* (or *short arc*). The two form a pair of *conjugate arcs*.
2. An edge of a directed *graph.
3. *See* path.

arccos, arcsin, arctan, etc. *See* inverse trigonometric functions.

arccosh, arcsinh, arctanh, etc. *See* inverse hyperbolic functions.

Archimedean property *See* Archimedes, axiom of.

Archimedean solid *See* polyhedron.

Archimedean spiral *See* spiral.

Archimedes of Syracuse (*c*.287–212 BC) Greek mathematician who in his *Measurement of a Circle, Quadrature of the Parabola*, and *On Spirals* tackled difficult problems of description and mensuration in plane geometry. Comparable work in solid geometry was displayed in his *On the Sphere and Cylinder* and *On Conoids and Spheroids*. Equally original was Archimedes' *On Floating Bodies*, the first application of mathematics to hydrostatics, and his work on the lever, specific gravity, and the centre of gravity of a variety of bodies. In pure mathematics he succeeded in solving cubic equations, squaring a parabola, and summing higher series as well as, in *The Sand Reckoner*, providing a notation for the representation of very large numbers. In *The Method*, an important work discovered only in 1906, Archimedes described how the use of mechanical principles first led him to consider such propositions as 'the area of any segment of a parabola is 4/3 times that of the triangle with the same base and height'. His mathematical proof, by the method of *exhaustion, came later.

Archimedes, axiom of In the introduction to his *Quadrature of the Parabola*, Archimedes formulated the following principle, since known as the axiom of Archimedes: 'The more by which the greater of two unequal areas exceeds the less can, by being added to itself, be made to exceed any given area.' This is equivalent to saying that, given any positive real numbers a and b such that $a < b$, there exists an integer n such that $na > b$. It is sometimes called the *Archimedean property* of the real numbers.

Archimedes' principle The principle that if a body is wholly or partially submerged in a fluid (liquid or gas) it experiences an upward force (upthrust) equal to the weight of fluid that it displaces.

Archytas of Tarentum (*c*.428–*c*.365 BC) Greek mathematician who, much to the disgust of Pythagoreans and Platonists alike, was one of the first to apply his skills to practical problems in mechanics. He was also the first to offer a solution to the problem of *duplicating the cube.

arc length *See* length.

are Symbol: a. A unit of area in the *metric system, equal to 100 square metres. It is most commonly used in the form of the *hectare. 1 are = 119.60 square yards.

area A measure of a surface. For a rectangle, the area is the product of two adjacent sides. A triangle has an area equal to the product of half its base and its altitude. Areas of other rectilinear figures can be found as a combination of triangles. For plane curved figures, the area is found by *integration. In Cartesian coordinates, it is possible to find the area between a curve $y = f(x)$ and the x-axis. The *element of area is $f(x)\,dx$, and the area between $x = a$ and $x = b$, where $a < b$, is given by the definite integral

$$\int_a^b f(x)\,dx$$

In using this integral, care must be taken if the curve $y = f(x)$ crosses the x-axis since the integral determines the algebraic value of the area, i.e. it gives negative values for areas below the axis. Area can also be determined by using a double integral (*see* multiple integral).

In polar coordinates, the area between the rays $\theta = \theta_1$ and $\theta = \theta_2$, and the curve $r = f(\theta)$, where $\theta_1 < \theta_2$, is given by

$$\frac{1}{2} \int_{\theta_1}^{\theta_2} r^2 \, d\theta$$

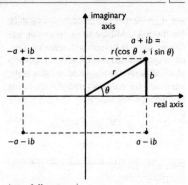

Argand diagram

area sampling A sampling method in which a geographical area is divided into small sub-areas by a grid of squares or rectangles, or by other definable means. Sub-areas are selected at random, and all units in them are included in the sample. For example, in a study of farming, a grid of 5×5 km squares might be used, and all farms lying completely within the selected squares included in the sample. *See* cluster sample; sample survey; sampling theory.

arg *See* argument (of a complex number).

Argand diagram (complex plane) Any plane with a pair of mutually perpendicular axes which is used to represent *complex numbers by identifying the complex number $a + ib$ with the point in the plane whose coordinates are (a, b) (*see* diagram). It is named after Jean Robert Argand (1768–1822), although the method's first exposition, in 1797, was by Caspar Wessel (1745–1818), and the idea can be found earlier in the work of John Wallis. *See also* extended complex plane.

argument 1. (amplitude) (of a complex number) The angle that the line representing a *complex number makes with the positive horizontal axis in an *Argand diagram. If the complex number is written in polar (modulus–argument) form as $r(\cos\theta + i\sin\theta)$, the argument is the angle θ. In the form $a + ib$, the argument is one of the values of $\tan^{-1}(b/a)$. The *principal value* of the argument of a complex number z is the value lying in the range

$$0 \leqslant \arg z < 2\pi$$

Some mathematicians use the range

$$-\pi < \arg z \leqslant \pi$$

2. A set of statements that serve as *premises*, together with a statement as the conclusion, such that the conclusion is supposed to follow from the premises. If the conclusion does so follow, the argument is said to be *valid*; otherwise, it is *invalid*. Note that the premises need not be true; an *indirect proof, for example, uses a false premise. *See* consequence; deduction; logic; logical form; syllogism; valid.
3. The *independent variable of a *function.
4. The *azimuth in a *polar coordinate system.

argumentum ad hominem *See* fallacy.

Aristarchus of Samos (*c*.310–*c*.250 BC) Greek mathematician and astronomer best known for his claim, in a lost work, that the sun and not the earth lies at the centre of the cosmos. He regarded astronomy as a mathematical rather than a descriptive science. In an extant work, *On the Sizes and Distances of the Sun and Moon*, Aristarchus used simple trigonometrical arguments to calculate that the sun is 18–20 times as far from the earth as the moon is. Although his argument was essentially sound, he relied upon inaccurate observational data and consequently reached inaccurate conclusions; the same applies to his estimates of the sizes of the sun and the moon.

arithmetic Computation using numbers and simple operations such as addition, subtraction, multiplication, and division.

arithmetic function A *function whose domain is the *natural numbers and whose range is a subset of the *complex numbers. An arithmetic function f is *multiplicative* if $f(ab) = f(a)f(b)$ whenever a and b are *relatively prime. It is *completely multiplicative* if $f(ab) = f(a)f(b)$ for every a and b. For example, *Euler's phi function is an arithmetic function that is multiplicative but not completely multiplicative. *See also* divisor function; Möbius function; partition function; sigma function.

arithmetic–geometric mean *See* mean.

arithmetic–geometric mean inequality The arithmetic *mean of any number of nonnegative numbers is never less than their geometric mean. Thus for two numbers a and b,

$$\tfrac{1}{2}(a + b) \geqslant \sqrt{ab}$$

and for n numbers $a_1, a_2, \ldots, a_n$,

$$\frac{1}{n}(a_1 + a_2 + \cdots + a_n) \geqslant (a_1 a_2 \cdots a_n)^{1/n}$$

The means are equal if and only if all the numbers are equal.

arithmetic mean *See* mean.

arithmetic modulo n Modular arithmetic; i.e. arithmetic involving *congruences modulo n.

arithmetic progression (arithmetic sequence) A *sequence in which each term (except the first) differs from the previous one by a constant amount, the *common difference*. If the first term is a and the common difference is d, then the progression takes the form

$$a, a + d, a + 2d, a + 3d, \ldots$$

and the nth term is

$$a + (n - 1)d$$

A sum of the terms of such a progression is an *arithmetic series*:

$$a + (a + d) + (a + 2d) + \cdots$$

The sum of the first n terms is given by

$$na + \tfrac{1}{2}n(n - 1)d$$

or alternatively by

$$\tfrac{1}{2}n[2a + (n - 1)d]$$

Compare geometric progression.

arithmetic series *See* arithmetic progression.

arm (side) One of the two lines forming an angle.

array 1. An ordered collection of elements of the same type (integers, real numbers, etc.). A one-dimensional array (e.g. a *vector) is a list of elements a_i, where i is the *index value* (an integer). A two-dimensional array (e.g. a *matrix) has elements a_{ij}, where i is the number of the row and j that of the column. Three- and higher-dimensional arrays are similarly defined. In computing, an array is often called a *subscripted variable*.
2. A display of a set of observations, often in some explicit order such as increasing or decreasing magnitude, or in increasing order of frequency. The observations 3, 7, 2, 9, 1 in increasing order give the array $\{1, 2, 3, 7, 9\}$.

Aryabhata ($c.475$–$c.550$) The first Indian mathematician of any consequence. In the section 'Ganita' (Calculation) of his astronomical treatise *Aryabhatiya* (AD 499), he made the fundamental advance, in finding the lengths of chords of circles, of working with the half-chord rather than the Greek custom of calculating on the basis of the full chord. His work also contains rules for finding π, extracting square roots, and summing arithmetic series.

ASCII *Acronym for* American Standard Code for Information Interchange. A code for converting common typographical symbols – letters, numerals, punctuation, etc. – into a binary code of length 7 for internal use in a computer, enabling 128 different symbols to be coded. For example, the ASCII codes of the letter A, the numeral 6, and the symbol = are 100 0001, 0110110, and 0111101, respectively.

assertion sign The sign $\vdash$ used to indicate assertion (i.e. to say that something is true). It is used between sentences to indicate that the sentence on the right of the sign can be deduced (*see* deduction) from that on the left. For instance, $X \vdash Y$ indicates that Y may be deduced from X. The notation $\vdash_S A$ indicates 'A is a logical *theorem of the system S'. The subscript S can be omitted if it is clear which system is intended.

associate matrix *See* Hermitian conjugate.

association In statistics, two or more variables that are not *independent are sometimes described as exhibiting association. The term is applied most frequently to *categorical variables, but for numerical variables the *correlation coefficient is sometimes described as a measure of association. Association need not imply a causal relationship. *See* contingency table; chi-squared test.

associative Describing a *binary operation ($\circ$) on a set S which satisfies the *associative law*, namely that $a \circ (b \circ c) = (a \circ b) \circ c$ for all elements $a, b, c \in S$, so that the result of combining a number of elements does not depend on how the elements are grouped. The associative law for addition is

$$a + (b + c) = (a + b) + c$$

and that for multiplication is

$$a(bc) = (ab)c$$

Subtraction and division are not associative, because, for example, $5 - (3 - 2) = 4$, whereas $(5 - 3) - 2 = 0$. The composition of mappings (*see* function) obeys the associative law.

associative algebra *See* algebra.

astroid A plane curve that is the *locus of a fixed point P on the circumference of a circle that rolls on the inside of a fixed circle of four times its radius. It has four *cusps and is an example of a *hypocycloid. If the fixed circle has radius a and its centre at the origin, then the curve has parametric equations $x = a \cos^3 \theta$ and $y = a \sin^3 \theta$. In Cartesian coordinates the equation is $x^{2/3} + y^{2/3} = a^{2/3}$.

The curve has the property that the part of the length of its tangent intercepted by the coordinate axes is constant and equal to a. It is thus the *envelope of a line segment of length a which moves so that its end points lie on the coordinate axes. *See also* deltoid.

astroid

astronomical coordinate system A system used for locating the positions of celestial objects in the sky. Astronomical coordinates define positions on the *celestial sphere; as such they do not indicate distances from the earth. The four main systems in use are the *ecliptic, *equatorial, *galactic, and *horizontal coordinate systems. Transformations between systems can be made by using spherical triangles (*see* parallactic angle).

astronomical triangle *See* parallactic angle.

astronomical unit Symbol: AU. A unit of length used in astronomy, equal to the mean distance between the centre of the earth and the centre of the sun. $1 \, \text{AU} = 1.496 \times 10^{11}$ metres or approximately 92.9×10^6 miles.

asymmetric relation *See* symmetric relation.

asymmetry *See* symmetry; skewness.

asymptote A line related to a given curve such that the distance from the line to a point on the curve approaches zero as the distance of the point from an origin increases without bound. In other words, the line gets closer and closer to the curve but does not touch it. For example, the asymptotes of the curve $y = x^2 + 1/x$ are the straight line $x = 0$ and the curve $y = x^2$. *See* hyperbola.

asymptotic 1. Describing a curve that has a line as an *asymptote.
2. Describing two *functions $f(x)$ and $g(x)$ such that $f(x)/g(x)$ tends to 1 as x tends to infinity or to a limit. The function $f(x)$ is said to be *asymptotic* to $g(x)$, and vice versa. This is sometimes written as $f(x) \sim g(x)$, using the symbol '$\sim$'. For example, $x^2 + 1 \sim x^2$ as $x \to \infty$, and $\sin x \sim x$ as $x \to 0$. The *prime number theorem, concerning $\pi(x)$, the number of primes less than or equal to x, can be stated as $\pi(x) \sim x/\ln x$. *See also* order properties.

asymptotic distribution The limiting *distribution of a random variable Z_n as $n \to \infty$. For example, if $X_1, X_2, \ldots X_n$ are n independent observations from *any*

distribution with mean μ and finite variance σ^2, then $Y_n = X_1 + X_2 + \cdots + X_n$ has mean $n\mu$ and variance $n\sigma^2$, both of which tend to infinity (unless $\mu = 0$). However, the *central limit theorem implies that

$$Z_n = \frac{Y_n - n\mu}{\sigma\sqrt{n}}$$

has the standard normal distribution as $n \to \infty$.

asymptotic relative efficiency *See* efficiency.

asymptotic series A *divergent series of the form

$$a_0 + \frac{a_1}{x} + \frac{a_2}{x^2} + \cdots + \frac{a_n}{x^n} + \cdots$$

where the coefficients $a_0, a_1, a_2, \ldots$ are constants. This is an asymptotic representation of a function $f(x)$ if

$$\lim_{|x| \to \infty} x^n[f(x) - s_n(x)] = 0$$

for any value of n, where s_n is the sum of the first $n + 1$ terms of the series.

Atiyah–Singer index theorem *See* Riemann–Roch theorem.

atmosphere Symbol: atm. A unit of pressure, equal to the pressure that will support a column of mercury 760 millimetres high at $0\,°C$ at sea level at latitude $45°$. 1 atm $= 101\,325$ pascals or approximately 14.7 pounds per square inch.

atom An element A of a *lattice such that if $B < A$ then $B = A$ (the null element). For example, in the lattice of all subsets of a given set X, the atoms are the sets $A = \{a\}$ consisting of exactly one element of X, the null element being the empty set $\varnothing$.

atomic sentence A sentence (*see* wff) that contains no logical *constants, such as & (and) or $\vee$ (or). For example, in the *propositional calculus the atomic sentences are those that do not contain any truth-functional connectives. *Compare* compound sentence.

atto- *See* SI units.

attractor *See* chaos.

augend *See* addition.

augmented matrix A *matrix corresponding to a system of inhomogeneous linear equations

$$a_{11}x_1 + \cdots + a_{1n}x_n = b_1$$
$$\vdots$$
$$a_{n1}x_1 + \cdots + a_{nn}x_n = b_n$$

and obtained by adjoining to the matrix of coefficients a_{ij} an additional column formed by the values b_i:

$$\begin{pmatrix} a_{11} & a_{12} & \cdots & a_{1n} & b_1 \\ a_{21} & a_{22} & \cdots & a_{2n} & b_2 \\ \vdots & \vdots & & \vdots & \vdots \\ a_{n1} & a_{n2} & \cdots & a_{nn} & b_n \end{pmatrix}$$

The matrix formed from the coefficients of the unknowns is the *matrix of coefficients* or *coefficient matrix*. The equations are solvable if and only if this has the same *rank as the augmented matrix.

Aut *See* automorphism.

autocorrelation If $x_1, x_2, \ldots, x_n$ is a sequence of n observations ordered in time or space, the product moment *correlation coefficient between all pairs $(x_i, x_{i-1}), i = 2, 3, \ldots, n$ is called the *autocorrelation of lag 1*. More generally, the correlation coefficient between all pairs $(x_i, x_{i-h}), i = h+1, h+2, \ldots, n$ is the *autocorrelation of lag h*.

automorphic function A *function f is said to be *automorphic* with respect to a group of *transformations if:

(1) the function is *analytic except for poles in a *domain D of the complex plane; and

(2) for every transformation T of the group, if z is in D, then $T(z)$ is also in D, and

$$f(T(z)) = T(f(z))$$

automorphism A *bijective mapping from an algebraic structure to itself that preserves all algebraic operations. That is, if M is an operation on the structure A and $\theta: A \rightarrow A$ is an automorphism, then it must satisfy

$$\theta(M(x_1, \ldots, x_n)) = M(\theta(x_1), \ldots, \theta(x_n))$$

For example, if A is the set of complex numbers considered as a field under the usual operations $+$ and $\times$, then complex conjugation $\theta(z) = \bar{z}$ is an automorphism, because

$$\overline{w+z} = \bar{w} + \bar{z} \quad \text{and} \quad \overline{wz} = \bar{w}\,\bar{z}$$

The set of all automorphisms of A forms a *group under composition of mappings, usually denoted by Aut(A). The group Aut(A) may be regarded as describing the symmetries possessed by the structure A. *See also* automorphic function.

autumnal equinox *See* equinoxes.

auxiliary circle One of the two eccentric circles of an *ellipse or *hyperbola. It is used in obtaining the parametric equations for the curve.

auxiliary equation An equation similar in form to a linear *differential equation, enabling the solutions of such an equation to be found. The quadratic equation

$$am^2 + bm + c = 0$$

is the auxiliary equation for the second-order differential equation

$$a\frac{d^2y}{dx^2} + b\frac{dy}{dx} + cy = 0$$

average *See* mean.

avoirdupois A British system of units of mass used in many English-speaking countries. It is based on the *pound (avoirdupois). In the UK the system is:

7000 grains = 1 pound (lb)
16 drams = 1 ounce
16 ounces = 1 pound
14 pounds = 1 stone
2 stones = 1 quarter = 28 lb
4 quarters = 1 hundredweight (cwt)
= 112 lb
20 cwt = 1 ton = 2240 lb

In the USA the system is:

100 pounds = 1 (short) hundredweight
5 cwt = 1 quarter (of 500 lb)
4 quarters = 1 (short) ton (of 2000 lb)

In the UK the system has been replaced by metric units for nearly all purposes, and by *SI units for scientific purposes.

axes *Plural of* axis.

axiom A statement used in the premises of arguments and assumed to be true without proof. In some cases axioms are held to be self-evident, as in *Euclidean geometry, while in others they are assumptions put forward for the sake of argument.

More precisely, an axiom is a *wff that is stipulated rather than proved to be so through the application of rules of inference (*see* proof). The axioms and the rules of inference jointly provide a basis for proving all other theorems. As different sets of axioms may generate the same set of theorems, there may be many *alternative axiomatizations* of the formal system. Axioms are often introduced through axiom *schemata.

Axioms are usually subdivided into *logical* and *nonlogical* axioms. The latter, but not the former, deal with some specific subject matter. For example, *Peano's postulates are nonlogical axioms whose symbols are interpreted with respect to a domain of numbers, while the axioms of the *propositional calculus are logical axioms whose symbols can be interpreted in a variety of ways. The word *postulate* is sometimes used as a synonym for 'axiom'. The postulates of *Euclidean geometry are nonlogical axioms.

axiom of abstraction Given any property, there exists a *set whose members are just those entities possessing that property, i.e.

$$(\exists y)(\forall x)(x \in y \leftrightarrow F(x))$$

First explicitly formulated by Frege in 1893, it was soon demonstrated by Russell to lead to a contradiction. Taking F in the axiom to be the property of not belonging to itself ($x \notin x$) leads to

$$(\exists y)(\forall x)(x \in y \leftrightarrow x \notin x)$$

which leads by simple steps to the contradiction

$$y \in y \,\&\, y \notin y$$

It was to avoid this paradox (*see* Russell's paradox) that Zermelo introduced in 1908 his axiom of separation:

$$(\exists y)(\forall x)(x \in y \leftrightarrow x \in z \,\&\, F(x))$$

in which the existence of the set y is no longer asserted unconditionally. *See also* Zermelo–Fraenkel set theory.

axiom of Archimedes *See* Archimedes, axiom of.

axiom of choice *See* choice, axiom of.

axiom of extensionality The axiom that two *sets are equal if they have exactly the same members, i.e. for sets A and B

$$(\forall x)(x \in A \leftrightarrow x \in B) \rightarrow A = B$$

For example, if $A = \{1, 2, 2, 3, 6, 4\}$ and $B = \{1, 2, 3, 4, 6\}$, then A and B contain exactly the same elements (repetition and order are irrelevant) and therefore the two sets are equal.

axiom of infinity Although earlier mathematicians had attempted to prove the
existence of an infinite set of objects, it
was left to Zermelo in 1908 to appreciate
that the existence of such a set needed to be
assumed axiomatically. He therefore
included in his system the axiom

$$(\exists A)(0 \in A \,\&\, (\forall y)(y \in A \rightarrow \{y\} \in A))$$

which allows the construction of the infinite set of natural numbers.

axis (*plural* **axes**) In general, a reference
line associated with a geometric figure or
an object.

1. (reference axis) A line from which distances or angles are taken in a *coordinate
system or *Argand diagram.

2. (axis of symmetry) A line associated
with a geometric figure such that every
point on one side of the line has a corresponding point on the other side. The axis
bisects at right angles the line segment joining the two points. Axes of figures such as
ellipses, parabolas, hyperbolas, and ellip-
soids are axes of symmetry. *See also* symmetry.

3. (axis of revolution *or* **rotation)** A line
about which a curve or figure is rotated
in forming a *surface of revolution or a
*solid of revolution.

A line about which an object rotates. *See*
moment of inertia.

4. A line joining a vertex of a cone or pyramid to the centre of the base, or joining the
centres of the bases of a frustum, truncated
solid, cylinder, or prism. Such an axis is not
necessarily an axis of symmetry (e.g. if the
solid is oblique).

5. A line along which a *pencil of planes
intersect.

See also Cartesian coordinate system;
helix.

azimuth 1. (amplitude; anomaly; argument)
Symbol: θ. The angle between the polar
axis and the radius vector in a *polar
coordinate system.

2. Symbol: A. The angular distance (measured from $0°$ to $360°$) of a point on the
*celestial sphere from the north point. It
is measured eastwards along the horizon
between the north point and the place at
which a great circle through the zenith and
the point intersects the horizon. *See* horizontal coordinate system.

B

Babbage, Charles (1792–1871) English mathematician best known for his work on the design and manufacture of the mechanical computer. Beginning in the 1820s, Babbage devoted much of his life to the construction of, first, his 'difference engine' and, later, his more ambitious 'analytical engine', which were in theory capable of performing mechanically any mathematical operation. Owing to a number of factors – personal, financial, and technological – Babbage failed to develop the machines as he intended; they did, however, contain in their design a number of essential features used in the modern electronic computer.

backward difference Given *function values $y_i = f(x_i)$, where $x_i = x_0 + ih$, $i = 0, 1, 2, \ldots$, the backward difference ∇y_i is defined by $\nabla y_i = y_i - y_{i-1}$. See finite differences.

backward difference formula See Gregory–Newton interpolation.

balanced incomplete block design An *experimental design in which t treatments are allocated to b blocks, each block containing k units, where $k < t$; blocks are thus 'incomplete' in the sense that there are not enough units to receive all treatments. The allocation is subject to the constraints that each treatment is replicated r times, and every pair of treatments occurs together in the same number (λ) of blocks. The total number of units is $bk = rt$, and $\lambda = r(k-1)/(t-1)$.

For four treatments A, B, C, and D, each replicated three times in six blocks of two units, an appropriate allocation of treatments to blocks would be AB, AC, AD, BC, BD, and CD. For a valid analysis the treatment pairs must be allocated to blocks at random, and within each block the treatments must be allocated to units at random. It is possible to compare treatments after eliminating block effects by an appropriate *analysis of variance.

ball The *n*-ball B^n ($n \geqslant 1$) is the subspace of *n*-dimensional *Euclidean space $\mathbb{R}^n$ of points $(x_1, \ldots, x_n)$ such that $\sqrt{(x_1^2 + \cdots + x_n^2)} \leqslant 1$. It contains the $(n-1)$-sphere S^{n-1} (see sphere) as a subspace. For example, B^2 is a circular disc and B^1 is the closed interval $[-1, 1]$.

ballistics The study of the propulsion, flight, and impact of projectiles. *Interior ballistics* is concerned with the motion of projectiles under propulsive power, e.g. with events occurring up to the instant that a bullet leaves the muzzle of a gun. The rate of chemical combustion of the propellant, the gas pressure behind the projectile, and the velocity imparted to the projectile are important factors. *Exterior ballistics* is concerned with the motion of projectiles that are no longer under propulsive power, i.e. with their motion through the air (or through water, say). Of prime interest is the way in which a projectile is affected by such factors as drag, cross-winds, and the *Coriolis effect (in long-distance flight), the maintenance of a stable trajectory, and the effects of varying initial velocity, angle of projection, etc.

Banach, Stefan (1892–1945) Polish mathematician noted for his work beginning in 1922 on a type of vector space, more general than Hilbert space, and since commonly known as *Banach space. He was also responsible, with Tarski, for the *Banach–Tarski paradox, which implies that any two spheres of different radii can be divided into the same number of congruent disjoint sets.

Banach contraction principle *See* contraction mapping.

Banach space A *complete normed *vector space over the real or complex field. Examples are all *Euclidean spaces (with the usual norm), and the space of all square-integrable real-valued functions. The major concepts of analysis, such as differentiation and integration, may be generalized to Banach spaces, giving the subject of *functional analysis*. This has important applications to the study of partial differential equations and integral equations, and appears to be a natural abstract setting for many general theories of analysis.

Banach–Tarski paradox A paradox based on the existence of sets which are *nonmeasurable*. Consider two sets in 3-dimensional space such as $B^3(1)$ and $B^3(2)$, the solid spheres of radii 1 and 2; they can be broken up into the disjoint union of a finite number of sets $B^3(1) = X_1 \cup X_2 \cup \cdots \cup X_m$ and $B^3(2) = Y_1 \cup Y_2 \cup \cdots \cup Y_m$ that have the remarkable property that, for each k, X_k is isometric to Y_k; that is, each Y_k can be obtained from X_k by a rigid motion (one that preserves all lengths). However, such decompositions cannot be constructed directly; their existence relies on the axiom of *choice. Named after S. Banach and A. Tarski. *See* measure.

band matrix A *matrix whose elements are zero outside a band of diagonals around the leading diagonal. An example of a band matrix is

$$\begin{pmatrix} 1 & 2 & 3 & 0 & 0 \\ 1 & 1 & 2 & 1 & 0 \\ 0 & 2 & 1 & 4 & 1 \\ 0 & 0 & 2 & 1 & 1 \\ 0 & 0 & 0 & 1 & 1 \end{pmatrix}$$

Another example of a band matrix is a tridiagonal matrix (*see* sparse matrix).

bandwidth *See* kernel density estimation; lowess.

bar A *c.g.s. unit of pressure, equal to a pressure of 10^6 dynes per square centimetre. $1\,\text{bar} = 10^5$ pascals or approximately 0.9869 atmosphere. The millibar is still in use for meteorological purposes.

bar chart A graph consisting of vertical bars with heights proportional to the frequencies of some event for each of several categories. The categories may be nominal, such as nationalities of passengers on an aircraft, or ordinal, such as days of the week (*see* diagram). The names *bar diagram* and *bar graph* are also used. *See also* histogram.

bar chart showing the number of items sold on different days.

bar diagram *See* bar chart.

bar graph *See* bar chart.

Barrow, Isaac (1630–77) English mathematician and theologian who published in his *Lectiones geometricae* (1670, Geometrical Lectures) a method of finding tangents similar to that now used in differential calculus. Barrow himself never developed the method – in his book he wrote that it is published in an appendix 'on the advice of a friend'. The friend, Isaac Newton, was later recommended by Barrow as his successor to the Lucasian chair of mathematics at Cambridge.

Bartlett's test *See* homogeneity of variance.

barycentre *See* centre of mass.

barycentric *See* centre of gravity.

barycentric coordinates A set of numbers that represent the position of a point in space relative to a set of fixed points. In three-dimensional space four points are used. p_0, p_1, p_2, and p_3 (p_0 is the point (x_0, y_0, z_0), etc.), and the four points are not coplanar. For any general point p there is a set of numbers $\lambda_0, \lambda_1, \lambda_2$, and λ_3 for which

$$p = \lambda_0 p_0 + \lambda_1 p_1 + \lambda_2 p_2 + \lambda_3 p_3$$

$$\text{and} \quad \lambda_0 + \lambda_1 + \lambda_2 + \lambda_3 = 1$$

The set $\{\lambda_0, \lambda_1, \lambda_2, \lambda_3\}$ consists of the barycentric coordinates of p. If point masses $\lambda_0, \lambda_1, \lambda_2, \lambda_3$ are placed at p_0, p_1, p_2, p_3, then p is the centre of mass of the system. The system can be generalized to n-dimensional space.

base 1. (of a number system) The number represented by the numeral '10' in a positional *number system. Thus, in the decimal system the base (ten) is represented by 10; in a binary system the base (two) is also represented by 10.
2. (of logarithms) The number which, raised to the power of a given *logarithm, produces a given number. Thus, if the logarithm of x to base b (written as $\log_b x$) is y, then $b^y = x$.

3. A line or plane in a geometric figure relative to which the altitude of the figure is measured.

base units A set of *units for dimensionally independent *physical quantities that form the basis of a system of units. There are seven base units in *SI units. *See also* coherent units; derived units; supplementary units.

basis (*plural* **bases**) A *subset of a *vector space that is linearly independent and spans the space. If $\mathbf{x}_1, \ldots, \mathbf{x}_t$ is a basis, then every element of the vector space has a unique representation as a linear combination

$$a_1 \mathbf{x}_1 + \cdots + a_j \mathbf{x}_j + \cdots + a_t \mathbf{x}_t$$

where the a_j are scalars. This permits the introduction of a coordinate system $(a_1, \ldots, a_t)$ on the vector space. Bases are in general not unique. Any two bases for a given vector space must contain the same number of elements: this number is the *dimension* of the vector space and is of fundamental importance.

Baudot code A *code invented by the French telegraph engineer Émile Baudot in the 1870s and which was widely used in telegraphy. Its *code length is 5 and allows transmission of letters and numerals. It superseded Morse code but has now been largely replaced by *ASCII. It is still used in specialist telephone devices for the deaf. *See also* Gray code.

Bayes, Rev. Thomas (1702–61) English mathematician and theologian best known for *An Essay Towards Solving a Problem in the Doctrine of Chances*, published posthumously (1763), which included both the uncontroversial *Bayes' theorem and a contentious postulate that is fundamental to *Bayesian inference. Works published in his lifetime dealt with the logical foundations of mathematics.

Bayes' factor (H. Jeffreys, 1935) In *Bayesian inference, suppose that we are given a set of data D assumed to have arisen under one of only two possible models or hypotheses M_1 or M_2, with prior probabilities $\Pr(M_1)$ and $\Pr(M_2) = 1 - \Pr(M_1)$. The *posterior probabilities are then $\Pr(M_1|D)$ and $\Pr(M_2|D) = 1 - \Pr(M_1|D)$. Applying *Bayes' theorem, it may be shown that

$$\frac{\Pr(M_1|D)}{\Pr(M_2|D)} = B_{12} \frac{\Pr(M_1)}{\Pr(M_2)}$$

where

$$B_{12} = \frac{\Pr(D|M_1)}{\Pr(D|M_2)}$$

is called the Bayes' factor.

This result implies that the posterior *odds on M_1 are obtained by multiplying the prior odds on M_1 by the Bayes factor, which is independent of the prior odds. The notion extends easily to pairwise comparisons of more than two alternative models.

Bayesian inference A method of statistical inference based on *Bayes' theorem. The unknown or unknowns (usually *parameters) to be estimated are assumed to have a *prior probability distribution. Using Bayes' theorem, this is combined with the information from observed data expressed in terms of the *likelihood to form a *posterior probability distribution for the unknown(s). If further data become available, this posterior distribution may be used as a prior distribution for a later analysis.

When the prior distribution is based on empirical data, as in the examples given in the entry for Bayes' theorem, the approach is not controversial. However, there are two other common approaches that involve subjective probabilities, and these are regarded by many as controversial.

In the first of these approaches, the prior probability distribution is based on a personal degree of belief. The precise nature of this distribution influences the posterior probability distribution. Proponents of this approach defend it on the grounds that the influence of any particular prior distribution becomes diluted in the posterior distribution by the impact of the data.

In the second common approach, the choice of a prior distribution is such as to indicate a noncommittal attitude about the unknown(s). Typically, a prior distribution may be in this situation a uniform distribution over a plausible range of values for the unknown(s).

Either approach, especially the latter, often but not always leads to inferences that differ little from those arrived at using a *frequential inference approach.

There are possible analytic or computational difficulties associated with some choices of prior distributions. Some of these difficulties are alleviated by choosing a *conjugate prior distribution*, which is a distribution that gives rise to a posterior distribution belonging to the same family as the prior; for example, both might be a *normal distribution or both a *gamma distribution. Bayesian inference often makes use of the *Gibbs sampler based on *Markov chain *Monte Carlo simulations. *See also* Bayes' factor.

Bayes' theorem (T. Bayes, published posthumously in 1763) A theorem on *conditional probability that evaluates the probability of an event (a cause) conditional upon another event (a consequence) of known probability having taken place. Suppose that $B_1, B_2, \ldots, B_n$ are a mutually exclusive and exhaustive set of events (i.e. a set of non-overlapping events covering the whole *sample space) and an event A is observed. The probability that the event B_j is the causal event giving rise to A, i.e. the probability of B_j conditional upon A, is given by Bayes' theorem, i.e.

$$\Pr(B_j|A) = \frac{\Pr(B_j)\,\Pr(A|B_j)}{\Sigma\Pr(B_i)\,\Pr(A|B_i)}$$

Suppose, for example, that box 1 contains 10 good screws and 2 unslotted screws, and box 2 contains 8 good screws and 4 unslotted screws. If a box is selected at random and a screw chosen from it is found to be unslotted, what is the probability that it came from box 2? If A is the event 'unslotted screw selected', and B_1 and B_2 are the events 'screw is selected from box 1, box 2' respectively, then $\Pr(B_1) = \Pr(B_2) = \frac{1}{2}$, $\Pr(A|B_1) = \frac{1}{6}$, and $\Pr(A|B_2) = \frac{1}{3}$, whence, by Bayes' theorem,

$$\Pr(B_2|A)$$
$$= \frac{\Pr(B_2)\Pr(A|B_2)}{\Pr(B_1)\Pr(A|B_1) + \Pr(B_2)\Pr(A|B_2)}$$
$$= \frac{\frac{1}{2} \cdot \frac{1}{3}}{\frac{1}{2} \cdot \frac{1}{6} + \frac{1}{2} \cdot \frac{1}{3}} = \frac{2}{3}$$

An important application arises in diagnostic testing. Suppose that in a population, 6 in every 1000 people have a disease or condition X. It is known that if a person has X, then there is a 92 percent chance that a blood test will be positive for X and also a 0.5 percent chance that a person without X will test positive. If an individual selected at random tests positive for X, what is the probability that they have X? If A is the event *tests positive for X*, and B_1 and B_2 the events *has X* and *does not have X*, we seek $\Pr(B_1|A)$, given that $\Pr(B_1) = 0.006$, $\Pr(B_2) = 0.994$, $\Pr(A|B_1) = 0.92$, and $\Pr(A|B_2) = 0.005$, whence

$$\Pr(B_1|A) = \frac{0.006 \times 0.92}{0.006 \times 0.92 + 0.994 \times 0.005}$$
$$= 0.526$$

So, before the test the probability of an individual having X is 0.006, but after a positive test the probability that an individual has X is 0.526.

bearing The angle between a course or direction and a northerly direction. The bearing of a point B from another point A (i.e. the bearing of the direction AB) is the angle of clockwise rotation of AB from a north-pointing line AN (*see* diagram). The angle is measured in degrees, and is usually stated in three-digit form (e.g. 045°, 229°).

bearing of B from A.

beats A phenomenon arising when two *waves of slightly different *frequency occur together: there is a slow fluctuation in the *amplitude of the resulting compos-

beats (a) Two waves of slightly different frequency, and (b) the resultant wave, showing beats.

ite wave as the two wavetrains continuously reinforce and then neutralize each other. Beats can be used in tuning musical instruments. Audible surges in volume are produced by playing two notes of approximately equal frequency; these surges are reduced to zero when the frequencies are made equal. The frequency at which the amplitude fluctuates is the *beat frequency*, which is equal to the difference in frequency of the combining waves, $|f_1 - f_2|$. If the two frequencies are close, the resulting beat frequency is low. If the two combining waves have equal amplitude a, the resulting amplitude is approximately given by

$$2a \cos[\pi(f_1 - f_2)t - \tfrac{1}{2}\theta]$$

where θ is the phase difference.

becquerel Symbol: Bq. The *SI unit of activity, equal to the number of atoms of a radioactive substance that disintegrate in 1 second. [After A.H. Becquerel (1852–1908)]

Bede, the Venerable (672–735) English scholar who produced works on the calculation of the date of Easter, finger-counting, the sphere, and division. These writings, in Latin, are probably the first mathematical works known to have been produced in England by an Englishman.

Behrens–Fisher test (W.V. Behrens, 1929; R.A. Fisher, 1937) A test involving two independent samples, which in essence extends the *t-test by relaxing the requirement of equal population variances. The test may be justified by *fiducial inference theory, but is often quoted as a case where fiducial and confidence-interval theory differ. The test is still a subject of controversy and its validity is not universally accepted.

bei function *See* Bessel functions.

bel Symbol: B. A unit for comparing two power levels, equal to the logarithm to the base ten of the ratio of the two powers. It is

most commonly used in the form of the *decibel. [After A.G. Bell (1847–1922)]

bell-shaped curve *See* Cauchy distribution; normal distribution.

Beltrami, Eugenio (1835–99) Italian mathematician who in his *Saggio di interpretazione della geometria non-euclidea* (1868, Studies in the Interpretation of Non-Euclidean Geometry) demonstrated how the various new geometries of Bolyai, Lobachevsky, and Riemann, as well as the traditional geometry of Euclid, can all be mapped on surfaces of constant curvature. Beltrami then succeeded in showing that if any of the new non-Euclidean geometries proved to be inconsistent, so too would be Euclidean geometry.

bending moment The algebraic sum of the *moments of the forces acting on one side of a cross-section of a beam or other structural member.

bend point A point on a curve at which the ordinate is a maximum or a minimum.

ber function *See* Bessel functions.

Bernoulli A family of Swiss mathematicians and physicists. About a dozen members of the family are remembered, the most important being:

Jacques Bernoulli (1654–1705; also known as *James* or *Jakob*) Noted for his work on calculus and probability. In 1690 he was the first to introduce the word 'integral'. Jacques was interested in applying the calculus to the study of curves, in particular the logarithmic spiral and the brachistochrone. The lemniscate of Bernoulli is named after him. He was one of the first to use polar coordinates, in 1691. He also wrote the first book concentrating on probability theory, *Ars conjectandi* (The Art of Conjecture, published posthumously in 1713). This contains an account of the *Bernoulli numbers and *Bernoulli's theorem.

Jean Bernoulli (1667–1748; also known as

John or *Johann*) The brother of Jacques, and also known for his work on the calculus. In 1694 he was the discoverer of L'Hôpital's rule. In 1696 he proposed the brachistochrone problem and, as a consequence, is often referred to as the originator of the calculus of variations. Jean Bernoulli had three sons, all of whom became professors of mathematics, the most prominent being Daniel.

Daniel Bernoulli (1700–1782) The son of Jean Bernoulli, noted for his book *Hydrodynamica* (1738, Hydrodynamics) in which he laid the foundations of the modern discipline of hydrodynamics and introduced *Bernoulli's equation (2). Daniel Bernoulli, like his uncle Jacques, worked on probability.

Bernoulli distribution *See* Bernoulli trial.

Bernoulli numbers Numbers originally introduced by Jacques Bernoulli in a formula for sums of the powers of integers. They are now often defined by taking the expansion of $x/(1 - e^{-x})$, giving

$$1 + \frac{1}{2}x + \frac{1}{6}\frac{x^2}{2!} - \frac{1}{30}\frac{x^4}{4!} + \frac{1}{42}\frac{x^6}{6!} - \cdots$$

The Bernoulli numbers are the coefficients of $x^n/n!$ when n is even, i.e. the values $1/6$, $-1/30$, $1/42$, ...

Bernoulli's equation 1. A first-order *differential equation of the form

$$\frac{dy}{dx} + y\,f(x) = y^n g(x)$$

where $f(x)$ and $g(x)$ are functions of x. It was first solved by Jacques and Jean Bernoulli and by Leibniz.
2. An equation in fluid mechanics:

$$\int \frac{dp}{\rho} + \frac{1}{2}v^2 + V = C$$

where p is the pressure of the fluid, ρ its density, v the velocity along a stream line, V the gravitational potential, and C a constant for a given stream line (*Bernoulli's constant*). The equation, which was first formulated by Daniel Bernoulli in 1738, is a statement of the principle of conservation of energy for nonviscous incompressible fluids.

Bernoulli's theorem A theorem in *probability introduced by Jacques Bernoulli in his book *Ars conjectandi* (1713). If p is the probability of a given event and m is the number of occurrences of the event in n trials, then the probability that, for any $\varepsilon > 0$,

$$|(m/n) - p| < \varepsilon$$

has a limit of 1 as $n \to \infty$.

Bernoulli trial (Jacques Bernoulli, 1713) A *trial or experiment with only two possible outcomes, often called success or failure, with probabilities p and $q = 1 - p$, respectively. A *Bernoulli variable* X takes the value $X = 1$ with probability p and $X = 0$ with probability q. Its distribution is the *Bernoulli distribution* which has *mean p and *variance pq. For example, if scoring either a five or a six when a fair die is cast is a success, denoted by $X = 1$, then $\Pr(X = 1) = \frac{1}{3}$ and $\Pr(X = 0) = \frac{2}{3}$, and X has a Bernoulli distribution with *mean $\frac{1}{3}$ and *variance $\frac{2}{9}$. *See also* binomial distribution; negative binomial distribution.

Bernoulli variable *See* Bernoulli trial.

Berry's paradox A *paradox stated by G.G. Berry in 1906. In general, the larger a number, the more syllables are needed to form English names of the number. Consider 'the least integer not nameable in fewer than nineteen syllables'. This expression appears to name a number (one hundred and eleven thousand, seven hundred and seventy-seven, 111 777), but it is also an expression of eighteen syllables that is itself a name of a number. So the least integer not nameable in fewer than nineteen syllables is nameable by an expression containing fewer than nineteen syllables, which is a contradiction.

Bertrand's postulate The *postulate that for any integer n greater than 3, there is always at least one *prime between n and $2n - 2$. The conjecture was first published in 1845 by the French mathematician Joseph Bertrand (1822–1900), and was proved in 1850 by Tchebyshev.

Bessel, Friedrich Wilhelm (1784–1846) German mathematician and astronomer noted for his introduction in 1824 of *Bessel functions into mathematics. Bessel's interest in them arose from his work on the perturbations observed in planetary motions.

Bessel functions *Functions that arise in the solution of the *wave equation expressed in *cylindrical coordinates and satisfy *Bessel's equation.

Bessel functions of the first kind are denoted by $J_n(z)$. For a nonzero integer n,

$$J_n(z) = \frac{1}{\pi} \int_0^\pi \cos(nt - z \sin t) dt$$

$$= \sum_{r=0}^\infty \frac{(-1)^r}{r! \, \Gamma(n+r+1)} \left(\frac{z}{2}\right)^{n+2r}$$

the series form being valid if n is a positive integer. The *order* of the function is n. The real and imaginary parts of $J_n[x \exp(3\pi i/4)]$, where x is real, are respectively the *ber* and *bei* functions of order n.

Bessel functions of the second kind (also called *Neumann functions*) are simple combinations of Bessel functions (written as $Y_n(z)$). *Bessel functions of the third kind* are called *Hankel functions* and have two forms:

$$H_n^{(1)}(z) = J_n(z) + iY_n(z)$$

$$H_n^{(2)}(z) = J_n(z) - iY_n(z)$$

The functions were originally introduced by Bessel in 1824. They arise in many problems in physics and engineering.

Bessel's equation A second-order *differential equation of the form

$$z^2 \frac{d^2y}{dz^2} + z \frac{dy}{dz} + (z^2 - n^2)y = 0$$

where n is a constant. This is Bessel's equation of order n.

The equation

$$z^2 \frac{d^2y}{dz^2} + z \frac{dy}{dz} - (z^2 + n^2)y = 0$$

is the *modified Bessel equation*. Bessel's equation occurs in many applications in physics and engineering. *See* Bessel functions.

beta distribution A distribution over [0, 1] with *frequency function

$$f(x) = \frac{1}{B(m, n)} x^{m-1}(1 - x)^{n-1}$$

where $B(m, n)$ is the *beta function and m, $n > 0$. The frequency function takes a wide range of shapes for different combinations of values of the parameters m and n, and includes *U-shaped distributions and the *uniform distribution over [0, 1] as special cases.

beta function The *function, denoted by $B(p, q)$, that is the integral of

$$x^{p-1}(1 - x)^{q-1}$$

from 0 to 1. It can be expressed in terms of the *gamma function as

$$B(p, q) = \frac{\Gamma(p)\Gamma(q)}{\Gamma(p + q)}$$

between-treatments sum of squares *See* analysis of variance.

Bézout's theorem Two algebraic plane curves whose degrees are d and e and which have no common part meet in exactly de points in the complex projective plane if the intersection points are counted with an appropriate multiplicity. In the projective plane geometry over any field, two such curves meet in at most *de* points. For example, two conics meet in at most four points. The theorem is named after the French mathematician Étienne Bézout (1730–83).

Bhaskara (1114–*c*.1185) Indian mathematician who published works on arithmetic, *Lilavati* (The Beautiful), and algebra, the *Bijaganita* (Seed Arithmetic).

bias *See* unbiased estimator; unbiased hypothesis test.

biconditional (in logic) A sentence of the form '*A* if and only if *B*' (*A* iff *B*). Such a statement is called biconditional because it is a joint assertion of two conditions: '*A* if *B*' and '*B* if *A*'. It is symbolized in *formal language by $A \equiv B$ or $A \leftrightarrow B$ (*see* equivalence).

bidiagonal matrix A square *matrix whose elements are zero except on the principal diagonal and the first superdiagonal or first subdiagonal. The matrix *A* is *upper bidiagonal* if $a_{ij} = 0$ when $i > j$ or $j > i + 1$, and *lower bidiagonal* if $a_{ij} = 0$ when $i < j$ or $i > j + 1$. The matrix

$$\begin{pmatrix} a & b & 0 \\ 0 & c & d \\ 0 & 0 & c \end{pmatrix}$$

is upper bidiagonal.

Bieberbach conjecture In 1916 the German mathematician Ludwig Georg Bieberbach (1886–1982) conjectured that if the *holomorphic function defined by the *power series

$$\mathrm{f}(z) = z + a_2 z^2 + a_3 z^3 + \cdots$$

where a_n are complex numbers is *injective for $|z| < 1$, then $|a_n| \leqslant n$ for all $n \geqslant 2$. After attempts by many mathematicians, the conjecture was proved by Louis de Branges in 1984.

Bienaymé–Tchebyshev inequality *See* Tchebyshev's inequality.

bifurcation A sudden change in the nature of an *attractor (or repellor) as the defining transformation or *flow changes with respect to changes in the defining equations. Important examples are:

(1) *Hopf bifurcation* (E. Hopf, 1942): a bifurcation in which a family of flows $x_\lambda(t)$, parametrized by a real number λ, has an attractor consisting of a fixed point replaced by a circle and a repelling fixed point for a small change in the parametrization (*see* diagram).

(2) *Flip* (or *period doubling*) *bifurcation*: a bifurcation in which a family of transformations T_λ, parametrized by a real number λ, has a repelling fixed point replaced by a pair of periodic points of period 2, forming an attractor. For example, if $T_\lambda(x) = x^2 - (1 + \lambda)$ the change occurs when $\lambda = 0$.

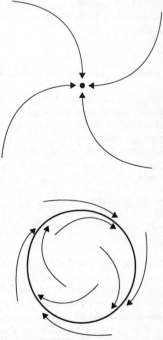

bifurcation Hopf bifurcation.

This is related to the *Feigenbaum number.

bijection A *mapping f: $X \rightarrow Y$, where X and Y are *sets, satisfying the properties

(1) if $x, y \in X$ and $f(x) = f(y)$ then $x = y$;
(2) if $y \in Y$ then $y = f(x)$ for some $x \in X$.

Any bijection has an inverse mapping f^{-1} such that $f(f^{-1}(y)) = y$ and $f^{-1}(f(x)) = x$ for all $x \in X$ and $y \in Y$; conversely any mapping f having such an inverse must be a bijection. Another name used for a bijection is *one-to-one correspondence. A bijection from set A to set B is a function that is both an *injection and a *surjection.

bilateral symmetry A geometric figure has bilateral symmetry if it has reflectional *symmetry in a line or plane. *See also* reflection.

bilinear Describing a mathematical expression that is *linear with respect to each of two variables considered separately. For example,

$$x + 2xy + y = 0$$

is a *bilinear equation*. $6xy$ is a *bilinear form*.

billion One thousand million (10^9). The term has long been established in this sense in the USA. In the UK 'billion' originally meant one million million (10^{12}), being a contraction of bi = two and *million*, but since the 1970s it has commonly been used to mean 10^9.

bimodal distribution A *distribution for which the *frequency function has two distinct maxima. *Compare* unimodal distribution.

binary code A set of *codewords, each a sequence of zeros and ones, such as 011001. Usually all the codewords in the set have the same length; *error-correcting codes are examples of such sets.

binary connective *See* connective.

binary notation The method of positional notation used in the *binary system.

binary number A number expressed in the *binary system.

binary operation A rule assigning, to two elements x and y of a *set, an element $x \circ y$ of the same set, often referred to as their *product* (although it need not be the product in the usual sense). For example, addition ($+$) is a binary operation on the set $\mathbb{Z}$ of integers, assigning to x and y the value $x + y$ and multiplication (.) is also a binary operation, assigning to x and y the usual product $x . y = xy$. If the set is S, a binary operation b on S can be regarded as a *function whose *domain is $S \times S$ and whose *codomain is S, and we can write $b: S \times S \rightarrow S$. *Compare* unary operation; *see also* Cartesian product; algebraic operation.

binary relation *See* relation.

binary string A string of elements such as 10001110 from the set $\{0, 1\}$. Information is often translated into binary strings before transmission because they are easy to transmit.

binary system A *number system using the base two. Two digits, 1 and 0, are used to denote binary numbers. Decimal 1 is 1 in binary, decimal 2 is 10, 3 is 11, 4 is 100, 5 is 101, etc. Binary numbers are used in computers because the two digits 1 and 0 can be represented by two alternative states of a component (e.g. the presence or absence of an electrical potential or magnetized region).

binary tree *See* tree.

binary variable A random variable that can take only two possible values; for example, the sex of a newborn child may be male or female, a bacterium may be present or absent in a clinical sample. For most analytic purposes, binary random variables may be coded by a random variable X that takes only the value 0 or 1.

For example, we might assign $X = 0$ to a male birth and $X = 1$ to a female birth; or in a coin tossing experiment we might assign $X = 0$ to the outcome 'tails' and $X = 1$ to the outcome 'heads'. *See also* dummy variable; Bernoulli trial.

binomial A *polynomial consisting of two terms, for example $1 + 2x$ or $p + q$.

binomial coefficients $\binom{n}{r}$, the coefficients of x^r in the expansion of $(1 + x)^n$, such that

$$\binom{n}{0} = 1$$

and

$$\binom{n}{r} = \frac{n(n-1)\cdots(n-r+1)}{r!}$$

for each positive integer r. When n is a positive integer, the binomial coefficients form a row of *Pascal's triangle. *See* binomial theorem.

binomial distribution The *distribution of the number of successes in a series of n independent *Bernoulli trials, at each of which the probability of success is p. The *frequency function is

$$\Pr(X = r) = \binom{n}{r} p^r q^{n-r}$$

where $q = 1 - p$, $0 \leqslant r \leqslant n$, and $\binom{n}{r}$ is a *binomial coefficient. The distribution is often denoted by $B(n, p)$. Successive terms in the frequency function are those in the binomial expansion of $(p + q)^n$. The mean is given by $E(X) = np$, and the variance by $\mathrm{Var}(X) = npq$. For example, if a die is cast four times and a score of 6 is a success, then $n = 4$ and $p = 1/6$, and the probability of two successes is

$$\binom{4}{2}(1/6)^2(5/6)^2 = 25/216$$

For large n, and both np and nq approximately 5 or more, the binomial distribution approaches the $N(np, npq)$ distribution (*see* normal distribution), and when $n \rightarrow \infty$

and $p \rightarrow 0$ in such a way that $np = \lambda$, it can be approximated by the *Poisson distribution with mean λ. *See also* continuity correction; negative binomial distribution.

binomial expansion The expansion given by the *binomial theorem.

binomial series An infinite *series that is the expansion of $(1 + x)^n$ or $(x + y)^n$ when n is not a positive integer or zero. *See* binomial theorem.

binomial theorem A theorem that gives the expansion for $(1 + x)^n$ as

$$1 + nx + \frac{n(n-1)}{2!}x^2$$
$$+ \frac{n(n-1)(n-2)}{3!}x^3 + \cdots$$

This is known as the *binomial expansion*.

When n is a positive integer the expansion is a finite series with $n + 1$ terms, the last term equalling x^n. When n is not a positive integer or zero, the expansion is an infinite series since the coefficients are all nonzero; it is known as the *binomial series*. It is convergent when $|x| < 1$ and divergent when $|x| > 1$. It is thus a valid expansion of $(1 + x)^n$ only when $|x| < 1$. (In the special case in which $x = 1$ the series is convergent if $n > -1$, and in the case in which $x = -1$ the series is convergent if $n > 0$.)

More generally, an expansion for $(x + y)^n$ is

$$x^n + nx^{n-1}y + \frac{n(n-1)}{2!}x^{n-2}y^2 + \cdots$$

This is valid for $|y| < |x|$. The coefficients of the terms are known as the *binomial coefficients*. Only when n is a positive integer is the expansion a finite series; it then has $n + 1$ terms, the last term equalling y^n.

biquadratic *See* quartic.

birectangular Having two right angles. *See* spherical triangle.

birth–death process A *stochastic process concerned with population changes due to births, deaths, immigration, and emigration.

bisect To divide into two equal parts. For instance, bisection of an angle involves drawing a line through the vertex that cuts the angle in half. A point, line, plane, etc. that bisects something is a *bisector*.

The point that bisects the line segment joining two other points is called their *mid-point*. If their position vectors are **a** and **b**, the mid-point has position vector $\frac{1}{2}(\mathbf{a} + \mathbf{b})$.

bisection method An interative method for solving a nonlinear equation $f(x) = 0$ in one variable. It is based on the fact that, if f is a *continuous function and f(a) and f(b) have opposite signs, then there is a zero of f between a and b. The bisection method takes an interval $[a, b]$ with $a < b$ and $f(a)f(b) < 0$, and evaluates f(c), where $c = \frac{1}{2}(a + b)$. If $f(c) = 0$, then c is a zero. If f(a) and f(c) have opposite signs, then there is a zero of f in $[a, c]$; otherwise f(c) and f(b) must have opposite signs and there is a zero of f in $[c, b]$. In either case, a new interval containing a zero has been constructed, of half the length of the original interval. This process is continued until the interval is so small that the zero is known to the desired accuracy. For example, if the interval is $[1.717, 1.724]$, then the zero is 1.72 to three significant figures.

bisector *See* bisect.

biserial correlation coefficient *See* correlation coefficient.

bit A unit of information, especially as used in digital computers, consisting of one binary digit; i.e. the amount of information required to specify one of two alternatives, such as the 0 and 1 in the binary system. *See* byte.

bitangent *See* double tangent.

bivalence, principle of The semantic principle which states that every sentence is either true or false. Under the standard interpretation of the logical connectives (*see* truth function) this principle is represented in a *formal system as the law of the *excluded middle: $A \lor \sim A$. *See* intuitionism; semantics.

bivariate data *See* data.

bivariate distribution The joint *distribution of two *random variables X and Y. The cumulative distribution function is

$$F(x, y) = \Pr(X \leqslant x, Y \leqslant y)$$

If X and Y are both discrete, the frequency function is

$$p_{ij} = \Pr(X = x_i, Y = y_j)$$

and

$$F(x, y) = \sum \sum p_{ij}$$

where the double summation is over all i and j such that $x_i \leqslant x$ and $y_j \leqslant y$.

If X and Y are both continuous, the frequency function is $f(x, y)$ and

$$F(x, y) = \int_{-\infty}^{x} \int_{-\infty}^{y} f(s, t) \, dt \, ds$$

In either case $F(x, y) \to 1$ as $x \to \infty$ and $y \to \infty$. For the continuous case the *marginal distribution* of X has marginal frequency function defined as

$$f_1(x) = \int_{-\infty}^{\infty} f(x, t) \, dt$$

with an analogous definition for $f_2(y)$, the marginal frequency function of Y. The marginal distribution functions are written as $F_1(x)$ and $F_2(y)$.

The *conditional distribution* of X, given $Y = y$, has conditional frequency function

$$g_1(x | Y = y) = \frac{f(x, y)}{f_2(y)}$$

with an analogous definition for

$g_2(y|X = x)$. Marginal and conditional frequency functions are defined on similar lines for the discrete cases, and it is possible to have one of X and Y continuous and the other discrete. If X and Y are independent variables, then

$$f(x, y) = f_1(x)\, f_2(y)$$

or

$$p_{ij} = \Pr(X = x_i)\, \Pr(Y = y_j)$$

for all x, y. For independence

$$F(x, y) = F_1(x)\, F_2(y)$$

also.

block 1. *See* randomized blocks.
2. *See* partition (of a matrix).

block design 1. *See* randomized blocks; balanced incomplete block design.
2. A collection of *subsets B_r of a finite set S such that each B_r has the same number k of elements; in addition, for every subset T of S with exactly t elements, T is a subset of exactly λ of the B_r. The numbers $k, t, \lambda,$ and n, the number of elements of S, are characteristic of the particular design. For example, the following is a block design with $n = 4$, $k = 3$, and $t = \lambda = 2$: $S = \{1, 2, 3, 4\}$ and the sets B_r are the subsets of S with exactly 3 elements; every subset of S with exactly $t = 2$ elements belongs to exactly $\lambda = 2$ of the sets B_r.

block diagonal matrix A square *block matrix with blocks A_{ij} such that A_{ij} is a zero matrix for $i \neq j$, so that the only non-zero blocks appear on the principal block diagonal.

block matrix The 4×3 matrix

$$\begin{pmatrix} 1 & 2 & 3 \\ 1 & 0 & 1 \\ -2 & 3 & -1 \\ 1 & 2 & 0 \end{pmatrix}$$

can be written as

$$\begin{pmatrix} A_{11} & A_{12} \\ A_{21} & A_{22} \end{pmatrix}$$

where A_{11} and A_{21} are 2×2 matrices and A_{12} and A_{22} are 2×1 matrices. This form of matrix, in which the elements are themselves matrices, is a *block matrix*, and this example illustrates a *block 2×2 matrix*. More generally, a block $p \times q$ matrix is of the form

$$A = \begin{pmatrix} A_{11} & \cdots & A_{1q} \\ \vdots & & \vdots \\ A_{p1} & \cdots & A_{pq} \end{pmatrix}$$

where the (i, j) element A_{ij} is a matrix of order $m_i \times n_j$

block triangular matrix A square *block matrix with blocks A_{ij} such that A_{ij} is a zero matrix for $i > j$ (*block upper triangular*) or $i < j$ (*block lower triangular*). The 4×4 matrix

$$\left(\begin{array}{cc|cc} 1 & 2 & 5 & 4 \\ 1 & 1 & 1 & 2 \\ \hline 0 & 0 & -1 & 1 \\ 0 & 0 & 0 & 1 \end{array} \right)$$

is block 2×2 upper triangular with 2×2 blocks when viewed as a block matrix as indicated by the lines, but it is not a *triangular matrix.

Bob The name conventionally used for the receiver of an encrypted message.

Boethius (*c*.475–524) Roman scholar whose *Geometry* and *Arithmetic* survived as standard texts in Europe for much of the medieval period. The former contained little more than Book I of Euclid, together with some elementary mensuration; the latter was based on the *Arithmetica* of Nicomachus (*c*. AD 100).

Bolyai, János (1802–60) Hungarian mathematician who demonstrated in 1823 that it was possible to develop an apparently consistent geometry in which the parallel postulate was rejected. Bolyai's system of hyperbolic geometry, published in 1832, was the first clear account of a *non-Euclidean geometry.

Bolzano, Bernard Placidus (1781–1848) Czech mathematician and philosopher who made an important contribution to analysis by offering in 1817 the first rigorous account of a *continuous function. He also published an influential work, *Paradoxes of the Infinite* (1850), in which he anticipated some of the later results of *Cantor.

Bombelli, Rafael (1526–72) Italian mathematician and author of the highly influential *L'Algebra* (1572). He published rules for the solution of quadratic, cubic, and quartic equations, and was one of the first mathematicians to accept imaginary numbers as roots of such equations.

Bonferroni inequalities Inequalities concerning the *probabilities of occurrence of combinations of *events. The best known of several inequalities given by Bonferroni states that if $E_1, E_2, \ldots, E_n$ is a set of n events and $\bar{E}_1, \bar{E}_2, \ldots, \bar{E}_n$ is the set of opposite events for which $\Pr(\bar{E}_i) = 1 - \Pr(E_i)$, then

$$\Pr(E_1 \cap E_2 \cap \cdots \cap E_n) > 1 - \sum_{i=1}^{n} \Pr(\bar{E}_i)$$

The inequalities are used in *multiple comparison tests when the tests are not independent of one another and in forming the associated simultaneous *confidence intervals. They are named after the Italian mathematician Carlo Emilio Bonferroni (1892–1960).

Boole, George (1815–64) English mathematician who in his *Mathematical Analysis of Logic* (1847) showed for the first time how algebraic formulae could be used to express logical relations. The *Boolean algebra developed in 1847 and in his *The Laws of Thought* (1854) has proved to have wide application in such diverse fields as computer design, topology, and probability theory.

Boolean algebra An algebraic system consisting of a *set of elements S together with two *binary operations, denoted by . (the *Boolean product*) and + (the *Boolean sum*) obeying certain *axioms (x, y, and z are members of S):

(1) The operations are commutative:

$$x \cdot y = y \cdot x, \quad x + y = y + x$$

(2) There are identity elements (0 and 1) for each of the operations, with the *identity laws*

$$x \cdot 1 = x, \quad x + 0 = x$$

(3) The distributive laws apply, each operation being distributive over the other:

$$x \cdot (y + z) = (x \cdot y) + (x \cdot z),$$
$$x + (y \cdot z) = (x + y) \cdot (x + z)$$

(4) Each member of S has an inverse (or complement) denoted by x', y', etc., with the *complement laws*

$$x \cdot x' = 0, \quad x + x' = 1$$

Note that the Boolean operations . and + are not the same as those in 'ordinary' algebra. For instance, in the algebra of numbers it is not true that

$$x + (y \cdot z) = (x + y) \cdot (x + z)$$

Various alternative axiomatizations of Boolean algebra can be given. Using the one shown here, certain other relationships can be proved, for example:
(a) The *duality principle* that if a given expression is valid, then the expression obtained by interchanging . and +, and 0 and 1, is also valid. This follows from the fact that the axioms are symmetrical with respect to . and + and to 0 and 1.

(b) The *idempotent laws*

$$x + x = x, \quad x \cdot x = x$$

(c) The *associative laws*

$$x \cdot (y \cdot z) = (x \cdot y) \cdot z$$

$$x + (y + z) = (x + y) + z$$

(d) The *absorption laws*

$$x \cdot (x + y) = x, \quad x + (x \cdot y) = x$$

(e) The *null laws*

$$x + 1 = 1, \quad x \cdot 0 = 0$$

There are two common examples of systems that are Boolean algebras:
(1) The algebra of sets, in which + is *union of sets ∪, . is *intersection ∩, 0 is the null set, and 1 is the universal set.
(2) The algebra of propositions (*see* propositional calculus) in which . is 'and' (&) and + is 'or' (∨).

Boolean algebras are applied extensively to logic design, switching theory, and other applications in computer science.

bootstrap A method for obtaining information about population parameters or characteristics by first taking a *random sample of n observations from a population, and then forming from this initial sample further random samples, called *bootstrap samples*. These are also of size n, and are obtained by sampling with replacement (*see* random sample). The method is especially useful when there is insufficient information to specify the population distribution, or when there is little analytic theory about properties of estimators. If B bootstrap samples are taken, they may be used to estimate standard errors of *estimators of parameters such as means, medians, or correlation coefficients obtained from the original sample without the need to make any assumption about the population distribution.

For example, if the sample median m is used to estimate the population median, then if the median of the bth bootstrap sample is m_b^*, with $b = 1, 2, \ldots, B$, then the bootstrap estimated standard error of m is given by

$$\mathrm{se}(m) = \sqrt{\left(\frac{1}{B-1} \sum_b (m_b^* - \overline{m}^*)^2 \right)}$$

where $\overline{m}^*$ is the mean of the m_b^*. Approximate 95 percent confidence limits for the population median are given by the 0.025 and 0.975 *quantiles of the set of B values m_b^*. In practice, good estimates of $\mathrm{se}(m)$ may be obtained with $B = 50$, but values of $B = 1000$ or $B = 2000$ are usually required for reliable estimates of confidence intervals.

Refinements are available to improve these estimates. Statistical software with a reliable random number generator is necessary for the practical implementation of the bootstrap.

Borda, Jean Charles (1733–99) French mathematician and astronomer who worked on problems in fluid mechanics, demonstrating that resistance is proportional to the square of the fluid velocity and to the sine of the angle of incidence. A naval captain, Borda also worked on geodesy, developing various measuring instruments, and helped to introduce the metric system into France.

bordering The process of adding a row and a column to a *determinant to increase its order. If the common element of the added row and column is 1, with other elements being zero, then the order is increased but the value of the determinant is unchanged.

Borel, Félix Edouard Émile (1871–1956) French mathematician noted for his work on set theory (*see* Borel set) and measure theory. Borel also introduced a definition for the sum of a divergent series.

Borel set A measurable set that can be obtained from *closed sets and *open sets on the real line by applying the operations of union and intersection repeatedly to *countable collections of sets.

Borromean rings An arrangement of three interlinked circles in space such that, if any one of them is removed, then the other two are unlinked. Named after the house of Borromeo, a wealthy Milanese banking family who incorporated the design in their family crest in the 15th century.

Borromean rings

Bouguer, Pierre (1698–1758) French mathematician and physicist who worked on problems of geodesy. He measured the *acceleration of free fall using a pendulum and was the first to observe that a pendulum could be affected by the gravitational pull of a high mountain.

bound **1.** (of a function) A restriction on the *range of a function. An *upper bound* is a number u such that $f(x) \leqslant u$ for all x in the domain, and a *lower bound* is a number l such that $f(x) \geqslant l$ for all x in the domain. For example, if $f(x) = \sin x$ then $+1$ is an upper bound and -1 is a lower bound. If a lower bound for $f(x)$ exists, f is said to be *bounded below*; if an upper bound exists, it is *bounded above*; if both exist it is *bounded*.

An upper bound u is a *least upper bound* (l.u.b.) if $u \leqslant v$ for any other upper bound v. A lower bound l is a *greatest lower bound* (g.l.b.) if $l \geqslant m$ for any other lower bound m.

See also unbounded function.

2. (of a sequence) *See* bounded sequence.
3. (of a set) *See* order properties.
4. *See* variable.

boundary *See* frontier.

boundary conditions If the solution to a *differential equation or *difference equation contains r arbitrary constants, these constants may be eliminated to give a unique solution to a problem if there are r given *conditions* that the solution must satisfy. Some of these may be *boundary* or *initial conditions*. Boundary conditions, which may be for the function and/or its derivatives at certain boundary points, may be used to obtain a solution which is valid over the region specified by the conditions. For systems evolving with time, initial conditions are those that must be satisfied by the solution function and its derivatives at the start.

For example, the differential equation

$$\frac{d^2 y}{dx^2} + 4\frac{dy}{dx} = 0$$

where $x \geqslant 0$, has the solution $y = A + Be^{-4x}$. If the boundary conditions are $y = 0$ and $dy/dx = 1$ when $x = 0$, then substituting $x = 0$ in the solution and its first derivative yields $A = \frac{1}{4}$ and $B = -\frac{1}{4}$.

bounded sequence A *sequence $\{a_n\}$ of *real numbers for which there is both an *upper bound* and a *lower bound*. If there is a number U that is greater than or equal to every number in the sequence, i.e. if $a_n \leqslant U$, then U is an upper bound of the sequence, which is then said to be *bounded above*. Similarly, if there is a number L such that $a_n \geqslant L$, then L is a lower bound of the sequence, which is then said to be *bounded below*.

The positive integers 1, 2, 3, … are bounded below since all exceed 0. The sequence $\{1 - 1/n\}$ for $n \geqslant 1$ is bounded above and below (i.e. is a bounded sequence) since all its terms lie between 0 and 1.

See also bound.

bounded set A *set A in a Euclidean space is bounded if there exists a number N such that $|x| \leqslant N$ for all $x \in A$. Otherwise the set is *unbounded*. A set A in a metric space X with metric d is bounded if there is a point $x \in X$ and a number N such that $d(a, x) \leqslant N$ for all $a \in A$.

bounded variation *See* variation.

Bourbaki, Nicolas A collective *nom de plume* of mainly French mathematicians whose aim has been to publish an ambitious *Éléments de mathématiques* in many volumes in which much of modern mathematics is treated rigorously, comprehensively, and in depth. Well over 40 volumes have appeared since 1939.

box-and-whisker diagram (boxplot) A useful graphical representation of the information contained in a *five-number summary. The box is a rectangle with length indicating the *interquartile range, and arbitrary breadth, divided lengthwise at the median. The whiskers are lines extending beyond the rectangle to indicate the range. Adjacent plots for data from two or more samples make it easy to see major differences between their characteristics. Box-and-whisker diagrams are shown below for two samples with the following five-number summaries:

Sample 1:   2   17   23   49   71

Sample 2:   6   11   25   56   93

box-and-whisker diagrams for two samples.

Box–Jenkins model (G.E.P. Box and G.M. Jenkins, 1967) A very general mathematical model for *time-series analysis in forecasting and prediction.

boxplot *See* box-and-whisker diagram.

brachistochrone A curve that is the path along which a particle will slide in the shortest time from one point to another, lower point (not directly beneath the first). The problem of finding the equation of such a curve was proposed in 1696 by Jean Bernoulli. The solution – that the curve is a *cycloid through the two points – was found by a number of mathematicians including Newton, Leibniz, and Jacques Bernoulli. *See also* calculus of variations.

Brahmagupta (*c*.598–*c*.665) Indian mathematician and astronomer noted for his introduction of negative numbers and zero into arithmetic. He also formulated the rule of three, gave the formula for the area of a cyclic quadrilateral in terms of its sides, and proposed rules for the solution of quadratic and simultaneous equations. His main work was an account in verse of Hindu astronomy and mathematics, *Brahmasphuta siddhanta* (The Revised System of Brahma).

Brahmagupta's formula A formula for the area A of a cyclic quadrilateral:

$$A = \sqrt{[(s - a)(s - b)(s - c)(s - d)]}$$

where a, b, c, and d are the lengths of the sides, and s is the semiperimeter, i.e.

$$s = \tfrac{1}{2}(a + b + c + d)$$

Compare Hero's formula.

braid A number of strings plaited together. The theory of braids studies the different ways in which a number of strings can be plaited. In the 1930s. E. Artin related the study of braids to group theory, and in the 1960s V.I. Arnol'd stressed the relevance of braid theory to the theory of functions. If n points are moved continuously in the plane back to the positions originally occupied, the movement can be recorded as a braid. Arnol'd used this idea to study how the roots of a polynomial vary as the coefficients of the polynomial are varied. A close relationship with *knot theory was established by J.W. Alexander in the 1930s, and exploited by V.F.R. Jones in the 1980s.

branch A part of a curve that is separated from another part by a *discontinuity or a *singular point.

branch point *See* singular point.

branching process A *stochastic process where individuals give rise to offspring, the distribution of descendants being likened to branches on a family tree. *See* tree diagram.

Bravais lattice *See* crystallography.

Brianchon, Charles Julien (1783–1864) French mathematician noted for his proof in 1806 of the dual version of Pascal's theorem.

Brianchon's theorem *See* Pascal's theorem.

Briggs, Henry (1561–1630) English mathematician who in his *Arithmetica logarithmica* (1624, The Arithmetic of Logarithms) published the first table of common *logarithms (formerly known as *Briggsian logarithms*).

Briggsian logarithm *See* logarithm.

British thermal unit Symbol: BTU. The *f.p.s. unit of energy, equal to the energy required to raise the temperature of one pound of water by $1\,^\circ$F. 1 BTU = 1055.06 joules or approximately 252 calories.

British units of length A system of *imperial units based on the *yard. In this system:

    12 inches = 1 foot
       3 feet = 1 yard
    22 yards = 1 chain
    10 chains = 1 furlong
    8 furlongs = 1 mile (of 1760 yards)

broken line A line formed of a number of discrete line segments joined together. A continuous curve can be approximated by a broken line – for example, a circle can be approximated by a polygon.

(a)

(b)

braid (a) The pigtail braid on three strings; (b) a braid on four strings.

Brouncker, William, Viscount (1620–85) English mathematician noted for his work on the early development of the calculus. He was one of the first mathematicians in Britain to use *continued fractions, and expressed $4/\pi$ as a continued fraction. He also published work on the rectification of the parabola and cycloid.

Brouwer, Luitzen Egbertus Jan (1881–1966) Dutch mathematician and philosopher. Beginning in 1912, Brouwer formulated the doctrine of *intuitionism and continued the attempt to construct a rigorous mathematics in accordance with its principles. He also worked on *fixed-point theorems in topology.

Brouwer's theorem *See* fixed-point theorem.

Brownian motion A physical phenomenon which may be modelled as a *stochastic process. The phenomenon was first noticed by the Scottish botanist Robert Brown in 1827 when he observed, under a microscope, the erratic motion of pollen grains suspended in water, these following a somewhat zigzag path. The term is now used more generally to describe the random movement of particles suspended in a fluid, or the stochastic model used to describe such random movements.

Buffon, Georges Louis Leclerc, Comte de (1707–88) French naturalist and mathematician best known as the creator of the immense *Histoire naturelle* (1794–1804, 44 vols). In mathematics he is still remembered for his work on probability and his famous needle problem, with which in 1777 he computed an approximation for π.

Buffon's needle problem A problem in *probability put forward by Buffon in 1777 in a supplement to *Histoire naturelle*. He considered a plane area ruled with parallel equidistant lines a distance d apart. The problem is to calculate the probability that a needle of length l ($l < d$), thrown at random onto the area, will come to rest across one of the lines. The answer, given correctly by Buffon, is $2l/\pi d$. Laplace in 1812 extended the problem to a rectangular grid of lines distances a and b apart, showing that the probability then becomes

$$\frac{2l(a+b)-l^2}{\pi ab}$$

This is sometimes known as the *Buffon – Laplace problem* or the *needle problem*.

bulk modulus A constant property of an elastic body, measuring the resistance to change in volume without change in shape. It is the ratio of compressive *stress per unit surface area of a body to the change in volume per unit volume associated with this stress, the pressure being uniform over the surface. *See also* elasticity.

bundle A topological notion more general than that of the product of two *spaces, and sometimes referred to as a *twisted product*. It is often described as a map f: $E \to B$ between topological spaces E and B, whose 'fibres' $f^{-1}(b)$ for $b \in B$ are all *homeomorphic to a single space. The simplest example, apart from products, is the *Möbius strip; in this case E is the Möbius strip, B is a circle, and the 'fibre' is an interval. *See also* sheaf.

buoyancy The upward *force exerted on a body by the fluid in which it is wholly or partly submerged. According to *Archimedes' principle, the magnitude of the force is equal to the weight of fluid displaced by the body; its line of action passes through the *centre of gravity of the displaced volume, a point known as the *centre of buoyancy*.

Burali-Forti's paradox Every *well-ordered set has an *ordinal number, and, as the set of all ordinals is well ordered, it too has an ordinal number, say A. But the set of all ordinals up to and including a given ordinal, say B, is itself well ordered and has ordinal number $B + 1$. So the set of all ordinals up to and including A has ordinal number $A + 1$, which is greater than A, so that A both is and is not the ordinal number of all ordinals. This *paradox is avoided in standard versions of set theory by denying that there exists a set of all ordinals. It was first stated in 1897, by the Italian mathematician Cesare Burali-Forti (1861–1931).

byte A unit of information, as used in digital computers, equal to eight *bits.

C

ℂ Symbol for the set of all *complex numbers.

ℂ∞ Symbol for the *extended complex plane.

Caesar cipher A *cipher where each letter is replaced by a letter which is a fixed number of places along in the alphabet. This fixed number is called the *Caesar shift* of the cipher. For example, if the number is 5 then ABY becomes FGD. (The alphabet should be regarded as written round a circle.) This type of cipher is one of the *substitution ciphers used and described by Julius Caesar (100–44 BC), after whom it is named.

Caesar shift *See* Caesar cipher.

calculus A branch of mathematics using the idea of a *limit, and generally divided into two parts: integral and differential calculus.

Integral calculus (*see* integration) can be used for finding areas, volumes, lengths of curves, centroids, and moments of inertia of curved figures. It can be traced back to Eudoxus of Cnidus and his method of *exhaustion (c.360 BC). Archimedes (in *The Method*) developed a way of finding the areas bounded by curves by considering them to be divided up by many parallel line segments, and extended it to determine the volumes of certain solids; for this, he is sometimes called the 'father of the integral calculus'.

In the early 17th century, interest again developed in measuring volumes by integration methods. Kepler used a procedure

for finding the volumes of solids by taking them to be composed of an infinite set of infinitesimally small elements (*Stereometria doliorum*, Measurement of the Volume of Barrels, 1615). These ideas were generalized by Cavalieri in his *Geometria indivisibilibus continuorum nova* (1635), in which he used the idea that an area is made up of indivisible lines and a volume of indivisible areas, i.e. the concept used by Archimedes in *The Method* (*see also* Cavalieri's principle). Cavalieri thus developed what became known as his *method of indivisibles*. John Wallis, in *Arithmetica infinitorum* (1655), arithmetized Cavalieri's ideas. In this period, infinitesimal methods were extensively used to find lengths and areas of curves.

Differential calculus (*see* differentiation) is concerned with the rates of change of functions with respect to changes in the independent variable. It came out of problems of finding tangents to curves, and an account of the method is published in Isaac Barrow's *Lectiones geometricae* (1670). Newton had discovered the method (1665–6) and suggested that Barrow include it in his book. In his original theory, Newton regarded a function as a changing quantity – a *fluent* – and the derivative, or rate of change, he called a *fluxion*. The slope of a curve at a point was found by taking a small element at the point and finding the gradient of a straight line through this element. The binomial theorem was used to find the limiting case, i.e. Newton's calculus was an application of infinite series. He used the notation $\dot{x}$ and $\dot{y}$ for fluxions and $\ddot{x}$ and $\ddot{y}$ for fluxions of fluxions. Thus, if $x = f(t)$, where x is distance and t time for a moving body, then $\dot{x}$ is the instantaneous velocity and $\ddot{x}$ the instantaneous acceleration. Leibniz had also discovered the method by 1676, publishing it in 1684. Newton did not publish until 1687 (in *Principia*). A bitter dispute arose over the priority for the discovery. In fact, it is now known that the two made their discoveries independently and that Newton made his about ten years before Leibniz, although Leibniz published first. The modern notation of dy/dx and the elongated S for integration are due to Leibniz.

From about this time, integration came to be regarded simply as the inverse process of differentiation. In the 1820s, Cauchy put the differential and integral calculus on a more secure footing by using the concept of a limit. Differentiation he defined by the limit of a ratio, and integration by the limit of a type of sum. The limit definition of an integral was made more general by Riemann.

In the 20th century, the idea of an integral was extended. Originally, integration was concerned with elementary ideas of measure (e.g. lengths, areas, and volumes), and with continuous functions. With the advent of set theory, functions came to be regarded as mappings, not necessarily continuous, and more general and abstract concepts of measure were introduced. Lebesgue put forward a definition of integration based on the Lebesgue measure of a set. Similar extensions of the concept have been made by other mathematicians. *See also* integrability; Lebesgue integral.

calculus of variations A branch of calculus concerned with finding the maximum or minimum values of *definite integrals. A well-known example of its use is the *brachistochrone problem, in which it is required to find the curve down which a particle will slide freely in the fastest time. If the equation of the curve is $y = f(x)$, it

can be shown that the time taken between two points is expressed by an integral of

$$\sqrt{\left(\frac{1 + (f'(x))^2}{2g\,f(x)}\right)}$$

where g is the *acceleration of free fall. The problem is then one of finding $f(x)$ such that the integral has a minimum value. This is often done by finding a differential equation (the *Euler–Lagrange equation*) which the function $f(x)$ must satisfy. Such a function, maximizing or minimizing a definite integral, is called an *extremal*. *See* minimal surface.

calorie Symbol: cal. A *c.g.s. unit of energy, equal to the energy required to raise the temperature of one gram of water by 1 °C. Various calories have been defined. The *15° calorie* specifies that the 1° rise in temperature should be from 14.5 °C to 15.5 °C; this calorie is equal to 4.1855 joules. The *IT* (international table) *calorie* is defined as 4.1868 joules. The *Calorie* (written with a capital C and also called the *large calorie*, *kilogram calorie*, or *kilocalorie*) is equal to 1000 calories. It is used in estimating the energy value of foods.

cancellation 1. The process of dividing the numerator and denominator of a fraction by the same number to produce a simpler fraction. Thus, 27/30 is $(9 \times 3)/(10 \times 3)$, which is 9/10. The 3 has been cancelled out of the fraction.
2. The process in which two equal quantities are removed from an equation. Thus, in $x + 3y = 7 + 3y$, the terms $3y$ can be cancelled out, leaving $x = 7$.

candela Symbol: cd. The *SI unit of luminous intensity, defined as the intensity, in a given direction, of a source that emits monochromatic radiation of a frequency of 540×10^{12} hertz and that has a radiant intensity in that direction of 1/683 watt per steradian.

Cantor set Construction of the set.

canonical When an expression can be expressed in a standard manner, usually at its most simple, it is said to be in *canonical* form. Thus the standard or canonical form for expressing a *quadratic equation is $ax^2 + bx + c = 0$, where a, b, and c are constants.

canonical form (normal form) (of a matrix) A form of *matrix to which all of a certain class of matrices can be reduced by transformations of a specified kind. For example, any symmetric matrix can be reduced to a *diagonal matrix by *similarity transformations, i.e. any symmetric matrix is *diagonalizable.

cantilever A beam or other structural member that is supported at one end only and supports a load along its length or at its free end. A cantilever can be horizontal or vertical. Cantilever construction is used, for example, in bridges and in the roofs and floors of buildings, permitting a large area to be spanned without obstructing supports.

Cantor, Georg (1845–1918) German mathematician who between 1874 and 1895 developed the first clear and comprehensive account of transfinite sets and numbers. He provided a precise definition of an infinite set, distinguishing between those which were denumerable and those which were not. *See* Cantor's theory of sets; diagonal argument.

Cantor–Bernstein theorem *See* Schröder–Bernstein theorem.

Cantor–Dedekind hypothesis *See* Dedekind cut.

Cantor set Any closed set topologically equivalent to the *Cantor middle third* (or *ternary*) *set*. This is constructed in the unit interval [0, 1] by deleting successive middle thirds of intervals. First $\left(\frac{1}{3}, \frac{2}{3}\right)$ is deleted, then $\left(\frac{1}{9}, \frac{2}{9}\right)$ and $\left(\frac{7}{9}, \frac{8}{9}\right)$ are removed, and so on. The process is continued indefinitely. The Cantor middle third set is a *fractal set with *similarity dimension $\ln 2 / \ln 3$.

Cantor's paradox A *paradox in *set theory. Is the cardinality of the set of all sets C greater than or equal to the cardinality of its *power set PC? The sets of PC must belong to the set of all sets ($PC \subset C$) and its *cardinal number must therefore be less than or equal to the cardinal number of C. But, by *Cantor's theorem, the cardinal number of C is less than that of PC.

Cantor's theorem The theorem that, for any set A, the *cardinal number of A is less than the cardinal number of the *power set PA. It follows that for any cardinal number n there is always a cardinal number greater than n.

Cantor's theory of sets A theory of sets developed by Georg Cantor in 1874. Dedekind had earlier defined an *infinite set as a set S that can be put into *one-to-one correspondence with a proper subset of S. Unlike Dedekind, Cantor realized that not all infinite sets are the same. He showed that the rational numbers are countable – i.e. they can be put into one-to-one correspondence with the positive integers (they have *cardinal number aleph-null, $\aleph_0$). He also showed that the algebraic numbers are countable. However, the set of all real numbers (algebraic plus transcendental) cannot be put into one-to-one correspondence with the positive integers. This infinite set has a higher cardinal number (c). In this way, Cantor built up a theory of *transfinite sets*. He showed, for instance, that the set of subsets of a set always has a higher cardinal number than that of the set itself, and consequently that there is an infinite number of these *transfinite numbers*. Cantor also developed an arithmetic of transfinite *ordinal numbers.

cap The symbol ∩, used to denote the *intersection of two sets A and B, as in $A \cap B$. *Compare* cup.

capture–recapture sampling A statistical procedure for estimating animal populations. In a simple form, a sample of n_1 animals is captured and each is tagged and released. A second sample of n_2 animals from the same population is captured at a later date and the number m of tagged animals is noted. If the unknown population size is N and any animal is equally likely to be captured, the proportion of tagged animals in the second sample, m/n_2, is an intuitively reasonable estimate of the unknown proportion of tagged animals in the whole population, n_1/N. Taking these proportions to be approximately equal leads to an estimate of N given by $N^* = (n_1 n_2)/m$ provided $m \neq 0$ (since $m = 0$ implies an infinite population). For example, if 25 squirrels are humanely trapped, marked, and released, and on a later date 40 squirrels are trapped and 5 of these are marked, we have $n_1 = 25$, $n_2 = 40$, and $m = 5$, implying that $N^* = (25 \times 40)/5 = 200$.

The *estimator N^* is biased, and tends to overestimate the population size, especially when m is small. The validity of N^* is strongly dependent on the assumptions that both the population size and the probability of capture remain constant between the sampling times. The former assumption ignores births, deaths, or migration between sampling times, and the latter

will not hold if animals find the first capture appealing, perhaps because of a food reward; alternatively if an animal is frightened by capture this may reduce the probability of recapture. More sophisticated estimators take such factors into account.

The method extends to the medical sciences, where it is used to estimate total numbers in the population exhibiting some abnormality, when there are two incomplete registers of patients with the abnormality, some patients (the 'recaptured units') appearing on both registers, and the remainder only on one.

Carathéodory's theorem (C. Carathéodory, 1911) If a point $x \in \mathbb{R}^n$ lies in the *convex hull of a set $S \subset \mathbb{R}^n$, then there is a subset T of S with no more than $n + 1$ points such that x lies in the convex hull of T.

Cardano, Girolamo (1501–76) Italian mathematician, physician, and astrologer noted for the first publication of the solution to the general *cubic equation in his book on algebra, *Ars magna* (1545, The Great Art). The solution was in fact found by Tartaglia and had been revealed in confidence. Although Cardano credited Tartaglia with the discovery, the revelation led to a bitter dispute between the two. *Ars magna* also contains the solution of the general quartic equation found by Cardano's former assistant, Ferrari. Cardano is also known for his speculations on philosophical and theological matters and, in mathematics, for early work in the theory of probability, published posthumously in *Liber de ludo aleae* (1663, A Book on Games of Chance).

Cardano's method *See* cubic.

cardinal number (cardinality) A number that indicates the number of elements in a *set. Thus the set of the members of a football team has cardinal number of 11 while, more generally, a set with n distinct elements has a cardinal number of n. If two sets can be put into a *one-to-one correspondence with each other then they have the same cardinal number. With regard to infinite sets, all *countable sets have a cardinal number $\aleph_0$ (aleph-null), while the cardinal number of the real numbers is denoted by c or by $\aleph_1$. Sets, whether finite or infinite, with the same cardinal number are described as being *equipollent, equipotent, equinumerable*, or *equivalent*. The cardinal number of a set is sometimes called its *power* or *potency*. Common notations for the cardinal number of a set A are $\bar{A}$, $n(A)$, and $|A|$.

cardinal points 1. The four directions on the earth's surface: north, south, east, and west.
2. Four points on the *celestial sphere lying on the horizon. The east and west points are the intersections of the horizon with the celestial equator. The north and south points are midway between these. The points are named so that the north point is the one closest to the north celestial pole, with the east point $90°$ clockwise from the north point.

cardioid A plane curve; the *locus of a fixed point P on a circle rolling on an equal fixed circle. A cardioid is a type of *epicycloid in which the two circles have equal radii. In polar coordinates it has the equation

$$r = a(1 + \cos\theta)$$

The name means 'heart-shaped'. *See also* limaçon of Pascal.

Carmichael numbers *See* pseudoprime.

Carnot, Lazare Nicolas Marguerite (1753–1823) French mathematician and politician best known for his work on the foundations of the calculus. Unhappy with the fluxions of Newton, the differentials of Leibniz, and the limits of d'Alembert, he argued in his *Réflexions sur la métaphysique du calcul infinitésimal* (1797) that infinitesimals should be regarded merely as convenient aids, introduced only to facilitate calculations, and should be eliminated

from the final result. In a later work, *Géométrie de position* (1803), Carnot helped lay the foundations of modern geometry.

Carnot, Nicolas Léonard Sadi (1796–1832) French mathematical physicist best known for his classic work *Réflexions sur la puissance motrice de feu* (1824, Reflections on the Motive Power of Fire), a cornerstone of the science of thermodynamics. It contained the crucial insight, *Carnot's theorem*, that all reversible heat engines operating between the same temperatures are equally efficient. As later developed by Lord Kelvin and Rudolf Clausius, Carnot's work led directly to the discovery of the second law of thermodynamics.

Cartesian coordinate system A *coordinate system in which the position of a point is determined by its relation to reference lines (*axes*). In two dimensions, two lines are used; commonly the lines are at right angles, forming a *rectangular coordinate system* (*see* diagram (a)). The horizontal axis is the x-axis and the vertical axis is the y-axis. The point of intersection O is the *origin* of the coordinate system. Distances along the x-axis to the right of the origin are usually taken as positive, distances to the left negative. Distances along the y-axis above the origin are positive; distances below are negative. The position

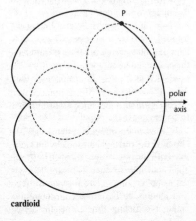

cardioid

of a point anywhere in the plane can then be specified by two numbers, the *coordinates* of the point, written as (x, y). The x-coordinate (or *abscissa*) is the distance of the point from the y-axis in a direction parallel to the x-axis (i.e. horizontally). The y-coordinate (or *ordinate*) is the distance from the x-axis in a direction parallel to the y-axis (vertically). The origin O is the point $(0, 0)$. The two axes divide the plane into four *quadrants*, numbered anticlockwise starting from the top right (positive) quadrant: the *first quadrant*.

Cartesian coordinates were first introduced in the 17th century by René Descartes. Their discovery allowed the application of algebraic methods to geometry and the study of hitherto unknown curves. As a point in Cartesian coordinates is represented by an ordered pair of numbers, so is a line represented by an equation. Thus, $y = x$ represents a set of points for which the x-coordinate equals the y-coordinate; i.e. $y = x$ is a straight line through the origin at $45°$ to the axes. Equations of higher degree represent curves; for example,

$$x^2 + y^2 = 4$$

is a circle of radius 2 with its centre at the origin. A curve drawn in a Cartesian coordinate system for a particular equation or function is a *graph* of the equation or function.

The axes in a planar Cartesian coordinate system need not necessarily be at right angles to each other. If the x- and y-axes make an angle other than $90°$ the system is said to be an *oblique coordinate system* (*see* diagram (b)). Distances from the axes are then measured along lines parallel to the axes.

Cartesian coordinate systems can also be used for three dimensions by including a third axis – the z-axis – through the origin perpendicular to the other two. The position of a point is then given by three coordinates (x, y, z). The coordinate axes may be left-handed or right-handed,

(a)

(b)

(c)

Cartesian coordinate system (a) Rectangular and (b) oblique coordinate systems; (c) a right-handed system of axes.

depending on the way positive directions are given to the axes. In a right-handed system (*see* diagram (c)), if the thumb of the right hand points in the positive direction of the x-axis, the first and second fingers can be pointed in the positive directions of the y- and z-axes respectively. The axes are said to form a *right-handed triad*. A left-handed system is the mirror image of this (i.e. determined using the left hand), the axes being said to form a *left-handed triad*.

See also rotation of axes; translation of axes.

Cartesian metric When X and Y are *metric spaces with metrics d_X and d_Y, respectively, the Cartesian metric d on the Cartesian product $X \times Y$ is defined by

$$d((x_1, y_1), (x_2, y_2))^2 = d_X (x_1, x_2)^2 + d_Y (y_1, y_2)^2$$

Cartesian product The Cartesian product of two *sets A and B, denoted by $A \times B$, consists of the set of all *ordered pairs (x, y) such that $x \in A$ and $y \in B$:

$$A \times B = \{(x, y): (x \in A) \,\&\, (y \in B)\}$$

For example, if $A = \{1, 2\}$ and $B = \{3, 4\}$ then

$$A \times B = \{(1, 3), (1, 4), (2, 3), (2, 4)\}$$

If each of the two sets has some particular structure (they might both be groups, vector spaces, or metric spaces), then the product can usually be given the same structure. For example, if V_1 and V_2 are both real vector spaces, then the product set $V_1 \times V_2$ is also a real vector space with addition defined by $(x_1, x_2) + (y_1, y_2) = (x_1 + y_1, x_2 + y_2)$, and scalar multiplication by $\lambda(x, y) = (\lambda x, \lambda y)$.

The product of two or more topological spaces is also a topological space. For example, the plane is the product $\mathbb{R} \times \mathbb{R}$.

The Cartesian product is named after Descartes since it generalizes the concept of a *Cartesian coordinate system.

Cartesian tensor *See* tensor.

Cassini's ovals Curves that are the *loci of the vertex of a triangle in which the side of the triangle opposite the vertex is fixed, and the product of the two sides adjacent to the vertex is constant. In Cartesian coordinates, the equation has the form

$$[(x + a)^2 + y^2][(x - a)^2 + y^2] = k^4$$

where a is half the length of the fixed side and k^2 the constant. If $k^2 > a^2$, there is a single curve. If $k^2 < a^2$, the curve consists of two ovals. If $k^2 = a^2$, the curve is a *lemniscate. The ovals are named after the Italian astronomer Giovanni Domenico Cassini (1625–1712).

Cassini's ovals: $PS \cdot PS' = k^2$.

casting out nines A simple method of manually checking any addition, subtraction, or multiplication of *natural numbers, by first finding the *digital root* of each of the relevant numbers. The digital root of a natural number is the single *digit obtained by adding the decimal digits of the number to give a smaller number, adding the digits of that number to give a still smaller number, and so on until a single digit is reached.

For example, for the number 38 247 we have $3 + 8 + 2 + 4 + 7 = 24$, and $2 + 4 = 6$. In this process we can leave out (or *cast out*) any 9 that appears. So for 61 934 we calculate $6 + 1 + 3 + 4 = 14$, and then $1 + 4 = 5$. Then, to check the addition $38\,247 + 61\,934 = 100\,181$, we see whether the sum of the digital roots on the left has the same digital root as 100 181. In this case $6 + 5 = 11$, which has digital root 2, and this is also the digital root of 10 081. Similarly, to check the subtraction $61\,934 - 38\,247$ and multiplication 61 934

× 38 247, we look to see whether the digital root of the answer is equal to that of $5 - 6$ and 5×6. If a difference of digital roots is negative, add 9; thus $5 - 6$ becomes 8.

The method works because $10 \equiv 1$ (mod 9), so any natural number is *congruent modulo 9 to the sum of its decimal digits. The digital root of n is the least non-negative integer congruent to n modulo 9, and if $n_1 \equiv d_1$ (mod 9) and $n_2 \equiv d_2$ (mod 9), then $n_1 \pm n_2 \equiv d_1 + d_2$ (mod 9) and $n_1 n_2 \equiv d_1 d_2$ (mod 9).

The method is mentioned in *Fibonacci's *Liber abaci* (1202) but is probably much older.

Catalan, Eugène Charles (1814–94) French mathematician noted for his work in number theory, geometry, and combinatorics.

Catalan numbers (E.C. Catalan, 1838) The sequence of numbers c_n (for $n \geqslant 3$) that gives the number of ways to divide a regular n-sided *polygon into triangles, using non-intersecting diagonals. So $c_3 = 1$, $c_4 = 2$, $c_5 = 5$, $c_6 = 14$, ..., and each c_n can be calculated as

$$c_n = \frac{1}{n+1} \binom{2n}{n}$$

Catalan's conjecture The conjecture (first proposed by Catalan in 1844) that when m and n are integers greater than 1, the only solution in positive integers x and y of the equation $x^m - y^n = 1$ is $x = 3$, $m = 2$, $y = 2$, $n = 3$, i.e. $3^2 - 2^3 = 1$. In 1342, Levi ben Gerson proved that the only powers of 2 and 3 above the first to differ by 1 were 8 and 9. The conjecture was proved to be true by P. Mihailescu in 2004.

Catalan's constant The number K defined by

$$K = \sum_{n=0}^{\infty} \frac{(-1)^n}{(2n+1)^2} = 1 - \frac{1}{3^2} + \frac{1}{5^2} - \cdots$$

Its numerical value is $0.915\,965\ldots$.

catastrophe theory A theory of dynamic systems developed by René Thom in 1972 to explain biological growth and differentiation, in which slow growth is accompanied by 'catastrophic' changes in form. It is based on the fact that the state of a system, which depends on a number of factors, can be represented by a set of points in n-dimensional space. It concentrates on the topological classification of these sets, connecting sudden discontinuous changes (i.e. 'catastrophies') with changes in topology. The theory has since been applied to many other fields, such as sociology, economics, engineering, and linguistics. *See* singularity theory.

categorical data Counts of the number of units or items falling into one or more categories. For a human population, categories might be gender (male or female), marital status (single, married, widowed, divorced), etc. If units can be allocated to only one of two categories, the data are described as *dichotomous data*.

categorical proposition A proposition which affirms or denies that a *predicate holds of a subject. Traditionally the following four *forms* were distinguished: *universal affirmative* (all A is B), *universal negative* (no A is B), *particular affirmative* (some A is B), and *particular negative* (some A is not B). Categorical propositions are to be distinguished from *conditional propositions* (if A then B) and from *modal propositions* (it is possible that A).

categorical variable A *random variable whose values are categories. For example, side effects of a drug classed as (none, slight, moderate, severe), character classifications (introvert, extravert), times to failure of a machine part (< 100 hours, 100–199 hours, 200–299 hours, $\geqslant 300$ hours), and marital status (single, married, widowed, divorced). Where there is no natural ordering of the categories, the variable is described as *nominal*. Thus marital status is a *nominal variable*. If there is a natural ordering of categories, the variable is described as *ordinal*. Thus the side effect of a drug is an *ordinal variable*. Categorical variables that can take only two possible values are called *dichotomous variables*.

category **1.** A classification of *sets into two types: sets are either of the first category or of the second category. A set X is of the first category if X is a *countable union of subsets that are nowhere *dense. An example is the set of rational numbers, since it can be represented as the countable union of *unit sets that are nowhere dense. All sets not of the first category are of the second category.
2. An entity that consists of *objects* and *morphisms*. A morphism can be regarded as a function between two of the objects and, when appropriate, morphisms can be composed. The two basic axioms for a category are associativity of composition of morphisms and the existence of an identity morphism for each object. An example is the category of real vector spaces, which has vector spaces as its objects and linear transformations as its morphisms.

category theory A language that unifies many concepts in mathematics. The theory aims to provide an understanding of properties of mathematical objects and mappings between them from a general point of view. *See* category (**2**).

catenary A plane curve with the equation, in Cartesian coordinates,

$$y = c\cosh(x/c) = \tfrac{1}{2}c(e^{x/c} + e^{-x/c})$$

It is symmetric about the y-axis; c is the intercept on the y-axis. The curve is the shape that a uniform flexible chain would assume if hung from two points. Huygens was the first to show that the catenary was nonalgebraic. Its equation was discovered by Jacques Bernoulli. *See also* intrinsic equation.

catenary: $y = c\cosh(x/c)$.

catenoid The surface formed by rotating the *catenary about the x-axis. Under certain conditions, the surface of minimal area bounded by two coaxial rings is a catenoid. *See* minimal surface.

Cauchy, Augustin-Louis, Baron (1789–1857) French mathematician who strove to introduce a more rigorous approach into analysis. In his *Cours d'analyse* (1821, A Course of Analysis) he introduced the modern notion of a limit and went on to use it to define the important concepts of continuity, convergence, and differentiability. In group theory, Cauchy proved in 1845 the fundamental theorem, since known as *Cauchy's theorem*, that every group whose order is divisible by a prime p contains a subgroup of order p. He also contributed to the calculus of variations, probability theory, and the study of differential equations. *See* Kovalevsky.

Cauchy convergence condition **1.** A condition for convergence stating that an *infinite sequence converges if and only if, beyond a certain point in the sequence, the numerical difference between any two terms is as small as desired. Thus the sequence $\{a_n\}$ converges if and only if, given any positive number ε, however small, there is an integer N, dependent on ε, such that

$$|a_i - a_j| < \varepsilon$$

for all $i, j > N$.

2. An infinite *series Σa_n converges if and only if, given any positive number ε, however small, there is an integer N, dependent on ε, such that

$$|a_{r+1} + a_{r+2} + \cdots + a_{r+i}| < \varepsilon$$

for all $r > N$ and $i > 0$. This is not normally used as a test for convergence but can be used to derive such tests. *See also* convergent series.

Cauchy convergence test Let

$$a_1 + a_2 + \cdots + a_n + \cdots$$

be an infinite *series of positive terms. If

$$\lim_{n \to \infty} (a_n)^{1/n} < 1$$

(*see* limit), then the series converges. If the limit exceeds 1, the series diverges. *See also* ratio test.

Cauchy distribution A continuous *distribution with *frequency function

$$f(x) = \frac{\lambda}{\pi\{\lambda^2 + (x - \theta)^2\}}$$

where $\theta, \lambda > 0$ are *parameters. The distribution has no finite moments, but the median is θ, and the curve is symmetric about the median and *bell-shaped*. The parameter λ determines the spread of the distribution and is often called a *scale parameter*. For a fixed λ, if we change θ the graph of the distribution will retain the same shape with the maximum shifted to correspond to the new value of θ.

If X has a Cauchy distribution, then

$$\Pr(\theta - \lambda < X \leqslant \theta + \lambda) = 0.5$$

The ordinate of $f(x)$ at $x = \theta - \lambda$ and at $x = \theta + \lambda$ is half the maximum ordinate at $x = \theta$, so λ is called the half-width at half-height. If $\theta = 0$ and $\lambda = 1$ the distribution is the *standard Cauchy distribution*. If X and Y are independent variables having *normal distributions $N(0, \sigma^2)$ and $U = X/Y$, then both U and $1/U$ have a

standard Cauchy distribution. This implies that the *t-distribution with one *degree of freedom also has a standard Cauchy distribution. If X has a *uniform distribution over the interval $[-\frac{1}{2}\pi, \frac{1}{2}\pi]$, then $Y = \tan X$ has a standard Cauchy distribution.

This last result and minor generalizations thereof have many applications in physics. A simple example is that where particles are emitted in a plane from a point source, A, in random directions (i.e. equally likely to be emitted in any direction) and follow a straight-line path, then the distribution of the points of impact of the particles on any straight line in the plane at unit distance from the point A has a standard Cauchy distribution.

Cauchy–Crompton formula *See* Radon transform.

Cauchy integral *See* integration.

Cauchy integral test A test for *convergence or divergence of a given infinite *series of positive terms,

$$a_1 + a_2 + \cdots + a_n + a_{n+1} + \cdots$$

where $a_{n+1} < a_n$. Suppose that the nth term can be expressed in the form $a_n = f(n)$, where $f(x)$ is a continuous function defined for all $x \geqslant 1$ (and not just for integral values, $x = n$). If $f(x) > 0$ for $x \geqslant 1$ and if $f(x)$ decreases steadily as x increases, then the series converges if the integral

$$\int_1^n f(x)\,dx$$

tends to a finite limit A as $n \to \infty$, i.e. if the integral

$$\int_1^\infty f(x)\,dx$$

exists; the sum of the series lies between A and $A + a_1$. The series diverges if the first integral tends to infinity as $n \to \infty$.

Cauchy ratio test (for convergence). *See* ratio test.

Cauchy–Riemann equations Given a function $f(z)$ of the complex variable $z = x + iy$ which is *analytic in some region of the complex plane, and expressible as $u + iv$, where u and v are real-valued functions of x and y, the following equations are satisfied in that region:

$$\frac{\partial u}{\partial x} = \frac{\partial v}{\partial y} \quad \text{and} \quad \frac{\partial u}{\partial y} = -\frac{\partial v}{\partial x}$$

These are the Cauchy–Riemann equations, named after A.-L. Cauchy and G.F.B. Riemann.

Cauchy–Schwarz inequality Related *inequalities are associated with the names of A.-L. Cauchy and K.H.A. Schwarz (1843–1921). Well-known forms are:
(1) For integrals: if $f(x)$ and $g(x)$ are real functions whose squares are integrable, then

$$\left(\int [f(x)g(x)] \, dx \right)^2$$
$$\leqslant \left(\int [f(x)]^2 dx \right) \left(\int [g(x)]^2 dx \right)$$

A statistical application in terms of *expectations is to two random variables X and Y with finite second moments, whence $[E(XY)]^2 \leqslant E(X^2) \, E(Y^2)$.
(2) For sums: if a_i and b_i, $i = 1, 2, \ldots, n$, are real numbers, then

$$\left(\sum (a_i b_i) \right)^2 \leqslant \left(\sum a_i^2 \right) \left(\sum b_i^2 \right)$$

which may be written in vector notation as

$$(\mathbf{a} \cdot \mathbf{b})^2 \leqslant (\mathbf{a} \cdot \mathbf{a})(\mathbf{b} \cdot \mathbf{b})$$

In statistics this result implies that the sample *correlation coefficient r satisfies the inequality $r^2 \leqslant 1$.
See also Hölder's inequality.

Cauchy sequence *See* metric space.

Cauchy's integral theorem The theorem that for a *closed curve C and an *analytic function of a complex variable $f(z)$,

$$\int_C f(z) \, dz = 0$$

there being no singular point of $f(z)$ in or on C. *See* contour integral.

Cauchy's residue theorem *See* residue.

Cauchy's theorem *See* Cauchy.

cause variable *See* regression.

caustic A curve formed by reflected light, for example in a teacup. It is the *envelope of the reflected rays. A caustic can also be formed by refracted light.

Cavalieri, Bonaventura Francesco (1598–1647) Italian mathematician. In his *Geometria indivisibilibus continuorum nova* (1635, A New Geometry of Continuous Indivisibles) Cavalieri introduced his method of indivisibles, a forerunner of the integral calculus, to determine the areas enclosed by certain curves.

Cavalieri's principle A principle used by Cavalieri in the early development of the calculus. If two solids have equal heights and their sections at equal distances from the base have areas that always have a given ratio, then the volumes of the solids are in the same ratio.

Cayley, Arthur (1821–95) English mathematician and a prolific writer, with 967 papers contained in his collected works. Of these, one of the most significant was his 'Memoir on the Theory of Matrices' (1858) which created a new mathematical discipline. Earlier, in collaboration with Sylvester, from 1843 onwards, Cayley had begun the development of the theory of invariants. A further innovation, dating from 1854, was his work on abstract groups. In algebraic geometry it was Cayley in 1843 who began the study of n-dimensional spaces where $n > 3$.

Cayley algebra (J.T. Graves, 1843) The *vector space of all 8-tuples (*see* n-tuple) of real numbers (with addition defined coordinate by coordinate) together with an internal multiplication which is not always *commutative or *associative, but which is *distributive over addition. There is a multiplicative *identity element, and every nonzero element has a multiplicative *inverse. The Cayley algebra is thus a *division algebra. It is denoted by $\mathbb{O}$ and its elements are called *octonions*. Like *complex numbers and *quaternions, each octonion c has a real *norm $\|c\|$ such that $\|c_1\|\|c_2\| = \|c_1 c_2\|$. The only normed division algebras over the real numbers are $\mathbb{R}, \mathbb{C}, \mathbb{H}$, and $\mathbb{O}$ (A. Hurwitz, 1898). John Graves first discovered the octonions, but Arthur Cayley was the first to mention them, in an article in 1845. *See* algebra; Frobenius's theorem.

Cayley–Hamilton theorem The theorem that every square matrix satisfies its own *characteristic equation. In other words, if

$$|x\mathbf{I} - \mathbf{A}| = x^n + a_{n-1} x^{n-1}$$
$$+ \cdots + a_1 x + a_0 = 0$$

is the characteristic equation of the $n \times n$ matrix $\mathbf{A}$, then

$$\mathbf{A}^n + a_{n-1}\mathbf{A}^{n-1} + \cdots + a_1\mathbf{A} + a_0\mathbf{I} = \mathbf{0}$$

where $\mathbf{0}$ is a *zero matrix and $\mathbf{I}$ is an *identity matrix.

Cayley's theorem The theorem that any *group is isomorphic (*see* isomorphism) to a group of *permutations. *See also* tree.

Cayley table *See* multiplication table.

c.d.f. *See* d.f.

ceiling function *See* integer part.

celestial axis *See* celestial equator.

celestial equator The *great circle that is the intersection of the plane of the earth's geographical equator with the *celestial

sphere. The poles of this circle are the north and south *celestial poles*. A line joining these is the *celestial axis*. *See* equatorial coordinate system.

celestial latitude (ecliptic latitude) Symbol: β. The angular distance of a point on the *celestial sphere from the ecliptic taken along a great circle passing through the ecliptic poles. Celestial latitude is measured from $0°$ to $90°$ north (taken as positive) or south (taken as negative) of the ecliptic. The complement of the celestial latitude, $90° - \beta$, the *colatitude*, is sometimes used. *See* ecliptic coordinate system.

celestial longitude (ecliptic longitude) Symbol: λ. The angular distance (measured from $0°$ to $360°$) of a point on the *celestial sphere from the vernal equinox. It is measured eastwards along the ecliptic between the vernal equinox and the place at which a great circle through the point and the ecliptic poles intersects the ecliptic. *See* ecliptic coordinate system.

celestial mechanics The study of the *dynamics of planets, satellites, comets, double stars, star clusters, etc.

celestial meridian A great circle on the *celestial sphere passing through the two celestial poles and the observer's zenith.

celestial pole *See* celestial equator.

celestial sphere An imaginary sphere of very large indeterminate radius with its centre at the centre of the earth, used in locating points in the sky. The positions of stars (and other celestial objects) can be taken as the radial projection of these objects onto the surface of the sphere. Since the radius of the sphere is large compared with that of the earth, observers on the earth can usually be considered to be at the earth's centre. Because of the rotation of the earth, the celestial sphere appears to make a full rotation in every 24-hour period.

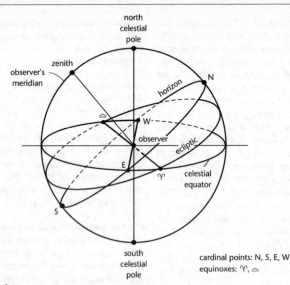

celestial sphere

cardinal points: N, S, E, W
equinoxes: ♈, ♎

celestial sphere

Positions on the celestial sphere are measured with respect to certain great circles and fixed points (*see* diagram):

(1) The *celestial equator*, which is the projection of the earth's equator onto the sphere. *See* equatorial coordinate system.
(2) The *ecliptic*, the circle which is the intersection of the earth's orbital plane with the celestial sphere. *See* ecliptic coordinate system.
(3) The *horizon*, the circle which is the intersection of a horizontal plane passing through the observer and perpendicular to the observer's *zenith with the celestial sphere. *See* horizontal coordinate system.
(4) The *galactic equator*, the circle which is the intersection of the plane of the Galaxy with the celestial sphere. *See* galactic coordinate system.

The principal points of the celestial sphere are the geometric poles of these circles and the points at which they intersect. Thus, the poles of the celestial equator are the north and south *celestial poles*; those of the galactic equator are the *galactic poles*; those of the ecliptic are called the *poles of the ecliptic*. The poles of the horizon are

the observer's *zenith* and *nadir*. The ecliptic and celestial equator intersect at the two *equinoxes. The horizon and the celestial equator intersect at two *cardinal points.

cell 1. A *topological space homeomorphic to the *n*-ball (*see* ball) is called an *n*-cell.
2. When *data are classified into categories, as in for example *grouped data for one variate or in a *contingency table for multivariate data, the subcategories are called cells. The frequency with which observations fall into a particular cell is the *cell frequency*.

Celsius degree Symbol: °C. A division of a temperature scale in which the melting point of ice is taken as 0 degrees and the boiling point of water is taken as 100 degrees. This degree and scale were formerly known as the *centigrade degree* and the *centigrade scale*. [After A. Celsius (1701–44)]. *See also* Fahrenheit degree; kelvin.

censored observations In statistical studies involving times to failure (e.g. the breakdown of machines, or the deaths of individuals), data may be incomplete in the sense that some 'units' may not have failed by the end of the study, or may have been withdrawn or lost before failure. Such data are said to be censored or, more specifically, *right censored*. In studies of a disease there is often a similar difficulty if the time of onset is of interest, for this may not be known if the disease is detected only when a patient is clinically examined and definite symptoms are apparent. In these circumstances data are described as *left censored*. Censoring may or may not depend on how patients are treated. A typical treatment-dependent situation is where severe side effects of a treatment cause patients receiving that treatment to withdraw from a study. Standard statistical techniques can be modified to take censoring into account. *See also* survival analysis.

census In *statistics, a survey of a complete population as distinct from a *sample survey.

centesimal measure *See* angular measure.

centi- *See* SI units.

centigrade degree *See* Celsius degree.

central angle An angle in a circle between two radii.

central conic A *conic that has a centre of symmetry, i.e. an *ellipse or a *hyperbola.

central difference Given *function values $y_i = f(x_i)$, where $x_i = x_0 + ih$, $i = 0, 1, 2, \ldots$, the central difference $\delta_{i+1/2}$ is defined by $\delta_{i+1/2} = y_{i+1} - y_i$. *See* finite differences.

central force A *force that is directed towards a fixed point. It commonly obeys an *inverse square law and may be a force of attraction or of repulsion. For instance, to a first approximation the motion of the planets is subject to a central force of gravitational attraction by the sun. In a central force field the force at every point acts

along the *position vector of that point relative to some point of reference.

centrality (central tendency) In statistics, a property measured by the *mean, *median, or *mode.

central limit theorem (P.-S. Laplace, 1818; A.M. Lyapunov, 1901) A theorem which states that, under very general conditions, the distribution of the mean of n *random variables tends to a *normal distribution as $n \to \infty$. The main condition is that the variance of any one variable should not dominate. An important application is to the mean of a random sample of n independently identically distributed random variables each with mean μ and standard deviation σ. For large n the theorem implies that this mean will be $N(\mu, \sigma^2/n)$. In practice, convergence is usually very rapid; for example, the means of samples of only ten observations from a continuous *uniform distribution over $[0, 1]$ are for all practical purposes normally distributed. The central limit theorem does not hold for samples from the *Cauchy distribution because that distribution has no mean.

central polyhedral angle A *polyhedral angle formed at the centre of a sphere, i.e. the polyhedral angle at the vertex of a *spherical pyramid.

central projection The central projection of a given set of points in one plane onto a second plane is the set of points produced by lines through a fixed point C and through the given points intersecting the given plane. C is the *centre of projection*. *See also* projective geometry.

central tendency *See* centrality.

centre (centre of symmetry) A point about which a geometric configuration is symmetrical. A geometric figure has a centre of symmetry if every point in the figure has a corresponding point such that the centre bisects the line segment joining the points. *See also* symmetry.

central projection of points A and B to give A′ and B′.

centre of a group *See* conjugacy class.

centre of buoyancy *See* buoyancy.

centre of curvature *See* curvature.

centre of gravity The fixed point through which the *resultant of the gravitational forces acting on all particles of a body can be considered to act, regardless of the orientation of the body. Since the resultant of the forces exerted on a body by the gravitational field constitutes the body's weight, the centre of gravity can be regarded as the point at which the weight of the body acts; if supported there, the body would remain balanced.

For a body in a *uniform gravitational field (e.g. a body in the earth's gravitational field which is small compared with the earth), the centre of gravity coincides with the *centre of mass. For a body in a non-uniform gravitational field, the forces on the body are reducible to a single force and a *couple whose plane is perpendicular to the line of action of the force. This force does not in general pass through a single point fixed with respect to the body as the body is turned in the field. However, if the matter in the body is distributed with spherical symmetry, the couple reduces to zero and the force always passes through the centre of mass. Only such a body has a centre of gravity in a non-uniform field, and is said to be *barycentric* or *centrobaric*.

centre of mass (CM; barycentre, mass centre) For a body of mass M made up of n particles m_i ($i = 1, 2, \ldots, n$) with position vectors $\mathbf{r}_i$, the centre of mass is the point with position vector $\bar{\mathbf{r}}$ given by

$$M\bar{\mathbf{r}} = \sum_{i=1}^{n} m_i \mathbf{r}_i$$

For a rod, lamina, or solid of mass M with density $\sigma(\mathbf{r})$, at the point with position vector $\mathbf{r}$ the equation becomes

$$M\bar{\mathbf{r}} = \int_V \sigma(\mathbf{r})\mathbf{r}\, dV$$

where the integral is over the region V occupied by the body. Thus, for a system of particles situated at points (x_i, y_i), in a plane, the coordinates of the centre of mass $(\bar{x}, \bar{y})$ can be found from the equations

$$M\bar{x} = \sum m_i x_i \quad \text{and} \quad M\bar{y} = \sum m_i y_i$$

For a uniform lamina of area A in the shape of the area under the curve $y = f(x)$ between $x = a$ and $x = b$, the position of the centre of mass $(\bar{x}, \bar{y})$ may be found from the equations

$$A\bar{x} = \int_a^b xy\, dx \quad \text{and} \quad A\bar{y} = \frac{1}{2}\int_a^b y^2 dx$$

When the *centre of gravity of a body exists, it coincides with the centre of mass. A table giving the centres of mass of various bodies is given in the Appendix.

centre of pressure The point on a plane surface immersed in liquid at which the *resultant pressure on the surface may be considered to act.

centre of rotation *See* rotation.

centre of symmetry *See* symmetry.

centrifugal force The *inertial force reacting against a *centripetal force.

centripetal component *See* acceleration; centripetal force.

centripetal force A *force that causes a body to deviate from motion in a straight line to motion along a curved path, or constrains a body to move in a curved path. The force at a point is directed inwards towards the centre of *curvature of this path at the point and, by Newton's laws of motion, has a magnitude equal to the mass of the body multiplied by its centripetal component of *acceleration. For motion around a circle of radius r, the centripetal force acts towards the centre of the circle and is equal to mv^2/r or $m\omega^2 r$, where m is the body's mass, v its speed, and ω its angular speed.

centrobaric *See* centre of gravity.

centroid The point in a geometrical figure whose coordinates are the arithmetical *means of the coordinates of the points making up the figure. If the figure represents a body of uniform density, the centroid coincides with the *centre of mass.

The centroid of a triangle with vertices (x_1, y_1), (x_2, y_2), and (x_3, y_3) is the point $\left(\frac{1}{3}(x_1 + x_2 + x_3), \frac{1}{3}(y_1 + y_2 + y_3)\right)$. It is the point of intersection of all three *medians, dividing each of them in the ratio $2:1$.

Ceva's theorem In a triangle ABC, L, M, and N are points on the sides AB, BC, and CA, respectively. The theorem, discovered by the Italian mathematician Giovanni Ceva (1647–1734) in 1678, states that the *necessary and sufficient condition for AM, BN, and CL to be concurrent is that

$$\left(\frac{AL}{LB}\right) \cdot \left(\frac{BM}{MC}\right) \cdot \left(\frac{CN}{NA}\right) = 1$$

An equivalent statement is that the *binary operation defined by taking the centre of mass of two (weighted) points is associative. It was in this context that it was originally studied by Ceva. *Compare* Menelaus' theorem.

cevian A line segment joining a vertex of a triangle to a point on the opposite side (or on its extension).

c.g.s. units A system of units based on the centimetre, gram, and second. It was formerly used for scientific purposes but has now been replaced by *SI units. The c.g.s. system used two different systems of electrical units: electrostatic units (e.s.u.) and electromagnetic units (e.m.u.).

chain 1. A totally ordered *set. *See* partial order.
2. *See* nested sets.

chain complex *See* homology group.

chain rule (for differentiation) A method of obtaining the *derivative of a *composite function. Differentiation is performed with respect to each function and the results are combined. If $y = f(u)$ and $u = g(x)$, then

$$\frac{dy}{dx} = \frac{dy}{du} \cdot \frac{du}{dx}$$

For example, $y = (3x + 1)^2$ can be regarded as $y = u^2$, where $u = 3x + 1$. Application of the chain rule gives

$$\frac{dy}{dx} = 2u \cdot 3 = 6(3x + 1)$$

For a greater number of functions the expression becomes

$$\frac{dy}{dx} = \frac{dy}{du} \cdot \frac{du}{dv} \cdot \frac{dv}{dx}$$

There is a version of the chain rule for functions of several variables which states that the *Jacobian matrix of a composite function fg is obtained by multiplying the Jacobian matrices of f and g. *See also* composite function.

Champernowne's number (D.G. Champernowne, 1933) The number

$$0.123\,456\,789\,101\,112\ldots$$

whose decimal digits are those of all the

*natural numbers in succession. It is an example of a *normal number.

change of variable (in integration) The transformation of an integral by substitution of a different variable. For the integration of a function $f(x)$ the method involves choosing a function $x = g(u)$, which is substituted in $f(x)$ to give a function of u, say $F(u)$. Differentiating $x = g(u)$ gives $dx = g'(u)du$. The change of variable is thus

$$\int f(x)\, dx = \int F(u)g'(u)\, du$$

For instance, the integral

$$\int \sqrt{(1-x)}\, dx$$

can be transformed by making the change of variable

$$x = 1 - u^2$$

so $1 - x = u^2$, and $dx = -2u\, du$. The integral then becomes

$$\int -2u^2 du$$

In the case of a definite integral, the limits ($x = a$ and $x = b$, say) are also changed using $x = g(u)$ to $u = g^{-1}(a)$ and $u = g^{-1}(b)$. The method is called *integration by substitution*.

channel A means of sending messages from one place to another. For example, a telephone line or a radio signal. A noisy channel is one that will corrupt or omit parts of the messages. *See* information theory.

chaos A general term for a type of behaviour found in certain *dynamical systems whose evolution, though deterministic, appears to be unpredictable and random.

There is no single accepted definition of chaos, although it is common to speak of a sensitive dependence on initial conditions,

i.e. the orbits of adjacent points evolve in markedly different ways. As the underlying equation of the map or *flow is usually nonlinear, chaos is an aspect of *nonlinear dynamics*. A chaotic dynamical system contains at least one point whose orbit is a *dense set. The term 'chaos' was introduced in 1975 by T.-Y. Li and J.A. Yorke in their work on periodic points for transformations of the real line. It is usually associated with the presence in the system of a strange attractor (*see* below).

For a dynamical system in a space X with an *iterative map $x_{n+1} = T(x_n)$, an *attractor* is an invariant set A in X towards which nearby points x converge, i.e. $T(A) = A$ and $x_n = T^n(x)$ approaches A as n increases for points close to A. For example, for the transformation $z \mapsto z^2$ on the complex plane, the single point $A = \{0\}$ is an attractor. However, the invariant set $\{z: |z| = 1\}$ is *not* an attractor. In fact this is a *repellor*, an invariant set from which nearby points diverge. Similarly, for a space X with a flow, an attractor is an invariant set towards which nearby points converge in time.

A *strange attractor* is an infinite invariant set A, usually an attractor, with additional properties. There is no single accepted definition, although the most commonly used is that orbits in A have a sensitive dependence on initial conditions and there is an open set of points attracted to A. The term was introduced by D. Ruelle and F. Takens in 1971. Two important examples are:

(1) The *Hénon attractor* (*see* diagram (a)). A strange attractor contained in the plane associated with the Hénon map $T(x,y) = (y + 1 - ax^2, bx)$ for some choice of real numbers a and b. It was studied in computer experiments by M. Hénon in 1976, and rigorously shown in 1991 to be a strange attractor by M. Benedicks and L. Carleson for many parameter values.
(2) The *Lorenz attractor* (*see* diagram (b)). A subset of three-dimensional space invari-

(a)

(b)

chaos (a) The Hénon attractor for $a = 1.3$, $b = 0.3$. (b) The Lorenz attractor.

ant under the flow which is the solution to the Lorenz equations:

$$\frac{dx}{dt} = 10(y - x)$$

$$\frac{dy}{dt} = -xz + 28x - y$$

$$\frac{dz}{dt} = xy - \frac{8}{3}z$$

The Lorenz attractor was originally studied by Edward N. Lorenz in 1963 as a model for weather; in 2002 the Swedish mathematician Warwick Tucker showed that it is a strange attractor.

characteristic 1. *See* logarithm.
2. (of a ring or field) For a given *ring R, if a positive integer n exists such that $na = 0$ for all a in R, then the least such positive integer is the characteristic of the ring. If no least positive integer exists, the characteristic is zero (or sometimes 'characteristic ∞' is used). The rings of integers, rational numbers, and real numbers all have characteristic 0. If the ring is an *integral domain, the characteristic is either zero or a prime.

See also Euler–Poincaré characteristic.

characteristic equation. *See* characteristic polynomial.

characteristic function 1. The expected value (*see* expectation) of the *function $g(X) = \exp(itX)$ of the *random variable X, for real t, written as $\phi(t) = E[\exp(itX)]$. It exists for, and uniquely defines, any distribution, hence the name. For the standard *normal distribution

$$\phi(t) = \exp(-\tfrac{1}{2}t^2)$$

See also moment generating function.
2. The characteristic function χ_A of a subset A of a set X is the function $\chi_A: X \to \{0, 1\}$ defined by $\chi_A(x) = 1$ if $x \in A$ and $\chi_A(x) = 0$ if $x \notin A$. It is also called the *indicator function* of the set A.

characteristic polynomial For an $n \times n$ *matrix $\mathbf{A}$ the polynomial $\det(x\mathbf{I} - \mathbf{A})$,

where $\mathbf{I}$ is the identity matrix of the same dimension as $\mathbf{A}$ and x is a *scalar variable. The characteristic polynomial is a *polynomial of degree n: $\det(x\mathbf{I} - \mathbf{A}) = x^n + a_{n-1}x^{n-1} + \cdots + a_0$, where a_{n-1} is minus the *trace of $\mathbf{A}$ and $a_0 = (-1)^n \det(\mathbf{A})$. The equation $\det(x\mathbf{I} - \mathbf{A}) = 0$ is the *characteristic equation* of $\mathbf{A}$ and its roots are the *eigenvalues of $\mathbf{A}$. Sometimes the characteristic polynomial is defined as $\det(\mathbf{A} - x\mathbf{I})$. See also Cayley–Hamilton theorem.

Chebyshev *See* Tchebyshev.

check digit A digit in a *codeword whose sole purpose is to check whether there has been an error in transmission or transcription. For example, if the codewords are written using 5 decimal digits, a 6th digit could be added, for example, it could be the last digit of the sum of the 5 digits; this extra information would enable a receiver to detect that one of the six digits has been transmitted incorrectly. More subtle check digits can detect transposition of digits in the original message. *See* ISBN.

Chinese postman problem Any problem equivalent to that of a city postman who wishes to visit all the streets in his area to deliver his letters and return to his starting point, having covered the least possible distance. In terms of *weighted graphs, the problem is to find a closed *walk which includes every edge (at least once) and has least total weight. *See also* travelling salesman problem; network analysis.

Chinese remainder theorem The theorem that if $m_1, \ldots, m_r$ are natural numbers every pair of which are *relatively prime, and $a_1, \ldots, a_r$ are any integers, then there is an integer x that simultaneously satisfies the *congruences

$$x \equiv a_1 \,(\text{mod}\, m_1), \ldots, x \equiv a_r \,(\text{mod}\, m_r)$$

Also, if $x = a$ is any solution then all other solutions are congruent to a modulo the product $m_1 m_2 \cdots m_r$.

Simultaneous congruences occur in such problems as that of finding a number that leaves the remainders 2, 3, and 2 when divided by 3, 5, and 7, respectively. This requires finding an integer x such that:

$$x \equiv 2 \bmod 3$$

$$x \equiv 3 \bmod 5$$

$$x \equiv 2 \bmod 7$$

The theorem is so named because it originates in the study of problems such as the one above in the *Sunzi suanjing* (Master Sun's Mathematical Manual), which circulated in China from the 3rd century AD onwards. It was first established by Qin Jiushao in 1247.

There is a more elaborate version that gives precise conditions under which a set of simultaneous congruence equations have a solution when the moduli m_k are not pair-wise relatively prime. For example, the equations $x \equiv 1 \bmod 2$ and $x \equiv 2 \bmod 4$ do not have a simultaneous solution.

chi-squared distribution The sum of squares of n independent standard normal variables (*see* normal distribution) has a chi-squared (χ^2) distribution with n *degrees of freedom. The distribution belongs to the *gamma distribution family, and has mean n and variance $2n$. Tables of percentiles of the distribution for various values of n are available for use in the *chi-squared test. Many statistical software packages provide exact one-tail *p-values.

chi-squared test 1. A test of goodness of fit of observations to a theoretical discrete *distribution. If a value x_i ($i = 1, 2, ..., n$) that is expected to occur E_i times for that distribution occurs O_i times, the statistic

$$\chi^2 = \sum_i \frac{(O_i - E_i)^2}{E_i}$$

has a chi-squared distribution with $n - p$

*degrees of freedom, where p is the number of distribution parameters estimated from the data and used to compute the E_i. Significantly high values of χ^2 lead to the rejection of the hypothesized distribution. Some modifications are needed if a few E_i are small (e.g. if several are 5 or less).

If a fair die is cast 96 times, then $x_i = 1$, 2, 3, 4, 5, or 6 are the possible scores for each throw, and $E_i = 16$ for all i. If $O_1 = 14$, $O_2 = 19$, $O_3 = 11$, $O_4 = 21$, $O_5 = 12$, and $O_6 = 19$, then $\chi^2 = 5.5$. The mean of 16 is computed from the data, so there are $6 - 1 = 5$ degrees of freedom. The χ^2 value is not significant at the 5 percent level, so the hypothesis that the die is fair is not rejected.

The test may be adapted and applied to grouped data from a continuous distribution, when these are the only data available, to see whether they are consistent with a specified distribution (e.g. a normal distribution). However, if individual observations are available these should not be arbitrarily grouped simply to allow the test to be applied because the outcome of the test is not independent of the choice of class intervals for grouping, and some groupings may lead to a significant value of the chi-squared statistic, while others may not. *See also* Kolmogorov–Smirnov test.

2. A test for lack of association (independence) between numbers in row and column categories in an $r \times c$ *contingency table. The expected numbers in any cell may be computed from the fixed marginal totals, and a statistic of the form of χ^2 given in **1** above is calculated by taking the observed and expected numbers in each cell and summing over all cells. The degrees of freedom are $(r - 1)(c - 1)$. Large values of χ^2 lead to the rejection of the independence hypothesis, since they indicate some association between row and column categories. In the special case of 2×2 tables, *Yates's correction* should be applied by subtracting 0.5 from the magnitude of each difference, $|(O_i - E_i)|$, before squaring. The test is an approximation to *Fisher's exact test for 2×2 tables. The cautions about small expected numbers in **1** also apply here.

Chiu-chang Suan-shu See: *Jiuzhang suanshu*.

choice, axiom of An axiom of set theory that states that for any *set S there is a *function f (called the *choice* or *selection function*) such that for any nonempty subset X of S, $f(X) \in X$. The set of values of f is called the *choice set*. A choice function for S may be regarded as selecting a member from each nonempty subset of S. For example, if $S = \{1, 2\}$, then nonempty subsets of S are $X_1 = \{1\}$, $X_2 = \{2\}$, and $X_3 = \{1, 2\}$. Two choice functions for S may then be defined:

$$f_1(X_1) = 1, \quad f_1(X_2) = 2, \quad f_1(X_3) = 1$$

and

$$f_2(X_1) = 1, \quad f_2(X_2) = 2, \quad f_2(X_3) = 2$$

Zermelo first used (an equivalent of) this axiom to prove that every ordered set can be *well ordered. The axiom has been thought to be counterintuitive, mainly on the grounds that it asserts the existence of (choice) sets independently of any property all the members of the set satisfy. In 1938 Gödel proved that the axiom is *consistent with the other axioms of set theory, and in 1963 Cohen proved its *independence.

Cholesky factorization A *matrix A that is symmetric and *positive definite can be factored as $A = R^T R$, where R is an *upper triangular matrix with positive diagonal entries, called the *Cholesky factor*. For example,

$$\begin{pmatrix} 1 & -1 \\ -1 & 5 \end{pmatrix} = \begin{pmatrix} 1 & 0 \\ -1 & 2 \end{pmatrix} \begin{pmatrix} 1 & -1 \\ 0 & 2 \end{pmatrix}$$

It is named after André-Louis Cholesky (1875–1918).

chord A straight-line segment joining any two points on a curve or surface. A chord is a segment of a *secant lying between two points of intersection of the secant and the curve. If two tangents are drawn to a circle from a point outside the circle, the chord joining the two points of contact of the tangents is called the *chord of contact*.

Chuquet, Nicolas (*c*.1440–*c*.1488) French mathematician and author of *Le Triparty* (*c*.1480), a treatise on algebra existing only in manuscript until published in 1880. The first part deals with the newly introduced Hindu–Arabic numbers, while later parts cover a number of notational innovations and, apparently for the first time, introduce an isolated negative number.

Church, Alonzo (1903–95) American mathematical logician and author of *Introduction to Mathematical Logic* (1956). In 1935 he proposed to identify effective computability with λ-definability or general recursiveness; in the following year he went on to show that the first-order functional calculus was undecidable.

Church's theorem (A. Church, 1936) The theorem that there is no *effective procedure for deciding whether or not a given *wff of the *predicate calculus is a theorem. In other words, the *decision problem for the predicate calculus has a negative solution (is *unsolvable*).

Church's thesis (A. Church, 1935) The principle, according to one formulation, that all effectively computable (*see* effective procedure) functions are *recursive. This thesis ties together an intuitive concept of effective computability and a precise mathematical concept, and is thus not susceptible to proof. However, evidence for the thesis can be adduced from the provable equivalence of many different attempts to characterize accurately the notion of effectiveness.

Chu Shih-chich See Zhu Shijie.

cipher (cypher) **1.** The symbol 0 for zero.

2. To calculate; to carry out computations using numbers.

3. An algorithm used to transform text (e.g. a message or credit card number), called the *plaintext*, into another text, called the *ciphertext*. This is done before transmission to a recipient, the aim being to prevent the plaintext from being recovered by a third party. The algorithm usually consists of replacing each character of the plaintext by a *word in some *alphabet, and its exact implementation depends on a chosen *key. A recipient who knows the cipher and the key can then recover the plaintext from the ciphertext.

ciphertext The text obtained after applying a *cipher to the plaintext.

circle A plane curve that is the *locus of a point which moves at a fixed distance (the radius r) from a fixed point (the centre). The area enclosed by a circle is πr^2 and the circumference is $2\pi r$. Theorems associated with circles include the following:

(1) Angles subtended by an arc at the circumference and lying in the same segment are equal.

(2) The angle that an arc subtends at the centre of a circle is twice the angle that the arc subtends at points on the remainder of the circumference.

(3) An angle subtended at the circumference by a semicircle is a right angle.

(4) If two tangents are drawn from an external point P to a circle, then:

(a) the tangents have equal length;

(b) the tangents subtend equal angles at the centre of the circle;

(c) the line from the point to the centre bisects the angle between the tangents.

(5) The *tangent–secant theorem*: if a tangent PA and a *secant PBC are drawn from an external point P, then $PA^2 = PB \cdot PC$.

(6) The *intersecting chords theorem*: if two chords AB and CD intersect at a point Y, then $AY \cdot BY = CY \cdot DY$.

(7) The *alternate segment theorem* (or *tangent–chord theorem*): if a straight line touches a circle, and from the point of contact a chord is drawn, the angles which the chord makes with the tangent are equal to the angles subtended by the chord at the circumference in the alternate segments.

A circle can be regarded as a *conic with an *eccentricity of 0 (i.e. a special case of an ellipse). In rectangular Cartesian coordinates its equation is

$$(x - a)^2 + (y - b)^2 = r^2$$

where r is the radius and (a, b) the centre. The *parametric equations of this circle are (*see* diagram)

$$x = a + r \cos \theta \quad \text{and} \quad y = b + r \sin \theta$$

circle of convergence *See* power series.

circle of curvature *See* curvature.

circuit *See* walk.

circulant A type of *matrix (or determinant) in which each row is a *cyclic permutation of the row above, and such that all the elements of the principal diagonal are

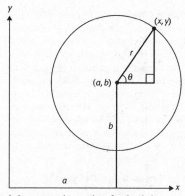

circle: parametric equations for the circle.

identical; for example,

$$\begin{pmatrix} a & b & c & d \\ d & a & b & c \\ c & d & a & b \\ b & c & d & a \end{pmatrix}$$

circular 1. Having the form of a circle.
2. Having a circle as base, as in a *circular cone* or *circular cylinder*.

circular data *See* directional data.

circular functions *See* trigonometric functions.

circular helix *See* helix.

circular measure *See* angular measure.

circular motion Motion along the circumference of a circle. For the motion to be uniform – at constant speed – there must be a continuous *acceleration towards the centre of the circle, i.e. a *centripetal force must be acting, and the tangential component of acceleration must be zero.

circular permutation *See* cyclic permutation.

circumcentre The centre of the circumcircle of a given polygon. *See* circumscribed.

circumcircle A circle *circumscribed about a given polygon.

circumference 1. The length of a circle, equal to $2\pi r$, where r is the radius.
2. (of a sphere) The length of a great circle on the sphere.
3. The length of any closed curve or figure (i.e. the perimeter).

circumferential mean *See* subharmonic function.

circumscribed Describing a relationship in which one figure encloses another. Most commonly it is used to describe the situation in which a *polygon can be completely enclosed by a circle (the *circumcircle*) that passes through all the vertices

of the polygon. The polygon is then said to be *circumscribed by* the circle; the circle is *circumscribed about* the polygon. Alternatively, a polygon completely enclosing a circle so that every side is a tangent to the circle is said to be circumscribed about the circle. The term can be extended to other figures, including solid figures. A polyhedron can be circumscribed by a sphere if all the vertices lie on the surface of the sphere. A prism can be circumscribed by a cylinder if all the edges of the prism lie on the cylinder's surface. *See also* inscribed.

cis *See* complex number.

cissoid A plane curve with the equation

$$r = 2a \tan \theta \sin \theta$$

in polar coordinates. In Cartesian coordinates, the equation is

$$y^2(2a - x) = x^3$$

The curve is symmetrical about the x-axis and has a *cusp at the origin. It can be generated by taking a circle (radius a) with a fixed point P on the circle. A tangent

cissoid

is drawn to the circle at the opposite end of the diameter through P. A variable line from P cuts the circle at Q and the tangent at R. The cissoid is the locus of points S, such that $PS = QR$.

Clairaut, Alexis-Claude (1713–65) French mathematician and physicist who worked on problems of geodesy and celestial mechanics, and on differential equations, in which field he established *Clairaut's equation. He also published important work on cubic curves.

Clairaut's equation A *differential equation of the form

$$y = x\frac{dy}{dx} + f\left(\frac{dy}{dx}\right)$$

where $f(dy/dx)$ is a function of (dy/dx) only.

class *See* set.

class equation An equation that counts the number of elements in the *conjugacy classes of a finite *group G. The group is the union of its distinct conjugacy classes, so the *order $|G|$ of G is the sum of the numbers of elements in the conjugacy classes. Since the union of the *singleton conjugacy classes is the centre, $Z(G)$, their number is its order $|Z(G)|$. So the class equation is usually written as

$$|G| = |Z(G)| + l + m + \cdots$$

where $l, m, \ldots$ are the numbers of elements in the different multi-element conjugacy classes.

class frequency The number of observations in a given *class interval.

class group In a *field of *algebraic numbers, a *group can be made out of *equivalence classes of the set of *ideals of the algebraic integers of the field. The equivalence *relation is that the ideals I and J are equivalent if and only if there are principal ideals S and T such that $IS = JT$; and the group operation is that the product of the

classes containing I and J is the class containing IJ. The *class number* of the algebraic number field is the number of elements in the class group. It can be shown that there is unique factorization in the ring of integers of the field if and only if the class number is 1.

classical mechanics (Newtonian mechanics) The study of the behaviour of systems under the action of forces, i.e. the study of the motions and states of *equilibrium of bodies, based on *Newton's laws of motion. Classical mechanics forms a basic and long-established part of physics and engineering. It can be divided into *dynamics (*kinematics plus *kinetics) and *statics, or into dynamics (including statics) and kinematics. It is usually concerned with the motions of solid bodies rather than fluids. Newton's laws are inadequate for the treatment of systems in which components move at speeds approaching that of light, or of systems of atoms, molecules, etc.; these systems are the subject matter of *relativistic mechanics and *quantum mechanics respectively. *See also* mechanics; hydrostatics.

class intervals Intervals in which data are grouped. For example, if employees' weekly wages are known we may count the number receiving between £0.00 and £199.99, between £200.00 and £299.99, etc. The class intervals are £0.00–£199.99, £200.00–£299.99, etc. The number of employees in each class interval gives the *class frequency for that interval. Class intervals often are, but need not necessarily be, all of the same length. *See also* grouped data; histogram.

class number *See* class group.

closed curve A curve that has no end points; i.e. one that is a continuous transformation of a *closed interval $[a, b]$ in which the images of a and b coincide. *Compare* open curve.

closed interval A *set of real numbers $\{x: a \leqslant x \leqslant b\}$, written as $[a, b]$. The interval contains the end points a and b. In n-dimensional space, if $a = (a_1, \ldots, a_n)$ and $b = (b_1, \ldots, b_n)$ are two distinct points with $a_j \leqslant b_j$ $(j = 1, \ldots, n)$, then the closed interval $[a, b]$ is given by

$$\{(x_1, \ldots, x_n): a_j \leqslant x_j \leqslant b_j, \quad j = 1, \ldots, n\}$$

An interval is partly open and partly closed if it contains just one of its end points, and is written as $[a, b)$ if it contains a and $(a, b]$ if it contains b. *Compare* open interval.

closed region *See* region.

closed set (of points) A *set A is closed if it contains all its *limit points. For example, the points corresponding to the real numbers equal to or greater than 0 and equal to or less than 1 constitute a closed set. A closed set is the complement of an *open set.

closure The closure of an *open set A is obtained by adding to it all *limit points of A. Thus, if A equalled the real numbers between 0 and 1, the closure of A would be obtained by adding to A the limit points 0 and 1. The closure of a set A is denoted by $\bar{A}$. *See also* derived set.

cluster analysis Statistical techniques for determining, on the basis of measurements of one or more characteristics for each of a number of items, whether the items fall into recognizable groups called *clusters*. For a chosen *metric, items in any one cluster will in general be closer to each other than they are to items in another cluster. Objective criteria are needed to determine the number of clusters and to allocate items to clusters.

Data on age, income, ownership of home or car, time spent out of the home each week, etc. for a number of people are likely to show evidence of several distinct clusters, each corresponding to a category such as employed, unemployed, pensioners, or students.

cluster point *See* limit point.

cluster sample A sample in which natural or artificial groups of sampling units (each called a *cluster*), rather than individual units, are selected from a population. Observations are made on all units in each selected cluster. For example, households may each form a cluster, and individuals in each selected household may be the units; or farms may each form a cluster, and individual fields on each farm the units. *See also* area sampling.

CM *Abbreviation for* *centre of mass.

coaltitude *See* zenith distance.

coaxial Having the same *axis, as in *co-axial cylinders*.

cobordism Two n-manifolds, M and N, are *cobordant* if there exists (*see* manifold) an $(n + 1)$-manifold-with-boundary, W, whose boundary is the disjoint union of M and N; W is called a *cobordism* between M and N.

The notion of cobordism is due to R. Thom (1954), who gave *necessary and sufficient conditions for two (differential) manifolds to be cobordant. Thom's work was extended by J.W. Milnor (1960) and C.T.C. Wall (1960). It is an important tool in the classification of manifolds.

Cocker, Edward (1631–75) English mathematician. As a London teacher and the author of the posthumous *Arithmetick* (1678), Cocker was sufficiently well known to endow the phrase 'according to Cocker' with an almost proverbial status.

code 1. *See* coding.
2. A particular method of encoding, for example *Reed–Solomon code.

codebreaking An attempt by someone other than the intended recipient (who does not know the *key) to reconstruct the original message from the coded message.

codeclination In *equatorial coordinates, the complement of the declination.

code correction A correction to a *codeword that has been corrupted (usually during transmission).

code length The length of the codewords in a *code (all codewords in a given code usually have the same length).

code weight The number of 1's in a particular *codeword from a *binary code.

codeword *See* coding.

coding 1. A method of rewriting a message before it is transmitted. Two common purposes for doing this are to make the message difficult to read, except by the intended recipients, and to introduce checks to correct errors that might be introduced during transmission.

More formally, a *code C* of length *n* is a set of *n*-tuples from a set *A*, its *alphabet. An element of *C* is called a *codeword*. A codeword is usually written as a *string $a_1a_2 \cdots a_n$, but sometimes as an *n*-tuple $(a_1, a_2, \ldots, a_n)$. Often the alphabet is a *field *F* such as $\mathbb{F}_2 = \{0, 1\}$; then *C* is a *linear code* if it is a subspace of the *vector space of all *n*-tuples from *F*. If *C* is linear and of dimension *k*, a $k \times n$ *matrix whose rows form a basis for the vector subspace *C* is called a *generator matrix* for *C*.
2. *See* data coding.

coding theory The theory of the encryption of messages, using *ciphers, for security during the transmission of data, or for the recovery of information from corrupted data (e.g. in reconstructing messages sent over long distances by space probes). One of the most secure ciphers is the *RSA cipher, in which information is stored using integers modulo a large number *N*, but which can be deciphered only if one knows the prime *factorization of *N*; *N* is usually chosen to be the product of two large primes. Many *error-correcting codes which identify and correct errors in messages corrupted during transmission have been constructed using *Galois fields.

codomain *See* function.

coefficient 1. In general, the product of all the factors in an expression except for a specified factor. Thus the coefficients of *x* in the expressions $3x$, $(a + b)x$, and $2xyz$ are respectively 3, $(a + b)$, and $2yz$. A coefficient is usually a constant.
2. A number that serves as a measure of some property or characteristic of a body, material, process, etc.

coefficient matrix *See* augmented matrix.

coefficient of concordance (M.G. Kendall, 1939) A test for the consistency of more than two sets of rankings, such as the merit ordering of competitors by several judges in a sporting contest. If *m* judges each award ranks 1 to *n* independently to competitors, the sum s_i of the ranks awarded to competitor *i* has mean $\frac{1}{2}m(n + 1)$. The sum of the squares of the deviations of the s_i from their mean is

$$S = \sum \left[s_i - \frac{1}{2}m(n + 1)\right]^2$$

and the coefficient of concordance is

$$W = \frac{12S}{m^2n(n^2 - 1)}$$

It can be shown that $0 \leqslant W \leqslant 1$. If all judges give identical rankings, then $W = 1$. The minimum possible value is $W = 0$, which corresponds to complete disagreement, and values near zero imply little agreement. If four judges give the ranks shown in the table to three competitors, clearly there is little agreement, and $S = 2$ and

$$W = \frac{12 \times 2}{16 \times 3 \times 8} = \frac{1}{16}$$

The test is equivalent to *Friedman's test. It will not detect patterns other than overall preference; so, for example, if two judges each rank four candidates in order of preference 1, 2, 3, 4, and another two rank them 4, 3, 2, 1, there is a reversal of preferences and $W = 0$. *See also* nonparametric methods; rank.

	Judge				
Competitor	A	B	C	D	s_i
I	1	1	2	3	7
II	3	2	3	1	9
III	2	3	1	2	8

coefficient of correlation *See* correlation coefficient.

coefficient of determination (index of determination) If data consist of n paired observations (x_i, y_i) and a least-squares linear *regression of y upon x is fitted, the proportion of the total variance of the y_i attributable to dependence on x (as opposed to independent variation) is r^2, the square of the product moment *correlation coefficient, and r^2 is called the coefficient (or index) of determination. The dependence is total if $r = \pm 1$, when the regression line accounts for all the variation in the y_i because the fit is perfect. The quantity $1 - r^2$ may be regarded as a measure of independence or lack of correlation because it takes a maximum value of 1 when $r = 0$ and there is zero correlation, and a minimum value of zero when $r = \pm 1$ and there is total dependence.

coefficient of friction *See* friction.

coefficient of kurtosis *See* kurtosis.

coefficient of multiple determination *See* multiple correlation coefficient.

coefficient of restitution *See* Newton's law of restitution.

coefficient of skewness *See* skewness.

coefficient of variation *See* variation, coefficient of.

cofactor A number associated with an element of a *determinant. If the element is in the ith row and jth column, its cofactor equals the determinant of lower order obtained by removing the row and the column in which the element appears, multiplied by $(-1)^{i+j}$. The determinant of lower order is called the *minor* of the original determinant, and the cofactor is sometimes called the *signed minor*. A cofactor of a *matrix is a cofactor of the determinant of the matrix.

For the determinant

$$\begin{vmatrix} a & b & c \\ d & e & f \\ g & h & i \end{vmatrix}$$

the cofactor of e is

$$(-1)^4 \begin{vmatrix} a & c \\ g & i \end{vmatrix} = ai - gc$$

and the cofactor of d is

$$(-1)^3 \begin{vmatrix} b & c \\ h & i \end{vmatrix} = ch - bi$$

cofunctions Pairs of *trigonometric functions that are equal when the variable in one function is the complement of the variable in the other. The sine and cosine functions are cofunctions:

$$\sin \theta = \cos(90° - \theta)$$

Other pairs of cofunctions are the tangent and cotangent, and the secant and cosecant.

Cohen, Paul Joseph (1934–2007) American mathematician who finally resolved (1963) the status of Cantor's continuum hypothesis. Gödel had shown in 1938 that the hypothesis could not be disproved in restricted set theory; Cohen went further and demonstrated that it could not be proved either, thus showing the hypothesis to be independent of the axioms of Cantor's set theory.

Cohen's kappa statistic (J. Cohen, 1960) A measure of agreement, κ, between two observers when they have independently classified each one of a set of items, there being two or more classes, and they are said to agree over an item if they assign it to the same class.

For example, two doctors each examine 80 patients claiming to suffer from depression, and independently classify each patient as to whether treatment with a specified antidepressant drug is appropriate or inappropriate. Both agree in 32 cases that treatment is appropriate and both agree in 35 cases that treatment is inappropriate. In the remaining 13 cases there is disagreement, one doctor thinking treatment is appropriate, while the other does not.

In general, if there are N items and n is the observed number of agreements over all classes then $p_{obs} = n/N$ is the observed proportion of agreements. If p_{exp} is the expected proportion of agreements over all classes under a random assignment, calculated in the usual manner for a *contingency table, then

$$\kappa = \frac{p_{obs} - p_{exp}}{1 - p_{exp}}$$

The coefficient takes values between $-p_{exp}/(1 - p_{exp})$ and 1. The value $+1$ corresponds to perfect agreement, zero to the level expected by chance, and negative values to apparent disagreement. Formulae are available for obtaining *confidence intervals for κ.

coherent units A system of units in which the *derived units are obtained from the *base units by multiplication or division without the introduction of numerical factors. *SI units form a coherent system of units.

cohomology The *cohomology groups* $H^n(X)$ $(n \geqslant 0)$ of a *topological space X are variants of the *homology groups of X, but with the characteristic property that, given a continuous map f: $X \to Y$, the corresponding homomorphisms f* run from $H^n(Y)$ to $H^n(X)$ rather than the other way round.

Cohomology groups arise naturally in the statement of the *Poincaré duality theorem for manifolds. They are important also because the cohomology groups of X can be given the additional structure of a ring, making them slightly more powerful tool in algebraic topology than homology. Cohomology has been adapted and usefully applied in several areas of mathematics. *See* differential form.

colatitude **1.** Symbol: θ. The angle between the polar axis and the radius vector in a *spherical coordinate system.
2. The complement of *celestial latitude in an *ecliptic coordinate system.

collinear Having a common line. Thus, *collinear points* are points that lie on a straight line. *Collinear planes* are planes that intersect in a common straight line. *See* concurrent.

collinearity transformation (collineation) A *transformation that takes collinear points into collinear points. *See* matrix.

collision Momentary point contact between two objects (e.g. snooker balls) and their resulting interaction, or the deflection of two particles (e.g. nuclear particles) from their original paths as a result of long-range interaction rather than direct contact. *Kinetic energy can be lost in a collision as a result of changes in the internal energies of the two objects, as by the heating up of a snooker ball or the excitation of an atom. If no change in kinetic energy occurs, i.e. if kinetic energy is conserved, then the collision is said to be *elastic*; otherwise it is described as *inelastic*. *See* Newton's law of restitution.

column A vertical line of elements in an array, as in a *determinant or *matrix.

column rank The dimension of the *column space of a matrix. It is equal to the *row rank and the *rank of the matrix.

column space The vector space of all *linear combinations of the columns of a matrix.

column vector (column matrix) A *matrix having a single column of elements.

combination The number of selections of r different items from n distinguishable items when order of selection is ignored. Denoted by $\binom{n}{r}$ or ^nC_r, it has the value $n!/[r!(n-r)!]$; and $\binom{n}{r}$ is the coefficient of $x^r y^{n-r}$ in the binomial expansion of $(x+y)^n$. *See* binomial distribution; permutation.

combinatorial theory *See* combinatorics.

combinatorial topology The study of *topological spaces that are constructed by piecing together elementary 'blocks' called *simplexes*, which are higher-dimensional analogues of points, line segments, and triangles.

More precisely, for $n \geqslant 0$ an *n-simplex* σ (or *simplex of dimension n*) is defined to be the *convex hull in some Euclidean space $\mathbb{R}^m$ of a set of $n+1$ points $a_0, a_1, \ldots, a_n \in \mathbb{R}^m$ (called *vertices*), provided these points a_i are 'independent' in the sense that the equations

$$\sum_0^n \lambda_i a_i = 0 \quad \text{and} \quad \sum_0^n \lambda_i = 0$$

(where $\lambda_0, \lambda_1, \ldots, \lambda_n$ are real numbers) imply that $\lambda_0 = \lambda_1 = \cdots = \lambda_n = 0$ (thus three points a_0, a_1, and a_2 are independent if they are not collinear). For example, a tetrahedron is a 3-simplex in $\mathbb{R}^3$.

A (nonempty) subset of $r+1$ vertices of σ determines an *r-simplex contained in σ, called a *face* of σ.

A *simplicial complex K* is a finite set of simplexes in some $\mathbb{R}^m$, with the property that all faces are included and any simplexes meet, if at all, in a common face.

The union of the simplexes in K is called the *polyhedron* of K. The *dimension* of K is the dimension of its simplex of highest dimension.

See also homology group.

combinatorics The branch of mathematics involved in the study of discrete objects – those where continuity plays no role. Enumeration and *graph theory are important examples of areas of combinatorics. The topic has applications in many branches of science, especially computer science.

commensurable Describing two quantities that are integral multiples of a common unit. 16 and 12 are commensurable since they are both integral multiples of 1, 2, or 4. Likewise, 3 feet and 2 inches are commensurable quantities since 3 feet contains 2 inches an integral number of times. Numbers are commensurable if their ratio is rational. $\sqrt{2}$ and 1 are incommensurable since $\sqrt{2}/1$ is not rational.

common denominator A common multiple of the denominator of two or more fractions, i.e. a number that each denominator divides exactly. For example, the fractions 1/2, 1/3, and 3/7 have common denominators of 42, 84, 126, 168, etc. The *least common denominator* (LCD) is the lowest such number, in this case 42. The common denominator is used in adding fractions.

common difference *See* arithmetic progression.

common factor (common divisor) A number that divides two or more given numbers exactly. For example, the numbers 20, 70, and 80 have 2 as a common factor; other common factors are 5 and 10. The largest number that is a common factor of the given numbers is the *highest common factor* (HCF), also called the *greatest common divisor* (GCD). In the case above the HCF is 10. *See also* Euclidean algorithm.

common fraction (simple fraction, vulgar fraction) A fraction in which both numerator and denominator are integers. *Compare* complex fraction.

common multiple A number that is a multiple of two or more other numbers. The lowest number that is a multiple of a given set of numbers is their *least common multiple* (LCM). For example 3, 9, and 11 have a LCM of 99; i.e. 99 is the smallest number that all three of the given numbers will divide exactly. The LCM can be found by splitting each number into prime factors. Thus, to find the LCM of 7, 9, 12, and 14:

$$7 = 7$$
$$9 = 3^2$$
$$12 = 3 \times 2^2$$
$$14 = 7 \times 2$$

The LCM is obtained by multiplying the prime factors together, taking each the maximum number of times it occurs in any of the numbers. In this case the LCM is $7 \times 3^2 \times 2^2 = 252$.

common ratio *See* geometric progression.

common tangent A line that is a *tangent to two separate curves. Two circles that lie outside each other have four common tangents: two *external tangents* (the circles lie on the same side of the tangent) and two *internal tangents* (the circles lie on opposite sides).

commutative Describing a *binary operation ∘ where the result of the operation does not depend on the order of the elements a and b it is applied to; that is,

$$a \circ b = b \circ a$$

Thus the commutative law of addition is

$$a + b = b + a$$

and the commutative law of multiplication is

$$a \times b = b \times a$$

Many mathematical systems contain noncommutative operations. *See* group; ring; vector product.

commutator (of elements in a group) The commutator c of two elements a and b in a *group is an element such that $bac = ab$.

commute If a and b are elements in a *set with a *binary operation ∘, they commute if $a \circ b = b \circ a$. *See also* commutative.

compact A subset A of a *topological space X is compact if, whenever A is contained in the *union of a collection of open sets U_i, then A is contained in the union of a finite number of the U_i.

When X is a *metric space, a subset A is compact if every *sequence has a *limit point in A. A closed interval $A = [a, b] \subset \mathbb{R}$ is compact, but the open interval $A = (0, 1)$ is not compact since the limit point 0 of the sequence $\{1/n\}, n > 1$, does not lie in A. The *Heine–Borel theorem* states that a subset $A \subset \mathbb{R}^n$ is compact if and only if A is closed and *bounded. It is named after Heinrich Eduard Heine (1821–81) and F.E.E. Borel.

companion matrix The companion matrix associated with the polynomial $p(x) = x^n - a_{n-1}x^{n-1} - \cdots - a_1 x - a_0$ is the $n \times n$ matrix

$$C = \begin{pmatrix} a_{n-1} & a_{n-2} & \dots & \dots & a_0 \\ 1 & 0 & \dots & \dots & 0 \\ 0 & 1 & \ddots & & 0 \\ \vdots & & \ddots & 0 & \vdots \\ 0 & \dots & \dots & 1 & 0 \end{pmatrix}$$

The *eigenvalues of C are the zeroes of p, so one method of finding the zeroes of a polynomial is to find the eigenvalues of the corresponding companion matrix.

comparative experiments *See* experimental design; factorial experiments.

comparison test A test for determining whether a given infinite *series is convergent or divergent by comparing it with another series of known convergence or divergence. Let

$$\sum a_n \quad \text{and} \quad \sum b_n$$

be two series of positive terms. Then one form of the test states that

(1) if $a_n \leqslant b_n$ for all n and if $\Sigma\, b_n$ converges then $\Sigma\, a_n$ converges;

(2) if $a_n \geqslant b_n$ for all n and if $\Sigma\, b_n$ diverges then $\Sigma\, a_n$ diverges.

In case (1) the value of the summation of a_n does not exceed that of b_n.

Another form of the comparison test states that if a_n/b_n tends to a nonzero (finite) limit as $n \to \infty$, then either both series converge or both diverge. *See* convergent series.

compass (compasses) An instrument for drawing circles.

complement 1. *See* complementary angles.
2. The complement of a *set A, denoted by A' or sometimes $\mathscr{C}A$, consists of all those elements that are not members of A:

$$A' = \{x : x \notin A\}$$

For example, in the domain of natural numbers, if A is the set of even numbers then its complement A' is the set of odd numbers. *See* relative complement.

complementary angles Two angles that have a sum of $90°$. Each angle is said to be the *complement* of the other.

complementary error function *See* error function.

complementary function A part of the general solution of a linear *differential equation with constant coefficients. If the equation has the form

$$a\frac{\mathrm{d}^2 y}{\mathrm{d}x^2} + b\frac{\mathrm{d}y}{\mathrm{d}x} + cy = \mathrm{f}(x)$$

where a, b, and c are constant, the complementary function is the general solution of

the equation

$$a\frac{\mathrm{d}^2 y}{\mathrm{d}x^2} + b\frac{\mathrm{d}y}{\mathrm{d}x} + cy = 0$$

See differential equation.

complete 1. A *formal system S is said to be simply complete if and only if, for every *wff A of S, either A or $\sim A$ is a theorem of S. This is a proof-theoretic notion of completeness, and it is in this sense that arithmetic was shown by Gödel to be incomplete if consistent (*see* proof theory; Gödel's proof).
2. An interpreted *logistic system (*see* interpretation) is said to be complete if and only if all *valid *wffs are theorems. Completeness in this sense is the converse of soundness (*see* sound). A *completeness theorem* for a logistic system S establishes that all the valid arguments that can be formulated in S are such that the conclusion is deducible from the premises (*see* deduction). Examples of complete systems of logic are the propositional calculus and the predicate calculus. *See also* logic.
3. *See* truth function.

complete bipartite graph *See* graph.

complete field *See* order properties.

complete graph *See* graph.

complete induction *See* induction.

complete lattice A *lattice in which every subset has a *greatest lower bound and a *lowest upper bound.

completeness property *See* order properties.

completeness theorem *See* complete.

complete quadrangle *See* quadrangle.

complete quadrilateral *See* quadrilateral.

complete space *See* metric space.

completing the square The process of writing a quadratic expression in a form in which the variable appears only in a squared term. Most commonly, 'complet-ing the square' refers to a method of solving *quadratic equations by putting an equation

$$ax^2 + bx + c = 0$$

in the form

$$(x + k)^2 + A = 0$$

where a, b, c, k, and A are constants. It can be used when it is not evidently or easily possible to factorize the left-hand side of the equation. For instance, the equation

$$3x^2 + 24x + 9 = 0$$

is divided through by 3 to give

$$x^2 + 8x + 3 = 0$$

To complete the square, this has to be put in the form

$$(x + 4)^2 + A = 0$$

where $A = -13$ and thus

$$(x + 4)^2 = 13$$

giving $x = -4 + \sqrt{13}$ or $-4 - \sqrt{13}$. See also quadratic formula.

complex analysis The study of *functions of a complex variable. The theory involves *holomorphic and *meromorphic functions. The theory can be used to study functions of a real variable, for example the integrals of certain functions can be successfully evaluated using *Cauchy's residue theorem.

complex conjugate 1. The complex conjugate of a *complex number $z \ (= a + ib)$ is the complex number $a - ib$ and is denoted by $\bar{z}$ or z^*. The number and its conjugate form a *conjugate pair*; each is the conjugate of the other.
2. The complex conjugate of a *matrix A is the matrix formed by replacing each element of A by its complex conjugate. It is

denoted by $\bar{A}$, or sometimes A^*. See also Hermitian conjugate.

complex fraction A fraction in which either the numerator or the denominator or both are themselves fractions. Compare common fraction.

complex function (function of a complex variable) A *function whose *domain and *codomain are sets of *complex numbers. For example, the function $f: z \mapsto z^2$ which maps the set of complex numbers $\mathbb{C}$ to $\mathbb{C}$ is a complex function. A complex function may also be regarded as a map of all or part of the complex plane to itself. Some complex functions take only real values. Examples include $z \mapsto \operatorname{Im} z$ and $z \mapsto \arg z$, where Im and arg denote the imaginary part and principal value of the *argument, respectively; they map $\mathbb{C}$ to $\mathbb{R}$. Compare real function.

complexity (of an algorithm) The number of discrete steps (such as addition and multiplication) needed to complete the execution of an *algorithm, expressed as a function of the size of the input. For example, to multiply two $n \times n$ matrices takes of the order of $2n^3$ additions and multiplications if done in the usual way, but the computation can be done in a number of operations proportional to $n^{\log_2 7}$, using *Strassen's method. Since $\log_2 7 < 3$, Strassen's method requires fewer operations than conventional multiplication for sufficiently large n, and it is said to have a lower computational complexity. Complexity has come into prominence because of the increase in the number, variety, and running costs of algorithms used by computer programmers. See also polynomial time; NP problem.

complex number A number of the type $a + ib$, where i is $\sqrt{-1}$ and where a and b are real numbers. a is said to be the *real part* of the complex number and b the *imaginary part*. The real and imaginary parts of a complex number z are denoted by Re z and Im z. (Sometimes j is used for $\sqrt{-1}$ in place of i.) If $b = 0$, the number has no imaginary part and is a *real number. The real numbers are considered to be a subset of the complex numbers. If b is nonzero then the number is an *imaginary number*; imaginary numbers in which $a = 0$ (i.e. ones with no real part) are said to be *pure imaginaries*. Complex numbers arise from attempts to solve equations that involve roots of negative numbers. For instance, the equation $x^2 + 4 = 0$ has roots of $\pm\sqrt{-4}$. These are pure imaginary numbers, written as $+2i$ and $-2i$, where i stands for $\sqrt{-1}$. The set of all complex numbers is usually denoted by $\mathbb{C}$.

Complex numbers can be represented on an *Argand diagram using two perpendicular axes. The real part is the x-coordinate and the imaginary part is the y-coordinate. A complex number $a + ib$ is then represented either by the point (a, b) or by a vector from the origin to this point. This gives an alternative method of expressing complex numbers, in the form $r(\cos\theta + i\sin\theta)$, where r is the length of the vector and θ is the angle between the vector and the positive direction of the x-axis. The value r is the *modulus (or absolute value) of the complex number; the angle θ is the *argument (or amplitude) of the number. This form of complex number is referred to as the *polar form* or *modulus–argument form*. Sometimes the expression $(\cos\theta + i\sin\theta)$ is abbreviated to cis θ.

Complex numbers can be added (or subtracted) by adding (or subtracting) their real and imaginary parts separately. For example:

$$(3 + 2i) + (5 + 4i) = 8 + 6i$$

In multiplication, the brackets are expanded:

$$(a + ib)(c + id) = ac + iad + ibc + i^2bd$$

Since $i^2 = -1$, this becomes

$$(ac - bd) + i(ad + bc)$$

If the complex numbers are in polar form they can be multiplied by multiplying their moduli and adding their amplitudes. Thus,

$$r_1(\cos\theta_1 + i\sin\theta_1)r_2(\cos\theta_2 + i\sin\theta_2)$$
$$= r_1r_2[\cos(\theta_1 + \theta_2) + i\sin(\theta_1 + \theta_2)]$$

See also complex conjugate; Euler's formula; quaternion.

complex plane *See* Argand diagram.

component (of a vector) One of a set of *vectors whose sum is the given vector. The component of a vector in a given direction is, however, the projection of the vector onto a line in that direction. The components of a vector are often taken at right angles to each other; if, for example, they are directed along x, y, z coordinate axes, then the components can be expressed as $a\mathbf{i}$, $b\mathbf{j}$, $c\mathbf{k}$, where $\mathbf{i}$, $\mathbf{j}$, $\mathbf{k}$ are *unit vectors.

component analysis *See* principal component analysis.

composite function (function of a function) A *function h such that $h(x) = g(f(x))$, where f and g are functions and the *domain of h is the set of x in the domain of f for which $f(x)$ is in the domain of g. For example, if $f(x) = x^2 + 1$ and $g(x) = x + 1$, and the domain of f is the set of real numbers, then

$$h(x) = g(f(x)) = g(x^2 + 1)$$
$$= (x^2 + 1) + 1$$
$$= x^2 + 2$$

The composite function can be written as gf or g ∘ f. In general, fg is not the same as gf. In the above example,

$$f(g(x)) = f(x + 1)$$
$$= (x + 1)^2 + 1$$
$$= x^2 + 2x + 2$$

If f is continuous at $x = a$ and g is continuous at $f(a)$, then h is continuous at $x = a$. A composite function can be differentiated using the *chain rule.

composite hypothesis *See* hypothesis testing.

composite number A number that is not *prime, i.e. one that has factors other than itself and 1. *See also* Gaussian integer.

composition (of vectors) *Vector addition, i.e. the process of determining the sum, or resultant, of vectors.

composition series A sequence $H_0, H_1, \ldots, H_n$ of *subgroups of a *group G with *identity element e, such that $H_0 = \{e\}$, H_i is a *maximal *normal subgroup of H_{i+1} for $0 \leqslant i < n$, and $H_n = G$. For $n > 2$, $n \neq 4$, the *symmetric group S_n has one composition series: $\{e\}, A_n, S_n$, where A_n is the *alternating group.

compound distribution A term used chiefly but not exclusively for the *distribution of a sum

$$S_N = X_1 + X_2 + \cdots + X_N$$

where the X_i are mutually independent discrete *random variables often with the same distribution. N may also be a random variable. In particular, if the X_i are all Bernoulli variables with $\Pr(X_i = 1) = p$ (*see* Bernoulli trial) and N has a *Poisson distribution with mean λ, then S_N has a Poisson distribution with mean λp.

compound interest *See* interest.

compound pendulum Any *rigid body that swings about a horizontal axis that passes through the body (but not through its centre of mass). *See* pendulum.

compound sentence (molecular sentence) A sentence (*see* wff) that contains logical constants such as & (and) or ∨ (or). For example, in the *propositional calculus the set of compound sentences can be identified with those sentences that contain a truth-functional connective. *Compare* atomic sentence.

compression A *force that compresses or tends to compress a body or structure, or the change in shape that results from the application of such a force. For example, a sphere under a uniform compression might decrease in radius (i.e. in volume). *Compressive stress* is set up within the body or structure in reaction to such a force. *See also* stress.

computable *See* effective procedure.

computer algebra The use of a computer program to carry out algebraic (as opposed to purely numeric) operations; also called *symbolic manipulation*. Computer algebra programs can be classified as either general purpose (examples being Maple® and Mathematica®), or specialized, the latter being dedicated to particular types of computation (e.g. *group theory).

concave function *See* convex function.

concave polygon A *polygon that has at least one interior angle greater than 180°. Interior angles greater than 180° are said to be *re-entrant angles*. *Compare* convex polygon.

concave polyhedron A *polyhedron for which at least one face lies in a plane that cuts other faces, i.e. the polyhedron does not lie completely on one side of that plane. *Compare* convex polyhedron.

concentric Having the same centre. The term can be applied to any two or more figures that have centres of symmetry. *Compare* eccentric.

conchoid A plane curve that can be generated by first taking a fixed point P outside a fixed line. A variable line through P cuts the fixed line at Q. If points R and R′ are chosen on this line such that RQ = QR′ = b (a constant), the conchoid is the locus of R and R′ as PQ varies. It has two branches on each side of the fixed line; both branches are asymptotic to this line. In Cartesian coordinates, if P is the origin and the fixed line is $x = a$, the equation is

$$(x - a)^2(x^2 + y^2) = b^2 x^2$$

The polar equation is

$$r = a \sec \theta \pm b$$

If $a = b$, one branch has a cusp; if $a < b$, it has a loop.

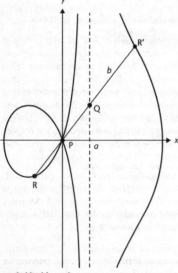

conchoid with $a < b$.

conclusion *See* argument; syllogism.

concomitant variable *See* analysis of covariance.

concordance *See* coefficient of concordance.

concurrent Having a common point. *Concurrent lines*, for example, are lines that all pass through a certain point. *See* collinear.

concyclic Describing points that lie on the same circle. For instance, the vertices of a *cyclic polygon are concyclic.

conditional A statement that something is true or will be true provided something else is also the case. It is a sentence of the form 'if A then B', often symbolized in a *formal language as $A \supset B$ or as $A \to B$. A is called the *antecedent* and B the *consequent* of the conditional. *See also* implication.

conditional distribution *See* bivariate distribution; multivariate distribution.

conditional equation *See* equation.

conditionally convergent series *See* convergent series.

conditional probability *See* probability.

condition number A number that measures the degree of ill-conditioning of a mathematical problem. For example, for a linear system of equations with *coefficient matrix A, a condition number is the *norm of A multiplied by the norm of A^{-1}; if the system is *ill-conditioned, A^{-1} will have relatively large elements and the condition number will be large.

cone A solid figure formed by a closed plane curve on a plane (the *base*) and all the lines joining points of the base to a fixed point (the *vertex*) not in the plane of the base. The closed curve is the *directrix* of the cone and the lines to the vertex are its *generators* (or *elements*). The curved area of the cone forms its *lateral surface*. Cones are named according to the base, e.g. a circular cone or an elliptical cone. If the base has a centre of symmetry, a line from the vertex to this centre is the *axis* of the cone. A cone that has its axis perpendicular to its base is a *right cone*; otherwise the cone is an *oblique cone*. The *altitude* of a cone (h) is the perpendicular distance from the plane of the base to the vertex.

The volume of any cone is $\frac{1}{3}hA$, where A is the area of the base. A right circular cone (circular base with perpendicular axis) has a *slant height* (s), equal to the distance from the edge of the base to the vertex (the length of a generator). The lateral area of a right circular cone is πrs, where r is the radius of the base. The term 'cone' is often used loosely for 'conical surface'. *See also* spherical sector (for *spherical cone*).

confidence interval If the distribution of a *random variable X contains an unknown parameter θ, a $100(1 - \alpha)$ percent confidence interval for θ is an interval formed by a rule which ensures that, in the long run, $100(1 - \alpha)$ percent of such intervals will include the parameter θ. Typically the interval is derived from the information obtained from a random sample of n observations from the distribution. If $\alpha = 0.05$, the interval is a 95 percent confidence interval. For example, if $\bar{x}$ is the mean of a sample of n observations from a normal distribution with unknown mean μ and known *standard deviation σ, then a 95 percent confidence interval for μ is

$$[\bar{x} - 1.96\sigma/\sqrt{n},\ \bar{x} + 1.96\sigma/\sqrt{n}]$$

The end points of the interval are called *confidence limits*.

A $100(1 - \alpha)$ percent confidence interval for a parameter θ derived from a given sample covers all values θ_0 of that parameter that would be accepted at significance level α in a *hypothesis test of $H_0: \theta = \theta_0$ against the alternative $H_1: \theta \neq \theta_0$.

For two or more parameters, the concept extends to a *confidence region*. *Bayesian confidence intervals* are conceptually different, but in many cases are numerically equivalent. *See also* Bayesian inference; confidence level.

confidence level The value of $100(1 - \alpha)$ percent associated with a *confidence interval or region. Common values are 90, 95, 99, and 99.9 percent, corresponding to $\alpha = 0.10, 0.05, 0.01,$ and 0.001, respectively.

confidence limits *See* confidence interval.

confidence region *See* confidence interval.

configuration A particular arrangement of points, lines, curves, etc.

configuration space An *abstract space describing all possible positions of a system (e.g. a mechanical system). The dimension of the configuration space is the number of *degrees of freedom of the system.

confocal conicoids *Conicoids that have the same principal planes and having sections by these planes that are *confocal conics.

confocal conics *Conics that have the same focus or foci. For example, families of confocal conics can be generated by an equation of the form

$$\frac{x^2}{a^2 - k} + \frac{y^2}{b^2 - k} = 1$$

where $a^2 > b^2$ and k is a parameter taking all real values, provided $a^2 > k$ and $b^2 \neq k$. Thus, for values of k less than b^2, ellipses are generated; values greater than b^2 generate hyperbolas. Confocal ellipses and hyperbolas intersect at right angles.

conformable matrices Two *matrices such that the number of columns in one equals the number of rows in the other. Matrices must be conformable for matrix multiplication to be possible.

conformal transformation A *transformation such that if two curves intersect at an angle θ, their images also intersect at angle θ. Thus, a conformal transformation (or map) is one that preserves angles, and is also called an *equiangular* or *isogonal* transformation. In Euclidean space, inversion, reflection, translation, and magnification are conformal transformations.

confounding *See* factorial experiments.

congruence 1. The property of being *congruent.
2. *See* Congruence modulo n.

congruence class (residue class) A *set consisting of all the integers that are *congruent modulo n to a given integer. For example, with respect to the modulus 7, some of the integers in the congruence class containing 2 are $\ldots$, $-19, -12, -5, 2, 9, 16, \ldots$. A congruence class modulo n can equally be regarded as the set of all integers that leave a particular remainder on division by n.

For any modulus n there are n different congruence classes and they form an *additive group, where the class obtained by adding the classes containing a and b respectively is the class containing $a + b$. This group, which is also a *cyclic group, is usually denoted by $\mathbb{Z}_n$.

congruence modulo n (C.F. Gauss, 1801) A relation, usually between integers, expressing the fact that two integers a and b differ by a multiple of a chosen natural number n. The two integers are said to be *congruent modulo n*, written as $a \equiv b \pmod{n}$. For example:

$$8 \equiv -1 \pmod 3$$
$$42 \equiv 18 \pmod 8$$
$$365 \equiv 1 \pmod 7$$

Integers c and d that are not congruent modulo n are said to be *incongruent modulo n*, written as $c \not\equiv d \pmod{n}$.

Two congruences with the same modulus (n above) can be added, subtracted, and multiplied just like ordinary equations. So $8 \equiv 18 \pmod{10}$ and $27 \equiv 7 \pmod{10}$ together imply

$$(8 + 27) \equiv (18 + 7) \pmod{10}$$
$$(8 - 27) \equiv (18 - 7) \pmod{10}$$
$$8 \times 27 \equiv 18 \times 7 \pmod{10}$$

This is not true for division; we definitely cannot conclude that

$$8/27 \equiv 18/7 \pmod{10}$$

nor even that

$$8/2 \equiv 18/2 \pmod{10}$$

A common factor can be cancelled from both sides of a congruence if as much as possible of the factor is also divided into the modulus. Thus

$$8 \equiv 18 \pmod{10}$$

does imply that

$$8/2 \equiv 18/2 \pmod{10/2}$$

and

$$6 \equiv 36 \pmod{15}$$

implies that

$$2 \equiv 12 \pmod 5 \quad \text{and} \quad 1 \equiv 6 \pmod 5$$

The arithmetic of congruences (or *modular arithmetic*) is useful in many 'cyclic situations' in everyday life. For instance, the problem of finding the day of the week for a certain date involves congruences modulo 7; and the fact that a 24-hour clock might say 21:00 when a 12-hour clock says 9:00 corresponds to the fact that $21 \equiv 9 \pmod{12}$.

The congruence notation $a \equiv b \pmod{n}$ is sometimes extended to include cases in which a and b are more general real numbers. It then means, as before, that $a - b$ is an integer that is an integer multiple of the natural number n. For example, $1.6 \equiv 0.6 \pmod 1$, $5.74 \equiv -3.26 \pmod 3$.

See also division modulo n; factor modulo n; congruence class.

congruence transformation *See* matrix.

congruent 1. Describing two or more geometric figures that differ only in location in space. The figures are congruent if one can be brought into coincidence with the other by a rigid motion in space (without changing any distances in the figure). Note that two plane figures can be congruent without being identical. For instance, two scalene triangles with identical sides and angles are not identical if one is drawn as a mirror image of the other. They are, however, congruent (on this definition) since one can be rotated through 180° about an axis in the plane (or 'picked up' off the plane and put down again the opposite way round). In the case of three-dimensional figures, this point is important since mirror images cannot be made coincident by a rigid motion in (three-dimensional) space. If two solid figures are identical, they are *directly congruent*. If each is identical to the mirror image of the other, they are *oppositely congruent*.

Two triangles are congruent if there is a correspondence between them satisfying one of the following conditions:

(1) All three pairs of corresponding sides are equal (the SSS condition).
(2) Two pairs of corresponding sides are equal, and the angles between them are equal (SAS).
(3) Two pairs of corresponding angles and a pair of corresponding sides are equal (AAS or ASA).
(4) The triangles are right-angled, and they have equal hypotenuses and a further pair of equal sides (RHS).

2. *See* congruence modulo *n*.

congruent matrices *See* matrix.

conic (conic section) A type of plane curve that is the *locus of all points such that the ratio of their distance from a fixed point (the *focus*) to their distance from a fixed line (the *directrix*) is a constant. The constant is the *eccentricity* (*e*) of the conic, and its value determines the type of curve:

ellipse $0 \leqslant e < 1$

parabola $e = 1$

hyperbola $e > 1$

A circle is a special case of an ellipse with zero eccentricity ($e = 0$).

Conics were first studied by the Greek Menaechmus (*c*.350 BC), who identified them as sections of different types of circular cone. Other early investigations were made by Conon of Samos (*c*.245 BC). Apollonius of Perga in around 225 BC produced an extensive study in his book *Conics*, and showed that they could be formed by different sections of any circular conical surface. Thus, an ellipse is formed by a plane cutting the surface at an angle (to the axis) greater than the *generating angle; a parabola is formed by a plane section at an angle equal to the generating angle; and a hyperbola by a plane at an angle less than the generating angle. A circle is a special case of an ellipse, formed by a section perpendicular to the axis of the conical surface (*see* diagram). Apollonius also recognized that a conical surface has two *nappes and was therefore able to show that the hyperbola has two branches.

The hyperbola and the ellipse, which have centres of symmetry, are known as *central conics*. There are certain limiting cases of conic sections that give rise to what are known as *degenerate conics*. Thus, a point is a limiting case of an ellipse in which the intersecting plane cuts the vertex of the cone. A single line is a limiting case of a parabola in which the plane is tangent to the surface. A pair of intersecting straight lines is a 'hyperbola' through the vertex of the cone.

The treatment of conic sections by coordinate geometry was begun in the 17th century, notably by Jan de Witt (1629–72), who gave the definition in terms of the focus and directrix, and independently in 1655 by John Wallis. In a Cartesian coordinate system, the equation for a conic can be expressed in various ways. For instance, if *e* is the eccentricity, the origin is the focus, and the directrix is a distance *k* from the origin, then

$$(1 - e^2)x^2 + 2e^2kx + y^2 = e^2k^2$$

The *general conic* is expressed by a general equation of the second degree:

$$ax^2 + 2hxy + by^2 + 2gx + 2fy + c = 0$$

The equation can, by translation of axes, be put in a form in which it contains no terms of the first degree:

$$ax^2 + 2hxy + by^2 - \frac{\Delta}{h^2 - ab} = 0$$

Here, Δ is the determinant (called the *discriminant* of the conic):

$$\Delta = \begin{vmatrix} a & h & g \\ h & b & f \\ g & f & c \end{vmatrix}$$

Its value is

$$abc + 2fgh - af^2 - bg^2 - ch^2$$

If the discriminant is nonzero, then the conic is an ellipse, parabola, or hyperbola:

$$h^2 - ab < 0, \quad \text{ellipse}$$

$$h^2 - ab = 0, \quad \text{parabola}$$

$$h^2 - ab > 0, \quad \text{hyperbola}$$

If the discriminant is zero, then the conic is degenerate:

$h^2 - ab < 0,$ point
$h^2 - ab = 0,$ a pair of parallel or coincident lines or an imaginary locus
$h^2 - ab > 0,$ a pair of intersecting lines

Conics can also be treated by *projective geometry. In 1604 Kepler introduced the idea that intersecting straight lines, hyperbolas, parabolas, ellipses, and circles all belong to the same family depending on

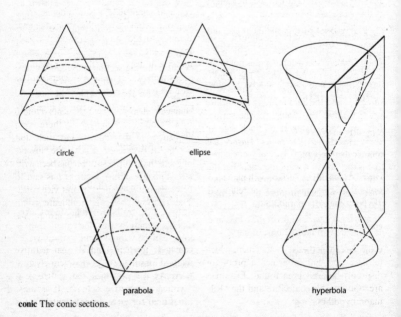

circle   ellipse

parabola   hyperbola

conic The conic sections.

the positions of the foci. In particular, he regarded the parabola as a curve with one focus at infinity. Desargues developed the projective geometry of conics, showing that a projection of any conic is also a conic. The work was largely disregarded at the time – partly because of Desargues's obscure terminology, but also because his treatment was overshadowed by the contemporary interest in analytic geometry.

See also circle; ellipse; hyperbola; parabola.

conical Denoting or concerning a *cone.

conical pendulum A simple *pendulum whose bob swings in a horizontal circle, i.e. the cord generates a right-circular conical surface with a vertical axis. The angular speed of the bob is constant. The period of revolution T is given by

$$T = 2\pi\sqrt{\left(\frac{h}{g}\right)}$$

where h is the height of the point of suspension above the centre of the circle and g is the acceleration of free fall.

conical surface A surface generated by all the straight lines that pass through a given point and intersect a curve that is not in the same plane as the given point. The point is the *vertex* of the surface, the curve is its *directrix*, and the lines forming the surface are *generators* (or *elements*). The surface has two parts (*nappes*) on each side of the vertex. A *circular conical surface* has a circle as directrix.

conicoid (conoid) A surface with plane sections that are conics, e.g. an *ellipsoid, *hyperboloid, or *paraboloid.

conic section *See* conic.

conjecture (hypothesis) A statement which may be true, but for which a proof (or disproof) has not been found. Examples are *Goldbach's conjecture and the *Riemann hypothesis.

conjugacy class A *set in a *group consisting of all the group elements that are *conjugate to a given element. Each element in a group belongs to a unique conjugacy class. The identity element is the only element in its conjugacy class, and also this is the only conjugacy class which is also a subgroup. There may be other elements that also form one-element conjugacy classes, and the set of all such elements of a group G is a subgroup of G called the *centre*, $Z(G)$, of G. The elements of the centre are precisely the elements that each *commute with every element of G. So the centre is the whole group when it is *Abelian. *See* class equation.

conjugate angles Two angles that have a sum of $360°$. Each angle is said to be the *explement* of the other.

conjugate arcs *See* arc.

conjugate axis *See* hyperbola.

conjugate complex numbers *See* complex conjugate.

conjugate diameters A pair of diameters of a given *conic, such that one diameter belongs to the family of parallel chords whose centres define the other. The major and minor axes of an ellipse, for example, are a pair of conjugate diameters (in this case they are perpendicular).

conjugate elements Two elements a and b in a *group are conjugate if there is an element g in the same group such that $b = g^{-1}ag$ (where g^{-1} denotes the *inverse of g and the group operation has been written as juxtaposition). Element b is said to be *conjugate to* element a, and vice versa. The relation of being conjugate is an *equivalence relation on the group. *See* conjugacy class.

conjugate gradient method An iterative method for solving a linear system of equations $\mathbf{Ax} = \mathbf{b}$ in which the matrix $\mathbf{A}$ is *symmetric positive definite. It is most often used for large *sparse matrices, for which it is particularly appropriate because each iteration involves a single product between the matrix and a vector. It is closely related to the *Lanczos method for computing *eigenvalues of a matrix.

conjugate lines (of a conic) Two lines such that each contains the *pole of the other.

conjugate pair *See* cross-ratio.

conjugate points (of a conic) Two points such that each lies on the *polar of the other.

conjugate prior distribution *See* Bayesian inference.

conjugate set *See* transform.

conjugate to *See* conjugate elements.

conjunct *See* conjunction.

conjunction A sentence of the form '*A* and *B*', often symbolized in a *formal language as '*A* & *B*' (*see* and). '*A*' and '*B*' are called *conjuncts*.

conjunctive normal form A formula is in conjunctive normal form if it consists entirely of a *conjunction of *disjunctions, with each disjunction formed only from *atomic sentences or their *negations. It can be shown that every *wff of the *propositional calculus can be expressed as an equivalent formula in canonical normal form. For example, the expression $(p \vee q)$ & $(p \vee \sim r)$ & $(q \vee r)$ is in conjunctive normal form. It is thus possible to see whether any formula is a *tautology of the propositional calculus by noting, as in the following case, that each disjunction contains an atomic sentence and its negation: $(p \vee q \vee \sim p)$ & $(q \vee \sim q \vee r)$ & $(r \vee p \vee \sim r)$. Using this approach it is possible to show that the propositional calculus is *complete. *Compare* disjunctive normal form.

connected graph *See* walk.

connected relation A *binary relation R on a *set *A* is connected if for all pairs of members *x* and *y*

$$x \neq y \rightarrow (x \ \mathbf{R} \ y) \vee (y \ \mathbf{R} \ x)$$

For example, in the domain of natural numbers the relation 'greater than' is connected.

connected set A *set *A* is a connected set if there do not exist disjoint nonempty subsets of *A* (*X* and *Y*) such that $X \cup Y = A$, and no *limit point of *X* is a member of *Y* and no limit point of *Y* is a member of *X*. *Compare* disconnected set.

connected space A *topological space *S* is a connected space if there do not exist disjoint, nonempty open sets of *S* (*X* and *Y*) such that $X \cup Y = S$. The real line $\mathbb{R}$ is connected, but $X = \mathbb{R} \backslash \{0\}$, the real line with the origin removed, is disconnected because it is the union of the two disjoint open subsets $(-\infty, 0)$ and $(0, \infty)$. A space *X* is called *path-wise connected* (or *arc-wise connected*) if any two points $a, b \in X$ can be joined by a *path lying entirely within *X*. For many spaces the definitions of connected and path-wise connected are equivalent. *Compare* disconnected space.

connection A connection on a differential *manifold *M* is a way of defining the parallelism of vectors. It allows consistent differentiation everywhere on *M*, since it 'connects' the various local coordinates on the manifold. A *Riemannian metric gives rise to a connection, but not vice versa. The idea of a connection was stressed by Weyl in his efforts to unify the theories of relativity and electromagnetism. Connections are also used in the study of *vector bundles and, in this context, they are basic in the development of *gauge theory.

connective In mathematical *logic, a symbol that can be combined with one or more sentences in order to form a new sentence. If a connective joins two sentences then it is called a *binary* (or *dyadic*) connective.

Examples are 'and', 'or', 'iff', and 'if … then'. *See also* truth function.

conoid *See* conicoid.

Conon of Samos (*fl.* 245 BC) Greek mathematician and astronomer responsible for early investigations into conics. His work was absorbed into the later work of Apollonius.

Conover squared rank test *See* homogeneity of variance.

consecutive angles Two angles in a *polygon that share a common side.

consecutive sides Two sides in a *polygon that share a common vertex; adjacent sides.

consequence **1. (logical consequence)** A *wff A is a logical consequence of a set of wffs $B_1, \ldots, B_n$ if and only if, given the truth of $B_1, \ldots, B_n$, A must also be true. Equivalently, A is a logical consequence of $B_1, \ldots, B_n$ if and only if

$$(B_1 \,\&\, \cdots \,\&\, B_n) \supset A$$

is a valid wff (*see* implication, material).
2. (formal consequence) A is a formal consequence in a *formal language S of the *wffs $B_1, \ldots, B_n$ if and only if A is deduced from $B_1, \ldots, B_n$ by use of the rules of inference of S (*see* deduction). In those logistic systems where the deduction theorem holds, A is a formal consequence of $B_1, \ldots, B_n$ if and only if

$$(B_1 \,\&\, \cdots \,\&\, B_n) \supset A$$

is a theorem. If a logistic system is also *sound and *complete then A is a logical consequence of $B_1, \ldots, B_n$ if and only if A is a formal consequence of $B_1, \ldots, B_n$. *See* logic.

consequent **1.** The second term in a ratio. Thus in the ratio $5:7$, 7 is the consequent (5 is the *antecedent*).

2. That part of a *conditional statement that indicates what is or would be the case given the initial condition. Thus in the conditional 'if p then q', q stands for the consequent (p is the *antecedent*).

conservation laws Laws requiring that, in an isolated or undisturbed system, the total amount of some *physical quantity does not change in the course of time; the quantity is said to be *conserved*. Such quantities include mass or mass–energy, momentum, and electric charge. The basis for such laws lies in the symmetry of space (and time): a given conserved quantity remains constant under a particular symmetry transformation.

conservation of energy The principle stating that in any isolated system the total *energy remains constant in time. There can be interconversion between different forms of energy – mechanical, heat, electrical, chemical, etc. – but the sum of these energies cannot change. Some components of the system may gain energy, but others must lose an equivalent amount.

In the theory of relativity, energy and mass are equivalent and interconvertible according to the *mass–energy equation, $E = mc^2$, where c is the speed of light. Thus a considerable amount of energy can be generated by the destruction of a small quantity of matter. In systems in which such conversion takes place, e.g. by nuclear reactions, the *conservation of mass–energy* must be invoked: the sum of the total (rest) mass plus the total energy remains constant.

conservation of mass The principle stating that in any isolated system the total *mass remains constant in time. Matter can change its form, as in combustion or metabolism, but the mass of all the products will equal that of the initial mass. According to the theory of relativity, however, mass and energy are equivalent. In addition, the mass of a body increases quite considerably as its speed approaches the speed of light. These changes in mass can normally be ignored, but they are significant in systems involving, for example, reactions of nuclear particles. In such systems there is conservation of mass–energy, where the mass is the particle's *rest mass. *See* conservation of energy; mass–energy equation.

conservation of momentum The principle stating that in a system in which components are undergoing *collisions or mutually attracting or repelling each other, then, in the absence of an external force, the sum of the momenta of the components in any particular direction remains constant: the *momentum gained by one component is balanced by a loss of momentum of one or more other components. For a body, or system of particles, rotating about a fixed axis, there is also conservation of *angular momentum, provided no external torque is applied.

conservative field The *field of force associated with a *conservative force.

conservative force A *force, such as gravitation, that acts on a particle in such a way that the work done in moving the particle from one point to another depends only on these end points and is independent of the path taken; the net *work evaluated around a closed loop is zero.

The work done by a conservative force in bringing a particle from a given point to some standard point is the *potential energy of the particle at the given point. Potential energy can be defined only for conservative forces. For motion under conservative forces, the total energy, *kinetic plus potential energy, remains constant, i.e. is conserved. *See also* potential; field.

consistent **1.** Describing equations that have a set of values that satisfies all the equations. For example, the equations

$$x + y = 10 \quad \text{and} \quad x + 2y = 15$$

are consistent, since they are satisfied by $x = 5$ and $y = 5$. The equations

$$x + y = 10 \quad \text{and} \quad x + y = 15$$

are *inconsistent* – there is no pair of values of x and y that satisfies both simultaneously.

2. Describing a *formal system in logic which is free from contradiction, i.e. one containing no *wff A such that both A and its negation $\sim A$ are provable (*see* proof; contradiction). A formal system is said to be *absolutely consistent* if not all wffs are *theorems. In many formal systems, consistency in the first sense is equivalent to absolute consistency. Although consistency is a proof-theoretic notion, its motivation is semantic in character: we are not interested in those systems that contain, as theorems, wffs that cannot be true. Inconsistent systems have no *models.

consistent estimator *See* estimator.

constant **1.** A fixed quantity or numerical value.

2. A symbol that is assigned a specific fixed entity under an *interpretation. Constants contrast with *variables, which range over a set of entities. An *individual constant* is an expression that is assigned an object under an interpretation. For example, a name such as 'Aristotle' would be treated as an individual constant. A *logical constant* is a logical expression that is used when giving the logical form of a sentence. Thus the logical form of 'some men are mortal' is

$$(\exists x)(M(x)) \ \& \ F(x))$$

and the logical constants are '∃' and '&'.

constant of integration *See* integration.

constant of proportionality *See* variaton.

constant term The term in a *polynomial that does not involve any power of the variable. For example, in the polynomials $x^3 - 6x + 2$ and $x^4 + 2x^3 - x$, the constant terms are 2 and 0, respectively. If the variable is x, as here, the constant term can be regarded as the coefficient of x^0.

construction The process of drawing a given geometric figure; for example, the construction of a line at right angles to a given line or the construction of a line bisecting a given angle. Usually, it is required that this be done using only compasses and straightedge. The three classical constructions of antiquity, dating from the 4th century BC, are *squaring the circle, *duplication of the cube, and *trisection of an angle. *Mascheroni constructions* are ones that require only compasses. *See also* Fermat numbers.

constructive Describing a definition or proof in which there is an *effective procedure for the construction of every object, such as a number or set, in it. The concept of a constructive definition is linked to the insistence of *intuitionism that before the existence of a mathematical object can be accepted, an effective procedure must be given for its construction. Thus the inference 'It is false that every number n lacks the property F, therefore at least one number n has the property F' would not be allowed in a constructive proof unless we could first identify that number or provide an effective procedure for its identification. *Compare* non-constructive.

consumer's risk *See* acceptance sampling.

contact, point of *See* tangent.

contingence, angle of The angle between the positive directions of two *tangents to a plane curve at two given points on the curve.

contingency table A table of r rows and c columns ($r, c \geqslant 2$) in which each row and each column is associated with a specified attribute, and the number n_{ij} lying in row i and column j (called the *cell* (i, j)) is a count of the number of units having that combination of attributes. The table shown here has $r = 2$ and $c = 3$, and refers to

the three qualities (superior, average, or poor) of items produced by two factories (A and B). Each of 40, 60 items from factory A, B were graded for quality. A *chi-squared test may be used to see whether proportions in each category differ significantly between factories.

	Superior	Average	Poor
Factory A	12	20	8
Factory B	14	39	7

Row and/or column categories may be *nominal* (e.g. married, single, widowed) or *ordinal* (e.g. aged under 10, aged 10–19, aged 20–29, aged 30 or over). While the chi-squared test for independence (lack of association between row and column attributes) is suitable for contingency tables with nominal categories, more powerful tests based on *loglinear models and *odds ratios are available when row and column categories are ordinal. *See also* Fisher's exact test.

continua *Plural of* continuum.

continued fraction A fraction in which the denominator is a number plus another fraction, which in turn may have a denominator consisting of a number plus another fraction, and so on:

$$a_0 + \cfrac{b_1}{a_1 + \cfrac{b_2}{a_2 + \cfrac{b_3}{a_3 + \cfrac{b_4}{a_4 \cdots}}}}$$

The series may be finite (*terminating fraction*) or infinite (*nonterminating fraction*).

Every positive real number has a *simple continued fraction* expansion in which each *partial quotient* $a_0, a_1, a_2, \ldots$ is an integer, $a_1, a_2, \ldots$ are positive integers, and $b_1, b_2, \ldots$ are each equal to 1. In this form, the expansion is often written as $[a_0, a_1, a_2, \ldots]$.

For example, $\sqrt{5}$ equals the infinite continued fraction

$$2 + \cfrac{1}{4 + \cfrac{1}{4 + \cfrac{1}{4 + \cfrac{1}{4 \dots}}}}$$

which can be written as $[2, 4, 4, 4, \dots]$.

All *rational numbers have finite simple continued fractions, and real *irrational numbers have infinite expansions. In each case the rational numbers $\dfrac{p_0}{q_0}, \dfrac{p_1}{q_1}, \dots$ obtained by truncating the expansion after a finite number of partial quotients are called the *convergents* of the continued fraction. Thus $\dfrac{p_0}{q_0} = [a_0] = a_0$, $\dfrac{p_1}{q_1} = [a_0, a_1]$ $= a_0 + \dfrac{1}{a_1}$, $\dfrac{p_2}{q_2} = [a_0, a_1, a_2]$, $\dots$. The convergents of $\sqrt{5}$ are 2, $2 + \frac{1}{4} = \frac{9}{4}$, $[2, 4, 4] = \frac{38}{17}$, $\dots$. The convergents of a number x tend to x and can be determined *recursively from the values $\dfrac{p_0}{q_0}, \dfrac{p_1}{q_1}$, and $\dfrac{p_n}{q_n} = \dfrac{a_n p_{n-1} + p_{n-2}}{a_n q_{n-1} + q_{n-2}}$ for $n \geqslant 2$.

continued product A product of factors; this is often written using the notation

$$\prod_1^m T_i$$

which signifies the product

$$T_1 \times T_2 \times \cdots \times T_m$$

A continued product may contain an infinite number of factors. *See also* infinite product.

continuity correction The addition or subtraction of 0.5 to values of a discrete *random variable taking integral values to obtain closer agreement to a continuous approximation. For example, when approximating to the binomial distribution by a normal distribution, $\Pr(X \leqslant 15)$, (binomial) is best approximated by $\Pr(X \leqslant 15.5)$ (normal).

continuity equation An equation that is used in many branches of physics and is applied to the continuous flow of a conserved quantity such as mass or electric charge. For mass, it equates the rate of increase of fluid mass in any volume in the fluid to the net rate of mass flow into this volume. This can be expressed as

$$\frac{\partial \rho}{\partial t} + \nabla \cdot (\rho \mathbf{v}) = 0$$

where ρ is the fluid density, $\partial \rho / \partial t$ is the rate of change of density at some point, and $\mathbf{v}$ is the velocity at that point; ∇ is the operator *del.

continuous distribution *See* distribution.

continuous function (continuous mapping, continuous map) A *function for which a small change in the independent variable causes only a small change, and not a sudden jump, in the dependent variable.

A *real function $f(x)$ is *continuous* at $x = c$ if the right- and left-hand limits of $f(x)$ at $x = c$ and $f(c)$ all exist and are equal. Otherwise, $f(x)$ is discontinuous at $x = c$. An equivalent definition is that for any positive number ε a positive number δ depending on ε and c can be found such that

$$|f(x) - f(c)| < \varepsilon$$

whenever $|x - c| < \delta$.

If f is continuous at every point of the open interval (a, b) it is said to be *continuous in* (a, b). If, in addition,

$$\lim_{x \to a^+} f(x) = f(a), \quad \lim_{x \to b^-} f(x) = f(b)$$

then f is continuous in the closed interval $[a, b]$. The function f is *sectionally* or *piecewise continuous* in (a, b) if the interval can be subdivided into a finite number of intervals in each of which the function is continuous with a finite right-hand limit at each lower end point and a finite left-hand limit at each upper end point.

Elementary functions such as polynomials, and trigonometric, logarithmic, and exponential functions, are continuous at all points of their domains.

A complex function $f(z)$ is continuous at $z = z_0$ if, as z tends to zero in any manner, the limit of $f(z)$ is $f(z_0)$. Alternatively, given a positive number ε, a positive number δ depending on ε and z_0 can be found such that $|f(z) - f(z_0)| < \varepsilon$, whenever $|z - z_0| < \delta$.

If a function is differentiable at $x = c$ it must be continuous at that point. The converse is false: for example, $|x|$ is continuous but not differentiable at $x = 0$. The sum, difference, and product of continuous functions are themselves continuous. The quotient of two continuous functions is continuous at points where the denominator is not equal to zero.

A function of two variables is continuous at (c, d) if, as x and y tend to c and d respectively in any manner whatsoever,

$$\text{Lim } f(x, y) = f(c, d)$$

provided $f(c, d)$ exists. If the function is continuous at every point of a region A in the x–y plane it is said to be continuous over A.

If its domain X and codomain Y are both *metric spaces or topological spaces, then a function f is continuous at $x \in X$ if for any *neighbourhood V of $f(x)$ in Y there is a neighbourhood U of $x \in X$ such that for all $u \in U$ then

$$f(u) \in V \text{ (or } f(U) \subseteq V)$$

If f is continuous at all points of X it is said to be *continuous on X*.

See also uniformly continuous function; topological space; *compare* discontinuous function.

continuous mapping, continuous map *See* continuous function.

continuous random variable *See* random variable.

continuum (*plural* **continua**) **1.** An entity or substance whose structure or distribution is continuous and unbroken: for example, the *real numbers $\mathbb{R}$, *time, a fluid or plasma, *spacetime.
2. A *compact *connected set. Examples are a closed *interval and, in higher-dimensional space, a *ball or a *sphere. In this sense the set of real numbers is not a continuum since it is not compact.

All continua (in senses **1** and **2**) have the same *cardinality, usually denoted by c.

continuum hypothesis A hypothesis in set theory first proposed by Cantor. The *set of all *natural numbers $\mathbb{N}$ has a *cardinal number $\aleph_0$. The *power set of $\mathbb{N}$ will therefore have a cardinality of $2^{\aleph_0}$, which is denoted by c – the cardinal number of the set of real numbers (the *continuum*). Cantor's hypothesis is that no infinite cardinal lies between $\aleph_0$ and c. He was unable to prove this as a theorem of set theory. Work by Gödel in 1938 and Cohen in 1963 demonstrated the independence of the continuum hypothesis by showing that the axioms of set theory would remain consistent, assuming that they were initially consistent, if either the continuum hypothesis or its negation were added. *See also* Cantor's theory of sets.

contour integral An *integral defined for a *function $f(z)$ in the *complex plane and for a curve or *contour C* in this plane. The integral of the function along the contour is written as

$$\int_C f(z) \, dz$$

and is defined as follows. C is divided into n segments by $n + 1$ points z_0, z_1, ..., z_n. Points on C are taken in each subinterval: t_1 between z_0 and z_1, t_2 between z_1 and z_2, and in general t_i between z_{i-1} and z_i. Numbers $|z_1 - z_0|$, $|z_2 - z_1|$, ..., $|z_n - z_{n-1}|$ are taken. If the largest of these is δ, then the contour integral is the limit of the sum

$$\sum f(t_i)(z_i - z_{i-1})$$

as n tends to infinity and δ tends to zero.

The limit exists if f(z) is continuous on C and C is a *rectifiable curve.

contour plot A contour plot of a function f(x, y) of two variables is a plot in the x–y plane in which points having equal f-values are joined by curves (*contour lines*), for a selection of different f-values. The diagram shows a contour plot for the function

$$f(x, y) = 4x^2 - 2.1x^4 + \frac{1}{3}x^6 + xy - 4y^2 + 4y^4$$

over the region $-3 \leqslant x \leqslant 3$, $-1.5 \leqslant y \leqslant 1.5$, with contour lines corresponding to function values between -10 and 10. The function has a number of minima, maxima, and *saddle points in the region.

contractible See homotopy.

contraction mapping A mapping f: X → X, where X is a metric space, is a contraction if it decreases distances in the sense that there is a positive constant α < 1 such that d(f(x), f(y)) ⩽ α d(x, y) for all x, y ∈ X. If X is *complete, then every contraction mapping has a unique fixed point: a point a ∈ X such that f(a) = a. This result is known as the *Banach contraction principle* or *contraction mapping theorem* in a metric space (S. Banach, 1922). This principle underlies the *Picard iteration method of solving differential equations numerically. *See also* fixed-point theorem.

contradiction A simultaneous assertion and denial of a proposition; i.e. a sentence of the form 'A and not A', often symbolized within a *formal language as 'A & ∼A'. Formal systems in which a contradiction is a theorem are said to be *inconsistent*. The *law of contradiction* is the logical principle that a proposition cannot be both asserted and denied; i.e. the theorem of the *propositional calculus ∼(A & ∼A).

contrapositive A statement that is related to a *conditional statement in the following way: the conditional statement 'if A then B' has a contrapositive 'if not B

contour plot Contour lines of a function f(x, y).

then not A'. Thus the contrapositive of the conditional '$A \supset B$' is '$\sim B \supset \sim A$'. A conditional and its contrapositive are materially equivalent (*see* equivalence), and this gives rise to a rule of inference (*contraposition*) whereby any occurrence of a conditional can be replaced by its contrapositive.

Thus we can derive from 'if the square of an integer is even then the integer is even' the contrapositive 'if an integer is not even then its square is not even'.

contravariant tensor *See* tensor.

control chart A graph used in *quality control to indicate whether or not a characteristic of mass-produced items, such as individual weight, mean lengths of items in batches, or variability of length of items in batches, falls within acceptable limits; and thus to indicate whether some deficiency in a production process is resulting in failure to meet such limits.

A common procedure for checking weight consistency is to take a specified number of units (four, say) from a production line at regular (e.g. hourly) intervals and record their total weight. This is plotted on a graph on which there is a *target line* representing the ideal total weight. Above and below this line at calculated distances are upper and lower control lines (often at about three *standard deviations above or below the target). If the total weight falls outside these lines, there is a need to check whether the process is out of control. There are sometimes additional warning lines indicating that a certain action may be needed if a specified number of consecutive sample values fall outside these lines. Modified charts are used to detect other undesirable features, such as increases in variability of output. *See also* cusum chart.

convergence A property of a *convergent series or *convergent sequence.

convergent fraction A nonterminating *continued fraction that has a *limit.

convergent integral An *infinite integral that has a definite limit.

convergent iteration An *iteration which generates a *convergent sequence.

convergent product An *infinite product that has a nonzero value.

convergents *See* continued fraction.

convergent sequence An infinite *sequence that has a *limit. *See* order (**12**).

convergent series An infinite *series

$$a_1 + a_2 + \cdots + a_n + \cdots$$

whose *partial sums, s_n, given by

$$s_n = a_1 + a_2 + \cdots + a_n$$

approach a limit S as the number of terms, n, approaches infinity; i.e. a series is convergent if

$$\lim_{n \to \infty} s_n = S$$

The series is then said to converge to the value S or to have the *sum S*. If s_n does not approach a limit as $n \to \infty$ the series is *divergent* (*see* divergent series).

Some of the terms in the infinite series $\Sigma\, a_n$ may be negative. If these terms are all made positive, i.e. if the absolute values $|a_n|$ are considered, and if the series $\Sigma\, |a_n|$ is also convergent, then the series $\Sigma\, a_n$ is said to be *absolutely convergent*. The series

$$\sum (-1)^{n-1}\left(\frac{1}{n^n}\right) = 1 - \left(\frac{1}{2}\right)^2 + \left(\frac{1}{3}\right)^3 - \left(\frac{1}{4}\right)^4 + \cdots$$

is absolutely convergent. If $\Sigma\, |a_n|$ is not convergent then $\Sigma\, a_n$ is said to be *conditionally convergent*. The series

$$\sum(-1)^{n-1}\left(\frac{1}{n}\right) = 1 - \left(\frac{1}{2}\right) + \left(\frac{1}{3}\right)$$

$$- \left(\frac{1}{4}\right) + \cdots$$

$$= \ln 2$$

is conditionally convergent since the *harmonic series $\Sigma(1/n)$ is divergent.

If Σa_n and Σb_n are two convergent series with sums S and T then

$\Sigma(a_n + b_n)$ converges; sum is $S + T$
$\Sigma(a_n - b_n)$ converges; sum is $S - T$
Σka_n converges, k constant; sum is kS
If $a_n \leqslant b_n$ for all n then $S \leqslant T$.

Since convergent series play a major role in mathematics it is necessary to be able to test a series for convergence. *See* comparison test; ratio test; Abel's test; Cauchy convergence test; Cauchy integral test; Dirichlet's test.

converse (of a theorem) A *theorem obtained by interchanging the premise and conclusion of a given theorem. For example, the theorem 'if two chords of a circle are equal distances from the centre, then the chords are equal' has the converse 'if two chords of a circle are equal then they are equidistant from the centre'. In this case the converse of the theorem is true, but this is not always so.

conversion period *See* interest.

convex combination A *linear combination in which the *scalar coefficients are non-negative and their sum is 1. For example, $0.1\mathbf{a} + 0.3\mathbf{b} + 0.6\mathbf{c}$ is a convex combination of the *vectors $\mathbf{a}$, $\mathbf{b}$, and $\mathbf{c}$.

convex function A *real function $f(x)$ is said to be *convex* if for every pair of points A and B on the curve $y = f(x)$ the line segment AB lies above the curve. It is said to be *concave* if the line segment always lies below the curve. For example, the function x^2 is convex, and the function x^3, for $x \leqslant 0$, is concave.

More formally, the real function $f(x)$ is said to be convex if, for any pair of x-values x_1 and x_2, and all numbers λ such that $0 \leqslant \lambda \leqslant 1$,

$$f(\lambda x_1 + (1 - \lambda)x_2) \leqslant \lambda f(x_1) + (1 - \lambda)f(x_2)$$

If the function $-f(x)$ is convex then $f(x)$ is a *concave function*. This formal definition extends to functions of more than one variable, the two numbers x_1 and x_2 then being replaced by a pair of *n-tuples.

If a function is both convex and concave then it is a *linear function.

convex hull The convex hull of a set of points X is the intersection of all *convex sets containing X, i.e. the smallest convex set containing X. Equivalently, the convex hull is the set of all *convex combinations of points of X.

convex polygon A *polygon that has all its angles less than or equal to $180°$. *Compare* concave polygon.

convex polyhedron A *polyhedron in which the plane of every face does not cut the polyhedron, i.e. the polyhedron lies completely on one side of the plane of each face. *Compare* concave polyhedron.

convex set A *set of points which, if it contains the points A and B, contains the line segment AB. *See also* convex hull.

convolution The convolution of two *functions $f(x)$ and $g(x)$ is the function

$$\int_0^x f(t)g(x - t)\,dt$$

coordinate One of a set of numbers (coordinates) specifying the position of a point relative to certain other lines or points. *See* coordinate system; *see also* abscissa; ordinate.

coordinate geometry (analytic geometry) A form of geometry in which lines, curves, etc. are represented by equations by using a coordinate system. Coordinate geometry was introduced in 1637 by René Descartes. *See also* Cartesian coordinate system.

coordinate system A system for locating points in space by using reference lines or points. The position of a point is given by a set of numbers (*coordinates*) that are distances or angles from the reference frame. *See* Cartesian coordinate system; polar coordinate system; astronomical coordinate system; geographical coordinates; inertial coordinates; *compare* intrinsic equation.

Copeland–Erdős number (A. Copeland and P. Erdős, 1946) the number

$$0.235\,711\,131\,7\ldots$$

whose decimal digits are those of all the *prime numbers in succession. It is *normal to base ten.

coplanar Lying in the same plane. Thus, coplanar lines (or curves) are lines (or curves) that lie in the same plane. Any three points are coplanar. Four points are coplanar if the *determinant which has the coordinates of the points as its first three columns, and a fourth column whose elements are unity, is zero:

$$\begin{vmatrix} x_1 & y_1 & z_1 & 1 \\ x_2 & y_2 & z_2 & 1 \\ x_3 & y_3 & z_3 & 1 \\ x_4 & y_4 & z_4 & 1 \end{vmatrix} = 0$$

coprime *See* relatively prime.

copunctal Having a common point. For instance, in a three-dimensional coordinate system the three coordinate axes are copunctal, the common point being the origin.

Coriolis force An *inertial force that arises when a body moves in a rotating *frame of reference. The force acts on the body at right angles to both the axis of rotation and the direction of motion of the body in the rotating frame, and vanishes when the velocity of the body is zero. It has a magnitude of $2mv\omega$, where m is the body's mass, v the magnitude of its velocity relative to the rotating frame, and ω the magnitude of the angular velocity of the rotating frame relative to an inertial frame. The *Coriolis acceleration* is the tangential acceleration experienced by the body as a result of this force: it acts in the same direction with magnitude $2v\omega$. The total force acting on the body is the sum of the 'real' force, the inertial *centrifugal force, and the Coriolis force.

The Coriolis force must be taken into account when considering motion relative to the earth's surface, e.g. the overall movement of winds or the trajectories of long-range weapons. It is named after the French engineer Gustav Gaspard de Coriolis (1792–1843).

Cornu spiral *See* spiral.

corollary *See* theorem.

correlation In a general sense, correlation between two or more quantities denotes an interdependence between them. The word is widely used in a more restrictive sense to indicate a degree of relationship, especially one more or less linear in nature, between two variables or between two sets of *ranks. Data pairs that show a close relationship are said to be *highly correlated*. High correlation need not imply causal relationship: for example, data for numbers of car owners and the average daily sales of alcohol in each of a number of cities are likely to be highly correlated, but this may simply reflect the influence of population size on both variables. *See* correlation coefficient; multiple correlation coefficient.

correlation coefficient **1.** The *product-moment correlation coefficient* (sometimes called the *Pearson correlation coefficient* after Karl *Pearson, who discovered many of its properties) between two random variables X and Y is defined as

$$\rho = \frac{\text{Cov}(X, Y)}{\sqrt{[\text{Var}(X)\,\text{Var}(Y)]}}$$

(*see* covariance, variance). For a straight-line relationship, $\rho = \pm 1$. If $\rho = 0$, then X and Y are said to be *uncorrelated*; this does not imply independence unless X and Y have a bivariate *normal distribution. For n paired observations (x_i, y_i), $i = 1, 2, \ldots, n$, the *sample correlation coefficient* r is given by

$$r = \frac{s_{xy}}{\sqrt{(s_{xx} s_{yy})}}$$

where s_{xy} denotes the sum of products of deviations of the x_i and y_i from their means, and s_{xx} and s_{yy} are sums of squares of deviations from their respective means. If $r = 1$, the points lie on a straight line of positive slope; if $r = -1$, they lie on a straight line of negative slope. If r is near zero there is virtually no linear association, but there may be some other form of association: for example, the points may be well scattered around the circumference of a circle.

2. *Spearman's rank correlation coefficient* (C. Spearman, 1904) is the product moment correlation coefficient between two sets of paired *ranks, such as the ranks assigned to each of the same set of candidates giving their order of merit on the basis of examination results for papers in (i) Mathematics and (ii) French. If no candidates are ranked equally (i.e. if there are no *tied ranks*) in any one subject, the Spearman's coefficient r_S is usually calculated by a formula that takes account of special properties of ranks:

$$r_S = 1 - \frac{6T}{n(n^2 - 1)}$$

where T is the sum of squares of the difference between the ranks for each of the n pairs. If measurements x and y have been replaced by ranks, a value of $r_S = \pm 1$ implies that y increases or decreases monotonically (but not necessarily linearly) as x increases.

3. *Kendall's rank correlation coefficient* (M.G. Kendall, 1938) is a measure of agreement between two sets of orderings of the same objects, and may be calculated using either *order statistics or ranks. The objects are arranged pair-wise in ascending order for the first set, and the number of objects out of natural order in the second set is counted. A coefficient is formed which may take values between -1 (complete disagreement) and $+1$ (complete agreement). Complete disagreement occurs when the rankings in the two sets are in reverse order.

4. The *biserial correlation coefficient* is an infrequently used measure of dependence between a *random variable X that varies continuously and a *random variable Y that may take only two values, y_1 or y_2.

correlation matrix A *matrix representation of all correlations between pairs of p ($\geqslant 2$) variables or sets of observations. The entry r_{ij} is the *correlation coefficient between the ith and jth variables. The matrix is symmetric with diagonal elements all unity and all *eigenvalues non-negative.

correspondence A *binary relation. *See also* one-to-one correspondence; many–one correspondence; one–many correspondence.

corresponding angles *See* transversal.

cos Cosine. *See* trigonometric functions.

cosecant (cosec) *See* trigonometric functions.

coset If H is a *subgroup of a *group G with group operation $\circ$, then to every element a of the group G there corresponds a *left coset*, denoted by $a \circ H$, which is the set of all elements of the form $a \circ h$, where $h \in H$. Similarly, there is a *right coset*, denoted by $H \circ a$, consisting of all elements of the form $h \circ a$, where $h \in H$. *See also* normal subgroup; ideal.

cosh Hyperbolic cosine. *See* hyperbolic functions.

cosine *See* trigonometric functions.

cosine curve A graph of a cosine function (*see* trigonometric functions). In rectangular Cartesian coordinates a graph of $y = \cos x$ is a regular undulating curve intersecting the y-axis at the point $(0, 1)$. It is the same shape as a *sine curve, displaced by $\frac{1}{2}\pi$ along the x-axis.

cosine rule (law of cosines) **1.** A formula used for solving triangles in plane trigonometry:

$$c^2 = a^2 + b^2 - 2ab \cos C$$

where C is the angle opposite side c (i.e. the angle included between sides a and b).
2. Formulae used in spherical trigonometry for solving *spherical triangles:

$$\cos c = \cos a \cos b + \sin a \sin b \cos C$$

$$\cos C = -\cos A \cos B + \sin A \sin B \cos c$$

where a is the side opposite angle A, b is opposite angle B, and c is opposite angle C.

cosine series **1.** The *series expansion for a cosine function:

$$\cos x = 1 - \frac{x^2}{2!} + \frac{x^4}{4!} - \frac{x^6}{6!} + \cdots$$

This is valid for all x. *See* trigonometric functions.
2. A *series in which the terms are cosine functions. *See* Fourier series.

cotangent (cot) *See* trigonometric functions.

coterminal **1.** *Coterminal angles* are angles which are rotations between the same two lines, i.e. angles that have the same initial and final lines. For example, $20°$, $-340°$, and $380°$ are coterminal angles.
2. *Coterminal edges* are edges of a geometric figure or *graph which have a common vertex.

Cotes, Roger (1682–1716) English mathematician and astronomer. Much of his short life was spent working with Newton on preparing the extensively revised second edition of Newton's *Principia* (1713). Cotes published just one mathematical paper of his own, *Logometria* (1714), in which he described new methods for computing logarithms and for converting logarithms from one base into another. His other mathematical papers, published posthumously in *Harmonia mensurarum* (1722), dealt mainly with problems on the integration of rational functions.

coth Hyperbolic cotangent. *See* hyperbolic functions.

coulomb Symbol: C. The *SI unit of electric charge, equal to the quantity of charge transferred by a current of 1 ampere flowing for 1 second. [After C.A. Coulomb (1736–1806)]

countable (denumerable, enumerable) Describing a *set that can be put into a *one-to-one correspondence with a subset of the positive integers. If the set is infinite it is described as *countably infinite*. Examples are the set of natural numbers and the set of rational numbers. The set of irrational numbers is not a countable set. *See also* Cantor's theory of sets.

counterexample An example clear enough to disprove a general statement. Thus the claim that there are no numbers other than 1 which are the sum of the cubes of their digits is refuted by the counterexample

$$153 = 1^3 + 5^3 + 3^3$$

counting number A number used in counting objects, i.e. one of the set of positive integers 1, 2, 3, 4, etc.

couple A system of two *forces that are equal in magnitude, act in exactly opposite directions, and do not have the same line of action. A couple has the same *moment about any point in the plane of the two forces. The moment is a *vector that acts at right angles to the plane of the forces: under the action of a couple, a rigid body rotates about an axis perpendicular to this plane. The magnitude of the vector is Fd, where d is the perpendicular distance between the forces and F is the magnitude of each force. A couple has no resultant force: it cannot be reduced to or balanced by a single force. It can be balanced by a couple having equal but opposite moment, applied in the same plane or in a parallel plane. In addition, two couples are together equivalent to a third couple whose moment is the vector sum of the separate moments.

covariance The first product *moment of two *variables about their means. If X and Y have means μ_x and μ_y then the covariance is

$$\mathrm{Cov}(X, Y) = \mathrm{E}[(X - \mu_x)(Y - \mu_y)]$$

$$= \mathrm{E}(XY) - \mu_x\mu_y$$

For a sample of n paired observations (x_i, y_i) the sample covariance is

$$c_{xy} = \sum_i \frac{1}{n}(x_i - \bar{x})(y_i - \bar{y})$$

The covariance of X and Y divided by the product of the standard deviations of X and Y is the product moment *correlation coefficient, which, unlike covariance, is not scale dependent.

covariance, analysis of *See* analysis of covariance.

covariance matrix The analogue of the *correlation matrix, with covariances in place of correlations and variances in place of unit elements on the main diagonal.

covariant tensor *See* tensor.

cover If A is a family of *sets, and if X is a set such that every element of X is included in at least one of the family of sets of A, then A is said to be a cover of X. For example, if

$$A = \{\{1,2\},\{3,4\}\} \quad \text{and} \quad X = \{1,3\}$$

then A is a cover of X. *See* compact.

CPA *Abbreviation for* *critical path analysis.

Cramer, Gabriel (1704–52) Swiss mathematician who in his *Introduction à l'analyse des lignes courbes algébriques* (1750, Introduction to the Analysis of Algebraic Curves) published a classification of algebraic curves. The book also contains *Cramer's rule for the solution of systems of linear algebraic equations.

Cramér, Harald (1893–1985) Swedish pure mathematician and statistician. His *Mathematical Methods of Statistics* (1945) was a definitive work linking the pure-mathematical theory of probability to statistical applications. As an adviser to the life assurance industry, he pioneered the statistical study of risk. His work on time-series analysis led to the development of an important method called Cramér–Wold decomposition; he was co-discoverer of the *Cramér–Rao inequality. His interest in mathematics owed much to a close friendship with G.H. Hardy.

Cramér–Rao inequality (C.R. Rao, 1945; H. Cramér, 1946) An *inequality giving a lower *bound to the *variance of an *estimator T of a parameter θ, extending that for an unbiased estimator to allow for bias. *See* information.

Cramer's rule (G. Cramer, 1750) A rule for solving systems of linear *simultaneous equations by *determinants. It requires the equations to be written in the form

$$a_1x + b_1y + c_1z = d_1$$

$$a_2x + b_2y + c_2z = d_2$$

$$a_3x + b_3y + c_3z = d_3$$

Then, the determinant of the coefficients of the unknowns is formed:

$$D = \begin{vmatrix} a_1 & b_1 & c_1 \\ a_2 & b_2 & c_2 \\ a_3 & b_3 & c_3 \end{vmatrix}$$

If $D \neq 0$ there is a unique solution. If $D = 0$, the equations are not independent and may either be *inconsistent or possess infinitely many solutions. For each unknown in the system of equations a determinant is formed in which the coefficients of that unknown are replaced by the constant terms appearing on the right-hand sides of the equations. Thus, to find x, the coefficients a_1, a_2, and a_3 are replaced by d_1, d_2, and d_3, to give

$$D_x = \begin{vmatrix} d_1 & b_1 & c_1 \\ d_2 & b_2 & c_2 \\ d_3 & b_3 & c_3 \end{vmatrix}$$

The value of x is then given by $x = D_x/D$. Similarly, for y

$$D_y = \begin{vmatrix} a_1 & d_1 & c_1 \\ a_2 & d_2 & c_2 \\ a_3 & d_3 & c_3 \end{vmatrix}$$

and $y = D_y/D$. The method can be used for any system of n linear equations in n unknowns. However, it requires many more arithmetic operations than *Gaussian elimination, and so is used only for small values of n.

Crelle, August Leopold (1780–1855) German mathematician and civil engineer noted for his founding in 1826 of the *Journal für die reine und angewandte Math-*

ematik (Journal of Pure and Applied Mathematics), known more familiarly as *Crelle's Journal*, one of the first journals to be devoted exclusively to mathematical research. He is also remembered for his publication in 1820 of extensive factor tables.

critical damping The situation occurring when a system, such as a pendulum, just fails to oscillate. *See* damped harmonic motion.

critical path analysis (CPA) A form of network analysis used to determine the optimum or permissible scheduling for meeting fixed constraints where there are interlocking and ordered operations, such as those involved in building a house or publishing a book. Characteristics are:

(1) there is an order of precedence for certain activities;

(2) some activities may be carried out simultaneously;

(3) the duration of each activity is known.

An *activity* is a task requiring time to perform it; a *dummy activity* is one that requires no action or time to complete and is included in a network to assist the solution of the problem. A CPA network is a logical combination of edges and nodes. Activities and times are indicated on edges of the network joining nodes which represent the start and finish of activities. In a simple situation the earliest possible time of completion is entered in each finish node, though further information may be added at nodes in larger networks.

The earliest completion time of an activity is also the earliest possible starting time for any activity that can be started only after that previous activity is completed. For some activities the completion time of the whole project will not be delayed if those activities are not started at the earliest possible time, but there will always be a latest time for starting any activity without delaying the completion of the whole project. The edges joining the nodes at which

Activity	Duration(minutes)	Precedents
A Disembark passengers	7	
B Unload baggage	10	
C Refuel aircraft	12	A
D Clean cabin	15	A
E Load catering requirements	6	D
F Load baggage	12	B
G Embark passengers	14	C, E
H Final loading check	2	F, G

critical path analysis for turning round an aircraft. The numbers represent time in minutes; times in nodes are cumulative. The dashed lines represent dummy activities. The critical path itself is via the edges A, D, E, G, H. Some activities not on the critical path may be started later without increasing the minimum turn-round time of 44 minutes given in the final node; for example, the start of activity C could be postponed for up to $28 - 19 = 9$ minutes without delaying take-off.

the earliest and latest possible starting times for an activity are equal form the *critical path* for the completion of the job on schedule. The method can be extended to work out how changes in the time taken to complete activities will delay or advance the completion date. Such studies are called *programme evaluation and review techniques*, commonly abbreviated to PERT.

The network in the diagram is for finding the shortest time (determined by the critical path) for turning round an aeroplane after landing if the activities, duration, and precedence requirements are as given in the table.

critical point 1. *See* stationary point.
2. A point on a graph at which a curve has a vertical *tangent.

critical region *See* hypothesis testing.

cross-cap *See* manifold.

cross product *See* vector product.

cross-ratio A particular ratio of ratios of lengths between four points A, B, C, and D on a line, defined as $(AC/CB)/(AD/DB)$ or $(AC . DB)/(AD . CB)$, and denoted by $\{A, B; C, D\}$. This equals the ratio of the ratios in which C and D divide AB.

If this cross-ratio is equal to -1, it is called a *harmonic ratio*, and the four points form a *harmonic range* or *set* in which (A, B) and (C, D) are *conjugate pairs*. In this case C and D divide AB in the same ratio (one internally, the other externally). The cross-ratio is harmonic if $\{A, B; C, D\} = \{B, A; C, D\}$.

If the line has a parametrized form in which the points have parameters a, b, c, and d, the cross-ratio is given by

$$\{A, B; C, D\} = \frac{(a - c)(b - d)}{(a - d)(b - c)}$$

See division in a given ratio; harmonic pencil.

cross-section *See* section.

cross-validation A procedure in statistical modelling in which data are first randomly divided into two or more *subsets. A model (e.g. some *regression model) is fitted to all but one of these subsets, and a *prediction error* of the fitted model when applied to the omitted subset is calculated. Each subset is omitted in turn, and a combined estimated prediction error is obtained. For small data sets an extreme but useful procedure is the *leave-one-out* cross-validation technique, where each of n observations is omitted in turn and a model is fitted to the remaining $n - 1$ data. A predicted value is obtained for the omitted observation using that model.

The method is useful both for assessing the overall goodness of fit of a model and for detecting outliers. The computations for a leave-one-out cross-validation are similar to those for the *jack-knife, but the aim is different.

cruciform curve A plane *curve with the equation

$$x^2y^2 = a^2(x^2 + y^2)$$

in Cartesian coordinates. The curve is cross-shaped, being symmetrical about the origin with four branches. The lines $x = \pm a$ and $y = \pm a$ are asymptotes.

crunode *See* node.

cryptanalysis The science of studying and developing methods of analysing *ciphertext in an attempt to find the *plaintext, usually by a recipient of an encrypted message who does not know the decryption *key.

cryptology (cryptography) The science of secret writing. It involves the development and understanding of methods of both encryption and decryption. The ultimate goal is to find reasonably quick methods of encryption that are difficult to decrypt without specific information and, on the other hand, to find methods that will decrypt *ciphertext when the decryption *key is not known.

crystallography The study of crystal structures. It is based largely on the algebraic properties of the *symmetry group* of the crystal *lattice, which is the group of all the *isometries of the crystal (regarded as being repeated indefinitely in all directions). The two main areas of crystallography are the study of the *Bravais lattice*, which gives information about the way in which the crystal structure is replicated throughout space, and the study of the *point group*, which gives information about the structure of the crystal in the neighbourhood of a particular atom. The *crystallographic restriction* is the theorem that any symmetry of a crystal that has finite *order must be of order 1, 2, 3, 4, or 6. There are 14 different Bravais lattices and 32 point groups; they can be combined in various ways to produce a total of 230 crystal *space groups*. For all but a few of these possibilities there are substances occurring naturally in crystalline forms having the corresponding crystal group symmetry.

Johann Kepler (1571–1630) was the first to study crystals and their symmetries. The theory was developed intensively by R.-J. Haüy (1743–1822) and A. Bravais (1811–63). In the 1980s *quasi-crystals* were discovered; unlike true crystals, they can have fivefold symmetry.

csc Cosecant. *See* trigonometric functions.

csch Hyperbolic cosecant. *See* hyperbolic functions.

ctn Cotangent. *See* trigonometric functions.

cube 1. The third power of a number. The cube of a is $a \times a \times a$, i.e. a^3.
2. A solid figure that has six identical square faces, all the face angles being right angles. The volume of a cube is a^3, where a

is the length of an edge. The cube is one of the five regular polyhedra. *See* polyhedron.

cube root A value or quantity that has a cube equal to a given quantity. The real cube root of 8, written as $\sqrt[3]{8}$, is 2 since $8 = 2^3$.

cube root of unity *See* root of unity.

cubic Describing a mathematical expression of the third *degree. Thus, a *cubic polynomial* in x is a polynomial of the type

$$ax^3 + bx^2 + cx + d$$

A cubic function of x is a function $f(x)$ whose value for a value of x is given by a cubic polynomial in x. A *cubic equation* is an equation of the general form

$$ax^3 + bx^2 + cx + d = 0$$

The first methods of obtaining a formula for x in terms of the coefficients were found by Scipione del Ferro (c.1465–1515) and later by Tartaglia. Del Ferro's solution was never published, but Tartaglia's appeared in 1545 (without his consent, but with acknowledgement) in *Cardano's book *Ars magna*.

This method (often known as Cardano's method) involved first recasting the equation by substituting $x = y - b/3a$. This removes the term in y^2. Dividing through by the coefficient of y^3 gives an equation of the form

$$y^3 + py + q = 0$$

This, the *reduced cubic*, is the starting point for the solution. Next, the substitution $y = u - v$ is made with the condition that $uv = \frac{1}{3}p$ (one-third of the y-coefficient). The equation becomes

$$u^3 - v^3 + q = 0$$

Substituting $v = p/3u$ gives

$$u^6 + qu^3 - (p/3)^3 = 0$$

which is a quadratic equation in u^3:

$$(u^3)^2 + qu^3 - (p/3)^3 = 0$$

Solving this for u^3 gives a value of u and hence a value of v. The general solution for y (in the reduced cubic) can then be found.

The nature of the roots of a cubic equation can be found from its *discriminant*. For the cubic equation

$$ax^3 + bx^2 + cx + d = 0$$

the discriminant can be found by dividing through by a so as to make the leading coefficient unity. The cubic then has the form

$$x^3 + px^2 + qx + r = 0$$

and the discriminant is

$$p^2q^2 + 18pqr - 4q^3 - 4p^3r - 27r^2$$

If the discriminant is negative, there are two conjugate imaginary roots and one real root. If it is zero, there are three real roots of which at least two are equal. If it is positive, there are three real roots that are not equal. This last case is a difficulty in Cardano's solution of the cubic because it leads to calculations that require the cube root of an imaginary number – the so-called *irreducible case* of the cubic.

The solution of such cases was first suggested by Viète, who noticed that it was linked with the geometrical problem of trisecting an angle. If the equation is put in the reduced form

$$y^3 + 3py + q = 0$$

and a substitution $my = x$ is made, to give

$$x^3 + 3m^2px + qm^3 = 0$$

then substituting $x = \cos\theta$ gives

$$\cos^3\theta + 3m^2p\cos\theta + qm^3 = 0$$

Viète compared this with his multiple angle formula for $\cos 3\theta$, which can be written as

$$\cos^3 \theta - \tfrac{1}{4}(3\cos\theta) - \tfrac{1}{4}(\cos 3\theta) = 0$$

It follows that if $3m^2p = -\tfrac{3}{4}$ then $\cos 3\theta = -4qm^3$. Since p and q are known, then m, and consequently $\cos 3\theta$, can be found. From the possible values of θ three values of $\cos\theta\ (= x)$ can be obtained and consequently the three values of y that satisfy the original cubic. In principle, solutions of the cubic equation can be found by such methods. In practice, however, it is usual to use numerical methods of solution (see numerical analysis).

A *cubic curve* is a curve with an algebraic equation of the third degree. A *cubic graph* is a regular *graph in which every vertex has degree three.

cubical parabola A plane *curve with the equation

$$y = ax^3$$

in Cartesian coordinates. It has a point of inflection at the origin. See also semicubical parabola.

cubic lattice A *lattice of points in three-dimensional space with coordinates (ax, ay, az), where x, y, and z integers and a is a positive constant (see diagram (a)). The cubic lattice with $a = 1$ is the three-dimensional integer lattice.

A lattice of points with coordinates $\left(\tfrac{1}{2}ax, \tfrac{1}{2}ay, \tfrac{1}{2}az\right)$ such that the integers x, y, and z are all even or all odd is a *body-centred cubic lattice*. This lattice comprises all the points of the cubic lattice with constant a and the points at the centre of each cube (see diagram (b)).

A lattice of points with coordinates $\left(\tfrac{1}{2}ax, \tfrac{1}{2}ay, \tfrac{1}{2}az\right)$ where the integers x, y, and z are such that $x + y + z$ is even is a *face-centred cubic lattice*. This lattice contains all the points of the cubic lattice with constant a and the points at the centres of the faces of each cube (see diagram (c)).

See also Kepler's conjecture.

cuboctahedron (*plural* **cuboctahedra**) A *polyhedron formed by truncating a

cubic lattice (a) cubic lattice, (b) body-centred cubic lattice, and (c) face-centred cubic lattice.

cube so that the vertices lie at the centre points of the cube edges. It can similarly be formed by truncating an octahedron (hence the name). The cuboctahedron is one of the Archimedean solids. It has 14 faces, 12 vertices, and 24 edges.

cuboid See parallelepiped.

cumulants The coefficient κ_r of $t^r/r!$ in the series *expansion of $\ln M(t)$, where $M(t)$ is the *moment generating function, is called the rth cumulant. For any distribution, κ_1 is the mean and κ_2 is the variance. Expressions for cumulants in terms of moments, and vice versa, are available.

cumulative frequency function An alternative name for a (cumulative) *distribution function. It is frequently applied to data to designate the proportion of a set of data less than a specified value. For example, if Y represents the number of teeth with dental caries in a group of 50 school-children (see table), then the cumulative frequency function $F(y) = $ frequency $Y \leqslant y$, where for these data $F(y)$ is a step function and the values at the steps are $F(0) = 27/50$, $F(1) = 39/50$, $F(2) = $

$45/50$, $F(3) = 49/50$, and $F(6) = 1$. F is a non-decreasing function such that, for all y, $0 \leqslant F(y) \leqslant 1$. The sums 27, $27 + 12$, ..., $27 + 12 + 6 + 4 + 1 = 50$ are sometimes called the *absolute cumulative frequencies*.

Number of teeth	Number of children
0	27
1	12
2	6
3	4
6	1

cup The symbol ∪ used to denote the *union of two sets A and B, as in the expression $A \cup B$. *Compare* cap.

curl For a vector function of position $\mathbf{V}(\mathbf{r})$, the curl of $\mathbf{V}$, written as curl $\mathbf{V}$, is given by $\nabla \times \mathbf{V}$, where ∇ is the operator *del. Thus

$$\text{curl } \mathbf{V} = \nabla \times \mathbf{V}$$
$$= \mathbf{i} \times \frac{\partial \mathbf{V}}{\partial x} + \mathbf{j} \times \frac{\partial \mathbf{V}}{\partial y} + \mathbf{k} \times \frac{\partial \mathbf{V}}{\partial z}$$

and

$$\mathbf{r} = x\mathbf{i} + y\mathbf{j} + z\mathbf{k}$$

An equivalent form for curl $\mathbf{V}$ is

$$\begin{vmatrix} \mathbf{i} & \mathbf{j} & \mathbf{k} \\ \partial/\partial x & \partial/\partial y & \partial/\partial z \\ V_x & V_y & V_z \end{vmatrix}$$

for which $\mathbf{V} = V_x\mathbf{i} + V_y\mathbf{j} + V_z\mathbf{k}$.

In fluid flow, $\frac{1}{2}$ curl $\mathbf{v}$ gives the angular velocity in an element of fluid ($\mathbf{v}$ is its velocity).

See divergence; gradient; Stokes's theorem.

curvature Symbol: κ. The rate of change of direction of a curve at a particular point on that curve. The angle $\delta\psi$ through which the tangent to a curve moves as the point of contact moves along an arc PQ is the *total curvature* of the arc PQ (*see* diagram (a)). The *mean curvature* of the arc PQ is defined as the total curvature divided by the arc length δs, i.e. $\delta\psi/s$, where s is the arc distance of P from a fixed point A. The curvature κ at the point P is the limiting value of the mean curvature of the arc PQ as $\delta s \to 0$, i.e. the derivative $d\psi/ds$.

If the curve is a circle with centre at C and radius R (*see* diagram (b)), then $\angle PCQ = \delta\psi$ and $\delta\psi/s = 1/R$. Thus at all points on a circle, the curvature is the reciprocal of the radius.

The *circle of curvature* at any point on a curve is the circle that is tangential to the curve at that point and whose curvature is the same as that of the curve at that point (*see* diagram (c)). The *centre of curvature* is the centre of this circle. The *radius of curvature* at P is the radius ρ of this circle, and $\rho = |ds/d\psi|$.

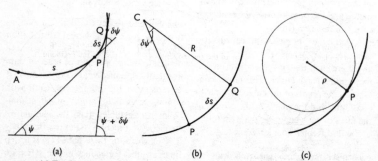

curvature (a) Total curvature of a curve; (b) curvature of a circle; (c) circle of curvature.

In Cartesian coordinates,

$$\rho = \left[1 + \left(\frac{dy}{dx}\right)^2\right]^{3/2} \Bigg/ \left|\frac{d^2y}{dx^2}\right|$$

In parametric form, if $x = f(t)$ and $y = g(t)$, then

$$\rho = \frac{(\dot{x}^2 + \dot{y}^2)^{3/2}}{|\dot{x}\ddot{y} - \ddot{x}\dot{y}|}$$

where x, $\ddot{x}$, y, and $\ddot{y}$ represent first and second derivatives with respect to t.

At a point on a surface, the curvature varies with direction. In general, there are two directions in which the radius of curvature has an absolute maximum and absolute minimum. These are the *principal directions*, and *Euler's theorem shows that they are perpendicular. The *principal curvatures* at the point are the curvatures in these directions. The *total* (or *Gaussian*) curvature of the surface at a given point is the product of the principal curvatures at that point. The curvature of higher-dimensional manifolds is also intensively studied, and is one of the main mathematical foundations for Einstein's theory of general relativity.

curve A line, either straight or continuously bending without angles. A curve can be considered as the path of a moving point (i.e. a point moving with only one degree of freedom). Alternatively it can be regarded as a set of points produced by a continuous transformation of a closed interval.

Curves are generally studied as graphs of equations using *coordinate systems. They are classified as *algebraic curves*, which have algebraic equations, and *transcendental curves*, which have equations containing transcendental functions. *Open curves* (or *arcs*) are curves that have end points. *Closed curves* have no end points, i.e. a closed curve is a transformation of a closed interval $[a, b]$ for which the images of a and b coincide. A curve that lies entirely in a plane is a *plane curve*. A curve that does not lie in a plane is a *skew* or *twisted curve* (e.g. a *helix). Any curve in three-dimensional space is described as a *space curve* (note that it need not also be a twisted curve).

curvilinear motion Motion along a curved path. *Circular motion is a special case of curvilinear motion.

cusp (spinode) A *singular point on a curve at which there are two different *tangents that coincide. A cusp is a special case of a *double point in which the tangents are

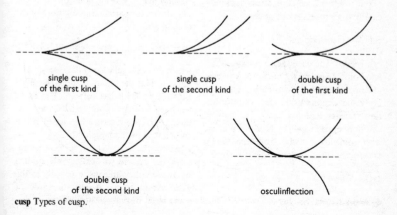

single cusp
of the first kind

single cusp
of the second kind

double cusp
of the first kind

double cusp
of the second kind

osculinflection

cusp Types of cusp.

coincident. In a *single cusp* the curve is not continuous through the point (i.e. two branches or parts of the curve meet at a point). A *double cusp* has both branches of the curve continuous through the point (i.e. the curve is tangential to itself). A double cusp is also called an *osculation* or a *tacnode*. Cusps (either single or double) are further classified into *cusps of the first kind* (in which both branches of the curve near the cusp lie on opposite sides of the tangent) and *cusps of the second kind* (in which the branches of the curve lie on the same side of the common tangent). Double cusps at which one or both branches of the curve have points of inflection are called points of *osculinflection* (i.e. both osculation and inflection).

cusum chart A *control chart designed to detect departures from acceptable operating standards in which a record is kept of the sum of deviations from an ideal or target value for successive samples. While the process is in control, the cumulative sum (abbreviated to *cusum*) of deviations should remain small, positive and negative deviations almost cancelling out. A run of samples in which either positive or negative deviations dominate indicates that the process may be out of control. Cusum charts are designed in a way that makes it easy for plant operatives to make a routine decision about whether such cumulative data indicate that a process is out of control.

cut *See* Dedekind cut.

cybernetics The science of communication and control applied to machines, animals, and organizations. Cybernetics attempts to unify such studies using ideas of information transfer and feedback. The subject was developed in 1946 by Norbert Wiener, who coined the name from the Greek *kubernētēs*, meaning 'pilot' or 'steersman'.

cycle *See* walk.

cycle per second *See* hertz.

cyclic Describing a polygon that can be *circumscribed by a circle, so that all its vertices lie on the circumference of the circle. Thus, a *cyclic quadrilateral* is a quadrilateral with its four vertices lying on a circle. *See also* Ptolemy's theorem.

cyclic code A *linear code in which every cyclic *permutation of a *codeword is also a codeword. For example, if 1011 is a codeword in a cyclic code, then so are 0111, 1110, and 1101.

cyclic group A *group, all of whose elements are powers of a single element. A finite cyclic group containing n elements will be generated by one element, say t, that satisfies $t_n = I$ (where I is the *identity element); it is denoted by $\mathbb{Z}_n$. When $n \geqslant 3$ it can be regarded as the group of all rotational *symmetries of a regular polygon with n sides. *See* dihedral group; generator.

cyclic permutation (circular permutation) A *permutation in which each member of a *set replaces a successive member or in which each member is replaced by a successive member. For example, $x \to y$, $y \to z$, $z \to x$ is a cyclic permutation of x, y, and z.

cycloid A plane *curve that is the *locus of a point on the circumference of a circle as the circle rolls (without slipping) along a straight line. If P is the point on the circle, of radius r, the parametric equations of the cycloid are

$$x = r(\theta - \sin\theta) \quad \text{and} \quad y = r(1 - \cos\theta)$$

where the circle rolls along the x-axis, starting with P at the origin, and θ is the angle through which P has rotated. The curve has a series of arches and touches the baseline at *cusps a distance $2\pi r$ apart. It is a special case of the *trochoid.

Although unknown to Greek geometers, the cycloid was extensively studied by later mathematicians, especially in the 17th century, when it was the cause of some bitter disputes about priority of dis-

cycloid

covery. It seems to have been recognized by Galileo, who attempted to find the area under an arch by experiment (weighing the shape). The curve was first studied extensively by Roberval, who proved (1634) that the area under an arch was three times the area of the generating circle. He also found (1638) the tangent at any point on the curve. If at a point P, a line PH is drawn parallel to the baseline and a tangent PT is drawn to the generating circle (PT indicating the direction of motion of P), then the tangent to the cycloid bisects the angle HPT (*see* diagram). Torricelli also discovered these results, publishing them first (in 1643 and 1644). Huygens, in 1658, considered the cycloid in his work on pendulum clocks. He showed that a simple pendulum in which the bob followed a cycloidal path would always have the same period of swing, irrespective of the amplitude, i.e. that the cycloid is an isochrone (or tautochrone). This is called the *pendulum property* of the cycloid.

cyclotomic polynomial The polynomial $\Phi_n x$ whose zeroes are the primitive nth roots of unity. For example, $\Phi_2(x) = x + 1, \Phi_4(x) = x^2 + 1$, and if n is a *prime number then $\Phi_n(x) = x^{n-1} + x^{n-2} + \cdots + x + 1$. *See* primitive roots.

cylinder A figure formed by cutting a *cylindrical surface by two parallel planes at an angle (> 0) to the generators. Usually, the term is used for a solid figure, i.e. one in which the directrix is a closed curve

such as a circle or ellipse. The cylinder then consists of two identical plane *bases* with a curved *lateral surface* formed by generators joining corresponding points on the bases. If the bases are perpendicular to the elements of the cylinder, the cylinder is a *right cylinder*; otherwise it is *oblique*.

The perpendicular distance between the bases of a cylinder is its *altitude* (h). The volume is Ah, where A is the area of a base; the area of the lateral surface is sp, where p is the perimeter of a section at right angles to the generators, and s is the slant height. For a right cylinder, $s = h$.

cylindrical coordinate system A *polar coordinate system in three dimensions. Cylindrical coordinates have the two coordinates (r, θ) of polar coordinates in a plane with an additional z-axis through

cylindrical coordinate system

the pole perpendicular to the plane. If r is constant and z and θ vary over all values, a cylindrical surface is generated.

It is possible to change between cylindrical and rectangular Cartesian coordinates. If the pole of the cylindrical system coincides with the origin of the Cartesian system, the polar axis coincides with the x-axis, and the z-axes coincide, then a point (r, θ, z) in cylindrical coordinates has Cartesian coordinates given by

$$x = r \cos \theta, \quad y = r \sin \theta, \quad z = z$$

Similarly, a point (x, y, z) in Cartesian coordinates has cylindrical coordinates given by

$$r = \sqrt{(x^2 + y^2)}, \quad \theta = \tan^{-1}(y/x), \quad z = z$$

the value of θ being chosen so that

$$x : y : r = \cos \theta : \sin \theta : 1$$

cylindrical surface A surface formed by all the straight lines that are parallel to a given line and that pass through a given curve which is not in the same plane as the reference line. The curve is the *directrix* of the cylindrical surface; the parallel straight lines are called *generators* (or *elements*) of the surface. If the directrix is a closed curve the surface is a *closed* cylindrical surface, otherwise it is an *open* cylindrical surface. Such surfaces are named according to the directrix, e.g. a circular cylindrical surface or a parabolic cylindrical surface.

cylindroid A *cylindrical surface. The term is often used for a cylindrical surface that has a circular or elliptical section.

cypher *See* cipher.

d'Alembert, Jean le Rond (1717–83) French mathematician, philosopher, and encyclopaedist. In his *Traité de dynamique* (1743, Treatise on Dynamics) he formulated what later became known as *d'Alembert's principle. He is also known for his work on the theory of vibrating strings, and partial differential equations.

d'Alembert's principle The principle that the internal forces in a system of *particles are in *equilibrium.

d'Alembert's test (for convergence) *See* ratio test.

d'Alembert's theorem *See* fundamental theorem of algebra.

damped harmonic motion The motion of a body that ideally would undergo simple *harmonic motion but in practice is subjected to some form of resistance. The damping is commonly due to viscous forces. For light damping, as occurs for a pendulum in air, the oscillations slowly die away; the decrease in amplitude is exponential in nature. For a heavily damped system, as would occur for a pendulum suspended in a very viscous fluid, there is no oscillation, although the decay is still exponential. When a system just fails to oscillate, *critical damping* is said to occur. The equation of motion can be written as

$$\frac{d^2 x}{dt^2} = -n^2 x - 2k\left(\frac{dx}{dt}\right)$$

where $k > 0$. It has three forms of solution:

(1) *Light damping* $(k < n)$, where the general solution is

$$x = e^{-kt}(A \sin n_1 t + B \cos n_1 t)$$

where $n_1 = \sqrt{(n^2 - k^2)}$.
(2) *Heavy damping* $(k > n)$, where

$$x = e^{-kt}(A e^{k_1 t} + B e^{-k_1 t})$$

and $k_1 = \sqrt{(k^2 - n^2)}$.
(3) *Critical damping* $(k = n)$, where

$$x = (A + Bt)e^{-kt}$$

Dandelin sphere If a circular conical surface is cut by a plane, the curve of intersection is a *conic. A Dandelin sphere is a sphere inside the conical surface that is *tangent to the conical surface along a circle, and is also tangent to the plane. The point of tangency to the plane is a focus of the conic (ellipses and hyperbolas have two Dandelin spheres; parabolas have one). It is named after the French mathematician Germinal Pierre Dandelin (1794–1847).

Darboux's theorem A theorem defining an integral in terms of upper and lower *bounds. For a function $f(x)$ which is bounded on the interval $[a, b]$, the interval is subdivided into n parts by points

$$a = x_0 < x_1 < \cdots < x_n = b$$

In the Riemann definition of *integration, intermediate points are considered on these subintervals. In Darboux's theorem, upper and lower bounds are taken for each interval. M_1 is the least upper bound of

$f(x)$ on $[x_0, x_1]$ and m_1 is its greatest lower bound; M_2 is the least upper bound of $f(x)$ on $[x_1, x_2]$ and m_2 is its greatest lower bound. The first (for upper bounds) is called the *upper Darboux integral*; the second (for lower bounds) is the *lower Darboux integral*. The function $f(x)$ has a Riemann integral if these two integrals are equal. The theorem is named after the French mathematician Jean Gaston Darboux (1842–1917).

bound; and in general M_i is the least upper bound of $f(x)$ on $[x_{i-1}, x_i]$ and m_i is its greatest lower bound. Two sums can then be formed:

$$\sum_1^n M_i(x_i - x_{i-1}) \quad \text{and} \quad \sum_1^n m_i(x_i - x_{i-1})$$

If the length of the largest subinterval is δ, then the limits of the above sums as δ tends to zero give two integrals.

dart *See* Penrose tiles.

data (*singular* **datum**) In statistics, information of a quantitative or qualitative nature. Data collected from records or by measurement in *sample surveys or designed experiments (*see* experimental design) or in observational studies are called *primary* or *raw data*. These may consist of measurements (e.g. weight, age, and height of each of a group of children) or counts (e.g. numbers of males and numbers of females living in a city) or of qualitative attributes (e.g. hair colour and eye colour for each of a number of individuals).

Summary statistics such as percentages, means, and standard deviations, derived from primary data are called *secondary data*.

Bivariate data consist of pairs of measurements or observations of two variables for each of a number of units (e.g. weight and age for each of 50 children). The concept extends to *multivariate data* when there are more than two measurements on each unit (e.g. sex, body weight, a respiratory rate measurement, and daily food intake for each of 20 mice). *See also* categorical data; grouped data; contingency table; scales of measurement.

data coding A way of simplifying manual calculations or reducing rounding errors when using pocket calculators or computers. In statistics, when calculating a *mean or *standard deviation for data sets, coding is based on two rules:

(1) If each datum is multiplied by a constant b, then the mean and standard deviation are each multiplied by b.
(2) If a constant a is added to each datum, then the mean is increased by a and the standard deviation is unaltered.

For example, to calculate the mean and standard deviation of 179.385, 179.387, and 179.392, we might multiply by 1000 and then subtract 179 387 (i.e. add $-179\,387$), giving $-2,0$, and 5. The mean and standard deviation of these numbers are easily computed as 1 and $\sqrt{(26/3)} = 2.94$. Adding 179 387 to 1 and multiplying by 1/1000 gives the mean of the original data as 179.388, and multiplying 2.94 by 1/1000 then gives the standard deviation as 0.002 94.

The addition of a constant a to each datum is equivalent to measuring from an *arbitrary origin*, $-a$, rather than from zero.

day Symbol: d. A unit of time based on the period of rotation of the earth about its axis. It can be defined in several ways. The *apparent solar day* is the interval between two successive meridian transits of the sun. It varies over the course of the year from 24 h 0 min 30 s to 23 h 59 min 39 s. The *mean solar day* is the interval between two successive meridian transits of an imaginary point in the sky (the mean sun) that moves along the celestial equator with a uniform rate of motion equal to the average rate of motion of the sun along the ecliptic. Its duration is exactly 24 hours. The *sidereal day* is the interval between two successive meridian transits of the vernal equinox. The *mean sidereal day*, which is very close in value to the *apparent sidereal day*, is 23 h 56 min 4.09 s.

death rate The number of deaths (in total or due to a specific cause) in a given period, divided by the population exposed to risk. For comparative purposes the rate is usually standardized for differences in age, sex, and exposure to risk in different populations. Also called a *mortality rate*. *See also* life tables.

dec *Abbreviation for* *declination.

deca- *See* SI units.

decade A group or series of ten numbers.

decagon A *polygon that has ten interior angles (and ten sides).

deceleration Negative *acceleration.

deci- *See* SI units.

decibel Symbol: dB. **1.** A unit for comparing two currents, voltages, or power levels, equal to one-tenth of a *bel.
2. A similar unit for measuring the intensity of sound, equal to ten times the logarithm to the base ten of the ratio of the intensity of the sound to be measured to the intensity of a reference sound, usually taken as the lowest audible sound of the same frequency.

decidable Describing a class of problems for which there is an *effective procedure (a *decision procedure*, or *algorithm*) for solving each problem in the class. *Formal systems are said to be decidable if there is an effective procedure for determining, for any *wff A of the system, whether or not A is a theorem of the system. In the case of the *propositional calculus, *truth tables provide an effective means of determining whether a wff is a *tautology. The completeness theorem (*see* complete) for the propositional calculus shows that every tautology is a theorem, and thus there is an effective method for determining

whether a wff is a theorem. In other words, the *decision problem* for the propositional calculus has a positive solution. *Church's theorem shows that the *predicate calculus is not similarly decidable.

decile *See* quantile.

decimal A number expressed using the decimal *number system. Commonly, the term is used for numbers that have fractional parts indicated by a decimal point. A number less than 1 is called a *decimal fraction*; for example, 0.537 is a way of writing

$$0 + \frac{5}{10} + \frac{3}{100} + \frac{7}{1000}$$

i.e. $(5 \times 10^{-1}) + (3 \times 10^{-2}) + (7 \times 10^{-3})$.

A *mixed decimal* is one consisting of an integer and a decimal fraction (e.g. 27.63). The first position to the right of the point (representing tenths) is the first *decimal place*; the second position is the second decimal place; etc.

A decimal fraction is a series of fractions, i.e. it is a number of tenths plus a number of hundredths plus a number of thousandths, etc. The decimal may have a fixed number of digits: for example, $\frac{5}{8}$ is 0.625; such numbers are called *finite* or *terminating decimals*. In other decimals the digits may continue indefinitely (they represent an infinite series); decimals of this type are called *infinite* or *nonterminating decimals*.

If the number is a rational number it may have an infinitely repeating digit or group of digits; decimals of this type are said to be *repeating* or *recurring decimals*. Thus $\frac{1}{3}$ is the decimal 0.333 33 . . . . This is sometimes written as $0.\dot{3}$ and referred to as 'nought point three recurring'. Another example of a repeating decimal is $\frac{5}{7}$, which is 0.714 285 714 . . . with the block of digits 714 285 repeated endlessly; this is written as $0.\dot{7}14\,28\dot{5}$. Such decimals are also called *periodic decimals*. Irrational numbers, such as π, $\sqrt{2}$, and e, are decimals that are infinite but do not repeat; such numbers are termed *nonrepeating* or *nonperiodic decimals*.

decimal fraction *See* decimal.

decimal notation The method of positional notation used in the decimal *number system.

decimal place *See* decimal.

decimal point A dot used to separate the integral part of a number from the fractional part in the decimal *number system. The point is either centred (as in 0·5) or, now more commonly, placed on the line (0.5), as in this Dictionary. In many European countries a comma is used (0,5).

decimal system The commonly used *number system using the base ten. *See also* decimal.

decision problem The problem of determining whether a class of problems is *decidable. *See* NP problem.

decision theory A framework for making decisions in the presence of uncertainty, which is appropriate in many situations in business or industry. A simple example of its application may be formulated as a two-person game (*see* game theory), with an entrepreneur as one player and nature (i.e. the world, a market, the competition) as the other. The entrepreneur may adopt any one of a finite number of strategies, but knows only the probabilities that nature will adopt each of its possible strategies. A *payoff matrix* gives the consequences (gain or loss to the entrepreneur) of each choice made by the entrepreneur for each choice made by nature. If the probabilities that nature will adopt each of its strategies are known, the entrepreneur may calculate the payoff expected from a particular action (i.e. a particular strategy). One criterion for choosing an action is to use the *minimax principle to select an action that minimizes the maximum expected loss. If an action a_2 produces losses that are always greater than the losses produced

by an action a_1, no matter what strategy nature adopts, a_2 is said to be *inadmissible* and may be excluded from further consideration in a decision problem. *See also* decision tree.

decision tree A form of *tree diagram useful in determining the optimum action to take when there are alternative strategies with uncertain outcomes. For example, a builder may be allowed to tender for only one of two contracts. If he tenders for contract A, there is a probability of 0.7 that he will win the contract and make a profit of £10 000. If he tenders for contract B, the probability that he wins is 0.2, but his profit will be £30 000. If he fails to win the contract he tenders for, he will do other work and make a profit of £5000. Which contract should he tender for to maximize his expected profit? The decision tree for deciding this is shown in the diagram.

Square nodes are called *decision nodes*; circular nodes are *random nodes*. Branches leaving random nodes have known attached probabilities. A vertical line represents a *terminal node*. Possible decisions or their consequences are indicated on each branch, with associated probabilities when relevant. A gain or other measured outcome is indicated at each terminal node. To determine the maximum expected gain, the expected gain is computed at each random node and is conventionally written in or below each node. For this example the expected gains at nodes A and B are computed as £10 000 × 0.7 +

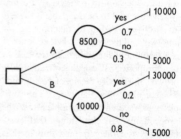

decision tree for deciding between two tenders.

$5000 \times 0.3 = £8500$ for A, and $£30\,000 \times 0.2 + 5000 \times 0.8 = £10\,000$ for B. To maximize his expected gain, the builder should thus tender for job B. In practice, most decision trees have many more branches and nodes than there are in this simple example.

declination (dec) Symbol: δ. The angular distance of a point on the *celestial sphere from the celestial equator, taken along a celestial meridian passing through the point. Declination is measured from $0°$ to $90°$ north (taken as positive) or south (taken as negative) of the celestial equator. Sometimes the complement $(90° - \delta)$, called the *north polar distance*, is used. *See* equatorial coordinate system.

decoding (decryption) In cryptography, the recovery of plaintext from ciphertext. *See* cipher.

decomposition 1. (of a fraction) The process of splitting a fraction into two or more *partial fractions.
2. (of a matrix) *See* factorization.

decreasing function *See* monotonic decreasing function.

decreasing sequence A *sequence $a_1, a_2, \ldots$ for which $a_n > a_n + 1$ for all n is said to be *strictly decreasing*. The sequence is described as *monotonic decreasing* if $a_n \geqslant a_{n+1}$ for all n.

If a monotonic decreasing sequence $\{a_n\}$ has a lower bound (*see* bounded sequence) then it tends to a finite limit; if no lower bound exists, then $a_n \to -\infty$ as $n \to \infty$.

Compare increasing sequence.

Dedekind, Julius Wilhelm Richard (1831–1916) German mathematician who in his *Was sind und was sollen die Zahlen?* (1888, The Nature and Meaning of Numbers) offered an axiomatic account of the natural numbers. He further defined the irrational numbers in terms of the *Dedekind cut.

Dedekind cut (J.W.R. Dedekind, 1872) A division of the *rational numbers into two (nonempty) *sets such that every number of the first set (A) is less than every number of the other set (B). If A has a largest member (or if B has a smallest member) the cut defines a rational number. If A has no largest member and B no smallest member, then the cut defines an irrational number. For example, the rational numbers could be put into two sets in which set A contains negative rational numbers together with those that have squares less than 2, and set B contains the positive rational numbers that have squares greater than 2. The cut itself defines the irrational number $\sqrt{2}$. A number can be indicated using the notation (A, B), where A and B are the sets in the cut. The real numbers are the set of all Dedekind cuts. The irrational numbers are the set of all Dedekind cuts for which the first set has no largest member and the other set no smallest member.

The method allows irrational numbers to be formally defined from rational numbers without geometric reasoning. The cut is equivalent to dividing a number line into two segments by a point, and the definition depends on the principle that the points on the line can be placed in *one-to-one correspondence with the real numbers. This idea is known as the *Cantor–Dedekind hypothesis*.

See real number.

deduction A valid argument in which the conclusion follows from the premises. Formally, it is a sequence of *wffs C_1, $\ldots$, C_m of a *formal language S such that for each C_i, $1 \leqslant i \leqslant m$, either

(1) C_i is an axiom of S (if there are such);
(2) C_i is a member of a set $B_1, \ldots, B_n$ (the *premises*, or *hypotheses* of the deduction); or
(3) C_i is *immediately inferred* from some previous wffs of the sequence by a single application of a rule of inference of S.

If we let $A = C_m$ then A is *deduced* from (or *proved* from) premises $B_1, \ldots, B_n$, or, equivalently, $B_1, \ldots, B_n \vdash_S A$ (sometimes the subscript S is omitted if it is clear which formal language is intended). Deductions are *proofs only if the set $B_1, \ldots, B_n$ is empty; proofs use only axioms and rules of inference. *See* argument; consequence; natural deduction.

deduction theorem The *theorem that if $B_1, \ldots, B_n \vdash A$ then

$$B_1, \ldots, B_{n-1} \vdash B_n \supset A$$

It holds in standard *formal systems such as the *predicate calculus.

deferred correction *See* Richardson extrapolation.

deficient number (defective number) *See* perfect number.

definite integral An expression for the difference between the values of an *integral when evaluated for two values of the *independent variable, written as

$$\int_a^b f(x)\,dx$$

The values $x = b$ and $x = a$ are called the *upper* and *lower limits* of the definite integral. If $F(x)$ is an integral of $f(x)$, then the value of the above definite integral is $F(b) - F(a)$. This is also written as $[F(x)]_a^b$.

The following example illustrates the evaluation of a definite integral:

$$\int_2^3 2x\,dx = [x^2]_2^3 = 3^2 - 2^2 = 9 - 4 = 5$$

Properties of definite integrals are:

$$\int_a^b f(x)\,dx = -\int_b^a f(x)\,dx$$

and

$$\int_a^b dx = b - a$$

If k is constant, then

$$\int_a^b kf(x)\,dx = k\int_a^b f(x)\,dx$$

If c is a point inside the range $a \leqslant x \leqslant b$, then

$$\int_a^c f(x)\,dx + \int_c^b f(x)\,dx = \int_a^b f(x)\,dx$$

If $f(x)$ and $g(x)$ are both integrable in the range $a \leqslant x \leqslant b$, then

$$\int_a^b \{f(x) + g(x)\}\,dx = \int_a^b f(x)\,dx + \int_a^b g(x)\,dx$$

Compare indefinite integral.

deformation Change in shape or size of a body as a result of the action of external *forces. The extent of the deformation depends on the material from which the body is made, the shape of the body, and the area of application of the force. Deformation is usually considered in terms of the *stress set up within a body and the *strain associated with such stress. *See also* elasticity.

degenerate conic A point, line, or pair of lines, regarded as a limiting case of a *conic section.

degree 1. The exponent of a variable in a term. For example, in $3x^3y^2z$, x has degree 3, y degree 2, and z degree 1. The degree of the whole term is the total of these exponents, in this case $3 + 2 + 1 \; (= 6)$. The degree of a polynomial or equation is the degree of its highest-degree term.

For instance,

$$x + 2xy + y = 0$$

is an equation of the second degree ($2xy$ is the highest-degree term). Its degree in x (or y) is 1.

2. (of a curve) The degree of an equation representing a plane algebraic *curve. For instance, $y = mx + c$, which represents a straight line, has degree 1. The equations $y^2 = 2x$ and $xy = 4$ both have degree 2, and the corresponding curves are *quadratic curves* (i.e. *conics). If the degree is 3, the curve is a *cubic curve*; if 4, a *quartic*; if 5, a *quintic*; etc.

3. (of a differential equation) The power to which the highest-order derivative is raised in a *differential equation.

4. (of a map of a sphere to itself) Let f: $S^n \rightarrow S^n$ be a continuous map from the n-sphere to itself ($n \geqslant 1$). Since the *homology group $H_n(S^n)$ is infinite and cyclic, the induced homomorphism f∗ must satisfy f∗$(x) = d \cdot x$ for all $x \in H_n(S^n)$ and some integer d. The integer d is called the degree of f.

It can be shown that two continuous maps f, g: $S^n \rightarrow S^n$ are *homotopic if and only if they have the same degree.

5. (of a vertex of a *graph) The number of edges joined to the vertex.

6. Symbol: °. A unit of angle equal to $1/360$ of a complete turn. *See* angular measure.

7. A subdivision of a scale of temperature measurement. *See* Celsius degree; Fahrenheit degree; kelvin.

degree measure *See* angular measure.

degree of arc A unit measuring the length of an arc, equal to the length of arc of a circle that subtends an angle of one degree at the centre of the circle. Note that the degree of arc is a measurement of length (not of angle) and is strictly defined only for a circular arc. It is used in astronomy to express distances on the celestial sphere. Similarly, the *minute of arc* and *second of arc* are defined as arc lengths that subtend an angle of a minute and a second, respectively.

degrees of freedom **1.** (in statistics) Degrees of freedom are in essence the number of independent units of information in a sample relevant to estimation of a *parameter or calculation of a *statistic. One approach is to regard the n observations as the initial units of information, one of which is used to determine the total or *mean. As the mean must be known before we can determine deviations from it, there are $n - 1$ degrees of freedom left for estimating the *variance in the sense that if the total is fixed, only $n - 1$ values can be assigned arbitrarily; the remaining one is then fixed to ensure the correct total. Likewise, in a 2×2 *contingency table with fixed marginal (i.e. row and column) totals there is only one degree of freedom, for once a value is assigned to any one of the four category cells the remaining values are determined by the constraint that they must add to the fixed marginal totals. For example, in the table below if we arbitrarily choose $a = 10$, it follows that $b = 2$, $c = 5$, and $d = 8$:

	Col. 1	Col. 2	Row totals
Row 1	a	b	12
Row 2	c	d	13
Column totals	15	10	

Similarly, if we put $a = 5$, then automatically $b = 7$, $c = 10$, and $d = 3$.

2. *See* normal modes.

del The operator

$$\nabla \equiv \mathbf{i}\frac{\partial}{\partial x} + \mathbf{j}\frac{\partial}{\partial y} + \mathbf{k}\frac{\partial}{\partial z}$$

where **i**, **j**, and **k** are *unit vectors along the x-, y-, and z-axes respectively. The symbol ∇ is also known as *nabla*. *See* curl; divergence; gradient; Laplace's equation; wave equation.

Delambre's analogies *See* Gauss's formulae.

de L'Hôpital's rule *See* L'Hôpital's rule.

Delian problem *See* duplication of the cube.

delta function *See* Dirac delta function.

deltoid 1. A concave *quadrilateral that has two pairs of equal adjacent sides. *See* concave polygon; *compare* kite.
2. A plane *curve that is the *locus of a point on the circumference of a circle that rolls on the inside of a fixed circle of three times its radius. It has three *cusps and is an example of a *hypocycloid. *See also* astroid.

demography The statistical study of human populations, in particular vital statistics (birth and mortality rates), movements of people, and other factors influencing population changes. *See* life tables.

de Moivre, Abraham (1667–1754) French mathematician and the author of *The Doctrine of Chances* (1718), one of the earliest works on probability theory. He is also known for *de Moivre's theorem, in which complex numbers were introduced into trigonometry for the first time.

de Moivre's theorem The relationship

$$(\cos\theta + i\sin\theta)^n = \cos n\theta + i\sin n\theta$$

involving the polar form of a *complex number. It was discovered by Abraham de Moivre around 1707.

De Morgan's laws Identities that hold for any two *sets A and B:

$$(A \cap B)' = A' \cup B'$$

$$(A \cup B)' = A' \cup B'$$

where A' denotes the *complement of A. By using these identities it is possible to convert any *intersection of sets into a *union of sets, or vice versa.

The name is also given to two theorems of the *propositional calculus:

$$\sim(A \& B) \equiv (\sim A) \vee (\sim B)$$

$$\sim(A \vee B) \equiv (\sim A) \& (\sim B)$$

where A and B are any two statements. These theorems may be used to replace a *disjunction in a *compound sentence by a *conjunction, and vice versa.

The formulae are named after the English mathematician and logician Augustus De Morgan (1806–71), who proposed them in 1847.

denary Pertaining to or based on the number ten.

denial of the antecedent The *fallacy of inferring from $A \supset B$ and $\sim A$ that $\sim B$, or an argument of this form. It is so called because the second premise, $\sim A$, is the *negation of the *antecedent of the *conditional statement forming the first premise. *See also* affirmation of the consequent.

denominate number A number that determines a unit of a *physical quantity, as in 5 metres or 6 volts. The unit involved (metre, volt, etc.) is the *denomination* of the number.

denomination *See* denominate number.

denominator The divisor in a fraction: i.e. the number on the bottom. In $\frac{3}{4}$, 4 is the denominator (3 is the numerator).

dense set A *subset $A \subset X$ is dense in the topological space X if every neighbourhood of every point of X contains at least one point of A. So, in a *metric space, every point of X is a *limit point of a sequence of points of A. The set of rational numbers is dense in the set of real numbers. *Weierstrass's theorem* states that the set of all polynomial functions is dense in the space of all continuous functions on a closed, bounded interval. *Compare* discrete set

density **1.** Symbol: ρ. The mass per unit volume of a material. It is usually expressed in grams per cubic centimetre or kilograms per cubic metre.
2. The value of some *physical quantity per unit volume (or area, or length). For example, surface charge density is the electric charge per unit area of surface.

denumerable *See* countable.

deontic logic The logic of obligation and permissibility. It was initially developed by the Finnish logician Georg von Wright (1916–2003) in the 1950s, since when many alternative systems have been proposed. Von Wright began by adding to the *propositional calculus the variables a, b, c, ... denoting acts, the two operators O (it is obligatory that) and P (it is permissible that), and two axioms:
A1. $Oa \rightarrow Pa$ (if a is obligatory then it is permitted);
A2. $P(a \lor b) \rightarrow (Pa \lor Pb)$ (if either a or b is permitted then either a is permitted or b is permitted).

The operator O can be defined in terms of P through the definition that 'Oa' is equivalent to and replaceable by '$\sim P \sim a$'; in other words, saying that a is obligatory is equivalent to saying that it is not permissible not to do a.

departure The length of arc cut off on a line of latitude by two *meridians. The value of the departure decreases with distance from the equator, falling to zero at the poles.

dependent equations An equation is dependent on a set of equations if it is satisfied by every set of values of the variables that satisfies the set of equations. A set of equations is dependent if one of them is dependent on the others. If the set contains no dependent equation it is *independent*.

For example, the set of equations

$$x + y = 3, \quad x(x + y) = 3x$$

is dependent since the second equation is dependent on the first, i.e. every pair of values (x, y) satisfying the first equation satisfies the second.

The equations

$$x + y = 3, \quad x + 2y = 6$$

are independent since, of all the pairs (x, y) satisfying one equation, only $(0, 3)$ satisfies the other.

In solving a system of *simultaneous equations, a dependent equation may be ignored.

dependent variable *See* function; regression; variable.

depression, angle of *See* angle.

derangement A *permutation with no *fixed points*, where a fixed point would be a point which is mapped to itself. So the permutation p from a *set X to itself is a derangement if $p(a) \neq a$ for every $a \in X$. The only derangements of the set $\{1, 2, 3\}$ are the permutations p_1 and p_2, where p_1 maps 1, 2, 3 to 2, 3, 1 and p_2 maps 1, 2, 3 to 3, 1, 2. The number d_n of derangements of an n-element set is given by $d_1 = 0$, $d_2 = 1$, and $d_n = (n - 1)[d_{n-1} + d_{n-2}]$ for $n \geqslant 3$.

de Rham cohomology *See* differential form.

derivative The rate of change of a *function with respect to the independent variable. It is also known as the *differential coefficient* or the *derived function*. For a function $y = f(x)$ the derivative can be written as dy/dx, y', $Df(x)$, $D_x y$, or $f'(x)$. For the function $y = f(x)$, a small change δx in x causes a change δy in y, where

$$\delta y = f(x + \delta x) - f(x)$$

and

$$\frac{\delta y}{\delta x} = \frac{1}{\delta x}[f(x + \delta x) - f(x)]$$

The derivative of the function, dy/dx, is the limit (if it exists) of this expression as δx

derivative As Q approaches P, the line PQ becomes the tangent at P.

approaches zero. A particular interpretation of the derivative at a point is that it is the slope of the tangent to the curve $y = f(x)$ at the point. Taking derivatives of derivatives gives derivatives of higher *order*. For instance, the function $y = x^4$ has a first derivative $dy/dx = 4x^3$. The second derivative, written as d^2y/dx^2, is obtained by differentiating this to give $12x^2$; the third derivative, d^3y/dx^3, is $24x$. For $n > 1$, the nth derivative is denoted by d^ny/dx^n. Other common notations for the second, third, and nth derivatives are y'', y''', $y^{(n)}$ and $f''(x)$, $f'''(x)$, $f^{(n)}(x)$.

When time t is the independent variable and $y = f(t)$, a common notation for the first and higher derivatives is $\dot{y}$, $\ddot{y}$, etc.

A table of derivatives is given in the Appendix. *See also* differentiation; partial derivative.

derived curve A curve obtained from an original curve by taking a *derivative. For instance, the first derived curve of the curve $y = f(x)$ is the curve $y = f'(x)$, where f' is the first derivative of f. For each point with a given abscissa on the second curve, the value of y equals the slope of the first curve at that value of x. A curve indicating the way distance changes with time for a moving body would have a derived curve show-

ing how velocity changes with time. The second derived curve would be produced by taking the second derivative of the function representing the original curve. In this example, it would show how acceleration changes with time.

derived equation 1. An equation obtained by an algebraic operation on a given equation, e.g. dividing both sides by the same factor or adding terms to both sides.
2. An equation obtained by *differentiation of both sides of a given equation.

derived function *See* derivative.

derived set The *set of all *limit points of a set. The derived set of a set A is usually denoted by A'. *See also* closure.

derived units A set of units derived from a set of *base units by multiplication or division without introducing numerical factors. For example, in *SI units the derived unit of velocity (metre per second) is obtained by dividing the base unit of length (metre) by the base unit of time (second).

DES *Abbreviation for* Data Encryption Standard. A method of *encryption that was recognized by US authorities as producing ciphertext that was very difficult to decipher and adopted as a standard in 1976. It has now been superseded by *AES, which is easier to implement and more secure.

Desargues, Girard (1591–1661) French mathematician and engineer who, in a work on the conic sections published in 1639, founded the discipline of *projective geometry. His work contained many original ideas, including what is now known as *Desargues's theorem, but it remained largely ignored until the 19th century.

Desargues's theorem A theorem of *projective geometry: if the lines joining corresponding vertices of two triangles pass through a common point, then the points of intersection of corresponding sides lie

Desargues's theorem

on a straight line. The dual theorem (*see* duality) is: if the corresponding sides of two triangles have points of intersection that lie on a straight line, then the lines joining corresponding vertices pass through a common point. The dual is also the *converse.

The theorem (as well as its converse) also holds in three dimensions.

Descartes, René (1596–1650) French mathematician and philosopher who in his *La Géométrie* (1637) introduced into mathematics the fundamental principles and techniques of coordinate geometry. He began with a solution to the problem of the four-line locus, went on to show how to draw tangents to curves and, in the final part, dealt with the solution of equations of degree higher than two, describing also the rule known as *Descartes's rule of signs. In the area of notation, it was Descartes who introduced the system of indices (x^2, x^3, etc.) and who began to employ the first letters of the alphabet to refer to known quantities and the last letters to represent unknowns. The adjective *Cartesian* is derived from his name.

Descartes's rule of signs A rule for finding the maximum number of positive *roots for a *polynomial equation. It depends on the number of *variations in sign* of the

coefficients of the polynomial, i.e. on the number of times the sign changes when the polynomial is written in descending order. Thus

$$x^5 + x^4 - 2x^3 + x^2 - 1 = 0$$

has three variations in sign. Descartes's rule states that the number of positive roots cannot be greater than the number of variations in sign (although it may be less). In the case above, it cannot exceed three. The rule can also be applied for negative roots by replacing x by $-x$. Thus, in the example the equation becomes

$$-x^5 + x^4 + 2x^3 + x^2 - 1 = 0$$

for which there are two variations in sign.

describe In geometry, to draw an arc or circle.

descriptive statistics *See* statistics.

designed experiment *See* experimental design.

det *See* determinant.

determinant A sum of certain products of numbers (*elements*) taken from a square array. It is commonly denoted by writing the array with a vertical line on each side, as in

$$\begin{vmatrix} a & b \\ c & d \end{vmatrix}$$

or as the 'det' function applied to a matrix, as in

$$\det \begin{pmatrix} a & b \\ c & d \end{pmatrix}$$

Horizontal lines of elements are *rows* and vertical lines are *columns*. The number of rows or columns is the *order* of the determinant (2 in the example above). A *diagonal* of the determinant is a diagonal line of elements. The one from top left to bottom right is the *main* or *principal diagonal* (a, d in the example); the other is the *main anti-diagonal* (b, c in the example). The second-order determinant above equals $ad - bc$. In general, the value of any determinant is obtained by taking any row or column, forming the products of each element and its *cofactor, and taking the algebraic sum of these products. Commonly, the elements of the first row are used in expanding determinants, as in the following example:

$$\begin{vmatrix} 1 & 2 & 3 \\ 4 & 1 & 2 \\ 6 & 5 & 4 \end{vmatrix} = 1 \times \begin{vmatrix} 1 & 2 \\ 5 & 4 \end{vmatrix} - 2 \times \begin{vmatrix} 4 & 2 \\ 6 & 4 \end{vmatrix}$$

$$+ 3 \times \begin{vmatrix} 4 & 1 \\ 6 & 5 \end{vmatrix}$$

$$= 1 \times (1 \times 4 - 2 \times 5)$$

$$- 2 \times (4 \times 4 - 2 \times 6)$$

$$+ 3 \times (4 \times 5 - 1 \times 6)$$

$$= -6 - 8 + 42$$

$$= 28$$

Various operations can be performed on determinants:

(1) If any two rows or columns are interchanged the value remains the same but the sign of the determinant changes.
(2) Multiplication of the determinant by a quantity is equivalent to multiplying all the elements of any row or column by that quantity.
(3) If all the elements of any row or column are multiplied by a quantity and added to the corresponding elements of another row or column, the value of the determinant is unchanged.
(4) Interchanging the rows and columns does not alter the value of the determinant. If any two rows (or columns) are equal or have proportional elements, the value of the determinant is zero.

Two determinants may also be multiplied, provided that they are of the same order, by the same method as in the multiplication of *matrices. The determinant of a square matrix A is written as $|A|$ or det A.

See also alternant; circulant; Cramer's rule; Jacobian; Vandermonde determinant.

determination See coefficient of determination.

deterministic model See model.

developable Describing a surface (e.g. a conical or cylindrical surface) that can be rolled out flat on a plane without any stretching or shrinking.

deviance (J.A. Nelder and R.W.M. Wedderburn, 1972) A measure for judging how well data fit a model using *maximum likelihood estimates in *generalized linear models.

deviation See mean absolute deviation; standard deviation.

d.f. Abbreviation for *degrees of freedom; less frequently for *distribution function. A preferred abbreviation for the latter is c.d.f. (cumulative distribution function).

diabolic square See magic square.

diagonal 1. A line segment that joins any two nonadjacent vertices of a *polygon.
2. A line segment that joins a vertex of a *polyhedron to another vertex that is not in the same face.
3. A set of diagonal elements forming part of a square array, as in a *determinant or square *matrix. The diagonal from top left to bottom right is the main or principal diagonal; that from top right to bottom left is the main antidiagonal.

diagonal argument An argument introduced into mathematics by Cantor in 1891 to prove that the *cardinal number of the set of *real numbers is greater than the cardinal number of the set of natural numbers: that the real numbers are, in fact, nondenumerable. Assume that the real numbers between 0 and 1 are *countable, and that they have been put into a *one-to-one correspondence with the natural numbers:

$$1 \equiv 0.a_{11}\, a_{12}\, a_{13} \cdots$$
$$2 \equiv 0.a_{21}\, a_{22}\, a_{23} \cdots$$
$$3 \equiv 0.a_{31}\, a_{32}\, a_{33} \cdots$$
$$\vdots$$

Cantor went on to show how to construct a real number X between 0 and 1 that differs from those already matched with the natural numbers. Thus, let X correspond to the nonterminating decimal

$$x = 0.x_1\, x_2\, x_3 \cdots$$

where $x_i = a_{ii} + 1$, unless $a = 9$, in which case let $x_i = a_{ii} - 1$. Clearly, the diagonal number X cannot be matched with any of the natural numbers. It follows that the set of real numbers is nondenumerable.

Diagonal arguments of this kind have proved to be very powerful and have been used by, amongst others, *Gödel and *Turing.

diagonalizable matrix A square matrix that can be reduced to a *diagonal matrix by a *similarity transformation, i.e. a square matrix A for which $X^{-1}AX$ is diagonal for some nonsingular matrix X.

diagonally dominant matrix A square *matrix is *diagonally dominant by rows* if, in each row, the absolute value of the element on the main diagonal exceeds the sum of the absolute values of the other elements. A matrix is *diagonally dominant by columns* if its *transpose is diagonally dominant by rows. For example, the matrix

$$\begin{pmatrix} 2 & 1 & 0 \\ 1 & -3 & 1 \\ -2 & 1 & 4 \end{pmatrix}$$

is diagonally dominant by rows, but not by columns. A diagonally dominant matrix (by rows or columns) is necessarily a *nonsingular matrix.

diagonal matrix A square *matrix in which all elements off the main diagonal are zero. A diagonal matrix in which all the elements on the main diagonal are equal is a *scalar matrix*.

diagonal point *See* quadrangle.

diameter 1. A *chord through the centre of a circle or sphere.
2. (of a conic) A straight line that is the *locus of the mid-points of any set of parallel chords of the conic. In the case of the ellipse and hyperbola, the diameters pass through the centre of the conic.

dichotomous data *See* categorical data.

dichotomous variable *See* categorical variable.

dichotomy Division of a population or sample into two groups based either on measurable variables (e.g. age under 18, age 18 or over) or on attributes (e.g. male, female).

Dido's problem The problem of determining the curve that encloses the maximum area for a given curve perimeter. The solution is that the curve is a circle. The problem is named after the mythological Queen Dido, who, according to legend, was given as much land as could be enclosed by a cow-hide. She cut the hide into narrow strips and, in an early application of the calculus of variations, laid them in a semicircle on the coastline, enclosing the land on which she founded the city-state of Carthage.

difference 1. A value or expression obtained by subtraction.
2. The difference of two *sets A and B (or the *relative complement* of B in A), denoted by $A \backslash B$ or $A - B$, consists of the set of those elements that are members of A but not members of B:

$$A \backslash B = \{x : x \in A\ \&\ x \notin B\}$$

For example, if A is $\{1, 2, 3\}$ and B is $\{1, 2, 4, 5\}$, then $A \backslash B$ is $\{3\}$.

difference equation An equation involving *finite differences. For example, the problem of finding a sequence y_0, y_1, y_2, ... such that $\Delta y_n = n$ has the solution

$$y_n = \tfrac{1}{2}n(n-1) + k$$

where k is an arbitrary constant. An *initial condition $y_0 = 1$ gives $k = 1$.

Since $\Delta y_n = y_{n+1} - y_n$, the above equation may be written as

$$y_{n+1} - y_n = n$$

Difference equations are often expressed in this way and are a type of *recurrence relation. They occur naturally in this form for *stochastic processes such as *random walks.

There are many analogies between difference equations and *differential equations, both in form and in methods of solution.

differentiable A *real function $f(x)$ is differentiable at $x = a$ if the limit

$$\operatorname*{Lim}_{h \to 0} \frac{f(a + h) - f(a)}{h}$$

exists; the value of the limit is then denoted by $f'(a)$. A real-valued function of several variables is differentiable at a point if all its *partial derivatives are defined at that point. The function is called differentiable on a set A if it is differentiable at every point of A. It can be shown that a differentiable function is *continuous.

differential The differential dx of an independent *variable x is any arbitrary change in the value of x, the corresponding differential dy being defined as $dy = f'(x)dx$, where $y = f(x)$ and $f'(x)$ is the derivative of $f(x)$. See also total differential.

differential calculus See calculus.

differential coefficient See derivative.

differential equation A relationship between an independent *variable x, a dependent variable y, and one or more of the *derivatives of y with respect to x. A simple example of a differential equation is

$$\frac{dy}{dx} = x$$

A *solution* of a differential equation is a function that, when substituted for the dependent variable in the equation, leads to an identity. Thus, for the equation above, $y = \tfrac{1}{2}x^2$ is a solution since substituting for dy/dx leads to $x = x$. Note that $y = \tfrac{1}{2}x^2 + C$, where C is a constant, is also a solution, in this case the *general solution* of the differential equation. A *particular solution* is one in which the constant(s) have particular values, e.g. $y = \tfrac{1}{2}x^2 + 5$ (see boundary conditions).

The *order* of a differential equation is the order of the highest derivative. The *degree* of the equation is the power to which the highest-order derivative is raised. Thus,

$$\frac{d^2 y}{dx^2} = kx$$

is a simple second-order equation of the first degree, and

$$\left(\frac{d^2 y}{dx^2} \right)^2 = kx$$

is a second-order equation of the second degree.

An equation involving more than one independent variable and *partial derivatives with respect to these variables is a *partial differential equation* (PDE). An important example is *Laplace's equation. A differential equation which does not contain partial derivatives is an *ordinary differential equation* (ODE).

Differential equations occur in numerous practical applications in science and engineering. There are various cases with standard methods of solution, as follows:

DIFFERENTIAL EQUATIONS OF THE FIRST ORDER AND FIRST DEGREE

(1) *Exact equations.* Equations of the form

$$P\left(\frac{dy}{dx}\right) + Q = 0$$

are exact if the left-hand side is the differential coefficient of some function $f(x, y)$ with respect to x. Integration gives the solution $f(x, y) = C$, where C is a constant. An exact equation is one in which the total differential of a function f is equal to zero, i.e.

$$\frac{\partial f}{\partial x}dx + \frac{\partial f}{\partial y}dy = 0$$

Thus an equation

$$A\,dx + B\,dy = 0$$

is exact if

$$\frac{\partial A}{\partial y} = \frac{\partial B}{\partial x}$$

(2) *Variables separable.* In this case, the equation can be put in the form

$$f(x) + g(y)\left(\frac{dy}{dx}\right) = 0$$

Rearrangement gives

$$f(x)\,dx = -g(y)\,dy$$

Both sides can then be integrated.

(3) *Homogeneous equations.* These can be written in the form

$$\frac{dy}{dx} = f\left(\frac{y}{x}\right)$$

The method of solution is to make the substitution $y = vx$, which reduces the equation to one in v and x only.

In the resulting equation, the variables are separable.

(4) *Equations reducible to homogeneous form.* Equations of the form

$$\frac{dy}{dx} = \frac{a_1 x + b_1 y + c_1}{a_2 x + b_2 y + c_2}$$

can be handled by substitution. Let $x = X + h$ and $y = Y + k$ where h and k are constants. Then,

$$\frac{dy}{dx} = \frac{dY}{dX}$$

$$= \frac{a_1(X + h) + b_1(Y + k) + c_1}{a_2(X + h) + b_2(Y + k) + c_2}$$

If h and k are chosen to be values of x and y, respectively, that satisfy the simultaneous equations

$$a_1 x + b_1 y + c_1 = 0$$

$$a_2 x + b_2 y + c_2 = 0$$

then the original equation becomes

$$\frac{dY}{dX} = \frac{a_1 X + b_1 Y}{a_2 X + b_2 Y}$$

which is homogeneous.

However, if $a_1/a_2 = b_1/b_2 \neq c_1/c_2$, then h and k cannot be chosen as above. In this case, let $a_2 = ma_1$ and $u = a_1 x + b_1 y$. The equation becomes

$$\frac{du}{dx} - a_1 = b_1\frac{u + c_1}{mu + c_2}$$

and the variables u and x can be separated.

(5) *Linear equations.* Equations of the form

$$\frac{dy}{dx} + Py = Q$$

where P and Q are functions of x, or constants, are said to be linear in y and can be solved by multiplying throughout by an *integrating factor $\exp(\int P\,dx)$. This makes the left-hand side of the equation an exact differential:

$$\exp\left(\int P\,dx\right)\frac{dy}{dx} + \exp\left(\int P\,dx\right)Py$$

$$= \exp\left(\int P\,dx\right)Q$$

$$\frac{d}{dx}\left[\exp\left(\int P\,dx\right)y\right] = \exp\left(\int P\,dx\right)Q$$

$$y\exp\left(\int P\,dx\right) = \int\exp\left(\int P\,dx\right)Q\,dx + C$$

where C is a constant.

DIFFERENTIAL EQUATIONS OF THE SECOND ORDER

(1) *Equations of the form*

$$\frac{d^2y}{dx^2} = f(x)$$

are immediately solvable by integrating twice.

(2) *Equations of the form*

$$\frac{d^2y}{dx^2} = f(y)$$

Here, the first integration is obtained by multiplying both sides of the equation by $2dy/dx$:

$$2\frac{dy}{dx}\frac{d^2y}{dx^2} = 2f(y)\frac{dy}{dx}$$

Integrating both sides with respect to x gives

$$\left(\frac{dy}{dx}\right)^2 = \int 2f(y)\,dy + C$$

where C is a constant.

The second integration is then accomplished by taking square roots and then separating the variables.

(3) *Linear equations with constant coefficients of the form*

$$a\frac{d^2y}{dx^2} + b\frac{dy}{dx} + cy = 0$$

This equation has a solution $y = e^{mx}$ if $am^2 + bm + c = 0$. This *auxiliary equation*

will have two roots which may be (a) real and different, (b) real and equal, or (c) imaginary. The cases are as follows:

(a) Real and different roots m and n. Here, e^{mx} and e^{nx} are solutions of the differential equation and the general solution will be

$$y = A\,e^{mx} + B\,e^{nx}$$

A and B being arbitrary constants.

(b) Real and equal roots m. The general solution is

$$y = e^{mx}(A + Bx)$$

(c) Imaginary roots of the form $m = p \pm iq$. The general solution is

$$y = e^{px}(A\cos qx + B\sin qx)$$

(4) *Linear equations with constant coefficients of the form*

$$a\frac{d^2y}{dx^2} + b\frac{dy}{dx} + cy = f(x)$$

If $y = u(x)$ is the general solution of the equation

$$a\frac{d^2y}{dx^2} + b\frac{dy}{dx} + cy = 0$$

and $y = v(x)$ is any particular solution of the given equation, obtained for example by inspection, it can easily be proved that the general solution of the given equation is

$$y = u(x) + v(x)$$

It follows that the general solution of the equation is made up of the sum of the two parts, one being the general solution of the allied equation

$$a\frac{d^2y}{dx^2} + b\frac{dy}{dx} + cy = 0$$

known as the *complementary function*, and the other being any particular solution of the given equation, known as a *particular integral*. The complementary function can be found by the methods given above. A particular integral of the equation

$$a\frac{\mathrm{d}^2 y}{\mathrm{d}x^2} + b\frac{\mathrm{d}y}{\mathrm{d}x} + cy = \mathrm{f}(x)$$

is found in one of the following ways.

If $\mathrm{f}(x)$ is a polynomial of degree n, a particular integral can be obtained by substituting

$$y = a_0 + a_1 x + a_2 x^2 + \cdots + a_n x^n$$

and determining the constants $a_0, a_1, \ldots, a_n$ by equating coefficients.

If $\mathrm{f}(x) = k e^{mx}$, a particular integral can be found by substituting $y = p\,e^{mx}$, p being determined by equating coefficients. If the function e^{mx} occurs in the complementary function p will be indeterminate, and it will be necessary to use the substitution $y = p\,x\,e^{mx}$ or possibly $y = p\,x^2\,e^{mx}$.

If $\mathrm{f}(x) = A\cos nx$ or $A\sin nx$, the substitution $y = p\cos nx + q\sin nx$ gives a particular integral. If functions of $\cos nx$ and $\sin nx$ occur in the complementary function, the substitution becomes

$$y = x(p\cos nx + q\sin nx)$$

A particular integral can be found for a wide range of functions using the method of *Laplace transforms. A particular integral can also be found by using methods involving the *differential operator D, meaning 'differentiate with respect to the independent variable involved', and its inverse.

If the differential equation is of the form $\mathrm{F(D)}y = \mathrm{f}(x)$ then a particular integral is

$$y = \frac{\mathrm{f}(x)}{\mathrm{F(D)}}$$

See also Bernoulli's equation; Bessel's equation; Clairaut's equation; Euler's equations; hypergeometric differential equation; Korteweg–de Vries equation; Laguerre's differential equation; Laplace's equation; Legendre's differential equation; Mathieu's equation; Maxwell's equations; Navier–Stokes equations; van der Pol's equation; wave equation.

differential form A differential form ω is an object which is defined on a differential *manifold M and which can be integrated over a k-dimensional set (as a *multiple integral), but does not depend on a particular choice of coordinate system. For example, a 1-dimensional form is (locally) the differential of a differentiable real-valued function, but it is not necessarily the differential of a (single-valued) function. Indeed, $\mathrm{d}\theta$ is not the differential of any single-valued differentiable function on the circle

$$\{(\cos\theta,\ \sin\theta)\colon 0 \leqslant \theta \leqslant 2\pi\}$$

The theory of differential forms allows the unification and generalization of a number of theorems of several-variable calculus. For example, the generalized Stokes's theorem

$$\int_{\partial K} \omega = \int_K \mathrm{d}\omega$$

(where ∂K denotes the boundary of the *domain K) includes *Green's theorem, *Stokes's theorem, and the *fundamental theorem of calculus as special cases. A more detailed analysis involves the study of the *homology groups of the manifold M; the *de Rham cohomology spaces* of M (G. de Rham, 1931) can be constructed using the set of all the differential forms on M, and are closely related to the *cohomology groups of M.

differential geometry The branch of geometry concerned with the intrinsic properties of curves and surfaces as found by differential *calculus. Gauss, in 1827, defined the total (or Gaussian) *curvature of a surface at a point and gave formulae for this in terms of the partial derivatives using different coordinate systems. This was later extended by Riemann (*see* Riemannian geometry) to a general differential geometry of any type of space in any number of dimensions.

differential manifold *See* manifold.

differential operator A symbol or letter indicating that *differentiation is to be performed, written as D or d/dx. Properties of the differential operator include the following:

$$\frac{f(x)}{D} = \int f(x)\, dx$$

$$\frac{x^n}{(D+p)^q} = \left(1 + \frac{D}{p}\right)^{-q} \frac{x^n}{p^q}$$

$$F(D)e^{ax} = e^{ax}F(a)$$

$$F(D)e^{ax}f(x) = e^{ax}F(D+a)f(x)$$

$$F(D^2)\sin ax = F(-a^2)\sin ax$$

$$F(D^2)\cos ax = F(-a^2)\cos ax$$

In general, any nonconstant polynomial in D is a differential operator. For example, $(D^3 + 2D + 6)y$ is $d^3y/dx^3 + 2dy/dx + 6y$.

differentiation The process of obtaining the *derivative of a *function by considering small changes in the function and in the independent variable, and finding the limiting value of the ratio of such changes. If $y = x^2$, for a small change δx in x,

$$y + \delta y = (x + \delta x)^2$$
$$= x^2 + 2x \cdot \delta x + (\delta x)^2$$
$$\delta y = 2x \cdot \delta x + (\delta x)^2$$
$$\frac{\delta y}{\delta x} = 2x + \delta x$$

As $\delta x \to 0$, $\delta y/\delta x \to dy/dx$. Thus

$$\frac{dy}{dx} = \lim_{\delta x \to 0}\left(\frac{\delta y}{\delta x}\right) = 2x$$

In general, if $y = x^n$, then $dy/dx = nx^{n-1}$.
Methods of differentiation include the use of the *chain rule, *product rule, and *quotient rule, and *implicit differentiation.

A table of derivatives is given in the Appendix. *See also* partial differentiation; numerical differentiation.

Diffie–Hellman–Merkle key exchange (B.W. Diffie, M.E. Hellman and R.C. Merkle, 1976) A safe method of exchanging a *key. To illustrate it, suppose that Alice and Bob wish to communicate securely. They decide to use the multiplicative *cyclic group G of order $p-1$ consisting of the nonzero elements of a *finite field of order p, and choose a *generator g for this group. Alice chooses an integer a and calculates $A = g^a$ modulo p; she sends this to Bob. Bob chooses an integer b and sends $B = g^b$ modulo p to Alice. Both can compute the key $K = g^{ab}$ modulo p since Alice can do $K = B^a$ and Bob can do $K = A^b$. An eavesdropper may know p, g, A, and B but cannot calculate K without solving the *discrete logarithm problem. If p is a large prime the problem is very difficult to solve, so K can be used as the basis of a private key for Alice and Bob.

digit A symbol used in writing numbers. For the decimal *number system ten digits are used: 0, 1, 2, 3, 4, 5, 6, 7, 8, and 9. *See also* duodecimal system; hexadecimal system.

digital root *See* casting out nines.

digraph *See* graph.

dihedral The configuration formed by two half-planes meeting at a common edge. The *dihedral angle* is the angle between these half-planes; i.e. the angle between two half-lines, one in each plane, drawn perpendicular to the edge from a common point.

dihedral group A *group generated by two elements, say r and t, that satisfy $r^2 = t^n = (tr)^2 = I$, where n is an *integer and is at least 2, and I is the *identity element. It is denoted by D_n and contains $2n$ elements. When $n \geqslant 3$ it can be regarded as the group of all rotational *symmetries of the solid figure produced by joining the bases of two identical regular *pyramids. Alternatively, it can be regarded as the group of all symmetries of a regular polygon with n sides. *See* cyclic group; generator.

Dijkstra's algorithm *See* network analysis.

dilatation (dilation) A *transformation that maps lines onto parallel lines. It is either a *translation or an *enlargement (a *central dilatation*).

dilation 1. The increase in volume per unit volume of a material.
2. *See* time dilation.

dim Dimension. *See* vector space.

dimension 1. Of space, the number of parameters needed to specify the position of a particular point. Space has n dimensions when n coordinates are required: points in one-dimensional space lie on a curve; points in two-dimensional space lie on a surface; points in three-dimensional space lie within a volume. Space is normally considered as three-dimensional. The dimensions of an object in three-dimensional space are given in terms of its volume.
2. The size of a *matrix expressed as $m \times n$, where m is the number of rows and n the number of columns. A square matrix of dimension $n \times n$ is sometimes said to be 'of dimension n'.
3. The number of elements in a *basis of a vector space.
4. (of a manifold) *See* manifold.
5. (of a simplex) *See* combinatorial topology.
6. The power of a fundamental *physical quantity, such as length, time, or mass, that is used in the description of the measure of any physical quantity. The physical quantity is represented by the product of particular powers of one or more fundamental quantities, without any numerical factor; this is known as its *dimensional formula*. (The same system is used in defining a coherent system of units, where the units of physical quantities can be derived from

fundamental units such as the metre, second, and kilogram.)

The dimensional formulae for mechanical quantities are usually expressed in terms of powers of length L, time T, and mass M. Although it is possible to give the dimensional formulae for non-mechanical (e.g. electrical) quantities in terms of these three quantities, fractional exponents are involved. However, an additional fundamental quantity can usually be introduced: current, for example, can be expressed as QT^{-1}, where Q is charge.

When two physical quantities are multiplied or divided, the exponents of their dimensional formulae are added or subtracted as appropriate. For two physical quantities to be added or subtracted, however, they must have the same dimensional formula. Furthermore, the arguments of trigonometric, exponential, and logarithmic functions must be dimensionless. It follows that in an equation involving physical quantities, the two sides of the equation must have the same dimensions.

See also dimensional analysis.

dimensional analysis A technique that makes use of the *dimensions of *physical quantities. It provides a means of checking equations that involve physical quantities: the terms on each side of an equation should have the same dimensional formulae; any numerical factors in the equation would have to be ignored. Equations can also be derived from a study of the dimensions of the physical quantities likely to be involved; again, numerical factors could not be obtained from such analysis. The technique can thus be used to obtain information about a system before a full analysis is undertaken.

Diocles (*c*.200 BC) Greek mathematician who wrote a work on conics, known only from extracts or from a much later dubious Arabic translation. He is reported to have invented the cissoid curve to solve the problem of duplicating the cube.

Diophantine analysis The study of *Diophantine equations.

Diophantine equation Any equation, usually in several unknowns, that is studied in a problem whose solutions are required to be integers, or sometimes more general *rational numbers. Examples of such problems are:

(1) To find all integers x and y that satisfy $11x + 3y = 1$.

(2) To find all rational numbers x, y, and z such that $x^3 + y^3 = z^3$.

Problems of this type are named after Diophantus of Alexandria, who investigated many similar questions in his book *Arithmetica*.

*Hilbert's 10th problem (1900) was to devise an *algorithm which would determine whether any given Diophantine equation is solvable in rational numbers. In 1970, Y. Matyasevic proved that no such algorithm can exist.

Diophantine problem A problem whose solutions are required to be integers or *rational numbers. *See* Diophantine equation.

Diophantus of Alexandria (*c*. AD 250) Greek mathematician and author of the *Arithmetica*, of which ten of the original thirteen books are extant. Over 180 problems are considered, some of which are surprisingly hard, in the field of what have since become known as *Diophantine equations.

Dirac delta function (delta function) A *generalized function, denoted by δ_a cf.

p191, first used by the English physicist Paul Adrien Maurice Dirac (1902–84).

directed angle An angle measured from an initial line to a final line. If the sense of rotation is anticlockwise, the angle is positive in sign; if the sense is clockwise the angle is negative.

directed graph *See* graph.

directed line A line along which one direction is specified as positive with the opposite direction specified as negative.

directed number (signed number) A number with a positive or negative sign, indicating that it is measured in a certain direction from the origin along a line.

directional data (angular data, circular data) Data consisting essentially of angular bearings from a point. Such data may be represented by points on the circumference of a circle having that initial point as centre.

Directional data have some characteristics that differ from those of linear data. For example, if angles are measured over the range $0°$ to $360°$ and observed values are 1, 5, 10, 350, 355, and 359, then the linear counterpart *mean and *median are both $180°$. However, if these angles are measured with the same zero, but with a range from $-180°$ to $180°$, the readings become 1, 5, 10, -10, -5, and -1, giving a mean or median zero. It is clear from diagram (a) that the latter is a more sensible interpretation of centrality.

If points are distributed on the circumference of a circle and the probability that

(a)

(b)

directional data

any point lies on any arc of fixed length is the same, there is a uniform directional distribution. A typical sample of 10 from this distribution is illustrated in diagram (b). There is no unique mean direction in this case.

Often one is interested in whether directional data are clustered in some incomplete arc on the circle, and several distributions, including the *wrapped Cauchy distribution*, may then be relevant. Nonparametric methods exist for testing the hypothesis of a uniform distribution against an alternative of clustering.

Any data having a periodic pattern may be expressed in a circular representation, for example the distribution of births over time of day on a 24-hour clock, or suicides on dates throughout a year on a circular scale from 1 January to 31 December.

direction angles For a line in a three-dimensional *coordinate system, the direction angles are the three positive angles the line makes with the three coordinate axes (usually denoted by α, β, and γ for the x-, y-, and z-axes, respectively). The cosines of these angles are the *direction cosines* of the line, l, m, and n. The angle θ between two lines with direction cosines l_1, m_1, n_1 and l_2, m_2, n_2 is given by

$$\cos\theta = l_1l_2 + m_1m_2 + n_1n_2$$

Direction ratios (or *direction numbers*) are numbers in the ratio $l : m : n$. Note that dir-

direction angles of a line.

ection angles (cosines or ratios) are not independent; if two are known, the third is fixed. The direction cosines are related by

$$l^2 + m^2 + n^2 = 1$$

direction cosines *See* direction angles.

direction ratios *See* direction angles.

direct iteration *See* iteration.

directly congruent *See* congruent.

directly proportional *See* variation.

director circle A circle that is the *locus of the point of intersection of pairs of perpendicular *tangents to an *ellipse or *hyperbola.

direct proof A method of proof in which conclusions are derived from *axioms and established laws in accordance with accepted definitions and rules of *inference. Thus, assuming that 'greater than' is a *transitive relation, and the two premises $12 > 10$ and $10 > 8$ hold, then we can directly infer the conclusion $12 > 8$. *Compare* indirect proof.

directrix (*plural* **directrices**) **1.** *See* conic. **2.** A curve defining the *generators of a *ruled surface. *See also* cone; conical surface; cylinder; cylindrical surface.

direct trigonometric function A *trigonometric function such as sine or cosine, as distinguished from an *inverse trigonometric function.

Dirichlet, Peter Gustav Lejeune (1805–59) German mathematician who first formulated the modern notion of a function. In number theory he demonstrated in 1825 that *Fermat's last theorem held for $n = 5$ and later proved what is now known as *Dirichlet's theorem. In other work, Dirichlet dealt with boundary problems and with Fourier series, in which latter field he was able to define (1829) the conditions sufficient for convergence.

Dirichlet's principle *See* pigeonhole principle.

Dirichlet's test A test for *convergence of a *series. Let $\Sigma\, a_n$ be an infinite series whose *partial sums

$$s_n = a_1 + a_2 + \cdots + a_n$$

are bounded, i.e. there is a positive number H such that

$$|s_n| < H \quad \text{for all } n$$

If the numbers $b_1, b_2, \ldots, b_n, \ldots$ constitute a monotonic *decreasing sequence that approaches zero, then the infinite series

$$a_1 b_1 + a_2 b_2 + \cdots + a_n b_n + \cdots$$

converges. The test can also be used to determine whether a functional series has *uniform convergence. *See also* Abel's test.

Dirichlet's theorem The theorem that in every *arithmetic progression a, $a+d$, $a+2d$, ..., where a and d are *relatively prime, there are an infinite number of primes.

disc (disk) The set of all points lying on a circle or within it is a *disc* or *closed disc*. The set of all points lying within the circle is an *open disc*. *See* ball.

disconnected graph *See* walk.

disconnected set A *set A is disconnected if there exist disjoint nonempty subsets of A (X and Y) such that $X \cup Y = A$, and no *limit point of X is a member of Y and no limit point of Y is a member of X. *Compare* connected set.

disconnected space A *topological space S is disconnected if there exist disjoint, nonempty open sets of S (X and Y) such that $X \cup Y = S$. *Compare* connected space.

discontinuity A point in the *domain of a *function at which the function is discontinuous. A real-valued function f has a *jump discontinuity* at $x = c$ if the right-hand and left-hand limits at $x = c$ exist

but are not equal; f(c) may equal one or neither of these limits. For example,

$$\text{f}(x) = 0 \quad \text{for } x > 1$$
$$\text{f}(x) = 1 \quad \text{for } x < 1$$
$$\text{f}(x) = \tfrac{1}{2} \quad \text{for } x = 1$$

has a jump discontinuity at $x = 1$.

The function f has a *removable discontinuity* at c if the left- and right-hand limits at $x = c$ are equal to each other but unequal to f(c). The function f can be made continuous by redefining f(c) to have the same value as the limits. For example,

$$\text{f}(x) = x \sin(1/x) \quad \text{for } x \neq 0$$
$$\text{f}(x) = 1 \quad \quad \quad \text{for } x = 0$$

has a removable discontinuity at $x = 0$, which can be removed by letting f$(0) = 0$. Removable and jump discontinuities are *simple discontinuities*.

A function f may not have a finite left- or right-hand limit at $x = c$, or the function may be undefined at $x = c$. Thus f$(x) = 1/(x-1)$ has a discontinuity at $x = 1$ as it is undefined there. If f remains finite at $x = c$ it is said to have a *finite discontinuity* at that point. For example, f$(x) = \cos(1/x)$ has a nonremovable finite discontinuity at $x = 0$. If, however, $|\text{f}(x)|$ becomes arbitrarily large near $x = c$ it is said to have an *infinite discontinuity* at that point. For example, f$(x) = 1/x$ has an infinite discontinuity at $x = 0$.

See also discontinuous function.

discontinuous function A function that is not a *continuous function. A function that is not continuous at $x = c$ is said to be discontinuous at $x = c$ and discontinuous on any interval containing c, if the *domain and *codomain are sets of real numbers. *See also* discontinuity.

discrete distribution *See* distribution.

discrete Fourier transform A technique similar to the Fourier transform that is used to study discrete phenomena. It can be used to speed up computer calculations. If $x_1, x_2, \ldots, x_n$ is a sequence of complex numbers, its discrete Fourier transform is the sequence $\hat{x}_1, \hat{x}_2, \ldots, \hat{x}_n$, where

$$\hat{x}_k = \sum_{r=1}^{n} x_r \exp\left(\frac{-2\pi i k r}{n}\right)$$

and its inverse is

$$x_k = \frac{1}{n} \sum_{r=1}^{n} \hat{x}_r \exp\left(\frac{2\pi i k r}{n}\right)$$

Evaluation directly from these formulae requires a multiple of n^2 operations. The *fast Fourier transform* is a way of evaluating the discrete Fourier transform in a number of operations proportional to $n \ln n$.

discrete logarithm problem If g is a *generator of a *cyclic group G (often the multiplicative group of nonzero elements of a *finite field), the discrete logarithm problem is to solve the equation $y = g^x$ for the unknown integer x. For large cyclic groups this is regarded as a very difficult problem. It is analogous to the problem of solving $y = e^x$ where x and y are real numbers and whose solution is $x = \ln y$.

discrete random variable *See* random variable.

discrete set A *set such as the set of *natural numbers is said to be *discrete* in the sense that there are elements of the set between which there are no other natural numbers. The set of rational numbers is not discrete since between any two members there is always at least one other rational number. In general, a set A is discrete if every point of A has a *neighbourhood containing no other point of A. *Compare* dense set.

discrete space A topological space in which every subset is an *open set.

discriminant (of a polynomial equation) A value obtained by taking the differences of all possible pairs of the *roots of an equation, squaring each difference, and taking the product of these squares. For example, if an equation has three roots, $r_1, r_2,$ and r_3, the discriminant is

$$(r_1 - r_2)^2 (r_2 - r_3)^2 (r_3 - r_1)^2$$

The discriminant can be found from the coefficients of the equation and can give information on the form of the roots. For instance, the quadratic equation with real coefficients

$$ax^2 + bx + c = 0$$

has a discriminant $b^2 - 4ac$. If this is zero, the roots are real and equal; if positive, the roots are real and unequal; if negative, the roots are imaginary. *See also* cubic; conic.

discriminant function A function that assigns an individual to one of two or more *populations on the basis of data for that individual. The function is based on measurements on individuals for whom the population to which each belongs is known. It is often linear, and is chosen to minimize the probabilities or costs of misclassification.

disjoint Describing *sets that have no common members. Two sets are disjoint if their *intersection is empty. For example, the sets $A = \{1, 2\}$ and $B = \{4, 5\}$ are disjoint sets.

disjunct *See* disjunction.

disjunction (alternation) A sentence of the form 'A or B', often symbolized in a formal language as '$A \vee B$' (*see* or). 'A' and 'B' are called *disjuncts*. If a disjunction 'A or B' is read as 'A or B but not both' then the disjunction is said to be *exclusive*; if 'A or B' is read as 'A or B or both' then the disjunction is said to be *inclusive*. In general, it is the inclusive sense that is implied by logicians in using the symbol $\vee$.

disjunctive normal form A formula is in disjunctive normal form if it consists of a *disjunction of *conjunctions, with each conjunction formed only from *atomic sentences or their *negations. It can be shown that every *wff of the propositional calculus can be expressed as an equivalent formula in disjunctive normal form. Thus the expression $(p\,\&\,q\,\&\,r\,\&\,\sim r) \vee (q\,\&\,\sim q) \vee (q\,\&\,p\,\&\,\sim p)$ is in disjunctive normal form. It is thus possible to see if any formula is a *contradiction of the propositional calculus by noting, as in the case above, that each disjunction contains both an atomic sentence and its negation. *Compare* conjunctive normal form.

disjunctive syllogism *See* implication.

dispersion The spread of a *random variable or a set of observations. Widely used measures of spread are *variance, *standard deviation, *range, *interquartile range, and semi-interquartile range. For random variables the range may be infinite. For random samples, the sample equivalents of their population counterparts may be used as *plug-in estimators, but sometimes modified estimators are preferred. For example,

$$s^2 = \frac{1}{n-1} \sum_i (x_i - \bar{x})^2$$

is an *unbiased estimator of population variance based on a random sample of n observations $x_1, x_2, \ldots, x_n$, but the plug-in estimator, which is the sample variance with divisor n in place of $n-1$, is a biased estimator.

dissection *See* partition.

dissection proof A proof that involves cutting a geometric object into a finite number of pieces. For example, one can show that two polygons have the same area if one of them can be broken into finitely many polygons that can be reassembled to form the other. In particular, a triangle can be dissected into three pieces which can be rearranged to form a rectangle; thus proving a formula for the area of a triangle (*see* diagram).

dissection proof

In 1807, the Scottish mathematician William Wallace formulated the theorem that if two planar rectilinear polygons have the same area then one of them can be dissected into pieces that can be reassembled to form the other. The corresponding problem in three dimensions was one of Hilbert's problems and the German mathematician Max Dehn showed in 1901 that there are two polyhedra of the same volume for which it is impossible to reconstruct one from the other by dissection. *See also* equidecomposable.

distance **1.** The length of a line segment between two points, lines, planes, etc. For example, the distance between two parallel lines or planes is the length of a line segment that is perpendicular to both. The distance of a point from a line, curve, plane, or surface is the length of the shortest line segment joining the point to the line, curve, plane, or surface.
2. (**angular distance**) The distance between two points as measured by the angle between two lines through the points and through a common reference point. For instance, the angular distance between points A and B with respect to point P is the angle APB.
3. (**arc distance**) The distance between two points on a curve as measured by the *length of the arc joining them.

distance function *See* metric.

distribution A *random variable that takes only a finite or countably infinite set of values has a *discrete distribution*. More formally, for a random variable X taking a finite or countably infinite set of values x_i, the discrete distribution of X is the set of pairs $(x_i, \Pr(X = x_i))$. In most practical cases the values taken are non-negative integers. The *binomial distribution takes integral values in $[0, n]$. For the *Poisson distribution, any non-negative integral value is possible.

A random variable that may take any value in a finite or infinite interval has a *continuous distribution*. More formally, for a random variable X taking a value between x and $x + \delta x$ with probability $f(x)\,\delta x$, the continuous distribution of X is the set of pairs $(x, f(x))$. The *normal and *gamma distributions are well-known examples.

See also Bernoulli distribution; bivariate distribution; Cauchy distribution; chi-squared distribution; extreme value distribution; F-distribution; geometric distribution; hypergeometric distribution; logarithmic distribution; multinomial distribution; multivariate distribution; negative binomial distribution; Pareto distribution; Pearson distributions; t-distribution; triangular distribution; uniform distribution; Weibull distribution.

distribution-free methods Methods for testing hypotheses or setting up *confidence intervals that include many *non-parametric methods. A distribution-free method for making *inferences from a sample does not depend on the form of the underlying population distribution; there may be requirements such as continuity or symmetry, but it is not assumed that the sample comes from any specific family such as that of the normal or exponential distributions.

The methods often depend only on the ranks of observations, and are therefore particularly useful when only the order of data is known but not precise values.

Many of the tests have analogues in which it is assumed that samples are from a given family of distributions, and when that assumption is valid the distribution-free tests are generally less efficient. However, a distribution-free test is often more efficient than a parametric test when assumptions for the latter break down.

Many sophisticated distribution-free methods once presented formidable computational problems, even for small samples, but these have been overcome by modern computer software developments. For larger samples the *central limit theorem allows the development of approximations based on the *normal distribution.

See coefficient of concordance; correlation coefficient; Friedman's test; Jonckheere–Terpstra test; Kolmogorov–Smirnov tests; Kruskal–Wallis test; median test; Page test; permutation test; sign test; Wilcoxon rank sum test; Wilcoxon signed rank test.

distribution function If X is a *random variable its (cumulative) distribution function is

$$F(x) = \Pr(X \leqslant x)$$

For discrete variables,

$$F(x) = \sum_{x_i \leqslant x} \Pr(X = x_i)$$

and for continuous variables,

$$F(x) = \int_{-\infty}^{x} f(t)\,dt$$

where $f(t)$ is the frequency function. $F(x)$ is *monotonic increasing, and $F(x) \to 1$ as $x \to \infty$. Also, $F(x) \to 0$ as $x \to -\infty$. The definition extends to *multivariate distributions. For a *bivariate distribution,

$$F(x, y) = \Pr(X \leqslant x, Y \leqslant y)$$

distributive Describing an operation on a combination in which the result is the same as that obtained by performing the operation on the individual members of the combination, and then combining them. For example,

$$2 \times (3 + 6) = (2 \times 3) + (2 \times 6)$$

is an example of the *distributive law* of arithmetic (or algebra), i.e.

$$a \times (b + c) = (a \times b) + (a \times c)$$

In this case it is said that 'multiplication is distributive over addition'. Note that addition is not distributive over multiplication:

$$2 + (3 \times 6) \neq (2 + 3) \times (2 + 6)$$

In the algebra of sets, intersection ($\cap$) is distributive over union ($\cup$). Thus for sets A, B, and C,

$$A \cap (B \cup C) = (A \cap B) \cup (A \cap C)$$

Also union is distributive over intersection, i.e.

$$A \cup (B \cap C) = (A \cup B) \cap (A \cup C)$$

See also Boolean algebra; field.

div *See* divergence.

divergence 1. A property of a *divergent series or *divergent sequence.
2. (div) For a vector function of position $V(r)$ the divergence of V, written as div V, is given by $\nabla.V$, where ∇ is the operator *del. Thus

$$\text{div } V = \nabla.V = i.\frac{\partial V}{\partial x} + j.\frac{\partial V}{\partial y} + k.\frac{\partial V}{\partial z}$$

and

$$r = x\mathbf{i} + y\mathbf{j} + z\mathbf{k}$$

The divergence is useful in certain physical applications. For example, ρ div v gives the rate of loss of mass of fluid per unit volume, ρ being the density and v the velocity; and div D gives electric charge density,

where D is electric displacement. The divergence of a vector is a scalar.

See curl; gradient; Green's theorem.

divergence theorem *See* Gauss's theorem.

divergent integral An *infinite integral that has no definite *limit.

divergent product An *infinite product that has a value of zero or infinity.

divergent sequence An infinite *sequence that has no *limit. A divergent sequence is either *properly divergent* or *oscillating* depending on whether it tends to infinity or oscillates in value. *See also* divergent series.

divergent series An infinite *series

$$a_1 + a_2 + a_3 + \cdots + a_n + \cdots$$

whose *partial sums

$$s_n = a_1 + a_2 + \cdots + a_n$$

do not approach a limit as the number of terms, n, becomes increasingly large (*compare* convergent series).

A series is *properly divergent* if s_n tends to infinity as n tends to infinity, i.e. if $s_n \to +\infty$ or $s_n \to -\infty$ as $n \to \infty$. An example is the series $1 + 2 + 3 + 4 + \cdots$. If a divergent series is not properly divergent it must be an *oscillating series*, i.e. s_n oscillates in value. An example is the series

$$\sum (-1)^n = -1 + 1 - 1 + 1 - \cdots$$

for which

$$s_n = 0 \quad \text{when } n \text{ is even}$$

$$s_n = -1 \quad \text{when } n \text{ is odd}$$

There are several methods by which a sum can be attributed to a divergent series.

divide To perform a division; to split into two or more parts.

divided difference interpolation formula A formula for *interpolation that makes use of divided differences; it is also called the *Newton divided difference interpolation formula*. If a function $y = f(x)$ has known values $y_0, y_1, \ldots, y_n$ at points $x_0, x_1, \ldots, x_n$, and a value y' is to be estimated at x', the formula is

$$y' = a_0 + a_1(x' - x_0)$$
$$+ a_2(x' - x_0)(x' - x_1) + \cdots$$
$$+ a_n(x' - x_0)(x' - x_1) \cdots$$
$$(x' - x_{n-1})$$

The numbers a_k are *divided differences*, given by $a_k = f[x_0, x_1, \ldots, x_k]$, where the divided differences are defined recursively by $f[x_k] = f(x_k)$ and

$$f[x_0, x_1, \ldots, x_{k+1}] =$$
$$\frac{f[x_1, x_2, \ldots, x_{k+1}] - f[x_0, x_1, \ldots, x_k]}{x_{k+1} - x_0}$$

Thus, for example,

$$f[x_0, x_1] = \frac{f(x_1) - f(x_0)}{x_1 - x_0}$$

The divided differences can be computed by forming a triangular divided difference table in which each entry is computed from the ones immediately west and north-west of it using the formula above. For $n = 3$, the table has the form

x_0 $f[x_0]$

x_1 $f[x_1]$ $f[x_0, x_1]$

x_2 $f[x_2]$ $f[x_1, x_2]$ $f[x_0, x_1, x_2]$

x_3 $f[x_3]$ $f[x_2, x_3]$ $f[x_1, x_2, x_3]$ $f[x_0, x_1, x_2, x_3]$

dividend A number or *polynomial that is divided by another number or polynomial. *See* division.

divisible Capable of being divided by a number or *polynomial an exact number of times (with zero remainder). Simple tests exist for the divisibility of numbers. Thus a number is divisible by:

2 if it is even
3 if the sum of the digits is divisible by 3
4 if the last two digits give a number divisible by 4
5 if the last digit is 5 or 0
6 if the number is even and divisible by 3
8 if the last three digits form a number divisible by 8
9 if the sum of the digits is divisible by 9
10 if the last digit is 0
11 if the sum of the digits in odd-numbered positions equals the sum of those in even-numbered positions, or if the two sums differ by a multiple of 11.

See also factor theorem

division The inverse operation to multiplication; the operation of finding – for two numbers or *polynomials – a third number or polynomial (the *quotient*) such that the first number or polynomial is equal to the quotient multiplied by the second. The operation is written as

$$q = a \div b$$

where q is the quotient, a is the *dividend*, and b is the *divisor*. Other common methods of indicating a quotient are $\frac{a}{b}$, a/b, and $a:b$.

The *division algorithm* is applied to integers and states that for any integers a and b there are two other integers q and r such that

$$a = bq + r$$

where $|r| < |b|$. Here, q is the quotient (sometimes called the *partial quotient* if $r \neq 0$) and r is the *remainder*. The algorithm also applies to polynomials, for which it states that for any polynomials $A(x)$ and $B(x)$, there are polynomials $Q(x)$ and $R(x)$ such that

$$A(x) = Q(x)B(x) + R(x)$$

where the *degree of $R(x)$ is less than the degree of $B(x)$.

division algebra *See* algebra.

division algorithm *See* division.

division in a given ratio **1.** Division of a quantity into parts which are in a certain ratio to one another. For instance, 100 divides in the ratio $2:3$ into parts of 40 and 60, while 150 divides in the ratio $1:2:3$ into parts of 25, 50, and 75.

2. Given three collinear points A, B, and P, then P divides AB in the ratio $m:n$ if $AP:PB = m:n$. The lengths are *directed numbers, so that if P lies between A and B, the ratio is positive and P divides AB *internally* in the ratio $m:n$. When P is not between A and B, the ratio is negative and P is said to divide AB *externally* in the ratio $|m|:|n|$.

If A, B, and P have position vectors $\mathbf{a}$, $\mathbf{b}$, and $\mathbf{r}$, then

$$\mathbf{r} = \frac{n\mathbf{a} + m\mathbf{b}}{m + n}$$

See also cross-ratio; golden section.

division modulo n The process of dividing an integer b by an integer a modulo n. This means finding an integer q such that $aq \equiv b \pmod{n}$, and is possible only with a q that is unique modulo n if a and n are relatively prime. For example, $7/2 \equiv 8 \pmod 9$ since $2 \times 8 \equiv 7 \pmod 9$, $1/3 \equiv 6 \pmod{17}$ and $4/5 \equiv 8 \pmod{12}$.

The concept can be extended to include the possibility of dividing an integral polynomial $f(x)$ (one with integral coefficients) by another integral polynomial $g(x)$ modulo n. As above, this amounts to finding a polynomial $h(x)$ such that

$$f(x) \equiv g(x)h(x) \pmod{n}$$

where the congruence means that the respective coefficients of $f(x)$ and $g(x)h(x)$ are congruent modulo n. For example, modulo 3, $x^3 - x^2 - 1$ is divisible by $x + 1$ since

$$(x+1)(x^2 + x - 1) = x^3 + 2x^2 - 1$$
$$\equiv (x^3 - x^2 - 1) \pmod 3$$

and modulo 11 the same polynomial $x^3 - x^2 - 1$ is not divisible by $x + 1$ but is divisible by $x - 5$ since

$$(x-5)(x^2 + 4x - 2) = x^3 - x^2 - 22x$$
$$+ 10 \equiv (x^3 - x^2 - 1) \pmod{11}$$

See also congruence modulo n; factor modulo n.

division ring (skew field) A *set D with operations of addition $(+)$ and multiplication $(\times)$ that is a *ring and that also has an identity element 1 (i.e. $1 \times a = a \times 1 = a$ for each a in D) and in which every nonzero element b has a corresponding inverse element b^{-1} (so that $b^{-1} \times b = b \times b^{-1} = 1$). The axioms for a division ring are just those for a field with the single exception that multiplication need not always be *commutative (i.e. there may be elements a and b in D for which $a \times b \neq b \times a$).

division sign The sign $\div$ denoting *division. It first appeared in 1659 in an algebra by Johann Heinrich Rahn. The alternative sign $/$, or *solidus*, is often used to separate the *numerator and *denominator of a fraction, as in $2/3$. Its use was recommended by William de Morgan in 1845 to simplify the printing of fractions. *See also* multiplication sign.

divisor **1.** *See* division.
2. An integer or *polynomial that divides another integer or polynomial exactly. The notation $m \mid n$ denotes that the integer m is a divisor of another integer n, i.e. that n is *divisible by m. *See* factor.

divisor function The *function $\tau(n)$ (or sometimes $d(n)$) that gives the number of positive *divisors of a *natural number n. For example, $\tau(6) = 4$. If n has the *prime factorization $p_1^{a_1} p_2^{a_2} \cdots p_r^{a_r}$, then

$$\tau(n) = (1 + a_1)(1 + a_2) \cdots (1 + a_r)$$

divisor of zero An element a in a *ring is a *left divisor of zero* if there is an element $b \neq 0$ in the ring such that $ab = 0$. The element a is a *right divisor of zero* if there is an element $c \neq 0$ such that $ca = 0$. It is just called a divisor of zero (or a zero divisor) if it is both a left and a right divisor of zero. If the ring is *commutative, any left divisor is also a right divisor and vice versa. An element is a proper divisor of zero if it is a divisor of zero and is not itself zero. *See also* integral domain.

dodecagon A *polygon that has 12 sides.

dodecahedron (*plural* **dodecahedra**) A *polyhedron that has 12 faces. A regular dodecahedron, in which all the faces are regular pentagons, is one of the five regular polyhedra.

d.o.f. *Abbreviation for* *degrees of freedom.

domain 1. (of a function) The set of values that can be assumed by the independent variable of a *function. Thus, if for every number in $0 \leqslant x \leqslant 2$

$$y = f(x) = x^3$$

then the domain of f is the closed interval $[0, 2]$. The *range is $[0, 8]$.
2. *See* integral domain.
3. (universe of discourse) The entities referred to by a language. More formally, the set of entities assigned as semantic values to the nonlogical expressions in the intepretation of a *formal language. *See* interpretation; semantics.
4. *See* region.

dot product *See* scalar product.

double-angle formulae Formulae in plane trigonometry that give trigonometric functions of double angles in terms of functions

of the angles, as follows:

$$\sin 2x = 2 \sin x \cos x$$

$$\cos 2x = \cos^2 x - \sin^2 x$$

$$\tan 2x = \frac{2 \tan x}{1 - \tan^2 x}$$

See also multiple-angle formulae.

double cusp *See* cusp.

double integral A *multiple integral involving two successive integrations. *See* area.

double negation (law of) The *theorem of the *propositional calculus stating that

$$A \equiv \sim\sim A$$

(A is equivalent to not not A). It follows from noncontradiction (not both A and not A) and the law of the excluded middle (either A or not A).

double point A *singular point on a curve at which the curve intersects itself so that there are two *tangents at the point. The tangents may be either coincident or non-coincident (in which case the point is a *node or *crunode*). *See also* cusp.

double root *See* multiple root.

double tangent 1. A *tangent that touches a curve at two separate points; a *bitangent*. **2.** A pair of coincident *tangents, as at a *cusp.

drag Resistance to the movement of a body through a fluid such as water or air; the fluid is set in motion by the moving body and the body thus experiences a *force opposing its motion. The amount of drag depends on the velocity of the body, its shape and size, and the density and viscosity of the fluid. There is no adequate formula giving the drag in every situation. One formula, used in aerodynamics, gives the drag force as $\frac{1}{2}k\rho Av^2$. A is a representative area of the body, v its speed, and ρ the fluid density; the coefficient k depends on the conditions and is a function of the Reynolds number, vl/v, where l is a representative length of the body and v is the coefficient of kinematic viscosity. Objects may be streamlined to reduce drag. *Compare* lift.

dual code The dual code to a *linear code C of length n is the set of all *binary strings $b_1b_2\dots b_n$ such that $a_1b_1 + a_2b_2 + \cdots + a_nb_n = 0$ modulo 2 for all *codewords $a_1a_2\dots a_n$ in C. The dual of a linear code C is also a linear code.

dual game *See* game theory.

duality **1.** The connection between lines and points in plane geometry (or between planes and points in solid geometry). A line can be defined by two points, and a point by the intersection of two lines; in this sense, the line and the point are said to be *dual elements* in plane geometry. Similarly, connection of points by lines and intersection of lines to give points are *dual operations*. A statement in which the name of each element is replaced by its dual element and the description of each operation is replaced by that of its dual operation leads to a *dual statement* (or *dual theorem*). The *principle of duality* in *projective geometry is that if a theorem is true, its dual is also true. *See also* Desargues's theorem; Pascal's theorem.
2. *See* Boolean algebra.

dual theorem, elements, etc. *See* duality.

dummy activity *See* critical path analysis.

dummy variable A quantity written in a mathematical expression in the form of a variable, although it represents a constant. Dummy variables typically take the value 1 or 0. For example, in statistics the *regression equation with two independent or *explanatory variables x_1 and x_2 is often written as $y = \beta_0 + \beta_1x_1 + \beta_2x_2$, but it is sometimes more convenient to write it as $y = \beta_0x_0 + \beta_1x_1 + \beta_2x_2$, where x_0 is a dummy variable always equal to 1. If we also wanted to consider the regression model in which $\beta_0 = 0$, this can be accom-

modated by setting the dummy variable $x_0 = 0$. The term is sometimes used when a variable characterizes attributes such as the presence of a property or characteristic ($x = 1$) or the absence of it ($x = 0$), but this usage is to be discouraged, for here x is what is properly called a *binary variable.

dunce hat The *topological space obtained from an equilateral triangle ABC by identifying together the three edges AB, AC, and BC.

duodecimal notation The method of positional notation used in the *duodecimal system.

duodecimal system A *number system using the base twelve. In the duodecimal system twelve different characters are needed; often the digits 0–9 are used together with the letters A and B: ten is A and eleven is B. Twelve is written as 10, thirteen as 11, etc. The number 7AB in duodecimal would, in the decimal system, be $(7 \times 12^2) + (10 \times 12^1) + (11 \times 12^0) = 1139$. The duodecimal system has an advantage over the decimal system for calculation because 12 has more factors than 10.

See also hexadecimal system.

duplication of the cube The problem of constructing, using only an unmarked straightedge and compasses, the edge of a cube having twice the volume of a given cube. It is also known as the *Delian problem*, since it is said that in 230 BC the oracle at Delos told the Athenians that, to rid themselves of a plague, they should construct an altar twice the size of the existing one to Apollo. It is one of the three classical problems of Greek geometry, along with *squaring the circle and *trisection of an angle. It is now known that the construction is impossible.

Durbin–Watson statistic (J. Durbin and G.S. Watson, 1950) A *statistic for testing in *time-series analysis (when the observations form an ordered sequence in time) whether there is a correlation between successive *errors.

dyadic connective *See* connective.

dynamical system An iterative procedure $x_{n+1} = T(x_n)$, where T is a mapping of a space X into itself, defines a dynamical system in X in which positions of points in X evolve iteratively under T. Thus the map $z \mapsto z^2 - 1$ and iteration $z_{n+1} = z_n^2 - 1$ together define a dynamical system in the complex plane. The term is also used to describe a point or set of points whose positions evolve with time according to a set of time-dependent equations or *flow*, typically the solution to a set of differential equations. For example, if the position of a point x at time t is denoted by $x(t)$, the equation $x(t) = t^2 + 1$ defines a flow on the real line which is a solution to the equation $x(t) = 2t$. Thus the term covers sets of states whose evolution is governed either by an iterated map or by a set of time-dependent equations (or flow).

For a map T and iteration $x_{n+1} = T(x_n)$, the ordered set of points $x_0, x_1, x_2, \ldots$ is called the *orbit* of x_0. For example, the set of points $1, \frac{1}{2}, \frac{1}{4}, \frac{1}{8}, \ldots$ is the orbit of the point 1 under the map $T(x) = \frac{1}{2}x$.

When T is *invertible, the orbit usually includes the points $x_{-1} = T^{-1}(x)$, $x_{-2} = T^{-1}(x_{-1})$, etc. For a flow, the orbit of a point x is the union of all points $x(t)$ for all t (*see also* trajectory).

dynamic programming A method of solving a wide range of optimization problems in *operational research where a step-wise approach to decisions is appropriate.

An example is illustrated in the diagram. The lines represent all possible routes for a three-stage bus journey from A to D (i.e. B_1, B_2, and B_3 are possible first-stage end points, C_1 and C_2 are second-stage end points, and D is the third-stage end point). The number on each line represents the stage fare for that route. The problem is to find the route from A to D with the lowest fare.

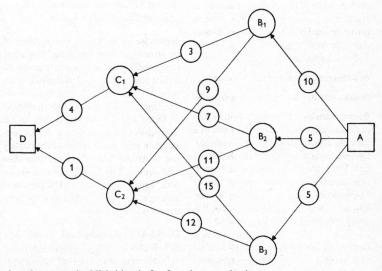

dynamic programming Minimizing the fare for a three-stage bus journey.

The dynamic programming approach first seeks the optimum one-stage strategy for the final (third) stage, i.e. C_1 or C_2 to D. This is a trivial problem as there is only one route from each. The next step is to find the optimal two-stage strategy from B_1, B_2, or B_3 to D. In this simple example it is easy to see that, starting this two-stage process from B_1, the optimal route to D is via C_1 (total fare 7, compared with 10 for travel via C_2). Similarly, from B_2 to D the route via C_1 is optimal (total fare 11) and from B_3 to D the route via C_2 is optimal (total fare 13).

The important point now is that, no matter how many additional *earlier* stages we add, the above routes are always optimal once we have arrived at B_1, B_2, or B_3. In this simple example it is now easy to see that, starting from A, the three-stage total fare via B_1 is $10 + 7 = 17$; via B_2, $5 + 11 = 16$; and via B_3, $5 + 13 = 18$. Thus the optimum route from A to D is via B_2 and C_1, with a fare of 16.

In more general problems of this type a *recursive approach is used along these lines to find the appropriate solution.

dynamics A field of *classical mechanics concerned with the study of the motion of material bodies under the influence of *forces. *Newton's laws of motion form the basis of this study. Dynamics can be divided into *kinetics, in which the relationships between force and motion are considered, i.e. the effects of forces are studied, and *kinematics, in which motion is described without regard to its cause, i.e. without considering the forces involved. Kinematics is, however, often treated as a separate field of classical mechanics. Dynamics and kinetics, then, are concerned with essentially the same subject matter and may be considered synonymous. *Statics deals with bodies in equilibrium under the action of forces.

dyne Symbol: dyn. A *c.g.s. unit of force, equal to the force required to impart to a mass of 1 gram an acceleration of 1 centimetre per second per second. 1 dyne = 10^{-5} newton.

E

e A transcendental *irrational number that is the base of natural *logarithms. It has the value $2.718\,281\,828\ldots$, the limit of $(1 + (1/n))^n$ as $n \to \infty$. It is the sum of the infinite series

$$1 + \frac{1}{1!} + \frac{1}{2!} + \frac{1}{3!} + \cdots$$

See also exponential function; exponential series.

eccentric Having centres that do not coincide. The term is applied to circles, ellipses, and other figures that have centres of symmetry. *Compare* concentric.

eccentric angle The angle that a radius of the *auxiliary circle makes with the positive x-axis, used in forming the parametric equations of an *ellipse or *hyperbola.

eccentric circle One of two circles that have the same centre as a *central conic and diameters equal to the conic's axes. *See* ellipse; hyperbola.

eccentricity Symbol: e. The ratio, for a point on a *conic, of its distance from a fixed point (the focus) to its distance from a fixed line (the directrix). *See also* ellipse; hyperbola; parabola.

ecentre *See* excentre.

echelon form *See* row echelon form.

ecliptic The *great circle that is the intersection of the plane of the earth's orbit with the *celestial sphere. The poles of this circle are the *poles of the ecliptic*. The line joining these poles is the *ecliptic axis*. The plane of

the ecliptic is inclined at an angle of $23° \, 26'$ to the plane of the celestial equator – an angle known as the *obliquity of the ecliptic*. *See* ecliptic coordinate system.

ecliptic axis *See* ecliptic.

ecliptic coordinate system An *astronomical coordinate system in which measurements are based on the ecliptic. In this system a point on the *celestial sphere is located by two angular measurements. The *celestial longitude (λ) is the angular distance measured eastwards along the ecliptic from the vernal equinox. The *celestial latitude (β) is the angular distance north or south of the ecliptic.

ecliptic latitude *See* celestial latitude.

ecliptic longitude *See* celestial longitude.

EDA *Abbreviation for* *exploratory data analysis.

edge 1. A line joining two vertices of a geometric figure or *graph.
2. A line or line segment from which one or more *half-planes extend; for example, the line segment between two faces in a *polyhedron, or the line between two planes in a *polyhedral angle.

effective procedure An *algorithm for determining whether or not a given object has a certain property. In particular, a function is said to be effectively *computable* if and only if, given the arguments of the function, there is an algorithm for determining its value. Many precise mathematical definitions have been provided that capture the intuitive notion of effectiveness, all of which are provably equivalent (*see* Church's thesis).

Examples of problems that are solvable by effective means are: whether or not a number c is the sum of numbers a and b; whether or not a given *wff is an axiom of the *predicate calculus (i.e. whether or not the predicate calculus is axiomatic); and whether or not a sequence of wffs is a *proof in the predicate calculus. *See* decidable; Church's theorem; recursive.

effect variable *See* regression.

efficiency 1. (R.A. Fisher, 1921) A *statistic T_0 used as an *estimator is more efficient than another estimator T_1 if it has smaller *variance. Efficiency has been studied principally for consistent and unbiased estimators. If two estimators of a parameter θ are such that sample sizes n_1 and n_2 are needed to give the same variance or, equivalently, the same power (*see* hypothesis testing), the relative efficiency of T_1 with respect to T_2 is n_2/n_1. Thus if $n_1 = 10$ and $n_2 = 20$, then T_1 is twice as efficient as T_2. The limit of the relative efficiency for large n is called the *asymptotic relative efficiency*.

Efficiency concepts extend to the comparision of tests, and relative efficiency may then depend on the *distribution from which a sample is obtained. While the *t-test is more efficient than the *Wilcoxon rank sum test for location comparisons of samples from a normal distribution, the reverse is true for samples from an exponential distribution.
2. In *experimental design one design is more efficient than another if the same precision can be obtained with less resources or if greater precision can be obtained with the same resources; a complication arises in that one design may be more efficient than another for some treatment comparisons but less efficient for others.
3. The ratio of the energy output to energy input over some time interval for a *machine or other energy-converting

mechanism, usually expressed as a percentage. This is equivalent to the ratio of work done on the load of a machine to work done by the effort. Efficiency is thus a measure of performance, and in practice is always less than 100 percent. Of the total energy available, some will always be lost in the form of unusable heat, e.g. through friction or exhaust fumes.

effort The *force applied to a particular part of a *machine, producing an effective force of different magnitude at some other part. The effective force is applied to the load of the machine.

eigenfunction A nontrivial solution of a differential equation subject to boundary conditions involving a parameter, for certain values of the parameter called *eigenvalues. For example, the differential equation with parameter λ

$$\frac{d^2y}{dx^2} + \lambda y = 0$$

subject to $y(0) = 0$ and $y(\pi) = 0$ has an eigenvalue m^2 with eigenfunction $\sin mx$, for any nonzero integer m.

eigenvalue In general, a characteristic value of some mathematical expression. The term comes from the German *eigen* ('characteristic' or 'own'). In some older texts the term 'latent root' is used.

For a square *matrix $\mathbf{A}$, the number λ is an eigenvalue of $\mathbf{A}$ if $\mathbf{Ax} = \lambda\mathbf{x}$ for some nonzero vector $\mathbf{x}$, the corresponding *eigenvector*. The eigenvalues are the zeroes of the *characteristic polynomial of $\mathbf{A}$. An $n \times n$ matrix of complex numbers always has n eigenvalues, but may have fewer than n *linearly independent eigenvectors. For a *symmetric matrix the eigenvalues are always real. For instance, the matrix

$$\begin{pmatrix} 1 & -2 & 5 \\ -1 & 6 & -1 \\ 5 & -2 & 1 \end{pmatrix}$$

has eigenvalues $\lambda_1 = -4$, $\lambda_2 = 4$, and $\lambda_3 = 8$, and corresponding eigenvectors

$$x_1 = \begin{pmatrix} 2 \\ 0 \\ -2 \end{pmatrix}, \; x_2 = \begin{pmatrix} 1 \\ 1 \\ 1 \end{pmatrix}, \; x_3 = \begin{pmatrix} 1 \\ -1 \\ 1 \end{pmatrix}$$

For a linear transformation on a *vector space $L: V \rightarrow V$, the number λ is an eigenvalue of L if $L(\mathbf{x}) = \lambda\mathbf{x}$ for some nonzero element $\mathbf{x}$ of V, the corresponding *eigenvector*.

eigenvector *See* eigenvalue.

Einstein, Albert (1879–1955) German-Swiss-American theoretical physicist who revolutionized modern physics with his special and general theories of relativity. Both needed for their expression mathematical techniques and insights not previously deployed in physics: in his special theory of 1905 Einstein substituted the Lorentz transformation for the classical Galilean transformation, while in his general theory he incorporated a curvature tensor based ultimately on the geometry of Riemann and the tensor calculus of Ricci-Curbastro.

Einstein's equation *See* mass–energy equation.

Eisenstein, Ferdinand Gotthold Max (1823–52) German mathematician who proposed the conjecture that numbers of the form $2^2 + 1$, $2^{2^2} + 1$, $2^{2^{2^2}} + 1$, …, are prime. This was disproved in 1953 when the *Fermat number $F_{16} = 2^{2^{16}} + 1$ was shown to be composite. He is, however, better known for his formulation of *Eisenstein's criterion.

Eisenstein's criterion A *sufficient condition for a polynomial with integer coefficients to be *irreducible (over the rational numbers) is that there exists a prime number p which divides all coefficients except the leading coefficient, but is such that p^2 does not divide the constant term. For example, $x^3 + 2x + 2$ is irreducible, but $x^3 + 2x + 4$ and $x^3 - 2x + 4$ may not be (the former is irreducible, the latter is reducible).

elastic collision *See* collision.

elastic constants Any of various constants that describe the behaviour of a homogeneous body under the action of a deforming *force. They include *Young's modulus, the *bulk modulus, and *Poisson's ratio. *See* elasticity.

elasticity The property of a solid body whereby it can resume its original shape and size once any deforming *forces are removed. The relations between a deforming force and the resulting change in shape or size of a homogeneous body can be described by its *elastic constants. The deforming force and the resulting deformation are considered in terms of *stress and *strain. For small stresses and strains, strain is proportional to stress. Above a certain stress not only will this proportionality no longer apply but also the body will not resume its original configuration. *See* Hooke's law.

elastic modulus *See* modulus.

elastic wave *See* wave.

electromagnetic wave *See* wave.

electronvolt Symbol: eV. A unit of energy used in atomic physics, equal to the energy acquired by an electron in passing through a potential difference of 1 volt. 1 electronvolt = 1.602×10^{-19} joule.

element 1. An expression following an integral sign in an *integration. For example, in the integration

$$\int_a^b f(x)\,dx$$

giving the area under the curve of $y = f(x)$ between $x = a$ and $x = b$, the expression $f(x)\,dx$ gives the *element of area* (i.e. the general formula for the area of the infinitesimal strips that are summed in finding the total area). Elements of length, volume, mass, etc. are similarly defined.

2. A *member of a *set or group.

3. *See* determinant; matrix.

4. *See* generator.

elementary matrix A square matrix obtained from the *identity matrix by applying an *elementary matrix operation. Multiplying a matrix A on the left (right) by an elementary matrix has the same effect as performing the corresponding elementary row (column) operation on A. An elementary matrix is *nonsingular.

For example, the matrix

$$\begin{pmatrix} 1 & 0 \\ 1 & 1 \end{pmatrix}$$

can be obtained from the identity matrix by the elementary row operation of adding the first row to the second. Multiplying the matrix

$$\begin{pmatrix} a & b \\ c & d \end{pmatrix}$$

on the left by this matrix has the same effect as the row operation, producing

$$\begin{pmatrix} a & b \\ a+c & b+d \end{pmatrix}$$

elementary matrix operation One of the operations of interchanging two rows or columns of a matrix, multiplying a row or column by a nonzero scalar, or adding a scalar multiple of one row or column to another. These are known as elementary row operations and elementary column operations, respectively. *Gaussian elimination makes use of elementary row operations.

elementary symmetric polynomial *See* invariant.

elevation 1. The height of a point above some baseline or plane.

2. (angle of) *See* angle.

eliminant (resultant) An expression formed from the coefficients of a set of *linear equations by eliminating the variables between the equations. In the case of a set of linear equations, the eliminant is a *determinant formed from the coefficients and constant terms. It is equal to zero for consistent simultaneous equations. For example, the three equations

$$a_1 x + b_1 y + c_1 = 0$$

$$a_2 x + b_2 y + c_2 = 0$$

$$a_3 x + b_3 y + c_3 = 0$$

have the eliminant

$$\begin{vmatrix} a_1 & b_1 & c_1 \\ a_2 & b_2 & c_2 \\ a_3 & b_3 & c_3 \end{vmatrix}$$

elimination The process of removing variables from a set of *simultaneous equations. There are various methods of elimination. For example, the equations

$$x + 3y = 7, \quad 2x + y = 9$$

can be handled by multiplying the second equation by 3, to give a third equation

$$6x + 3y = 27$$

Subtracting the left- and right-hand sides of the first from the third then gives

$$5x = 20$$

i.e. $x = 4$. The same result can be obtained by substitution. The first equation is put in the form

$$x = 7 - 3y$$

and this value of x is then substituted in the second:

$$2(7 - 3y) + y = 9$$

giving $y = 1$ and $x = 4$. *See also* Gaussian elimination.

ellipse A type of *conic that has an *eccentricity between 0 and 1 ($0 < e < 1$). It is a closed symmetrical curve like an elongated circle – the higher the eccentricity, the greater the elongation. Any chord through the centre is a diameter. The ellipse has two diameters that are axes of symmetry: the longest diameter is the *major axis* and the shortest the *minor axis*. A line segment from the centre to the ellipse along the major axis is a *semimajor axis*; one along the minor axis is a *semiminor axis*. Each of the two points at which the major axis meets the ellipse is a *vertex* of the ellipse. The area of an ellipse is πab, where a is the length of the semimajor axis and b the length of the semiminor axis.

The ellipse has two foci on the major axis and two directrices perpendicular to the major axis (*see* diagram (a)). Each

ellipse (a) F_1 and F_2 are foci.

focus is a distance ae from the centre, where e is the eccentricity. Each directrix is a distance a/e from the centre. The standard equation of an ellipse in Cartesian coordinates is

$$\frac{x^2}{a^2} + \frac{y^2}{b^2} = 1$$

The eccentricity of an ellipse is given by $c/2a$, where c is the distance between foci. Alternatively it is given by

$$e^2 - 1 - \frac{b^2}{a^2}$$

Either of the two chords through a focus of the ellipse and perpendicular to the major axis is a *latus rectum*. The length of the latus rectum is $2b^2/a$.

A circle with its centre at the centre of the ellipse and passing through the vertices (i.e. with radius a) is an *eccentric circle* of the ellipse. The circle with radius b is also an eccentric circle. The larger one (radius a) is called the *auxiliary circle* of the ellipse. If the ellipse has its centre at the origin and its major axis along the x-axis, the *eccentric angle*, α, is defined as follows (*see* diagram (b)). At a particular point P on the ellipse

ellipse (b) The eccentric angle of P is α.

the ordinate is extended to meet the auxiliary circle (at P′); α is then the positive angle between the x-axis and the radius OP′. The parametric equations of the ellipse are

$$x = a \cos \alpha, \quad y = b \sin \alpha$$

The ellipse has two well-known properties connected with its foci F_1 and F_2. For any point P on the ellipse, the sum of the distances PF_1 and PF_2 is constant (equal to $2a$). This is made use of in drawing ellipses by looping a string around two pins at the foci. The *focal property* of the ellipse is that if at any point P the tangent APB is drawn, then the lines from the foci to the point make equal angles with the tangent; i.e. $\angle APF_1 = \angle BPF_2$ *See* diagram (c)). This is also called the *reflection property*, since a

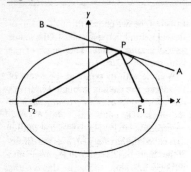

ellipse (c) $PF_1 + PF_2 = 2a$, and angle APF_1 = angle BPF_2.

reflector shaped like an ellipse would focus light from a source at one focus onto the other focus (the *optical property*), or would similarly focus sound (the *acoustical property*).

See also Kepler's laws.

ellipsoid A closed surface such that its plane sections are ellipses or circles. In Cartesian coordinates, it has the standard equation

$$\frac{x^2}{a^2} + \frac{y^2}{b^2} + \frac{z^2}{c^2} = 1$$

The centre of the ellipsoid is its centre of symmetry, and any chord through the centre is a diameter of the ellipsoid. Three of these chords are axes of symmetry – as with the ellipse, the largest is the *major axis* and the smallest the *minor axis*. The third axis, perpendicular to the other two, is the *mean axis*. The *semimajor* (a), *semiminor*

ellipsoid

(b), and *semimean* (c) axes are defined as for the ellipse. The volume of an ellipsoid is $\frac{4}{3}\pi abc$.

A special case of an ellipsoid is an *ellipsoid of revolution* (also called a *spheroid*) obtained by rotating an ellipse about one of its axes. In this case, plane sections perpendicular to the axis of revolution are circles. Revolution about the major axis gives a *prolate* ellipsoid (shaped like a rugby ball). Rotation about the minor axis gives an *oblate* ellipsoid. The earth, for instance, has a shape approximately that of an oblate ellipsoid.

ellipsoidal Denoting or concerned with an *ellipsoid.

elliptical Denoting, concerning, or connected with an *ellipse. For example, an elliptical cone (or cylinder) is a cone (or cylinder) having an ellipse as base.

elliptical paraboloid *See* paraboloid.

elliptic curve A curve, defined by a *polynomial equation with *rational *coefficients, having *genus 1 and containing at least one point, a *rational point*, with rational coordinates. The standard form of such a curve may be taken to be $y^2 = x^3 + ax + b$, where a and b are rational numbers, and the cubic polynomial $x^3 + ax + b$ has distinct *zeroes. For example, the elliptic curve $y^2 = x^3 + 17$ has a rational point $(-2, 3)$, but it also has many others, e.g. $(2, 3)$, $\left(\frac{1}{4}, \frac{33}{8}\right)$, and $(-1, 4)$.

The study of elliptic curves and their rational points was of great importance in the solution of *Fermat's last theorem.

elliptic functions Functions first derived from *elliptic integrals by Abel in 1826. For example, if the elliptic integral of the first kind is written in the form

$$u = \int_0^x \frac{\mathrm{d}t}{\sqrt{[(1 - t^2)(1 - k^2 t^2)]}}$$

then the elliptic functions sn, cn, and dn are defined by

$$x = \text{sn } u$$

$$\sqrt{(1 - x^2)} = \text{cn } u$$

$$\sqrt{(1 - k^2 x^2)} = \text{dn } u$$

This definition of the elliptic function sn is analogous to the definition of the *circular function sin by

$$x = \sin u \quad \text{and} \quad u = \int_0^x \frac{dt}{\sqrt{(1 - t^2)}}$$

Elliptic functions and integrals have many applications in mathematics. They were used by Hermite in 1858 in his solution of the general quintic equation.

elliptic geometry *See* Riemannian geometry.

elliptic integral An integral of the form

$$\int f(x, \sqrt{R}) \, dx$$

where R is a *cubic or *quartic function of x and f is any *rational function of x and $\sqrt{R}$. The name comes from the fact that integrals of this type were first met in the determination of the circumference of an ellipse. Any elliptic integral can be reduced to the sum of an elementary function and constant multiples of integrals in three standard forms involving x and constants k and n, known respectively as the *modulus* and *parameter*. These three integrals are called *Legendre's standard elliptic integrals* of the first, second, and third kinds respectively:

$$\int_0^x \frac{dt}{\sqrt{[(1 - t^2)(1 - k^2 t^2)]}}$$

$$\int_0^x \sqrt{\left(\frac{1 - k^2 t^2}{1 - t^2} \right)} \, dt$$

$$\int_0^x \frac{dt}{(1 + m^2)\sqrt{[(1 - t^2)(1 - k^2 t^2)]}}$$

elongation The total increase in length in the direction of a tensile *stress, or the increase in length per unit length caused by such a stress.

embedding A continuous map f: $X \rightarrow Y$ between *topological spaces is called an *embedding* if it is a *homeomorphism onto a subspace of Y.

empirical Based on observation or experiment rather than deduction from basic laws or postulates. An *empirical formula*, for example, is a formula that is devised to fit known data. An *empirical curve* is a curve drawn as the best approximation to fit a set of points.

empirical distribution function *See* sample distribution function.

empty set *See* null set.

encoding (encryption) In cryptography, the construction of ciphertext from plaintext. *See* cipher.

endogenous variables Variables such as price and demand in an economic *model that are an inherent part of the system, as distinct from *exogenous variables*, which impinge on the system from outside (e.g. exogenous variables such as rainfall may affect demand for certain products).

endomorphism A *mapping from a set to itself.

end point One of the numbers defining an *interval.

energy Symbol: E. A measure of the capacity of a body or system to do *work, i.e. to change the state of another body or system. Energy is measured in joules. Any

enlargement

body or system that is subject to a *conservative force, such as gravitation, has two forms of energy: *kinetic energy due to its motion and *potential energy due to its position; although two bodies can exchange kinetic and potential energy, the total energy remains constant in an isolated system. There is thus *conservation of energy. There are many kinds of energy, which are interconvertible: in a power station, the chemical energy stored in coal is converted by combustion to heat, which in turn is used to produce a jet of steam that drives a rotor whose mechanical energy is converted to electrical energy. Again, for a closed system energy is conserved.

enlargement (central dilatation, homothety, similitude) A *transformation involving a fixed point C, which is called the *centre of enlargement*. The image P' of a point P is the point on the *directed line CP such that $CP' = kCP$, where k, the *scale factor* of the enlargement, is a nonzero constant. When k is negative, C lies between P and P'. If C is the origin, a point with position vector $\mathbf{r}$ is mapped onto the point with position vector $k\mathbf{r}$. In the plane, an enlargement with centre at the origin and scale factor k maps the point with Cartesian coordinates (x, y) onto the point (kx, ky). Enlargement multiplies the distance between any two points by $|k|$.

enneagon *See* nonagon.

entailment *See* implication.

entire function (integral function) A function f of a *complex variable z that is an *analytic function for all finite values of z. Examples are e^z, $\sin z$, and $\cos z$. *Liouville's theorem* states that if $f(z)$ is entire and bounded, then it is constant.

enumerable *See* countable.

enumeration theory The theory of methods used to count or estimate the number of objects of a given type. An example of its application is in chemistry, to count the

number of different possible molecules of isomers.

envelope 1. A curve that touches (is *tangent to) every member of a given *family of curves. For instance, a family of circles with radius a, each of which has its centre on another circle of radius r, has an envelope consisting of a circle of radius $r + a$ and a smaller circle of radius $|r - a|$.

In general, a family of curves is defined by a parameter m, and members that differ by a small amount δm will intersect. The locus of these points of intersection as δm tends to zero becomes the envelope. The equation of the envelope can be found by equating to zero the partial derivative with respect to m of the equation of the family. For instance, the equation $y = 2mx + m^2$ represents a family of intersecting straight lines. Taking the partial derivative with respect to m (holding y and x constant) and equating to zero gives

$$0 = 2x + 2m$$

The equation of the envelope is then found by eliminating m between this equation and the original equation, to give $y = -x^2$; i.e. the envelope is a parabola.

2. A surface that is *tangent to all the members of a given *family of surfaces. For instance, the envelope of a family of spherical surfaces, each of which has its centre on a sphere, is itself a sphere. Such envelopes are used in constructing new wavefronts from secondary wavelets in wave theory.

epicycle *See* epicycloid.

envelope of a family of circles.

epicyclic Denoting, concerning, or connected with an *epicycloid. For example, an epicyclic gear is one in which one gear wheel moves around another.

epicycloid The *locus of a point on the circumference of a circle that rolls on the outside of a fixed circle (both circles being in the same plane). The parametric equations of the epicycloid are

$$x = (R + r)\cos\theta - r\cos[(R + r)\theta/r]$$

$$y = (R + r)\sin\theta - r\sin[(R + r)\theta/r]$$

where the fixed circle has its centre at the origin and has radius R, the rolling circle has radius r, and θ is the angle between the x-axis and a line through the centres of the two circles. The epicycloid was known to Apollonius of Perga, who used it in his method of representing planetary motion. In this, the moving circle was called the *epicycle* and the fixed circle the *deferent*. The system was later used in the Ptolemaic system of astronomy.

The epicycloid has *cusps at the points at which the moving point touches the fixed circle. If the ratio R/r is a rational number, the epicycloid is a closed curve. If $R = r$, the curve is a *cardioid; if $R = 2r$, the curve is a *nephroid. The epicycloid is a special case of an *epitrochoid. *See also* roulette.

epitrochoid The *locus of a point on the radius (or radius extended) of a circle that rolls on the outside of a fixed circle (both circles being in the same plane). The epitrochoid is thus a more general case of the *epicycloid and it has similar parametric equations:

$$x = (R + r)\cos\theta - d\cos[(R + r)\theta/r]$$

$$y = (R + r)\sin\theta - d\sin[(R + r)\theta/r]$$

where d is the distance of the point from the centre of the circle. *See also* roulette.

equals sign The sign used to represent equality, and to form an *equation. The

modern sign =, a pair of equal parallel line segments, was introduced in 1557 by Robert *Recorde, 'bicause noe 2 thynges can be moare equalle'.

equate To state that one expression is equal to another; to form an equation.

equate coefficients To conclude from the fact that two *polynomials are *identical that their *coefficients must be the same. This technique is frequently used to obtain information about the *roots of a polynomial equation. For example, if the roots of $x^2 + ax + b = 0$ are α and β, then

$$x^2 + ax + b = (x - \alpha)(x - \beta)$$

$$= x^2 - (\alpha + \beta)x + \alpha\beta$$

Since the coefficients of x must be the same, we conclude that $\alpha + \beta = -a$, and since the coefficients of $x^0 = 1$ (the *constant terms) are the same, we conclude that $\alpha\beta = b$.

equation A statement that two mathematical expressions are equal. A *conditional equation* is true only for certain values of the variables. Thus,

$$3x + y = 7$$

is true only for certain values of x and y. Such equations are distinguished from *identities*, which are true for all values of the variables. Thus,

$$(x + y)^2 = x^2 + 2xy + y^2$$

which is true for all values of x and y, is an identity. Sometimes the symbol $\equiv$ is used to distinguish an identity from a conditional equation.

equation of continuity *See* continuity equation.

equation of motion A *differential equation of the type

$$m\frac{d^2\mathbf{r}}{dt^2} = \mathbf{F}(\mathbf{r})$$

that gives the *position vector **r** of a particle of mass m moving under a *force **F**, as a function of time: the force is a function of position, i.e. it varies from point to point. Integration of this equation gives the velocity $d\mathbf{r}/dt$ at some particular time, and a second integration gives the position at some particular time. Two constants of integration are introduced, usually determined by the initial conditions, i.e. the velocity $\mathbf{v}_0$ and position $\mathbf{r}_0$ at time $t = 0$.

In the simple case of motion under a constant force, the equation reduces to $d^2\mathbf{r}/dt^2 = \mathbf{a}$, i.e. motion with constant acceleration **a**. Integration gives $d\mathbf{r}/dt = \mathbf{v}_0 + \mathbf{a}t$, and a second integration gives $\mathbf{r} = \mathbf{r}_0 + \mathbf{v}_0 t + \frac{1}{2}\mathbf{a}t^2$. *See also* Newton's laws of motion; Euler's equations.

equator *See* geographical equator; celestial equator; galactic equator.

equatorial coordinate system An *astronomical coordinate system in which measurements are based on the celestial equator. A point on the *celestial sphere is located by two angular measurements. The *right ascension (RA, α) is the angular distance measured eastwards along the celestial equator from the vernal equinox. The *declination (dec, δ) is the angular distance north or south of the terrestrial equator. Alternatively, *hour angle (t) can be used instead of right ascension. This is the angular distance measured westwards along the celestial equator. North polar distance, which is the complement of declination (i.e. $90° - \delta$), sometimes replaces declination. The equatorial system is the most widely used system of astronomical coordinates.

equiangular Having equal angles. The term is usually applied to geometric figures (for example, polygons) that have all their angles equal.

equiangular hyperbola *See* hyperbola.

equiangular spiral *See* spiral.

equiangular transformation *See* conformal transformation.

equicontinuous functions A family of *functions $\{f_i\}$ with the same *domain such that for all $\varepsilon > 0$ there exists a δ depending only on ε, and such that whenever

$$|x_1 - x_2| < \delta$$

then

$$|f_i(x_1) - f_i(x_2)| < \varepsilon$$

for all i.

equidecomposable Describing objects that can be broken down into identical sets of component pieces. More formally, two *polyhedra K_1 and K_2 in a *Euclidean space which are the union of n-dimensional simplexes are equidecomposable if each polyhedron is the union of a finite set of polyhedra

$$K_1 = A_1 \cup A_2 \cup \cdots \cup A_k$$

$$K_2 = B_1 \cup B_2 \cup \cdots \cup B_k$$

where any pair of the polyhedra A_i and A_j intersect only in lower-dimensional faces (and also for any pair B_i and B_j), and each A_i can be obtained from the corresponding B_i by a rigid motion. If K_1 and K_2 are equidecomposable, then they have the same (n-dimensional) volume. For planar polygons, i.e. the case $n = 2$, William Wallace proved in 1807 that if two polygons have the same area then they are equidecomposable (*see* diagram). In 1901 Max Dehn proved that the corresponding result is false in three dimensions; for example, a cube and a regular tetrahedron with equal volume are not equidecomposable.

equidistant Having equal distances from some specified point, line, etc.

equilateral Having equal sides or equal lengths. The term is usually applied to geometric figures (for example, polygons) that

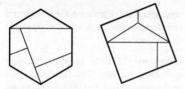

equidecomposable Two equidecomposable polygons.

have all their sides equal. It can also be used to denote two figures in which the corresponding sides are equal.

equilateral hyperbola *See* hyperbola.

equilateral polygon A *polygon that has all its sides equal. An *equilateral triangle* also has equal interior angles (60°).

equilibrant A *force or system of forces that can balance a given force or system of forces.

equilibrium A state attained or maintained by a particle or system of particles (a body) when it has no acceleration, neither translational nor rotational; the *resultant of the *external forces acting on the particle or body is zero, as is the sum of the *moments of all these forces. The equilibrium is said to be *stable* if, when slightly displaced, a particle or body returns to its original position; if the particle or body remains in its displaced position it is said to be in *neutral* equilibrium; if it moves to a different position, away from both the original and the displaced position, the equilibrium is described as *unstable*.

equinoxes (equinoctial points) Two points on the *celestial sphere at which the ecliptic intersects the celestial equator. The sun in its apparent annual motion crosses the celestial equator at these two points, crossing from south to north at the *vernal equinox* (♈) and from north to south at the *autumnal equinox* (♎). In the northern hemisphere these crossings occur around 21 March (vernal) and 23 September (autumnal), and they are marked by days on which the hours of daylight and dark-ness are equal. Points midway between these are the two *solstices* (or *solstitial points*).

equinumerable (equipollent, equipotent) Two *sets A and B are equinumerable if they can be put into a *one-to-one correspondence. The two sets are also described as *equivalent*. *See also* cardinal number.

equipotential surface An imaginary surface in a *conservative field on which all points have the same *potential.

equivalence 1. (material) Statements A and B are materially equivalent when both A and B are true, or both A and B are false. The material equivalence of A and B is symbolized by $A \equiv B$ (or $A \leftrightarrow B$, or A if and only if B) and is defined in a formal system as

$$(A \supset B) \,\&\, (B \supset A)$$

See also truth function.

2. (strict *or* logical) Statements A and B are strictly equivalent if they must have the same *truth value (i.e. if it is impossible for them to have different truth values). The strict equivalence of A and B is symbolized by $A \Leftrightarrow B$ and defined within a modal logic as $\Box(A \equiv B)$. *See* implication.

equivalence class If R is an *equivalence relation defined on the *set A then the equivalence class of any element $x \in A$, denoted by $[x]$, is the set of elements to which x is related by the equivalence relation R:

$$[x] = \{y : x \,R\, y\}$$

For example, if R is the equivalence relation 'the same height as', then the equivalence class of the element $x \in A$ consists of all elements of A with the same height as x. The equivalence relation R will also *partition the set A into the equivalence classes of A. Thus, if $A = \{u, v, w, x, y, z\}$, and if u, v, and w are of the same height, and x, y, and z are of the same height but different from u, v, and w, then $\{u, v, w\}$ and

$\{x, y, z\}$ constitute the equivalence classes of A.

equivalence principle The principle stating that, on a local scale, the physical effects of a uniform acceleration of some *frame of reference imitate completely the behaviour in a uniform gravitational field. For those on board a spacecraft far out in space, isolated from any gravitational field, everything (including themselves) would be weightless. If the spacecraft were given a uniform acceleration, corresponding to the *acceleration of free fall on earth, then everything in it would behave as if the spacecraft were stationary on earth. The principle of equivalence of these two frames of reference was introduced by Albert Einstein in his general theory of *relativity.

equivalence relation A *relation that is *reflexive, *symmetric, and *transitive on a set is an equivalence relation on that set. Examples of equivalence relations are parallelism between straight lines, congruence between figures, equality between numbers, and congruence modulo n.

equivalent (of sets) *See* equinumerable.

equivalent matrix *See* matrix.

eradius *See* exradius.

Eratosthenes of Cyrene (*c*.275–194 BC) Greek astronomer and mathematician who proposed as a means of collecting prime numbers the so-called *sieve of Eratosthenes. He is also remembered for his ingenious determination of the circumference of the earth. This he based on the observation that at midday at Syene (now Aswan) the sun is vertically overhead, while at the same time at Alexandria the rays make an angle of $7.2°$ with the vertical. He estimated the distance between Syene and Alexandria from the time taken for a camel train to make the journey, and thereby calculated the circumference of the earth. It is uncertain just how accurate his result was because the exact size of the unit used (the stade) is unknown. Eratosthenes also measured the angle of the obliquity of the *ecliptic.

Erdős, Paul (1913–96) Prolific Hungarian mathematician who was one of the best problem-solvers of the 20th century and made important contributions to number theory and combinatorics. He led a peripatetic life and wrote papers with very many different collaborators. In 1949, he and Atle Selberg found an elementary, if complicated, proof of the *prime number theorem. Most of his work came from his fascination with problems that were easy to state but difficult to solve.

erf *See* error function.

erg Symbol: erg. A *c.g.s. unit of work, equal to the work done by a force of 1 dyne acting through a distance of 1 centimetre. $1 \text{ erg} = 10^{-7}$ joule.

ergodicity The property of many time-dependent processes, such as certain *Markov chains, that the eventual (limiting) distribution of states of the system is independent of the initial state.

error 1. (in numerical computation) Errors are of three main types:

Rounding (or *roundoff*) *errors* are caused by approximating numbers by ones with fewer digits, and are an inevitable consequence of working with finitely many digits. These errors are dangerous for two reasons. First, in a computer calculation consisting of thousands or even millions of elementary operations (additions, subtractions, multiplications, divisions), small rounding errors can accumulate to produce large errors. Second, and more insidiously, a single rounding error can give rise to a large error. For example, if we evaluate $1/(1 - \cos 1°)$ in four-decimal-digit arithmetic, we obtain 5000, whereas the correct answer is 6565.8 (to one decimal place); the rounding error in the computed value of $\cos 1°$ is amplified by the subtraction and division.

Truncation errors are associated with essential limitations in the construction of approximations. They may arise from the use of an approximation rule, terminating an iterative method before it has converged, approximating a derivative by a difference (*see* numerical differentiation), or taking only finitely many terms of a series expansion such as a Taylor series. A truncation error also arises from the *truncation of a number.

Expressions and upper bounds for truncation errors are often available. For example, if the *trapezoidal rule is used to integrate f(x) over $[a, b]$ using n subintervals of length h, the truncation error E (integral minus approximation) is given by

$$E = -h^2(b - a)f''(\alpha)/12$$

where the second derivative f$''$ is evaluated at some unknown point α in $[a, b]$. If the trapezoidal rule is used to calculate the integral of $\cos x$ over $[0, 0.8]$ with $n = 8$ and $h = 0.1$, then, since $|f''(x)| \leqslant 1$, the truncation error cannot exceed $(0.1)^2 \times 0.8/12 = 0.000\,67$ in magnitude.

Data errors are errors that are inherent in the numbers that form the input to a computation. They may arise from errors in measuring physical quantities, or from rounding errors incurred in storing the numbers on a computer.

2. (in statistics) A *random error* is the discrepancy between an observed value and the value predicted by some appropriate *model, and represents uncontrolled *variation. In many practical situations errors are assumed to be independent and to have a *normal distribution with zero mean. *See also* hypothesis testing; residuals.

error-correcting code A form of encryption that identifies errors and corrects messages corrupted during transmission.

A *code C is k-error-correcting if up to k errors can be corrected. For example, if the *Hamming distance between any two codewords of C is at least $2k + 1$, then C

is k-error-correcting. A well-constructed code can be smaller and still be k-error-correcting. *See* coding theory.

error-detecting code A form of encryption which allows errors in messages to be identified during transmission, but not necessarily corrected. Such a *code is useful if the message can be resent easily. *See also* error-correcting code.

error function The function

$$\text{erf}(x) = \frac{2}{\sqrt{\pi}} \int_0^x \exp(-t^2)\,\mathrm{d}t$$

It is closely related to the standard *normal distribution cumulative distribution function $\Phi(x)$, since

$$\Phi(x) = \frac{1}{2}\left[1 + \text{erf}\left(\frac{x}{\sqrt{2}}\right)\right]$$

In applied mathematics, physics, and astronomy, the error function notation is widely used, while in probability and statistics the normal distribution is preferred.

The function

$$\text{erfc}(x) = 1 - \text{erf}(x)$$

$$= \frac{2}{\sqrt{\pi}} \int_x^\infty \exp(-t^2)\,\mathrm{d}t$$

is called the *complementary error function.* A *Taylor expansion of erf(x) valid for all real x takes the form

$$\text{erf}(x) = \frac{2}{\sqrt{\pi}}\left(x - \frac{x^3}{3 \times 1!} + \frac{x^5}{5 \times 2!}\right.$$
$$\left. - \frac{x^7}{7 \times 3!} + \frac{x^9}{9 \times 4!} - \cdots\right)$$

If measurement errors are normally distributed with *mean zero and *standard deviation σ, then erf $(x/\sigma\sqrt{2})$ is the probability that a single measurement error will lie in the interval $(-x, x)$.

error mean square In *analysis of variance, an *unbiased estimator of error *variance. It is the error (or residual) sum of squares divided by the error degrees of freedom. It is used as the denominator in *F-tests and is an estimator of the true error variance σ^2 which is used to construct *confidence intervals.

errors of the first and second kind *See* hypothesis testing.

error sum of squares *See* analysis of variance.

escape speed (escape velocity) The minimum speed at which an object must be propelled from a celestial body (such as the earth) in order to escape its vicinity, i.e. to avoid going into orbit around it or returning to its surface under the action of the body's gravitational field. It is equal to $\sqrt{(2MG/R)}$, where G is the gravitational constant and M and R are the mass and radius of the celestial body (assumed to be spherical). The escape speed from the earth's surface is about $11.2 \, \text{km s}^{-1}$.

escribed circle *See* excircle.

essentially bounded function A *function f for which there exists a number K such that the *set $\{x: |f(x)| > K\}$ has *measure zero. The *essential supremum* of $|f(x)|$ is the *greatest lower bound of all possible K, and is written as essup$|f(x)|$.

essential map *See* homotopy.

essential singularity *See* singular point.

essential supremum (essup) *See* essentially bounded function.

estimation The use of an *estimator to estimate a population parameter. The numerical value of an estimator calculated from a particular sample is called a *point estimate*; a *confidence interval is an *interval estimate*. A $100(1 - \alpha)$ percent confidence interval for a parameter contains all the values of that parameter that would be accepted under a (usually two-tail) hypothesis test (*see* hypothesis testing) at significance level α if the test were made using the given sample.

estimator A *statistic used to provide an estimate of a *parameter. For example, the sample mean $\bar{x}$ is an unbiased estimator of the normal population mean μ. The term *estimator* refers to the statistic $\bar{x}$; its value, 12.37, say, in a specific case is called an *estimate*. For a sample of size n, an estimator T_n of a population parameter θ is *consistent* if, for large n, T_n converges in probability to θ. i.e. if

$$\lim_{n \to \infty} \Pr\{|T_n - \theta| \geqslant \varepsilon\} = 0 \quad \text{for all } \varepsilon > 0$$

See unbiased estimator.

Euclid (c.300–260 BC) Greek mathematician and author of one of the most famous texts in the whole of mathematics, *Stoicheion* or *Elements*. In 13 books it covers the geometry of the triangle, the circle, various quadrilaterals, Eudoxus' theory of proportion, elementary number theory, irrational numbers, and solid geometry. The treatment throughout is axiomatic and based upon definitions, postulates, and 'common notions'. Important results established include the infinity of the primes (Book IX: 20), the fundamental theorem of arithmetic (Book IX: 14), Pythagoras' theorem (Book I: 47), the *Euclidean algorithm (Book VII), the existence of irrational numbers (Book X), and the construction of the five Platonic solids (Book XIII). Despite difficulties with the fifth postulate, the so-called *Euclidean geometry of the *Elements* survived unquestioned until the 19th century when the *non-Euclidean geometry of Bolyai and Lobachevsky was formulated. In addition to several other mathematical works, most of which are lost (including a work on conics), Euclid also wrote on astronomy, optics, and music.

Euclidean algorithm A systematic procedure for finding the highest *common factor (HCF) of two given natural numbers:

(1) If the two numbers are equal, their common value is also their HCF. Otherwise apply step 2.

(2) Divide the smaller number into the larger (possibly with a remainder).

(3) If the division at step 2 is exact then the divisor is the HCF of the original two numbers.

(4) If the division at step 2 is not exact, ensure that the remainder is smaller in absolute value than the divisor. The HCF of the original two numbers is the same as the HCF of the current divisor and the absolute value of the current remainder; so begin again at step 2 with these smaller numbers.

A very easy application of the algorithm is to find the HCF of 34 and 102. Here the algorithm stops after the first application of step 2 above, and the HCF is 34 since $102 = 3 \times 34$.

Another example, where two divisions are needed, is to find the HCF of 52 and 273. In this case

$$273 = 5 \times 52 + 13$$

$$52 = 4 \times 13$$

so the required HCF is 13.

The process always terminates, although several repetitions of steps 2 and 4 may sometimes be needed. For instance, the calculation of the HCF of 595 and 721 proceeds thus:

$$721 = 1 \times 595 + 126$$
$$595 = 5 \times 126 - 35$$
$$126 = 4 \times 35 - 14$$
$$35 = 2 \times 14 + 7$$
$$14 = 2 \times 7$$

so the HCF of the original two numbers is 7.

When the successive divisions are set out in order as above, the desired HCF is always the (absolute value of) the last non-zero remainder. At each division with remainder there is a choice between a positive and a negative remainder, but it is quicker always to choose the one that has the smallest absolute value. When positive remainders are always chosen, *Lamé's theorem* (G. Lamé, 1844) asserts that the number of steps taken by the algorithm never exceeds 5 times the number of decimal digits in the smaller of the original numbers.

Some variations on the algorithm have been discovered which allow faster computer evaluation.

Euclidean construction A geometrical construction that may be carried out using only an unmarked straightedge and compasses. For example, there is a Euclidean construction for the bisection of an angle, but not for its *trisection. *See* Mascheroni; Fermat numbers; duplication of the cube; squaring the circle.

Euclidean geometry *Geometry based on the definitions and axioms set out in Euclid's *Elements*. Book I starts out with 23 'definitions' of the type 'a point is that which has no part' and 'a line is a length without breadth'. Then follow ten axioms, which Euclid divided into five 'common notions' and five propositions. His common notions were:

(1) Things that are equal to the same thing are equal to one another.

(2) If equals are added to equals, the wholes are equal.

(3) If equals are subtracted from equals, the remainders are equal.

(4) Things that coincide with one another are equal to one another.

(5) The whole is greater than the part.

Euclid's postulates were:

(1) A straight line can be drawn from any point to any other point.

(2) A straight line can be extended indefinitely in any direction.

(3) It is possible to describe a circle with any centre and radius.

(4) All right angles are equal.

(5) If a straight line falling on two straight lines makes the interior angles on the same side less than two right angles, then the two straight lines, if produced indefinitely, will meet on that side on which the angles are less than two right angles.

With these basic assumptions, Euclid went on to prove propositions (theorems) about geometrical figures. Euclid's system of geometry was regarded as logically sound for 2000 years, although in fact it contained many concealed assumptions. In 1899, Hilbert, in *Grundlagen der Geometrie* (Foundations of Geometry), recast Euclidean geometry using three undefined entities (point, line, and plane). He introduced 28 assumptions, known as *Hilbert's axioms*. *See also* non-Euclidean geometry; parallel postulate.

Euclidean metric *See* Euclidean space.

Euclidean norm *See* norm (of a vector space).

Euclidean plane *See* Euclidean space.

Euclidean space Symbol: $\mathbb{R}^n$. For a fixed natural number n, $\mathbb{R}^n$ is the *set of all *n-tuples* $(x_1, \ldots, x_n)$ of real numbers $x_1, \ldots, x_n$, together with the operations of addition of pairs of n-tuples and multiplication of any n-tuple by any real number k, and a *norm for each n-tuple. These are defined by

$$(x_1, \ldots, x_n) + (y_1, \ldots, y_n)$$
$$= (x_1 + y_1, \ldots, x_n + y_n)$$
$$k(x_1, \ldots, x_n) = (kx_1, \ldots, kx_n)$$
$$\|(x_1, \ldots, x_n)\| = \sqrt{(x_1^2 + \cdots + x_n^2)}$$

The first two operations make $\mathbb{R}^n$ an n-dimensional *vector space, and the norm leads to a distance function

$d(\mathbf{x}, \mathbf{y}) = \|\mathbf{x} - \mathbf{y}\|$, where $\mathbf{x}$ denotes the n-tuple $(x_1, \ldots, x_n)$ and $\mathbf{y}$ denotes the n-tuple $(y_1, \ldots, y_n)$. The distance function $d(\mathbf{x}, \mathbf{y})$ is called the *Euclidean metric*.

The ordered pairs in $\mathbb{R}^2$ can be identified with geometrical points in a plane relative to fixed coordinate axes, so $\mathbb{R}^2$ is often called the *Euclidean plane*.

Euclid's proof (of the infinity of primes) Suppose that $p_1, p_2, \ldots, p_n$ is any finite list of *primes, and then form the number

$$N = 1 + p_1 \times p_2 \times \cdots \times p_n$$

Then N cannot be divisible by any of the primes $p_1, \ldots, p_n$, for a remainder of 1 is left whenever we try to divide by one of them. On the other hand, N is bigger than 1 and is either a prime number itself or is divisible by primes not in the given list. In either case, this demonstrates the existence of a prime p not in the original list. So the set of all primes cannot be contained in any finite list, and this is the required result.

Eudoxus of Cnidus (*c.*400–*c.*350 BC) Greek mathematician and astronomer noted for his introduction of the method of *exhaustion to determine areas bounded by curves. The theory of proportion in Book V of Euclid's *Elements* is also supposed to have been derived from the lost work of Eudoxus.

Euler, Leonhard (1707–83) Swiss mathematician who in his numerous works made major contributions to virtually every branch of the mathematics of his day. He published works on analysis (1748), the differential calculus (1755), the integral calculus (1768–70), the calculus of variations (1744), planetary motion (1744), and the moon's orbit (1753), as well as writing hundreds of memoirs. Amongst the many new symbols Euler introduced were the signs i for $\sqrt{-1}$, Σ for summation, and the functional notation f(x). Specific achievements were his theorem on polyhedra, his work on graph theory, his method for solving biquadratic equations, and his phi function for determining the number of positive integers not greater than and prime to a given number n. Not the least of Euler's achievements was his work in mechanics, notably his treatise of 1736, with which began the long struggle to introduce analytically rigorous methods into the discipline.

Euler characteristic For a surface, the number $V - E + F$, where V is the number of vertices, E the number of edges and F the number of faces of a *triangulation of the surface. The Euler characteristic is a property of the surface and not of the particular triangulation (as long as the faces are homeomorphic to discs). In particular, for the surface of a convex polyhedron, it is always the case that $V - E + F = 2$. If a surface is made up of h *handles and c *cross-caps, then its Euler characteristic is $2 - 2h - c$.

Eulerian graph A connected *graph is Eulerian if it contains a closed *walk which includes every one of its edges once and once only, i.e. a closed trail or circuit including every edge (*see* diagram (a)). The term arises from Euler's negative solution to the problem of finding a walk around Königsberg which crossed each of seven bridges exactly once and returned to its starting point (*see* Königsberg bridge problem).

A connected graph that contains a trail (not necessarily closed) which includes every one of its edges is called *semi-Eulerian* or *traversable* (*see* diagram (b)). This is equivalent to saying that one can draw the graph without lifting the pen from the paper and not retracing any edge. A connected graph is Eulerian if and only if the *degree of every vertex is even; a connected graph is semi-Eulerian if and only if there are 0 or 2 vertices of odd degree.

Compare Hamiltonian graph.

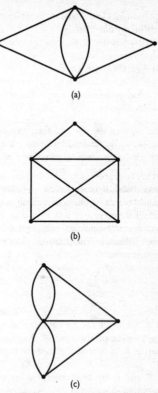

(a)

(b)

(c)

Eulerian graph (a) Eulerian, (b) semi-Eulerian, and (c) non-Eulerian graphs.

Euler–Lagrange equation *See* calculus of variations.

Euler line In any triangle that is not equilateral, the *circumcentre O, *centroid G, and *orthocentre H lie on a straight line, the *Euler line* of the triangle, and $OG = 2GH$.

Euler–Maclaurin summation formula A formula for the error in the repeated *trapezoidal rule. If $T(h)$ denotes the repeated trapezoidal rule approximation to $\int_a^b f(x)\,dx$ based on n subintervals of width h, then

$$T(h) - \int_a^b f(x)\,dx = \frac{h^2}{12}(f'(b) - f'(a))$$

$$- \frac{h^4}{720}(f'''(b) - f'''(a)) - \cdots$$

$$- \frac{B_{2r}}{(2r)!}h^{2r}(f^{(2r-1)}(b) - f^{(2r-1)}(a)) - \cdots$$

The coefficients B_{2r} are the *Bernoulli numbers. The Euler–Maclaurin summation formula has various uses, one of which is to estimate sums of series.

Euler–Poincaré characteristic A generalization of the Euler characteristic (*see* Euler's theorem).

Given a *simplicial complex K, the Euler–Poincaré characteristic $\chi(K)$ is defined to be

$$\sum_{n \geqslant 0} (-1)^n \alpha_n$$

where α_n is the number of n-simplexes of K. Since $\chi(K)$ is the Lefschetz number (*see* fixed-point theorem) of the identity map of K, it depends only on the homology groups of K, and so $\chi(K) = \chi(L)$ if K and L are homeomorphic (or even homotopy-equivalent).

In essence, the Euler–Poincaré characteristic is due to Euler, who observed that $\chi(K) = 2$ for regular polyhedra K in $\mathbb{R}^3$. Euler's original definition was extended by Cauchy (1813) and Poincaré (1895).

See combinatorial topology.

Euler's constant Symbol: γ. The limit of

$$\sum_1^n \frac{1}{r} - \ln n$$

as $n \to \infty$. To four decimal places, its value is 0.5772.

Euler's criterion *See* residue.

Euler's equations Three *differential equations expressing the motion of a *rigid body rotating about a fixed point, O, with *angular velocity $\boldsymbol{\omega}$. The forces on the body have *moment $\mathbf{M}$ about O. Euler's equations involve the components of moment along the principal axes of the body:

$$I_1 \frac{\partial \omega_1}{\partial t} - (I_2 - I_3)\omega_2 \omega_3 = M_1$$

$$I_2 \frac{\partial \omega_2}{\partial t} - (I_3 - I_1)\omega_3 \omega_1 = M_2$$

$$I_3 \frac{\partial \omega_3}{\partial t} - (I_1 - I_2)\omega_1 \omega_2 = M_3$$

where I_1, I_2, and I_3 are the principal *moments of inertia at O, and ω_1, ω_2, and ω_3 are the components of angular velocity along the principal axes.

Euler's formula The formula

$$e^{ix} = \cos x + i \sin x$$

It was introduced by Euler in 1748, and is used as a method of expressing *complex numbers. The special case in which $x = \pi$ leads to the formula $e^{i\pi} = -1$.

Euler's identities Three identities (*c*.1748) relating the trigonometric functions, exponential function, and i, the square root of -1:

$$\sin x = (e^{ix} - e^{-ix})/2i$$

$$\cos x = (e^{ix} + e^{-ix})/2$$

$$e^{ix} = \cos x + i \sin x$$

They are derived from the series for $\cos x$, $\sin x$, and e^x. *See also* hyperbolic functions.

Euler's method A numerical method for solving differential equations of the form

$$\frac{dy}{dx} = f(x, y)$$

given an *initial condition $y(a) = y_a$. Euler's method generates a sequence of approximations $y_n \approx y(x_n)$, in which y_{n+1} is obtained from y_n by the formula

$$y_{n+1} = y_n + h f(x_n, y_n), \quad n = 1, 2, \ldots$$

where $y_1 = y_a$ and $x_{n+1} = x_n + h$, and h is a positive step-size. For example, for the problem

$$y' = (1-x)y + \cos x, \quad y(0) = 1$$

Euler's method generates $y_1 = 1 + 2h$ as an approximation to $y(h)$.

Euler's phi function (phi function, totient function) (L. Euler, 1760) For a given natural number n the notation $\phi(n)$ indicates the number of natural numbers not exceeding n and *relatively prime to n. For example, $\phi(20) = 8$.

Euler's theorem uses this function to generalize *Fermat's theorem as follows. If a is any integer that is relatively prime to the natural number n, then $a^{\phi(n)} - 1$ is divisible by n (or equivalently, $a^{\phi(n)} \equiv 1 \pmod{n}$). Fermat's theorem results when n is a prime, since then $\phi(n) = n - 1$.

Foremost among the many other properties of Euler's function is the fact that it is *multiplicative: if m and n are relatively prime, then $\phi(mn) = \phi(m) \cdot \phi(n)$. This leads to the formula that if $p_1, \ldots, p_r$ are the distinct primes dividing n, then

$$\phi(n) = n \left(1 - \frac{1}{p_1}\right)\left(1 - \frac{1}{p_2}\right) \cdots \left(1 - \frac{1}{p_r}\right)$$

Euler's theorem 1. (for polyhedra) The relationship

$$V - E + F = 2$$

for any simple closed *polyhedron, where V is the number of vertices, E the number of edges, and F the number of faces. (A simple closed polyhedron is one that is topologically equivalent to a sphere.) *See also* Euler characteristic.
2. *See* curvature.
3. *See* Euler's phi function.

Euler's theorem For this triangular prism $V = 6$, $E = 9$, $F = 5$; and $V - E + F = 2$.

Eve The name conventionally used for a third party who might intercept and try to decode a message (an eavesdropper).

even function A *function f such that $f(-x) = f(x)$ for every x in the *domain. For example, $f(x) = x^2$ is an even function. The *graph of the function is symmetrical about the y-axis. *Compare* odd function.

even number An integer that is divisible by 2.

even permutation A *permutation that is equivalent to an even number of *transpositions. For example, 312 is an even permutation of 123 since it is equivalent to two transpositions: (13) and then (12). *Compare* odd permutation.

event A *subset of the *sample space of all possible outcomes of an experiment. If the outcome of a particular experiment belongs to a subset A, then A has occurred. If a die is cast and the sample space S represents all possible scores, and A the event 'score is even', then $S = \{1, 2, 3, 4, 5, 6\}$ and $A = \{2, 4, 6\}$. If we cast a die and score 4, the event A has occurred, but if we score 5 the event A has not occurred. The complement of A, denoted by A' or $\bar{A}$, represents the event 'A has not occurred'. The whole space S represents a *certain* or *sure* event, and $\Pr(S) = 1$.

evolute A curve that is the *locus of the *centres of curvature of a given curve. The given curve is the *involute of the evolute. The evolute of a circle is a point (its centre), and this is regarded as a degenerate case. The semicubical parabola $4x^3 = 27ay^2$ is the evolute of the parabola $y^2 = 4ax + 8a^2$ (*see* diagram).

evolution The process of extracting a *root of a number or equation. *Compare* involution.

exa- *See* SI units.

exact division Division in which there is zero remainder.

exact equation A type of *differential equation in which the *total differential of a function is equal to zero. Thus, if $z = \mathrm{f}(x, y)$,

$$\frac{\partial z}{\partial x}\,\mathrm{d}x + \frac{\partial z}{\partial y}\,\mathrm{d}y = 0$$

is an exact equation. An equation

$$A\,\mathrm{d}x + B\,\mathrm{d}y = 0$$

is exact if $\partial A/\partial y = \partial B/\partial x$.

excentre (ecentre) The centre of an *excircle of a triangle. *Compare* incentre.

excircle (escribed circle) A circle lying outside a given triangle *tangent to one of the sides and to the other two sides extended. The bisector of the angle between the two extended sides passes through the centre of the excircle. *Compare* incircle.

excluded middle, law or **principle of the** The *theorem of the *propositional calculus $A \lor \sim A$, i.e. the principle that for any statement A, the statement 'A or not A' is always true. *See also* bivalence.

exhaustion A method of treating areas and volumes of curved figures, dating back to Eudoxus of Cnidus (*c.*360 BC). Earlier mathematicians had considered the idea of finding areas of curved figures by approximating them by rectilinear figures. For example, if a circle is taken with an inscribed polygon and an escribed poly-

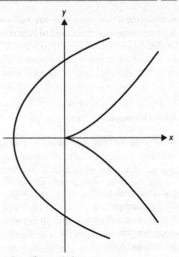

evolute of a parabola.

gon, the area of the circle must lie between the areas of the two polygons. Moreover, the more sides are taken for the polygons, the nearer they approximate the true area of the circle. Before Eudoxus, Greek mathematicians had no way of using this approach as they did not have the concept of a limit. Eudoxus is generally credited with the idea that, given a magnitude, if at least half the magnitude is subtracted and at least half subtracted from the remainder, and so on, then ultimately the remainder will be less than any preassigned magnitude. In modern notation, for a magnitude a and a ratio $0.5 \leqslant r < 1$, the limit of $a(1 - r)^n$ is zero as $n \to \infty$. Eudoxus used this concept to prove theorems about areas and volumes, for example to show that the volume of a cone is one-third of the volume of a cylinder with the same base and height.

existential import The existential commitment of particular kinds of proposition. A singular proposition such as 'Some fractions are reducible' is assumed to have existential import, and to be interpreted to mean that 'There exists at least one fraction which is reducible.' Universal propositions of the form 'All A is B' are held to carry no existential import and to be translated as 'If anything is an A then it is a B.' As a consequence, as there are no French kings, the universal statement 'All French kings are bald' will have an empty subject term, and will be true. *See* syllogism.

existential quantifier *See* quantifier.

exogenous variables *See* endogenous variables.

exp *See* exponential function.

expanded number A number written as a sum of multiples of powers of its *base. For instance, the number 163 (in decimal) can be written as $(1 \times 10^2) + (6 \times 10^1) + 3$ in expanded form.

expansion 1. A mathematical expression that is written as the sum of a number of terms. Expansion is also the process of putting an expression in this form, for example the expansion of $(x + 1)^3$ to give $x^3 + 3x^2 + 3x + 1$. The method of expanding such brackets is to take them in pairs and use the distributive law, thus:

$$(x + 1)^2 = (x + 1)](x + 1)$$

$$= x(x + 1) + 1(x + 1)$$

$$= x^2 + 2x + 1$$

$$(x + 1)^3 = (x + 1)(x^2 + 2x + 1)$$

$$= x(x^2 + 2x + 1)$$

$$+ 1(x^2 + 2x + 1)$$

$$= x^3 + 3x^2 + 3x + 1$$

The expansion of a function is the form of a function when it is represented as a sum of terms, i.e. as a finite series or as an infinite series that converges to the function for certain values of the variables (*see* convergent series). For example,

$$\cos 4x = 1 - 8 \cos^2 x + 8 \cos^4 x$$

is the expansion of $\cos 4x$ in terms of $\cos x$,

$$\sin x = \frac{x}{1!} - \frac{x^3}{3!} + \frac{x^5}{5!} - \frac{x^7}{7!} + \cdots$$

for all x (in radians), and

$$\ln(1 + x) = x - \frac{x^2}{2} + \frac{x^3}{3} - \frac{x^4}{4} + \cdots$$

for $-1 < x \leqslant 1$

The expansion of a *determinant is the conversion of the determinant to an expression containing determinants of lower order.

2. (of a number) The expansion of a number to a given *base is its representation in the *number system with that base. For example, the expansion of the number 5 to base 2, its *binary expansion*, is 101; and the expansion of 3/4 to base 10, its *decimal expansion*, is 0.75, while its binary expansion is the binary number 0.11.

expectation (expected value) The first *moment about the origin for a *random variable. Denoted by $E(X)$, it is also called the *mean value* of X. For a discrete random variable taking a finite or countably infinite set of values x_i with probabilities p_i,

$$E(X) = \sum_i p_i x_i$$

and for a continuous random variable with *frequency function $f(x)$,

$$E(X) = \int_{-\infty}^{\infty} x f(x) \, dx$$

The expected value of a function $g(X)$ of X is defined as

$$E(g(X)) = \sum_i g(x_i) p_i$$

or

$$E(g(X)) = \int_{-\infty}^{\infty} g(x) f(x) \, dx$$

as appropriate. *See also* moment; characteristic function.

experimental design   In comparative experiments where the aim is to compare the effect of administering two or more treatments such as different medicines, fertilizers, or lubricants to units such as patients, plots of land, or machines, the *efficiency and *precision of an experiment may often be improved by certain groupings of the units. The set of rules used for grouping units defines an experimental design. Any one group should consist of units that are as alike as possible in characteristics that may affect responses other than the applied treatments. Treatments are then compared within each group, often by means of the *analysis of variance. The relevant analysis removes the effect of differences – other than those resulting from the applied treatments – between units in different groups. Well-known and widely used designs include *randomized blocks, *balanced incomplete blocks, one or more *Latin squares, and *Youden squares. *See also* factorial experiments.

explanatory variable   *See* generalized linear model; regression.

explement   *See* conjugate angles.

explicit function   A *function defined by $y = f(x_1, x_2, \ldots, x_n)$ where y is the *dependent variable. An example is

$$y = x_1^2 + 2x_2 + x_2 x_3^2$$

Compare implicit function.

exploratory data analysis (EDA)   (J.W. Tukey, 1977) A term used to describe a preliminary examination of *data by operations such as grouping, graphing, and tabulation in a way that will highlight their general structure and characteristics and which will often detect *outliers. Tools such as *box-and-whisker diagrams, *five-number summaries, and histograms are widely used EDA tools. For multivariate data or large collections of data, methods such as *kernel density estimation are often used. EDA is often helpful in deciding which formal statistical analyses may be appropriate.

exponent (index)   A number placed in a superscript position to the right of another number or variable to indicate repeated multiplication: a^2 indicates $a \times a$, a^3 indicates $a \times a \times a$, etc. Sometimes 'power' is used instead of exponent; more strictly, *power* is the result of the multiplication – for instance, 4 is the second power of 2 (i.e. 2^2). If the exponent is negative then the expression is the reciprocal of the number with a positive value of the exponent: for example, $x^{-n} = 1/x^n$. Any number with an exponent of zero is equal to 1 ($x^0 = 1$). Certain *laws of exponents* (*laws of indices*) apply:

(1) *multiplication*: $a^n a^m = a^{n+m}$;
(2) *division*: $a^n/a^m = a^{n-m}$;
(3) *raising to a power*: $(a^n)^m = a^{nm}$.

Fractional exponents are defined by $a^{m/n} = \sqrt[n]{(a^m)}$.

exponential curve   A curve with an equation of the form $y = a^x$.

exponential distribution   The distribution of a random variable X with frequency function $f(x) = k e^{-kx}, x \geqslant 0$. The mean is $E(X) = 1/k$ and the variance $Var(X) = 1/k^2$. For a *Poisson process in which events occur at a rate of k per unit time, the intervals between successive events (the *waiting times*) are exponentially distributed with mean $1/k$. For example, if events occur at a mean rate of 4 per hour, the times between events are exponentially distributed with mean $\frac{1}{4}$, i.e. the expected time between events is 15 minutes. The distribution is a special case of the *gamma distribution.

exponential family of distributions   Distributions are said to belong to the exponential family if they have *frequency functions of the form

$$f(x, h) = \exp(a(x)b(h) + c(h) + d(x))$$

where h is a parameter. Distributions of this form are important in *generalized linear models, and the *normal distribution, *Poisson distribution, and *binomial distribution all belong to this family. Any parameters other than h are regarded as known.

exponential function The function $\exp x$ or e^x (*see* e). The term is also used for functions of the type a^x (where a is a constant) or, more generally, a function having variables expressed as exponents. For all x,

$$e^x = \lim_{n \to \infty} \left(1 + \frac{x}{n}\right)^n$$

See also exponential series.

exponential growth *or* **decay** A growth pattern for which some quantity y (e.g. weight, length of an organism, or numbers of individuals in an animal population) at time t is given by the equation $y = ae^{bt}$, where a and b are constants and $a > 0$. The rate of growth or decay is given by $dy/dt = abe^{bt} = by$, and is thus proportional to size. For growth $b > 0$; for decay $b < 0$.

exponential notation (standard form, index notation, scientific notation) A method of writing numbers as a product of a number between 1 and 10 and a power of 10. For instance, 1056 in exponential notation is 1.056×10^3.

exponential series The *series

$$\sum \frac{x^n}{n!} = 1 + x + \frac{x^2}{2} + \frac{x^3}{6} + \cdots$$

The series converges (absolutely) for all x. Its sum is a function of x: the exponential function, e^x. The exponential series is therefore an *expansion of the exponential function.

exponential time *See* polynomial time.

expression Any mathematical form expressed symbolically, as in an equation, polynomial, etc.

exradius (eradius) The radius of an *excircle of a triangle. *Compare* inradius.

extended complex plane Symbol: $\mathbb{C}_\infty$. A *set consisting of $\mathbb{C}$, the set of *complex numbers, and a symbol, denoted by '∞', which is not in $\mathbb{C}$.

The elements of $\mathbb{C}_\infty$ may be represented by the points on a sphere, as follows. Consider a sphere resting on a *complex plane with its south pole at the origin of the plane. A line is drawn from the point (a, b) in the plane to the north pole of the sphere. This line meets the sphere at a point which is uniquely determined by the point (a, b), and so by the complex number $a + ib$.

Every complex number thus corresponds to a point on the sphere below the north pole, and conversely every point on the sphere, apart from the north pole, corresponds to a complex number. The north pole is then identified with the symbol ∞, and the whole sphere, identified with $\mathbb{C}_\infty$ in this manner, is called a *Riemann sphere*.

The extended complex plane is a topological space in which the neighbourhoods of ∞ are defined as the complements in $\mathbb{C}_\infty$ of the closed and bounded subsets of $\mathbb{C}$.

extension field If a *field L contains another field K, and the field operations in K are the same as those in L (when they are just applied to elements of K), then L is an extension field of K and K is a *subfield of L. For instance, the field of *real numbers is an extension field of the field of *rational numbers, and is itself a subfield of the field of *complex numbers.

extensive definition An attempt to define a term by listing the elements to which it correctly applies. Thus the set of regular convex polyhedra is {tetrahedron, cube, octahedron, dodecahedron, icosahedron}. *Compare* intensive definition.

exterior (of a set) *See* frontier.

exterior angle 1. An angle formed outside a *polygon between one side and another side extended.
2. *See* transversal.

external angle An *exterior angle of a polygon.

external force Any *force that originates outside a particular system of particles considered as a whole. External forces can be distinguished from *internal forces, which arise from mutual interactions between the particles of the system and cancel each other out when the whole system is considered.

external tangent *See* common tangent.

extraction (of roots) The process of finding a *root or roots. For example, extracting the cube root of 27 is the process of finding its cube root (3). Extracting the root of an equation is the process of finding a number or numbers that satisfy the equation.

extrapolation If the values $y_1, y_2, \ldots, y_n$ of a *function $f(x)$ are known for values $x_1, x_2, \ldots, x_n$ of the independent variable, extrapolation is the process of estimating, from the given data, the value of the function for a further value of x lying outside the given range of x. *See also* interpolation.

extremal A point at which a *function attains an *extremum. The term is often used in the *calculus of variations for a function (thought of as a point in a space of functions) at which a *functional attains an extremum. For example, the *brachistochrone is the extremal for the functional

$$\phi(f) = \int_a^b \sqrt{\left(\frac{1 + (f'(x))^2}{2gf(x)} \right)} \, dx$$

where g is the *acceleration of free fall.

extreme value distribution A distribution associated with the least or greatest values in a *sample, i.e. the *order statistics $x_{(1)}$ and $x_{(n)}$ for a sample of size n. For example, these distributions are relevant to estimating the probability of future floods exceeding a certain magnitude on the basis of records for past floods, or estimating the earliest likely failure time of a specified machine component. The choice of an appropriate extreme value distribution depends on the type of population from which samples are drawn. A well-known extreme value distribution is the *Weibull distribution. Others include the *Frechet* and *Gumbel distributions*.

extremum (*plural* **extrema**) Greatest or least possible. An extremum of a *function is a *maximum or *minimum value of the function. The problem of maximizing or minimizing a function is an *extremum problem*. *See* Fermat point.

F

$\mathbb{F}_q$ Symbol for the *finite field with q elements.

face One of the plane regions bounding a *polyhedron, or the planes forming a *polyhedral angle.

face angle A plane angle between two adjacent edges in a *polyhedral angle.

factor 1. An integer or *polynomial that divides a given integer or polynomial exactly is called a *factor* or *divisor*. Thus, 1, 2, 3, and 6 are all factors of 6; and $x - 1$ and $x + 2$ are factors of $x^2 + x - 2$, since $(x - 1)(x + 2) = x^2 + x - 2$.

In a restricted sense of the definition, the factors of polynomials with rational coefficients must themselves be nonconstant polynomials with coefficients that are rational numbers (as in the above example). More generally, sometimes the factors are taken to include constants, e.g.

$$2x^2 + 2 = 2(x^2 + 1)$$

or to include irrational numbers, e.g.

$$x^2 - 2 = (x + \sqrt{2})(x - \sqrt{2})$$

$$x^2 + y^2 = (x + iy)(x - iy)$$

See also common factor; divisible; factor theorem.
2. *See* factorial experiments.

factorable 1. Of an integer, containing factors other than itself or unity. For instance, $8 \ (= 4 \times 2)$ is factorable. Prime numbers are not factorable.

2. Of a *polynomial, containing factors other than itself or a constant. For example, $x^2 + x - 2$ is factorable into $(x + 2)(x - 1)$.

factor analysis (L.L. Thurstone, 1935) A statistical technique that aims to express p observed *random variables as *linear functions of $m \ (<p)$ factors plus a term representing error (or residual) variation. There are several specifications of the basic problem, and estimation requires a knowledge of the *covariance matrix of the observations. Factor analysis was used originally in psychological experiments to try to explain individual test scores in terms of factors such as verbal ability, arithmetic ability, and manual skill. *See also* principal component analysis.

factor formulae Formulae from plane trigonometry expressing the sums and differences of sines and cosines as products of trigonometric functions:

$$\sin x + \sin y = 2 \sin \tfrac{1}{2}(x + y) \cos \tfrac{1}{2}(x - y)$$

$$\sin x - \sin y = 2 \cos \tfrac{1}{2}(x + y) \sin \tfrac{1}{2}(x - y)$$

$$\cos x + \cos y = 2 \cos \tfrac{1}{2}(x + y) \cos \tfrac{1}{2}(x - y)$$

$$\cos x - \cos y = -2 \sin \tfrac{1}{2}(x + y) \sin \tfrac{1}{2}(x - y)$$

See also product formulae.

factor group *See* normal subgroup.

factorial A number obtained by multiplying all the positive integers less than or equal to a given positive integer. The factorial of a given integer n is usually written as $n!$ (an old notation is $\lfloor n$), i.e.

$$n! = n \times (n-1) \times (n-2) \times \cdots$$
$$\times 3 \times 2 \times 1$$

By convention factorial zero, $0!$, is taken to be unity. *See* gamma function; factorial series; hypergeometric series; Stirling's formula.

factorial experiments (F. Yates, 1934) Experiments in which the treatment structure allows comparisons of several types of treatment, each called a *factor*, at different quantitative or qualitative levels. In an experiment measuring the yield of a chemical process, factor A might represent three different temperatures, $120\,°C$, $150\,°C$, and $180\,°C$, and factor B two different pressures, 1 and 2 atmospheres. The design allows the experimenter to assess whether the effects of each factor are simply additive or whether they *interact* (i.e. are not directly additive). There would be interaction if, for example, yield increased as temperature increased at the lower pressure, but yield decreased as temperature increased at the higher pressure.

The results are analysed by partitioning the between-treatments sum of squares in an *analysis of variance into *main effects* and *interactions*. Designs may involve any number of factors and any number of levels of each factor. In a *randomized block design every factor–level combination appears once in each block. *Efficiency can sometimes be increased by using a device known as *confounding*, which allows the use of blocks containing selected subsets of factor–level combinations. Certain components of interaction, usually assumed to be negligible, then become 'confounded' with differences between blocks. Special analyses are needed for sophisticated factorial designs, some of which may not include all factor–level combinations.

factorial series The infinite series

$$\sum \frac{1}{n!} = 1 + \frac{1}{1!} + \frac{1}{2!} + \frac{1}{3!} + \cdots$$

This is a *convergent series whose sum is the number e, i.e. $2.718\,28\ldots.$

factorization 1. The representation of a number or *polynomial as a product of *factors. If the factors of a number are all prime numbers, then their product is the *prime factorization* of that number. *See* fundamental theorem of arithmetic. 2. (of a matrix) The representation of a matrix as the product of two or more matrices. Important examples are *LU factorization, *Cholesky factorization, and *polar decomposition. *See also* Schur decomposition; singular value decomposition; QR factorization.

factor modulo n A number or *polynomial that divides another number or polynomial modulo n, i.e. a factor of it modulo n (*see* division modulo n). Thus modulo 12, 5 has the two factorizations 1×5 and 7×11, while 8 has the factorizations 1×8, 2×4, 4×5, 2×10, 4×8, 4×11, 7×8, 8×10. Modulo 7, $2x^4 - 4x - 3$ has the factors $2x^2 + 3x + 3$, $x - 2$, and $x + 4$ since

$$(2x^2 + 3x + 3)(x - 2)(x + 4)$$
$$= 2x^4 + 7x^3 - 7x^2 - 18x - 24$$
$$\equiv (2x^4 - 4x - 3) \pmod 7$$

Integers that are *coprime to n are the only numbers that, modulo n, are factors of every integer. These same coprime numbers, regarded as constant polynomials, are the only polynomials that have the similar universal property of dividing every polynomial modulo n.

factor theorem The theorem that for a given *polynomial in x, $x - a$ is a factor if the value of the polynomial is zero when a replaces x throughout. The *remainder theorem reduces to the factor theorem when the remainder is zero.

Fahrenheit degree Symbol: °F. A division of a temperature scale in which the melting point of ice is taken as 32 degrees and the boiling point of water is taken as 212 degrees. This scale has now been largely replaced by the *Celsius scale and, for many scientific purposes, by the *kelvin scale. To convert a Fahrenheit temperature to Celsius the formula used is $C = 5(F - 32)/9$. [After G.D. Fahrenheit (1686–1736)]

fair game A game in which the entry cost or stake equals the expected gain (*see* expectation). In a sequence of fair games between two adversaries the one with the larger capital has the better chance of ruining his opponent. *See* random walk; St Petersburg paradox.

fallacy An *argument or form of argument that is invalid. For example, the argument 'Given $x \geqslant y$ and $y \geqslant z$ it follows that $x > z$' is a fallacy. Since the time of Aristotle, logicians have sought to identify and classify persistent and systematic errors of reasoning usually described as fallacies. The major division is between *formal* and *informal* fallacies. Two formal fallacies when reasoning with *conditionals are *affirmation of the consequent and *denial of the antecedent. Similarly with the *syllogism, amongst a number of formal fallacies there is the *fallacy of four terms*, as in 'All metals are elements. Brass is a metal ∴. Brass is an element', where the term 'metal' is used ambiguously to refer to a pure substance and to an alloy. There are also a large number of informal fallacies whose force is more rhetorical than logical. These include *argumentum ad hominem*, where the man rather than his argument is attacked, for example 'No butcher can be expected to argue honestly about the meat trade'.

false position, rule of (*regula falsi*) In general, a method of successively approximat-

ing a *root of an equation $f(x) = 0$ from an initial estimate or estimates of the root.

In a method of *simple position*, a single estimate a_0 is made and an *iteration of the form $a_{n+1} = g(a_n)$ is used for $n = 1, 2, \ldots$. Examples are the direct iteration and Newton–Raphson methods.

In a method of *double position*, such as successive *linear interpolation, two estimates a_0 and b_0 are found such that $f(a_0)$ and $f(b_0)$ are close to zero but of opposite sign (*see* diagram). These estimates are then used as starting values in the formula

$$a_{n+1} = a_n - \frac{(b_n - a_n)f(a_n)}{f(b_n) - f(a_n)}$$

where, for $n = 0, 1, 2, \ldots, b_{n+1}$ is chosen from a_n and b_n so that $f(b_{n+1})$ is of opposite sign to $f(a_{n+1})$.

false position: linear interpolation.

family 1. A set of curves that are related by a common equation, so that all the curves can be generated by varying one or more parameters in the equation. For example, the equation

$$x^2 + y^2 = r^2$$

represents a family of concentric circles with centres at the origin and different values of r. The equation

$$(x - h)^2 + y^2 = a^2$$

where a is constant, represents a family of circles of equal radius (a) with centres along the x-axis (as h varies). The above cases are both examples of *one-parameter families*. Families of curves can also be generated by varying two or more parameters. Thus, in the second equation above both h and a can be varied to produce the two-parameter family of *all* circles that have their centre on the x-axis. The family of all circles in the plane is a three-parameter family obtained by varying h, k, and r in the equation

$$(x - h)^2 + (y - k)^2 = r^2$$

See also confocal conics.

2. A set of surfaces related by a common equation. As with curves, families of surfaces can be one-parameter, two-parameter, etc. For example, the equation

$$x^2 + y^2 + z^2 = r^2$$

represents a one-parameter family of concentric spheres for different values of r.

farad Symbol: F. The *SI unit of capacitance, equal to the capacitance of a capacitor between the plates of which a potential difference of 1 volt will appear when it is storing 1 coulomb of charge. [After M. Faraday (1791–1867)]

Farey sequence (of order n) (C. Haros, 1802; J. Farey, 1816) The finite *increasing sequence $\mathcal{F}_n$ of *irreducible fractions, between 0 and 1 inclusive, whose denominators do not exceed the *natural number n. Thus $\mathcal{F}_5$ is

$$\frac{0}{1}, \frac{1}{5}, \frac{1}{4}, \frac{1}{3}, \frac{2}{5}, \frac{1}{2}, \frac{3}{5}, \frac{2}{3}, \frac{3}{4}, \frac{4}{5}, \frac{1}{1}$$

One of the main properties of any Farey sequence is that if a/b and a'/b' are two adjacent terms with $a/b < a'/b'$, then $a'b - ab' = 1$. For instance, in $\mathcal{F}_5$ $a/b = 3/5$ and $a'/b' = 2/3$ are two such terms, and $2 \times 5 - 3 \times 3 = 1$.

fast Fourier transform (FFT) *See* discrete Fourier transform.

F-distribution (R.A. Fisher, 1922) The *distribution of the ratio of two independent chi-squared variables (*see* chi-squared

distribution), each divided by its *degrees of freedom. It is used in the *F-test* or *variance ratio test* in an *analysis of variance to test the null hypothesis that two components estimate the same variance against the alternative that the numerator component estimates a greater variance, the latter being indicated by a high F-value. The F-test may also be used to test the acceptability of the hypothesis that two samples are from normal distributions with the same variance. Tables giving critical values at the 0.05, 0.01, and 0.001 significance levels are widely available, and most computer programs for the analysis of variance give *p-values associated with calculated F-values. The distribution was first tabulated in 1934 by the American statistician George Waddel Snedecor (1881–1974), who named it the F-distribution in honour of Fisher. It is sometimes referred to as *Snedecor's F-distribution*.

feasible region, feasible solution *See* linear programming.

Feigenbaum number (M.J. Feigenbaum, 1979) A real number that characterizes the parameter values for which the *logistic map experiences *period doubling. If period doubling takes place at successive parameter values $a_1 = 3.5$, $a_2 = 3.56, \ldots$, $a_n, \ldots$, then

$$\lim_{n \to \infty} \frac{a_n - a_{n-1}}{a_{n+1} - a_n} = 4.669\,201\,660\,9\ldots$$

This is the Feigenbaum number. It is a universal constant, in that for similar maps with period doubling the limit has the same value.

Feit–Thompson theorem A finite *group of *prime order has no subgroups apart from the whole group and the subgroup consisting of the identity element, and is thus a *simple group. W. Feit and J.G. Thompson proved in 1963 that a finite simple group that does not have prime order must have even order.

femto- *See* SI units.

Fermat, Pierre de (1601–65) French mathematician who in his posthumously published *Arithmetica* (1670) established a number of important results in number theory. He was also responsible for some pioneering work on the calculus and devised a general procedure for finding tangents to curves. Further work in his *Isagoge ad locus planos et solidos* (1679, On the Plane and Solid Locus) foreshadowed the later analytic geometry of Descartes and allowed him to define such important curves as the hyperbola and parabola, the spiral of Fermat, and the cubic curve known as the witch of Agnesi. In optics, Fermat formulated the principle of least time. With Pascal, he laid the foundations of probability theory. *See also* Fermat's last theorem.

Fermat numbers (P. de Fermat, 1640) Numbers F_n of the form $2^{2^n} + 1$ where n is zero or a positive integer. The first few are

$$F_0 = 2^{2^0} + 1 = 2^1 + 1 = 3$$

$$F_1 = 2^{2^1} + 1 = 5$$

$$F_2 = 17$$

$$F_3 = 257$$

$$F_4 = 65\,537$$

A Fermat number that is prime is called a *Fermat prime*.

Each Fermat number is *relatively prime to every other Fermat number, and Fermat thought that they are actually all prime, as is the case for the examples above. However, in 1732 Euler found that F_5 is divisible by 641. To this date no one knows whether there are any Fermat primes after F_4.

In 1796 Gauss showed that Fermat numbers have a remarkable connection with geometry, since a regular polygon can be constructed with just unmarked straightedge and compasses if and only if the number of sides of the polygon is a power of 2, or a product of distinct Fermat primes, or a power of 2 multiplied by such a product.

Fermat point For a triangle ABC, the point P in its plane such that PA + PB + PC is a minimum. If all the angles of the triangle are less than 120°, then P lies where the angles BPC, CPA, and APB are all 120°. If the angle at one of the vertices of the triangle is greater than or equal to 120°, then P lies at that vertex.

This problem of finding the point in the plane that minimizes the sum of its distances from the vertices of a given triangle is sometimes called the *Fermat problem* or *Steiner's problem*. It is one of the first instances of an *extremum problem*.

Fermat's last theorem The result that if the *integer n is at least 3, then there are no integers $x, y,$ and z, none of which is zero, satisfying the equation

$$x^n + y^n = z^n$$

This was first conjectured by *Fermat in about 1637, and he said that he had a proof, though it was never found. For over 350 years, work on Fermat's conjecture provided much stimulus to the development of algebraic number theory; after a period of intense activity by the English mathematician Andrew Wiles it was eventually proved in 1995 by Wiles, with assistance from the English mathematician Richard Taylor.

Fermat's spiral *See* spiral.

Fermat's theorem The theorem (P. de Fermat, 1640) that if a is an integer and p is a *prime that does not divide a, then p does divide $a^{p-1} - 1$; or, in *congruence notation, $a^{p-1} \equiv 1 \pmod{p}$. For example, $8^4 - 1$ is divisible by 5. A simple corollary is that, whether p divides a or not, it must divide $a^p - a$: equivalently $a^p \equiv a \pmod{p}$. Chinese mathematicians 2500 years ago were aware that if p is prime then p divides

$2^p - 2$, which is the case $a = 2$. Leibniz was able to prove Fermat's theorem by 1683, but the first published proof was given in 1736 by Euler, who subsequently generalized the result (*see* Euler's phi function). The theorem is sometimes known as *Fermat's little theorem* to distinguish it from his celebrated last theorem.

Ferrari, Ludovico (1522–65) Italian mathematician who was the first to solve the *quartic equation. He was assistant to Cardano, who published the solution in his *Ars magna* (1545).

FFT *Abbreviation for* fast Fourier transform (*see* discrete Fourier transform).

Fibonacci, also known as **Leonardo of Pisa** (*c.*1175–*c.*1250) Italian mathematician who in his treatise on arithmetic and algebra, *Liber abaci* (1202, The Book of the Abacus), championed the Hindu–Arabic number system. One of its large collection of problems gives rise to the *Fibonacci sequence. A later work, *Liber quadratorum* (1225, The Book of Square Numbers) contains the first Western advances to be made in arithmetic since Diophantus.

Fibonacci sequence (Fibonacci, 1202) The *sequence 1, 1, 2, 3, 5, 8, 13, 21, ... where each term, after the first two, is the sum of the preceding pair of terms. Sometimes the sequence is begun 0, 1, 1, ... These Fibonacci numbers originally arose from a problem about the breeding of rabbits posed by Fibonacci in his *Liber abaci*. But they also occur elsewhere in the natural world, for example as the numbers of ancestors of a male honeybee in different generations. The sequence also has several interesting mathematical properties: for example, every two adjacent terms are relatively prime; any natural number is a sum of distinct Fibonacci numbers; and the ratios of successive terms, $1/1$, $2/1$, $3/2$, $5/3$, ... get closer and closer to the golden ratio (*see* golden section).

fictitious force *See* inertial force.

fiducial inference R.A. Fisher (1935) introduced the concept of a *fiducial distribution* to make probabilistic inferences about unknown parameter values. Fiducial theory very often gives similar results to theories leading to *confidence intervals, but the logical basis is distinct. The theory has some very subtle aspects and is not widely used. *See* Behrens–Fisher test.

field 1. A *set (of numbers or functions, for instance), together with ways of adding and multiplying together members of the set, that satisfy rules similar to the rules for the addition and multiplication of *rational numbers. In particular, given any element in the set, we must be able to add any element to it, subtract any element from it, multiply it by any element, or divide it by any nonzero element, and in each case obtain a result in the same set of elements. In detail, a set F will be a field if and only if the operations $+$ and $\times$ on F satisfy the following properties:

(1) for any a and b in F, $a + b$ and $a \times b$ must also be in F;
(2) for any a and b in F, $a + b = b + a$ and $a \times b = b \times a$;
(3) for any a, b, and c in F, $a + (b + c) = (a + b) + c$ and $a \times (b \times c) = (a \times b) \times c$;
(4) there is a special number 0 in F such that $0 + a = a$ for every a in F, and there is a special number 1 ($\neq 0$) in F such that $1 \times a = a$ for every a in F;
(5) to every element a there corresponds an element $-a$ in F such that $a + (-a) = 0$, and if $a \neq 0$ there is an element a^{-1} in F such that $a \times a^{-1} = 1$;
(6) for any a, b, c in F, $a \times (b + c) = (a \times b) + (a \times c)$.

Although there is no explicit mention of subtraction or division in properties (1)–(6), they are there implicitly since subtracting a is the same as adding $-a$, and dividing by a is the same as multiplying by a^{-1}.

The set of all rational numbers with their usual addition and multiplication is an example of a field. The real numbers and the complex numbers (with the appropriate addition and multiplication each time) are also fields. However, the set of integers, for example, is not a field. It fails to satisfy the last part of (5) as it has many elements (e.g. the integer 2) that do not have integer reciprocals.

There are also examples of fields that have only a finite number of elements. In these cases the easiest way to see how their elements are to be added and multiplied is to write down the addition and multiplication tables. The smallest possible field has just the two elements 0 and 1, which are added and multiplied together as in the following tables:

+	0	1		×	0	1
0	0	1		0	0	0
1	1	0		1	0	1

There are also fields with 3, 4, 5, 7, 8, 9, ... elements, but not 6 or 10, because the number of elements in any finite field must be a power of a prime. Conversely, if p^n is any prime power there is a unique finite field with p^n elements, often called the *Galois field, $GF(p^n)$. *See also* ring.

2. (field of force, force field) A phenomenon associated with a *conservative force: it is the force that would be experienced by a particle of unit mass (unit charge, etc.) due to some distribution of matter (charge, etc.). For example, a particle of mass m will experience a *gravitational force GMm/d^2 when it is a distance d from some body of mass M; the gravitational field at that position is GM/d^2. The field therefore depends on the distribution of matter that causes it; its effect is on another distribution of matter. The same applies to an electrostatic field arising from a distribution of charge.

A field of force is an example of a *vector field* or a *vector function of position, $\mathbf{g}(\mathbf{r})$, i.e. at every point there is specified a vector $\mathbf{g}$, the magnitude and direction of which varies from point to point. A relationship can be established between field and

*potential, which is a scalar function of position and an example of a *scalar field*.

field extension *See* extension field.

Fields Medal The Canadian mathematician John Charles Fields (1863–1932) sought to provide for mathematicians an award comparable in stature to the Nobel prizes. Consequently, he proposed to award quadrennially at least two gold medals for 'outstanding achievement in mathematics', normally to mathematicians under 40, at successive International Congresses of Mathematicians (ICM). In his will he left sufficient funds and suggestions for the organization of the award. The first awards were made in 1936; the recipients were Lars Ahlfors for his work on *complex analysis and Jesse Douglas for his work on the *Plateau problem. Following a wartime hiatus the awards were resumed in 1950. Fields wished to stress the international nature of mathematics, and consequently urged that there should not be attached to the medals 'the name of any country, institution, or person'. The award is administered by a Board of Trustees set up by the University of Toronto; the medals themselves are awarded by a committee of mathematicians appointed by the ICM.

figurate number An integer that can be represented by an *array forming a regular geometric figure. *See* triangular number.

figure 1. (geometric figure) A combination of lines, points, curves, surfaces, etc.
2. Any character or combination of characters representing a number.
3. A digit.
4. *See* syllogism.

filter Let X be any *set and F a collection of nonempty *subsets of X. F is a filter on X if and only if

(1) $(A \in F) \& (B \in F) \rightarrow (A \cap B) \in F$;
(2) $\varnothing \notin F$;
(3) $(A \in F) \& (A \subseteq B) \rightarrow B \in F$, where $B \subseteq X$.

For example, the set F of all closed intervals $[x, y]$ where $0 < x < \frac{1}{2} < y < 1$ is a filter on $[0, 1]$.

finite decimal (terminating decimal) *See* decimal.

finite differences When a *function is tabulated at equal intervals in the argument, the differences between successive function values, differences between successive differences, etc. are called *finite differences*.

There are several notations for finite differences. If $y = f(x)$ has known values $y_0, y_1, y_2, \ldots, y_n$ at $x_0, x_1, = x_0 + h, x_2 = x_0 + 2h, \ldots, x_n = x_0 + nh$, then

$$\Delta y_r = y_{r+1} - y_r, \quad r = 0, 1, 2, \ldots, n-1$$

is called the first *forward difference* of $f(x)$ at $x = x_r$; the difference

$$\Delta^2 y_r = \Delta(\Delta y_r) = \Delta y_{r+1} - \Delta y_r$$
$$= y_{r+2} - 2y_{r+1} + y_r$$

is called the second forward difference at $x = x_r$. The first *backward difference* is

$$\nabla y_i = y_i - y_{i-1}$$

and the first *central difference* is

$$\delta_{i+1/2} = y_{i+1} - y_i$$

More generally, the kth forward difference is

$$\Delta^k y_r = \sum_{s=0}^{k} (-1)^{k-s} \binom{k}{s} y_{r+s}$$

where $\binom{k}{s}$ is the *binomial coefficient. Note also that

$$y_1 = y_0 + \Delta y_0, \quad y_2 = y_0 + 2\Delta y_0 + \Delta^2 y_0$$

and in general

$$y_r = y_0 + \sum_{s=1}^{r} \binom{r}{s} \Delta^s y_0, \quad \text{for all } r \geqslant 1$$

If $y = x^n$ it is easily seen that, for all y, Δy is a polynomial of degree $n-1$, $\Delta^n y$ is a constant, and, for all $m > n$, $\Delta^m y = 0$. Thus if

y is *polynomial of degree n, then all its nth forward differences $\Delta^n y$ are constant. The converse is also true.

Finite differences are commonly set out in a table of the form

x	y	Δ	Δ^2	Δ^3
x_0	y_0			
		Δy_0		
x_1	y_1		$\Delta^2 y_0$	
		Δy_1		$\Delta^3 y_0$
x_2	y_2		$\Delta^2 y_1$	
		Δy_2		$\Delta^3 y_1$
x_3	y_3		$\Delta^2 y_2$	
		Δy_3		
x_4	y_4			

For example, if $x_0 = -2$, $y = x^2 - 2$, and $h = 1$ then

x	y	Δ	Δ^2	Δ^3
-2	2			
		-3		
-1	-1		2	
		-1		0
0	-2		2	
		1		0
1	-1		2	
		3		
2	2			

Each entry in the last three columns is obtained by subtracting the entry in the previous column immediately above it from that immediately below it.

Finite differences are important in *interpolation and *difference equations, and for many other problems such as *numerical integration and *numerical differentiation.

finite discontinuity *See* discontinuity.

finite element method A method for approximating the solution of a *partial differential equation with boundary conditions over a given domain. The domain is partitioned into elements (typically triangles for a two-dimensional region or tetrahedra in three dimensions) and on each element the solution is approximated by a suitable *function, usually a low-degree *polynomial. The coefficients that define the polynomials are chosen to satisfy a best approximation criterion.

finite field See Galois field.

finite Fourier transform Alternative name for a *discrete Fourier transform.

finite group A *group with a finite number of elements.

finite sequence A *sequence that has a finite number of terms.

finite series A *series that has a finite number of terms.

finite set A *set that is not infinite, i.e. one that cannot be put into a *one-to-one correspondence with a proper *subset of itself.

finite variation See variation.

first kind The terms *first kind, second kind, third kind*, etc. are sometimes used to distinguish two or more classes of mathematical object of some common overall type. For example, there are *elliptic integrals of the first, second, and third kinds. See also Airy functions; Bessel functions; cusp; hypothesis testing; integral equation; separation.

first-order convergence See order.

first-order differential equation A *differential equation containing only the first differential coefficient dy/dx.

Fisher, Sir Ronald Aylmer (1890–1962) English mathematician, statistician, and geneticist who in his *Statistical Methods for Research Workers* (1925) provided the basic statistical techniques and designs used by subsequent workers.

Fisher's exact test A test for lack of association in a 2×2 *contingency table. The test is based on the *hypergeometric distribution, and is available in many statistical software packages. If expected numbers in all cells are not too low, the *chi-squared test provides a good approximation. It is sometimes called the *Fisher–Irwin* or the *Fisher–Yates test*. An extension to $r \times c$ tables is called the *Fisher–Freeman–Halton test*.

Fisher's z-distribution A *distribution based on the *logarithm of the ratio of two *estimators of a common *variance. In practice the *F-distribution is used instead.

Fisher's z-transformation A *transformation of the sample estimate, r, of a bivariate normal *correlation coefficient to $z = \tanh^{-1} r$, giving a better approximation to a normal distribution.

Fitzgerald–Lorentz contraction See Lorentz–Fitzgerald contraction.

five-number summary (J.W. Tukey, 1977) For a set of observations the least value, first *quartile, *median, third quartile, and greatest value form a five-number summary of *order statistics, providing a rapid means of assessing the location, dispersion, and asymmetry (if any) of the observations. In association with a *stem-and-leaf display, this summary is generally superior to a *histogram in descriptive statistics, and it is widely used in *exploratory data analysis.

fixed-point iteration See iteration.

fixed-point theorem Any theorem that gives conditions which ensure that a *continuous mapping $f : X \to X$ has a *fixed point*, i.e. that there is an $a \in X$ such that $f(a) = a$. Examples are the *Banach contraction principle* (see contraction mapping) and *Brouwer's theorem* (L.E.J. Brouwer, 1912) which states that if X is the n-ball B^n (see ball) then any such f has a fixed point. Thus for $n = 2$, any continuous map of a circular disc onto itself has a fixed point.

There is a more general version called the *Lefschetz theorem* (S. Lefschetz, 1926; H. Hopf, 1928) that applies when X is any (compact) *polyhedron.

fixed variable *See* regression.

flat angle (straight angle) An angle equal to one-half of a complete turn ($180°$ or π radians).

flecnode A *node on a curve at which one or both branches of the curve have points of *inflection.

floating-point representation A floating-point system represents a real number x in a given *base β as $x = f \times \beta^e$, where f is a real number (the *mantissa) and e an integer (the *exponent or *index). For example, $105.7 = 0.1057 \times 10^{-3}$. The representation can be made unique by requiring that $1/\beta \leqslant |f| < 1$. *Exponential notation is a form of floating-point representation.

Computers use a floating-point system in which f and e have a limited range: f is a number with t base-β digits, and e lies in an interval $[L, U]$. In such a system there are finitely many numbers, and x can be written as

$$x = \pm(0.f_1 f_2 \ldots f_t) \times \beta^e$$

$$= \pm\left(\frac{f_1}{\beta} + \frac{f_2}{\beta^2} + \cdots + \frac{f_t}{\beta^t}\right) \times \beta^e$$

where each digit f_i satisfies $0 \leqslant f_i \leqslant \beta - 1$. Numbers for which $f_1 \neq 0$ are *normalized numbers*. On most computers, the base $\beta = 2$.

When a *floating-point operation* (addition, subtraction, multiplication, or division) is performed between two floating-point numbers, the result may not be a floating-point number, most likely because it has more than t digits in the mantissa; the result is then rounded (*see* rounding off) to

produce a floating-point number, thereby committing a rounding *error.

floor function *See* integer part.

flow *See* dynamical system.

flow chart (flow diagram) A diagram to indicate the relationships between logical steps in a well-defined procedure, such as an algorithm for a computer program to calculate the sum of N numbers X. A suitable flow chart would be as shown in the diagram. The standard convention is to use elliptical frames for 'start' and 'end', rectangles for operation boxes, and diamonds for decision boxes. There is one route out

flow chart for computing a sum of numbers.

from a start box. An operation box must have one or more routes in and exactly one route out; a decision box at least one route in and always two routes out, depending on the decision. An end box has at least one route in and no routes out. Flow charts are used in other contexts. Financial journals often use them to indicate optimal investment strategies in the light of different options facing investors in differing circumstances.

fluent *See* calculus.

fluid mechanics The study of the mechanical and flow properties of liquids and gases. *See also* hydrostatics.

fluxion *See* calculus.

focal chord A *chord that passes through the focus of a *conic.

focal property The property of a *conic in which lines from the foci to a point on the curve make equal angles with the *tangent at that point. It is also called the *reflection property*, since it shows how light (*optical property*) or sound (*acoustical property*) would be reflected by a reflector with the shape of the conic. *See* ellipse; hyperbola; parabola.

focal radius A line segment between the focus of a *conic and any point on the conic.

focus (*plural* **foci**) *See* conic.

folium of Descartes A plane *curve with the equation (in Cartesian coordinates) $x^3 + y^3 = 3axy$. It passes through the origin, has a single loop (hence *folium*, 'leaf'), and has two branches that are asymptotic to the straight line $x + y + a = 0$. It was proposed by René Descartes in 1638, who used it to cast doubt on a method of finding tangents invented by Pierre de Fermat.

foot **1.** Symbol: ft. A *British unit of length equal to one-third of a yard. 1 foot = 0.3048 metre.

2. The point at which a line perpendicular to another line or to a plane meets that line or plane.

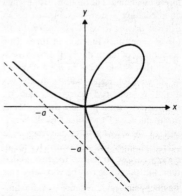

folium of Descartes

foot-pound Symbol: ft-lb. A unit of work in the *f.p.s. system, equal to the work done by 1 pound-force acting through 1 foot. 1 foot-pound = 1.35582 joule.

foot-poundal Symbol: ft-pdl. A unit of work in the *f.p.s. system, equal to the work done by a force of 1 poundal acting through a distance of 1 foot. 1 foot-poundal = 0.042 14 joule.

force Symbol: **F**. A dynamic influence that, when acting on a particle or system of particles (a body), causes or tends to cause it to accelerate. The particle or body can be moving or stationary. Force is expressed in newtons and is a vector quantity. Its magnitude can be given by the product of the magnitude of the acceleration, a, given to the particle or body, and the mass, m, of the particle or body. This is Newton's second law of motion. The direction of the force is the direction in which the acceleration is imparted (for linear motion). A body will be deformed by the action of a force. The deformation is usually ignored when studying the motion of the body as a whole. *See also* Newton's laws of motion; central force; centrifugal force; centripetal force; conservative force;

Coriolis force; external force; inertial force; internal force.

forced oscillation The motion arising when an oscillating system is subjected to an external driving force that is itself periodic (or is some other function of time). One component of the motion is the *free oscillation that would occur in the absence of the driving force and that eventually dies away (*see* damped harmonic motion). Ultimately, with a periodic driving force, the frequency of the forced oscillation is that of the driving force but there is a change in amplitude and phase. If the frequency is close to that of the free oscillation, the amplitude can be very large (*see* resonance).

force field *See* field.

force polygon A graphical representation of a system of forces acting at a point and their *resultant. Let the forces $\mathbf{F}_1$, $\mathbf{F}_2$, ..., $\mathbf{F}_n$ be represented by directed line segments A_0A_1, A_1A_2, ..., $A_{n-1}A_n$. The resultant of the system is represented by the directed line segment A_0A_n. If A_0 and A_n coincide, the resultant is zero and the system is in *equilibrium. *See also* polygon of forces.

force polygon for four forces, $\mathbf{F}_1$, $\mathbf{F}_2$, $\mathbf{F}_3$, and $\mathbf{F}_4$. The directed line segment A_0A_4 represents the resultant.

forest *See* tree.

form A homogeneous *polynomial in two or more variables. The form is said to be *linear* if the variables are separately of the first degree. The number of sets of variables is the *order* of the form. For instance, if there are two sets of variables, $x_1, \ldots, x_n$ and $y_1, \ldots, y_n$, the sum

$$a_{11}x_1y_1 + a_{12}x_1y_2 + \cdots + a_{1n}x_1y_n +$$
$$a_{21}x_2y_1 + a_{22}x_2y_2 + \cdots + a_{2n}x_2y_n +$$
$$\vdots$$
$$a_{n1}x_ny_1 + a_{n2}x_ny_2 + \cdots + a_{nn}x_ny_n$$

is a *bilinear form* (of order 2). A *quadratic form* is a form of the second degree, for example

$$ax^2 + bxy + cy^2$$

Quadratic forms in two variables (as above) represent *conics when put equal to a constant. Quadratic forms in three variables equal to a constant represent *conicoids. The study of forms was developed in the mid 19th century by A. Cayley and J.J. Sylvester, who called them *quantics*. In particular, they studied the *invariants of forms. For instance, if the form above is part of an equation representing a conic, then the expression $b^2 - 4ac$ is an invariant for translation or rotation of the coordinate axes.

formal calculation A calculation (often involving *power series) that is purely algebraic and disregards questions of *convergence. Euler used formal calculations to discover many new theorems.

formal consequence *See* consequence.

formalism The view, often associated with David Hilbert, that mathematics can be regarded as manipulation of symbols independently of their meaning or interpretation. To the formalist, mathematics is not true in the sense that it describes an independent reality, but is rather like a game played in accordance with certain rules that allow the construction of sequences of symbols from other sequences of symbols. Thus the formalist's concern is proof-theoretic, one of the tasks of mathematics being to provide consistency proofs (*see* consistent) that prevent contradictory claims from being made. Without consistency, mathematics would be useless.

The formalists, like the intuitionists, find acceptable only those proofs that do not require an infinite number of steps.

See intuitionism; logicism.

formal language In logic, a set of symbols together with a set of *formation rules* that designate certain sequences of symbols as *wffs, and a set of rules of inference (*transformation rules*) that, given a certain sequence of wffs, permit the construction of another wff. The symbols chosen vary from language to language, but typically they contain both logical *constants and nonlogical vocabulary. For example, in the language of the *propositional calculus the logical constants are truth-functional connectives, and the nonlogical vocabulary consists solely of sentence letters. In the *predicate calculus, variables, predicates, and quantifiers are needed. The formation rules will naturally reflect the chosen vocabulary. The rules of inference are to be thought of as governing only the manipulation of symbols, independently of any interpretation they might have.

Although formal languages do not require at any stage the notion of an interpretation, they are nevertheless constructed with interpretations in mind, and rules of inference that do not preserve truth, although not formally unsatisfactory, are of no interest. The term 'formal language' is also sometimes used as a synonym for 'formal system'.

See also proof theory; logic.

formal power series *See* formal calculation.

formal system (formal theory) A *formal language together with a set of *axioms.

formation rules In logic, the rules of a *formal language for constructing *wffs from symbols. For example, the rule in the *propositional calculus that if 'A' is a wff, then '$\sim A$' (i.e. 'not A') is a wff.

formula Any identity, general rule, or law of mathematics. *See also* well-formed formula.

forward difference Given *function values $y_i = \mathrm{f}(x_i)$, where $x_i = x_0 + ih$, and $i = 0, 1, 2, \ldots$, the forward difference Δy_i is defined by $\Delta y_i = y_{i+1} - y_i$. *See* finite differences.

forward difference formula *See* Gregory–Newton interpolation.

Foucault's pendulum A means of showing the rotation of the earth about its axis. It was demonstrated in 1851 by the French physicist Jean Bernard Léon Foucault (1819–68), who suspended a 28 kg ball on a 67 m length of wire inside the dome of the Panthéon in Paris. When such a pendulum is set in motion, with small displacements about its equilibrium position, the suspended weight swings in a plane (tracing a straight line on the floor beneath), and this plane slowly rotates about the vertical.

The maximum rate of rotation occurs at the earth's poles. The pendulum maintains a constant plane of oscillation in space (relative to the fixed stars) while the earth rotates. To an observer on earth the plane of oscillation makes one rotation every 24 hours (approximately). In general, the angular speed of rotation is $\omega \sin \lambda$, where ω is the earth's angular speed of rotation, $7.3 \times 10^{-5}\,\mathrm{rad\,s^{-1}}$ or $15°$ per (sidereal) hour, and λ is the local latitude; the direction of rotation is clockwise in the northern hemisphere, anticlockwise in the southern.

four-colour problem The problem of finding the minimum number of colours needed to colour a geographical map so that adjacent regions are distinguished by different colours. (Adjacent regions are ones with common boundary line segments.) It is clear that three colours will not suffice. It was proved in 1890 by P.J. Heawood that five colours are always enough; however, the problem of demonstrating that four is the minimum number of colours was resolved only as recently as 1976, by K. Appel and W. Haken.

As Appel and Haken had used some 1200 hours of computer time and in the process accepted a number of complicated computations uncheckable by human hand, some critics have objected that such an approach does not amount to an acceptable mathematical proof. Most mathematicians, however, have accepted the proof while continuing to hope for a more accessible demonstration of the theorem.

The problem applies to maps on a plane or sphere. For a torus, it has been proved that seven is the minimum number of colours required.

four group See Klein's four group.

Fourier, Jean-Baptiste Joseph, Baron (1768–1830) French mathematician who, in his *Théorie analytique de la chaleur* (1822, Analytical Theory of Heat), developed the technique since known as *Fourier analysis, which has proved to have wide application in a number of apparently unrelated disciplines.

Fourier analysis The use of *Fourier series and *Fourier transforms in analysis.

Fourier coefficients See Fourier series.

Fourier series The infinite *series

$$\frac{1}{2}a_0 + \sum (a_n \cos nx + b_n \sin nx)$$

Since the sine and cosine each have a period of 2π, the Fourier series also has a period of 2π. By a suitable choice of the coefficients a_n and b_n, the series can be made to converge to (i.e. the sum of the series can be made equal to) any periodic function of x defined on the interval $(-\pi, \pi)$. If f is such a function, the *Fourier coefficients* are

$$a_n = \frac{1}{\pi} \int_{-\pi}^{\pi} f(x) \cos nx \, dx$$

$$b_n = \frac{1}{\pi} \int_{-\pi}^{\pi} f(x) \sin nx \, dx$$

for $n = 1, 2, 3, \ldots$. The Fourier series is used in the analysis of a waveform into its constituent sine waves of different frequencies and amplitudes (see wave).

Fourier series were first used to study heat conduction but are now very widely used in electrical engineering, vibration analysis, acoustics, optics, data compression, signal processing and other areas. For example, the decomposition of light sources using spectroscopy relies on Fourier analysis and is used to obtain information about the chemical composition of stars.

Fourier's half-range series A *Fourier series that can take two forms:

$$\frac{1}{2}a_0 + \sum a_n \cos nx \quad \text{or} \quad \sum b_n \sin nx$$

The cosine is an *even function while the sine is an *odd function, i.e.

$$\cos x = \cos(-x), \quad \sin x = -\sin(-x)$$

The former (cosine) series can therefore be made to converge to any even function of x defined on the interval $(-\pi, \pi)$, and the latter (sine) series to any odd function of x defined on $(-\pi, \pi)$.

Fourier transform An *integral transform of the type

$$F(t) = \frac{1}{\sqrt{(2\pi)}} \int_{-\infty}^{\infty} e^{itx} f(x) \, dx$$

The function F is said to be the Fourier transform of the function f. It follows that

$$f(x) = \frac{1}{\sqrt{(2\pi)}} \int_{-\infty}^{\infty} e^{-itx} F(t) \, dt$$

F and f are said to be a pair of Fourier transforms. *See also* discrete Fourier transform.

four squares theorem *See* Lagrange's theorem (**1**).

f.p.s. units A system of units that was formerly used in English-speaking countries for scientific, engineering, and general purposes. A noncoherent system, based on the foot, pound, and second, it has been replaced for scientific purposes by *SI units.

fractal A term introduced in 1975 by the French mathematician B.B. Mandelbrot to describe geometric objects that, in a certain sense, have 'fractional dimension'. It includes sets such as the *snowflake curve and *Cantor set generated by some infinitely repeated process and possessing *self-similarity*, i.e. every point of the set is contained in a scaled-down copy of the entire set. In general, a fractal is a set of points with a similarity dimension or Hausdorff dimension which is not an integer (*see* below). In many cases, the attractor or strange attractor (*see* chaos) associated with a *transformation or *flow is a fractal.

A self-similar fractal S in a space of dimension d can be given a *similarity dimension* D, where $0 \leqslant D \leqslant d$. If there are N similarities with scale factors $r_1, r_2, \ldots, r_N$ that map S into itself, then D satisfies the equation $(r_1)^D + \cdots + (r_N)^D = 1$. For example, the Cantor set C has self-similarities $x \mapsto \frac{1}{3}x$ and $x \mapsto \frac{2}{3} + \frac{1}{3}x$. In this case, $r_1 = r_2 = \frac{1}{3}$, and so $D = \ln 2/\ln 3$. The interval $[0, 1]$, the snowflake curve, and the unit square have similarity dimension 1, $\ln 4/\ln 3$, and 2, respectively.

The *Hausdorff dimension* is a more general definition introduced by F. Hausdorff in 1919. It can be defined for any set in n-dimensional Euclidean space. For sets with self-similarities, the Hausdorff dimension and similarity dimension coincide.

Fractal curves are used in producing designs in computer graphics. Many of the designs have a natural form (e.g. the snowflake curve – *see* diagram). Fractal geometry has been used to study crystal formation, electrical discharges, coagulation of particles, urban growth, and many other areas.

fractal An early stage in the generation of the snowflake curve.

fraction A quotient of one number (or a expression) by another, indicated by $\frac{a}{b}$ (or a/b). The dividend a is the *numerator* and the nonzero divisor b is the *denominator*. Fractions are classified as:

Common (or *simple* or *vulgar*) *fraction* – the numerator and denominator are both integers.
Complex fraction – the numerator and denominator are themselves fractions.
Proper fraction – the numerator is less than the denominator, as in $\frac{7}{8}$.
Improper fraction – the numerator is greater than the denominator, as in $\frac{8}{7}$.
Mixed fraction – an integer together with a proper fraction, as in $1\frac{1}{2}$.

Rules for combining fractions are:

Addition. The fractions are put in a form in which their denominators are equal. For example, to add $\frac{1}{2}$ and $\frac{1}{3}$, write $\frac{1}{2} = \frac{3}{6}$ and $\frac{1}{3} = \frac{2}{6}$ (6 is the lowest *common denominator of the two fractions). Then,

$$\frac{1}{2} + \frac{1}{3} = \frac{3}{6} + \frac{2}{6} = \frac{3+2}{6} = \frac{5}{6}$$

Subtraction. The same method as addition, except that the numerators are subtracted rather than added.

Multiplication. The numerators are multiplied and the denominators are also multiplied. For example,

$$\frac{2}{3} \times \frac{4}{7} = \frac{2 \times 4}{3 \times 7} = \frac{8}{21}$$

Division. The divisor is inverted and the two fractions are then multiplied. Thus

$$\frac{1}{3} \div \frac{3}{4} = \frac{1}{3} \times \frac{4}{3} = \frac{4}{9}$$

See also continued fraction; decimal; partial fraction; reducible fraction.

frame In statistics, a specification of all units in a *population in sufficient detail for the selection of a random sample, including, where appropriate, information for selection of *stratified samples, etc. The UK Register of Electors forms a frame of all people in each district qualified to vote in parliamentary elections; unfortunately it rapidly becomes out of date through deaths, people moving to other districts, etc. *See* sampling theory.

frame of reference A means by which the position of a point or the time of an event can be defined in relation to an arbitrary point and an arbitrary pointer reading on a clock. These reference entities, which together form the frame of reference, are described in terms of some coordinate system and a linear timescale. If it is assumed that time is absolute, that observers all experience the same flow of time, then a particular frame of reference can be described merely in terms of a particular set of axes.

An *inertial* (or *Newtonian*) *frame of reference* is a frame of reference in which a body will remain at rest or move at constant velocity as long as no force is acting on it, i.e. Newton's first law of motion is valid. Any frame of reference moving at constant velocity relative to an inertial frame is also an inertial frame. A set of axes fixed in space relative to the positions of distant stars is a standard inertial frame. A set of axes on the earth's surface can be considered a good approximation to an inertial frame.

If a particle is fixed in a given frame of reference but is accelerated with respect to an inertial frame, then the given frame is a *noninertial* or *accelerated frame of reference*. A rotating frame of reference is noninertial: a particle fixed in such a frame will have a *centripetal component of acceleration relative to an inertial frame.

See also inertial force; relativity.

Fredholm's integral equations Certain types of *integral equation. A Fredholm integral equation of the first kind has the form

$$f(x) = \lambda \int_a^b K(x, y)\, g(y)\, \mathrm{d}y$$

g being the unknown function. A Fredholm integral equation of the second kind is

$$g(x) = f(x) + \lambda \int_a^b K(x, y)\, g(y)\, \mathrm{d}y$$

They are named after the Swedish mathematician Erik Ivar Fredholm (1866–1927).

free group A *group with no relations between its *generators a, b, ... except the trivial relations $aa^{-1} = I$, ... and their consequences (such as $b^{-1}aa^{-1}b = I$). Here I is the identity of the group and the operation has been written as juxtaposition. In such a free group, every element other than the identity can be written uniquely as a finite product $a^\alpha b^\beta \dots r^\sigma$ of powers of generators, where adjacent generators a, b, ... in the product are distinct and the exponents α, β, ... are nonzero integers. Every subgroup of a free group, apart from the identity, is also free; and every group is a homomorphic image of some free group.

free oscillation The motion of an oscillating system that occurs when it is displaced from its *equilibrium position and released. The system oscillates about this point with a frequency characteristic of the system. In practice there is some resistance to the motion, i.e. the oscillations are damped, and the free oscillations gradually die away (*see* damped harmonic motion). When it is necessary to maintain an oscillation, a compensating mechanism is used to overcome the resistance. This mechanism can be regarded as an external driving force, and the system will assume *forced oscillation.

free variable *See* variable.

Frege, Friedrich Ludwig Gottlob (1848–1925) German mathematician, logician, and philosopher who in his *Begriffsschrift* (1879, Concept-writing) developed the first adequate notation for mathematical logic and provided the first formalization of the propositional and predicate calculus. In his *Die Grundlagen der Arithmetik* (1884, The Foundations of Arithmetic) Frege offered a definition of number based on set theory, while his abortive *Grundgesetze der Arithmetik* (1903, Basic Laws of Arithmetic) tried to complete the logicist programme of deriving arithmetic from logic.

frequency 1. Symbol: v or f. The number of complete *oscillations or *cycles that occur in unit time, i.e. the rate of repetition of a periodic phenomenon. The various forms of wave motion have some value of frequency, as do pendulums. In one complete oscillation or cycle there is a displacement or variation from an equilibrium position or value, a return to equilibrium, a displacement or variation in the opposite sense, and a further return to equilibrium. Frequency is measured in hertz. *See also* angular frequency.
2. The *absolute frequency* of an observed value is the number of times that value

appears in a sample. In the sample 2, 5, 3, 3, 3, 5, 3, 6, 2, 3, 9, 5 the absolute frequency of the observation 3 is 5, and that of 9 is 1. The *relative frequency* of an observation is determined by dividing the absolute frequency by the total number of observations. There are 12 observations in the above sample, so the relative frequency of 3 is 5/12 and that of 9 is 1/12. The *cumulative frequency* of observations less than or equal to a given value is the sum of all frequencies of observations at or below that value. In the above sample the cumulative absolute and relative frequencies of observations less than or equal to 5 are respectively 10 and 5/6. The frequency for a range is the sum of all frequencies of observations in that range. In the above sample the absolute and relative frequencies of observations in the range 3 to 5 inclusive are 8 and 2/3.

frequency analysis An analysis of *ciphertext that relies on the *frequency of occurrence of certain elements. A classic example is in *substitution ciphers applied to English words, for which use is made of the fact that E is the most common letter in normal English.

frequency curve A smooth curve approximating a *frequency polygon for a large data set. The term is also used for the curve representing the *frequency function.

frequency distribution A specification of the frequencies with which values of a variable occur. For observed data, the distribution usually takes the form of a *frequency table, and for a continuous variable the data will need to be grouped (*see* grouped data). A theoretical distribution is usually specified by a *frequency function or a *distribution function.

frequency function For a discrete probability *distribution, the frequency function specifies for each realizable value x of a random variable X the *probability that X attains that value. Thus, for a *binomial distribution with parameters n and p, the frequency function is specified by

$$f(x) = \Pr(X = x) = \binom{n}{x} p^x (1-p)^{n-x}$$

$$\text{for } x = 0, 1, 2, \ldots, n$$

For a continuous random variable X, if $f(x)\delta x$ is the probability that X takes a value between x and $x + \delta x$, where $\delta x \to 0$, then $f(x)$ is the frequency function of X, and $f(x)$ gives the *probability density* of X at x. For the *exponential distribution with mean 1,

$$f(x) = 0 \qquad \text{if } x < 0$$

$$f(x) = e^{-x} \qquad \text{if } x \geqslant 0$$

For continuous distributions, if $F(x)$ is the *distribution function then $f(x) = F'(x)$. Alternative names for the frequency function are *probability mass function*, or *probability function* for discrete X and *probability density function* for continuous X.

For discrete data, the frequency function (sometimes called the *relative frequency function*) gives the relative frequency of each x-value. For example, if for a group of 100 children the numbers having 0, 1, 2, 3, 4 decayed teeth are 53, 29, 14, 1, 3 respectively, then the frequency function is such that $f(0) = 0.53$, $f(1) = 0.29$, $f(2) = 0.14$, $f(3) = 0.01$, $f(4) = 0.03$.

The concept may be extended to more than one random variable. *See* bivariate distribution; multivariate distribution.

frequency polygon A figure obtained by joining the mid-points of the tops of the rectangles forming a *histogram.

frequency table A table that summarizes the absolute or relative frequencies for a set of observations. The concept extends to cumulative frequencies; for example, for the observations 2, 3, 5, 5, 5, 7, 9, 9, 9, 10, the table gives absolute frequencies and cumulative frequencies. The idea extends to the frequencies of observations of *grouped data, where groups correspond to nonoverlapping ranges.

Observation	2	3	5	7	9	10
Absolute frequency	1	1	3	1	3	1
Cumulative frequency	1	2	5	6	9	10

frequential inference Statistical inference based on the frequential theory of probability, which regards the *probability of an event as the limit of the frequency of occurrence of that event in a series of n trials as $n \to \infty$. In *hypothesis testing this leads to familiar tests, e.g. the *t-test, as to whether data provide evidence against a null hypothesis that it comes from a population with a parameter θ, say, having some preassigned fixed value θ_0. The question that is posed and answered by this approach is 'What is the probability (given by a *p-value) of obtaining this or a more extreme (i.e. more unlikely) value of a relevant statistic obtained from the observed data when the null hypothesis is true?' *Bayesian inference, on the other hand, poses and attempts to answer the question 'How do the data influence our (prior) beliefs about what values of a parameter are plausible?'

In estimation problems a *confidence interval approach is a frequential inference in the sense that we interpret confidence intervals on the basis that a confidence interval of, say, 95 percent is such that if we form these for repeated samples from the same population, in the limit 95 percent of such intervals will cover (or include) the true parameter value.

Both Bayesian and frequential inference methods extend to more than one parameter.

Fresnel integrals The integrals

$$S(x) = \int_0^x \sin t^2 \, dt$$

$$C(x) = \int_0^x \cos t^2 \, dt$$

They are named after the French physicist Augustin Jean Fresnel (1788–1827), and are used for analysing light diffraction. *See* spiral.

friction A *force that opposes the relative motion between two surfaces in contact, and is encountered when an object slides on a surface or when motion is first initiated. It acts within the plane of contact and is independent of the apparent area of contact of the sliding surfaces. (The true area of contact is considerably smaller owing to the roughness of the surfaces.) Most of the energy used in overcoming friction is dissipated as heat.

In addition to the frictional force of magnitude F, two surfaces in contact experience a force of magnitude P that is due to their mutual reactions and acts perpendicular to the plane of contact. With no relative motion, F can take any value up to some limiting value which is roughly proportional to P. Thus for equilibrium

$$F < \mu_s P$$

where μ_s is the *coefficient of static* (or *limiting*) *friction*. When there is relative motion,

$$F < \mu_k P$$

where μ_k, the *coefficient of kinetic friction*, is approximately constant. In general, μ_k is less than μ_s.

If a rolling rather than a sliding motion can be used, as with ball bearings, there is much less friction (*see* rolling friction).

Friedman's test (M. Friedman, 1937) A nonparametric test using *ranks for testing equality of *means in a *randomized block experiment. The statistic used is similar to *Kendall's coefficient of concordance. *See* nonparametric methods.

frieze group A *symmetry group* of a strip pattern (a *frieze*). It is the one-dimensional analogue of the symmetry group of a crystal. It can be shown that there are exactly 7 different frieze groups. *See* crystallography; *compare* wallpaper group.

Frobenius's theorem The *theorem that a finite-dimensional associative *division algebra over the field of *real numbers must consist of either the real numbers themselves, or the *complex numbers, or the *quaternions. It is named after the German mathematician Georg Ferdinand Frobenius (1849–1917).

Since real multiplication and complex multiplication are *commutative operations, whereas quaternion multiplication is not always commutative, it follows from Frobenius's theorem that the quaternions form the only noncommutative, finite-dimensional associative division algebra over $\mathbb{R}$.

frontier (boundary) The *interior* of a *set A is the *union of all open *subsets of A. The *exterior* of set A is the interior of the complement of A. The *frontier* of A is the set of points that belong to neither the interior nor the exterior of A.

The frontier of a set A in a topological space X consists of those points $x \in X$ which are limits of a sequence of points in A and of a sequence of a set of points not in A. In symbols, $\mathrm{Fr}(A) = \bar{A} \cap \overline{(X \backslash A)}$, where $\bar{A}$ denotes the *closure of A.

frustum A part of a solid figure cut off by two parallel planes. The *altitude* (h) of the frustum is the distance between the planes. The volume of a frustum of any cone or pyramid is given by

$$\frac{1}{3} h [A_1 + A_2 + \sqrt{(A_1 A_2)}]$$

where A_1 and A_2 are the areas of the bases. If the cone is a right circular cone, the lateral area of the frustum is

$$\pi s(r_1 + r_2)$$

where r_1 and r_2 are the radii of the bases and s is the slant height of the frustum. The volume of the frustum of a pyramid can also be obtained from the *prismoid formula.

frustum of a cone.

F-test A statistical test based on the *F-distribution. *See* variance ratio.

fulcrum The pivot about which a *lever turns, and about which the moments of the applied force and the weight are calculated.

function (map, mapping) A rule that assigns to every element x of a *set X a unique element y of a set Y, written as $y = f(x)$ where f denotes the function. X is called the *domain* and Y the *codomain*. The set of all the elements $f(x)$ is called the *range* or *image* of f, and is denoted by R_f, Im f, or f(X). It is a subset of the codomain. For example, the area of a circle, y, is a function of the radius, x, written as $y = f(x) = \pi x^2$. x is called the *independent variable* or *argument*, and y is called the *dependent variable* or the *image* of x. If a function can be expressed algebraically the value of y can be calculated for any particular value of x. For example, a circle of radius 2 has area $f(2) = 4\pi$. However, some functions cannot be expressed algebraically: for example, the function 'is the birthday of', which has domain the set of all individuals and range the set of all days in a year.

A function can also be defined as the set of all *ordered pairs (x, y), with x belonging to the domain X and y belonging to the codomain Y, where there is a *many-to-one correspondence between the members of X and the members of Y.

A *multiple-valued function is not a function as defined above because for each value of x the corresponding y is not necessarily unique, but it can be considered as being made up of several branches, each of which is a (single-valued) function.

A function $y = f(x)$ can be graphically represented if (x, y) is plotted on rectangular coordinate axes for every x in X.

A function of two variables assigns to every element (x_1, x_2) of a set of ordered pairs a single element $y = f(x_1, x_2)$. Here, x_1 and x_2 are the independent variables and y is the dependent variable. For example, the volume of a right circular cylinder y is $\pi x_1^2 x_2$ where x_1 is the radius of the base and x_2 is the height. Similarly, a function of several variables assigns to every ordered n-tuple $(x_1, \ldots, x_n)$ a single element $y = f(x_1, \ldots, x_n)$. $x_1, \ldots, x_n$ are independent variables and y is the dependent variable.

A function with domain X and codomain Y is also called a *mapping* or *map* from X to Y, written as f: $X \rightarrow Y$; if, for example, for all $x \in X$ the function maps x onto x^2, this can be specified by using the notation f: $x \mapsto x^2$.

The image of an element $x \in X$ is the element f(x). The image of a subset $A \subset X$ is the set of all the images of the elements of A, and is denoted by f(A).

The *pre-image* of a subset B of the range of f is the set of all elements in X whose images are in B, and is denoted by $f^{-1}(B)$. Thus, for the function f: $x \mapsto x^2$ with domain $\mathbb{R}$ and codomain $\mathbb{R}$, the range of f is the non-negative real numbers, the image of 3 is 9, the image of the interval $[-2, 3]$ is the interval $[0, 9]$, and the pre-image of the interval [4, 9] is the union of two intervals $[-3, -2]$ and $[2, 3]$.

A *polynomial function* (or *rational integral function*) has the form

$$f(x) = a_0 + a_1 x + \cdots + a_n x^n$$

where $a_0, a_1, \ldots, a_n$ are constants.

An *algebraic function* $y = f(x)$ is one that can be defined by a relation of the form

$$p_0(x) + p_1(x)y + \cdots + p_{n-1}(x)y^{n-1}$$
$$+ p_n(x)y^n = 0$$

where $p_0(x), \ldots, p_n(x)$ are polynomials in x.

A *transcendental function* is a function that is not an algebraic function. Examples are the *trigonometric, *logarithmic, and *exponential functions.

See also analytic function; complex function; continuous function; discontinuous function; even function; inverse; limit; mean value; monotonic decreasing function; monotonic increasing function; odd function; rational function; turning point.

functional A *function that has a *domain that is a set of functions and a *range belonging to another set of functions. For example, the *differential operator d/dx is a functional of differentiable functions $f(x)$. The range of the functional may be a set of numbers. An example of this is a *definite integral of $f(x)$ with respect to x.

functional analysis *See* Banach space.

functional series A *series of the form

$$\sum f_n(x)$$

in which the terms are *functions of an independent variable x. The set of values of x for which the series converges constitutes the *region of convergence* of the series. *See also* power series.

function of a complex variable *See* complex function.

function of a function *See* composite function.

fundamental group *See* homotopy group.

fundamental theorem of algebra The theorem that every polynomial equation having complex coefficients and of degree greater than or equal to 1 has at least one complex root. The theorem was first conjectured by Albert Girard who, in 1629, published an account of the roots of equations in which he recognized the existence of imaginary roots. The name 'fundamental theorem of algebra' is due to Gauss, who first investigated the problem in his doctoral thesis (1799), showing that earlier 'proofs' were not sufficient. This proof of Gauss's was geometric, based on the then novel idea that the real and imaginary parts of a complex number could be interpreted as coordinates in a plane. Gauss later tried to prove the theorem by purely algebraic means, but failed. In France, the theorem is known as *d'Alembert's theorem* in recognition of d'Alembert's many attempts to prove it.

fundamental theorem of arithmetic The statement (known to Euclid) that every *natural number other than 1 can be uniquely expressed as a product of *primes. (A prime number itself is expressed as a product with one term in it.) The analogous result for all the integers is that every integer, apart from 0 and ± 1, can be expressed essentially uniquely as a product of prime integers. This means that it will be possible to express an integer in several different ways as a product of prime integers, e.g. $18 = 2 \times 3 \times 3 = (-2) \times (-3) \times 3$. However it is only possible if, as here, the individual prime integers in the two products differ only by factors that are unit integers (± 1).

fundamental theorem of calculus The theorem expressing the relationship between *integration and *differentiation, namely that if the integral

$$\int f(x)\,dx$$

exists, and a *function F(x) also exists for which $F'(x) = f(x)$ in $a \leqslant x \leqslant b$, then

$$\int_{a}^{b} f(x)\,dx = F(b) - F(a)$$

fuzzy logic A system of logic proposed in 1965 by Lofti Zadeh, an Iranian electrical engineer, based on *fuzzy set theory*. In classical set theory an object either is or is not a member of a given set; in fuzzy set theory membership is represented by a real number between 0 and 1. Thus whether someone is tall or not is not simply true or false, but more a matter of degree. Systems of fuzzy logic have been used, amongst other applications, to control elevators, dishwashers, and assembly line strategies in factories.

G

galactic axis *See* galactic equator.

galactic centre A point on the galactic equator taken as the centre of the Galaxy and used as the zero point in a *galactic coordinate system. It has an agreed position (in equatorial coordinates) of right ascension 17 h 46 min, declination −28°56′.

galactic coordinate system An *astronomical coordinate system in which measurements are based on the galactic equator. A point on the *celestial sphere is located by two angular measurements. The *galactic longitude (*l*) is the angular distance measured eastwards from the *galactic centre. The *galactic latitude (*b*) is the angular distance north or south of the galactic equator.

galactic equator (galactic circle) The *great circle that represents the intersection of the plane of the Galaxy with the *celestial sphere. The poles of this circle are the north and south *galactic poles*. The line joining these poles is the *galactic axis*. *See* galactic coordinate system.

galactic latitude Symbol: *b*. The angular distance of a point on the *celestial sphere from the galactic equator taken along a *great circle passing through the point and through the galactic poles. Galactic latitude is measured from 0° to 90° north (taken as positive) or south (taken as negative) of the galactic equator. *See* galactic coordinate system.

galactic longitude Symbol: *l*. The angular distance (measured from 0° to 360°) of a

point on the *celestial sphere from the *galactic centre. It is measured eastwards along the galactic equator between the galactic centre and the place at which a great circle through the point and the galactic poles intersects the galactic equator. *See* galactic coordinate system.

galactic pole *See* galactic equator.

Galilean transformation *See* relativity.

Galileo Galilei (1564–1642) Italian astronomer and physicist who, in *Discorsi e dimostrazione matematiche intorno a due nuove scienze* (1638, Dialogues on Two New Sciences) and other works, attempted to present a mathematically exact and experimentally based kinematics. He correctly formulated the law of acceleration ($s = \frac{1}{2}at^2$) and was the first to note the isochrony of the pendulum. The transformation of the parameters of position and motion is named after Galileo as the *Galilean transformation*.

gallon Symbol: gal. **1.** An *imperial unit of capacity or volume, equal to the volume occupied by ten pounds of distilled water. 1 gallon = $4.546\,09 \times 10^{-3}$ cubic metre. **2.** A unit of liquid volume in the *United States customary system equal to 231 cubic inches.   1 US gallon = $3.785\,411 \times 10^{-3}$ cubic metre = 0.832 674 imperial gallon. 6 US gallons ≈ 5 UK gallons.

Galois, Évariste (1811–32) French mathematician noted for his fundamental discovery in 1829 of group theory, although full details of his work were published only posthumously in 1846. His discovery arose from his realization that the general quintic equation was insoluble by the traditional method of extracting roots. Galois went on to establish precisely under what conditions such traditional methods would work.

Galois field (finite field) Any *field that contains only a finite number of elements. For example, the integers $0, 1, 2, \ldots, p - 1$ added and multiplied modulo a *prime p. The study of such fields was initiated by Galois in 1830. A finite field with q elements is denoted by GF(q) or $\mathbb{F}_q$.

Galois group A *group of *automorphisms associated with a pair of *fields E and F where one of the fields, say F here, is a *subfield of the other. It is denoted by G(E/F) and consists of all the automorphisms of E that leave each element of F fixed. That is, an automorphism σ of E is in G(E/F) precisely when $\sigma(a) = a$ for every a in F. If f(x) is a polynomial with all its coefficients in F, then the Galois group of the polynomial is G(K/F) where K is the smallest field containing F and all the roots of the equation f(x) = 0. *See* soluble group; simple group.

Galois theory The theory that reduces the study of *fields containing a given field to the study of the associated *Galois groups. Galois's powerful ideas can be used to produce explicit examples of polynomial equations (e.g. $x^5 - 10x + 2 = 0$) whose roots cannot be obtained from the coefficients by using (in any order and any number of times) just the operations of addition, subtraction, multiplication, division, and raising to powers of the form $1/n$ (where n is any natural number). The roots of a polynomial equation can be written in this way if and only if the corresponding Galois group is soluble. So the roots of any polynomial equation whose Galois group is not soluble can be written only by using functions that are more complicated than those described.

Galton, Sir Francis (1822–1911) English anthropologist and pioneer in the application of statistical techniques to the analysis of biological problems. He discovered the phenomenon of regression in 1875 and formulated his law of ancestral heredity shortly afterwards. In 1888 he introduced his index of correlation.

gambler's ruin A classic problem determining the probability that a gambler becomes bankrupt in a series of games at each of which he gains 1 unit of capital with probability p, or loses 1 unit with probability $q = 1 - p$. If the gambler has initial capital of C units, he is ruined if he loses all C units. There is usually also a condition that the game stops if the gambler attains a total fortune of a fixed number of units N ($> C$), implying either that the opponent is then ruined, or that one player or the other does not wish to continue. When $p = 0.5$ the probability of ruin is $(N - C)/N$, otherwise it is

$$q^C \frac{p^{N-C} - q^{N-C}}{p^N - q^N}$$

The problem in essence involves a sequence of *Bernoulli trials and is an example of a *random walk.

game theory In competitive situations different parties may make different decisions when their interests conflict, and the outcome is then determined by these decisions. Such conflicting situations may arise in business competition, politics, military operations, etc.

Game theory, first considered by Borel in 1921, was developed by von Neumann to cover conflicting situations where:

(1) there may be any finite number of players;

(2) each player may take one of a finite number of actions (and different players may take different actions);

(3) at each contest (play of a game) players do not know what action will be taken by the other players; and

(4) the outcome of a game determines a set of payments (positive, zero, or negative) to each player.

If the sum of payments to all players is zero the game is called a *zero-sum game. A game with two participants is a *two-person* or *dual game*.

The simplest game is a two-person zero-sum game, in which the win (loss) for player A equals the loss (win) for player B. Crucial to the theory is the *payoff matrix*. If player B may take any of four actions and player A any of three actions, the payoff matrix for player A takes the form:

		B		
A	1	2	3	4
1	a_{11}	a_{12}	a_{13}	a_{14}
2	a_{21}	a_{22}	a_{23}	a_{24}
3	a_{31}	a_{32}	a_{33}	a_{34}

Here a_{ij} is the amount (positive, zero, or negative) A wins if he takes action i and his opponent takes action j. Player B's payoff matrix has each a_{ij} replaced by $-a_{ij}$ (to conform with the zero-sum property).

If a player elects always to take the same action, this is a *pure strategy*. If he selects an action each time using a probabilistic or random choice, this is a *mixed strategy*. For optimality a player should list each of his strategies together with the worst outcome (from his viewpoint) that can result from his opponent's strategies, and then choose a strategy corresponding to the best of these worst possible outcomes; this is the *maximin criterion*.

If there is an entry in the payoff matrix that is a minimum in its row and a maximum in its column, it is called a *saddle point*. The optimum policy for each player is then to take the actions (pure strategies) corresponding to the saddle point.

If there is no saddle point, mixed strategies are appropriate and one can only maximize expected minimum gain over a series of contests. Von Neumann's *minimax theorem* (1928) shows that if each player adopts his best mixed strategy, then one player's expected gain will exactly equal the other's expected loss. This is called the *value* of the game. Although at any play of a game neither player knows what action the other will take, it is assumed that players will behave rationally and may use information about their opponent's strategies from previous games to assess their likely strategy in later games.

The theory of games has been extended to *n*-person nonzero-sum games and to games in which a continuous range of strategies is possible. In 1944, von Neumann and Morgenstern applied game theory to economic competition. Since then it has found many applications in commerce, politics, military strategy, etc. *See* prisoner's dilemma.

gamma distribution The gamma *distribution for a positive-valued *random variable has *frequency function

$$f(x) = \frac{x^{a-1} e^{-a/b}}{b^a \Gamma(a)}$$

where $x, a, b > 0$ and $\Gamma(a)$ is the *gamma function. It is an asymmetric distribution exhibiting positive *skewness, and the probability density function takes a wide range of shapes for different values of the parameters a and b. The case $a = 1$ gives the *exponential distribution important in *waiting time* problems (the distribution of the time from zero to the first occurrence of an event and of the interval between future occurrences).

gamma function The *function Γ defined by

$$\Gamma(x) = \int_0^\infty t^{x-1} e^{-t} \, dt$$

where x is real and greater than zero. The recurrence relation $\Gamma(x+1) = x\Gamma(x)$ is true for all x. Hence if n is a positive integer, $\Gamma(n+1) = n!\Gamma(1) = n!$, and if n is also odd, $\Gamma(\frac{1}{2}n)$ can also be derived since $\Gamma(\frac{1}{2}) = \sqrt{\pi}$. $\Gamma(x)$ for $x \leqslant 0$ can also be obtained using the recurrence relation. If z is a complex variable, then

$$\Gamma(z) = \int_0^\infty t^{z-1} e^{-t} \, dt$$

for $\mathrm{Re}(z) > 0$. The function was also defined by Weierstrass as

$$\frac{1}{\Gamma(x)} = x \exp(\gamma x) \prod_{n=1}^\infty \left[\left(1 + \frac{x}{n}\right) \exp\left(-\frac{x}{n}\right) \right]$$

where γ is *Euler's constant.

Gantt chart A diagram used in scheduling problems in operational research where, for example, separate time axes are allocated to each of two machines, and blocks placed on the time axes are used to indicate the time periods when each machine is performing specified jobs. It is useful to indicate when machines are necessarily idle if operations that require more than one machine must be performed in sequence, and later operations requiring one machine cannot be started until certain earlier operations requiring another machine are completed – and perhaps then only after additional time delays (e.g. to allow paint to dry or adhesives to set). It is named after the American management scientist Henry Laurence Gantt (1861–1919).

gauge theory A theory developed by mathematical physicists to study *fields, and which uses the theories of *groups and *bundles. The theory involves a group G, and in the simplest form of the theory it is an Abelian group; in this case the theory is a modern form of Maxwell's electromagnetic theory.

The ideas behind the theory have been used by the English mathematician Simon Donaldson to study the geometric and topological properties of 4-dimensional spaces. He has shown that their properties are fundamentally different from what would be expected by analogy with the study of spaces of either lower or higher dimension. A key ingredient in his work is the study of the set of all solutions of certain (nonlinear) differential equations; such sets are called *moduli spaces*.

gauge transformation A mathematical reformulation of a physical theory that does not change the physical interpretation. For example, the magnetic fields $\mathbf{E}$ and $\mathbf{B}$ can be written in terms of scalar and vector potentials ϕ and $\mathbf{A}$ as

$$\mathbf{E} = -\nabla\phi - \frac{\partial \mathbf{A}}{\partial t}, \quad \mathbf{B} = \nabla \times \mathbf{A}$$

However, ϕ and $\mathbf{A}$ are not unique: they can be changed by the gauge transformation

$$\phi \to \phi + \frac{\partial \psi}{\partial t}; \quad \mathbf{A} \to \mathbf{A} - \nabla\psi$$

in which the physical quantities $\mathbf{E}$ and $\mathbf{B}$ are not changed, but the purely mathematical ϕ and $\mathbf{A}$ have changed.

Gauss, Carl Friedrich (1777–1855) German mathematician who began a lifetime of prodigious mathematical creativity by proving in 1799 the fundamental theorem of algebra. This was followed in 1801 by his masterpiece, *Disquisitiones arithmeticae* (Arithmetic Disquisitions), in which he introduced into mathematics modular arithmetic and presented his results on the construction of regular polygons as well as proving the law of quadratic reciprocity. Later work by Gauss in astronomy led him in his *Theoria motus corporum coelestium* (1809, Theory of the Motion of Heavenly Bodies) to propose general solutions to the problem of determining planetary orbits, while in geometry he worked out the principles of hyperbolic geometry, independently of Bolyai and Lobachevsky. Other achievements were his method of least squares, and work in electricity, geodesy, complex numbers, and the convergence of series.

Gaussian curvature *See* curvature.

Gaussian distribution *See* normal distribution.

Gaussian elimination A formalization of the method of solving n linear equations in n unknowns by successive *elimination of variables. The equations are first written in *matrix form as $Ax = b$, where A is an $n \times n$ nonsingular matrix and x and b are column vectors of n components, x representing the unknowns. The procedure is to multiply the first equation by a_{21}/a_{11} and subtract it from the second equation, multiply the first equation by a_{31}/a_{11} and subtract it from the third equation, and so on. The effect is to eliminate x_1, the first element of x, from all equations except the first. The element a_{11} used as the first divisor is called the *pivot*, and the numbers a_{i1}/a_{11} are the *multipliers*.

The process is repeated with the new set of $n - 1$ equations (omitting the first) in $n - 1$ unknowns, then for $n - 2$ equations, and so on, until after $n - 1$ stages there results an equation in one unknown only. This equation is immediately solved for that unknown, and a process of substitution in the last but one equation in two unknowns, and so on, is used to obtain all the unknowns in the reverse order to that in which they were eliminated. The effect of the process on the matrix A is to reduce it to an *upper triangular matrix.

The same procedure can be applied to a system of m linear equations in n unknowns. After $\min(m - 1, n)$ stages the $m \times n$ matrix A is reduced to *row echelon form.

The process as described breaks down if one of the pivot elements used as a divisor is zero. If, for example, a_{11} is zero, row and/or column permutation can be used

to bring a nonzero element to the $(1, 1)$ position, and the process continues as usual. To reduce the effect of rounding errors in a computer calculation, it is usual to choose as the pivot the element of greatest absolute value in the column. Gaussian elimination is sometimes called *pivotal condensation*.

Gaussian field The *field of all those *complex numbers whose real and imaginary parts are both *rational, the field operations being the usual complex addition and multiplication. *See also* Gaussian integer.

Gaussian integer A *complex number whose real and imaginary parts are both ordinary integers, as in $2 - 3i$, 5, $-i$, and $1 + 2i$. Using complex arithmetic, Gaussian integers can always be added, subtracted, and multiplied, and sometimes divided, with results that are themselves Gaussian integers. With respect to these operations Gaussian integers behave much like ordinary integers. There are four Gaussian integers (± 1, $\pm i$) that divide 1 and so divide into every Gaussian integer. They are called *units*, and apart from them each Gaussian integer can be classified as *composite* if it is a product of two factors, neither of which is a unit, or *prime* otherwise. Thus

$$2 = (1 + i)(1 - i)$$

$$46 + 9i = (5 + 12i)(2 - 3i)$$

$$5 + 12i = (3 + 2i)^2$$

are composite Gaussian integers, whereas $1 + i$, $4 - i$, and $7 + 2i$ are Gaussian primes. Apart from the four units, every Gaussian integer has an (essentially unique) expression as a product of Gaussian primes. *See also* fundamental theorem of arithmetic.

Gaussian integration rule A *numerical integration rule of the form

$$\int_a^b w(x)\mathrm{f}(x)\,\mathrm{d}x \approx \sum_{i-1}^n w_i\,\mathrm{f}(x_i)$$

where $w(x)$ is a non-negative weight function on the interval $[a, b]$ in which both the n nodes x_i and the weights w_i are chosen to make the approximation exact when f is a polynomial of degree less than or equal to $2n - 1$. The purpose of the weight function is to build into the rule any special behaviour of the integrand; common choices include $w(x) = 1$ with $[a, b] = [-1, 1]$, and $w(x) = \mathrm{e}^{-x}$ with $[a, b] = [0, \infty]$.

An example of a Gaussian integration rule is the *three-point Gauss–Tchebyshev rule*:

$$\int_{-1}^1 \frac{\mathrm{f}(x)}{\sqrt{(1 - x^2)}}\,\mathrm{d}x$$

$$\approx \tfrac{1}{3}\pi[\mathrm{f}(-\sqrt{3/2}) + \mathrm{f}(0) + \mathrm{f}(\sqrt{3/2})]$$

The theory of Gaussian quadrature is intimately connected with the theory of *orthogonal polynomials.

Gauss interpolation formula *See* Gregory–Newton interpolation.

Gauss–Jordan elimination A variant of *Gaussian elimination, due to the German geodesist Wilhelm Jordan (1842–99), for solving a system $\mathbf{Ax} = \mathbf{b}$ of n linear equations in n unknowns in which the kth unknown is eliminated from all the other $n - 1$ equations at the kth stage. The effect of the process on the *matrix $\mathbf{A}$ is to reduce it to a *diagonal matrix. As a means of solving a linear system $\mathbf{Ax} = \mathbf{b}$, Gaussian elimination is preferred as its requires much less work. Gauss–Jordan elimination is sometimes used to invert a square matrix, for which task it has the same cost as Gaussian elimination. With a final row scaling in which each nonzero row is divided by its first nonzero element, Gauss–Jordan elimination applied to an $m \times n$ matrix produces its *reduced row echelon form.

Gauss–Markov theorem The theorem that the *least-squares estimator gives the *unbiased (linear) estimator of a parameter having minimum *variance. Here 'linear' means linear in the sample values. It is named after Gauss and A.A. Markov.

Gauss–Ostrogradsky theorem *See* Gauss's theorem.

Gauss–Seidel method An *iterative method of solving a system of linear equations $\mathbf{Ax} = \mathbf{b}$, published by P.L. von Seidel in 1874 but based on earlier work by Gauss. For three equations in three unknowns,

$$a_{11}x_1 + a_{12}x_2 + a_{13}x_3 = b_1$$

$$a_{21}x_1 + a_{22}x_2 + a_{23}x_3 = b_2$$

$$a_{31}x_1 + a_{32}x_2 + a_{33}x_3 = b_3$$

we may start with arbitrary solutions. In many practical problems a_{11}, a_{22}, and a_{33} are large compared with a_{ij}, $i \neq j$, and it is then convenient to take $x_1 = b_1/a_{11}$, $x_2 = b_2/a_{22}$, and $x_3 = b_3/a_{33}$ as starting values. Now, writing $\mathbf{x}_n$ for the column vector of values of x_1, x_2, x_3 after the nth iteration, and

$$\mathbf{L} = \begin{pmatrix} a_{11} & 0 & 0 \\ a_{21} & a_{23} & 0 \\ a_{31} & a_{32} & a_{33} \end{pmatrix}, \mathbf{U} = \begin{pmatrix} 0 & a_{12} & a_{13} \\ 0 & 0 & a_{23} \\ 0 & 0 & 0 \end{pmatrix}$$

the iterative relationship is

$$\mathbf{x}_{n+1} = \mathbf{L}^{-1}(\mathbf{b} - \mathbf{U}\mathbf{x}_n)$$

where $n = 0, 1, 2, \ldots$. Note that $\mathbf{L} + \mathbf{U} = \mathbf{A}$, and that $\mathbf{x}_0$ is the column vector of starting values. The iterations are continued to convergence.

Modifications of the Gauss–Seidel method produce a class of procedures called *successive over-relaxation* methods.

Gauss's formulae (Delambre's analogies) Formulae relating the angles (A, B, and C) and sides (a, b, and c, where a is opposite A, etc.) of a *spherical triangle:

$$\sin\tfrac{1}{2}c\sin\tfrac{1}{2}(A-B) = \cos\tfrac{1}{2}C\sin\tfrac{1}{2}(a-b)$$

$$\sin\tfrac{1}{2}c\cos\tfrac{1}{2}(A-B) = \sin\tfrac{1}{2}C\sin\tfrac{1}{2}(a+b)$$

$$\cos\tfrac{1}{2}c\sin\tfrac{1}{2}(A+B) = \cos\tfrac{1}{2}C\cos\tfrac{1}{2}(a-b)$$

$$\cos\tfrac{1}{2}c\cos\tfrac{1}{2}(A+B) = \sin\tfrac{1}{2}C\cos\tfrac{1}{2}(a+b)$$

Gauss's proof See fundamental theorem of algebra.

Gauss's theorem For a *vector field $\mathbf{F}$ and a volume V enclosed by a surface S,

$$\int_S (\mathbf{F}.\mathbf{n})\,\mathrm{d}A = \int_V (\mathrm{div}\,\mathbf{F})\,\mathrm{d}V$$

where $\mathbf{n}$ is a *unit vector normal to S. Intuitively, if the vector function $\mathbf{F}$ denotes the magnitude and direction of the flow of a fluid at a point, its *divergence is the net change at sources and sinks. The total of these is the net flow in or out of the region. The theorem is also known as the *divergence theorem* or the *Gauss–Ostrogradsky theorem*, after Gauss and Mikhail Vasilievich Ostrogradsky (1801–62). See also Green's theorem.

Gauss–Tchebyshev rule See Gaussian integration rule.

GCD Abbreviation for greatest common divisor. See common factor.

Gelfond–Schneider theorem (A.O. Gelfond, 1934; T. Schneider, 1934) The theorem that if a and b are *algebraic numbers, with $a \neq 0$ or 1 and b not rational, then a^b is a *transcendental number. For example, $2^{\sqrt{2}}$ and 3^i are transcendental.

generalized coordinates Any set of coordinates

$$q_1, q_2, q_3, \ldots, q_n$$

that is sufficient to specify the configuration of a mechanical system. A knowledge of the generalized coordinates implies a knowledge of the position of every particle of the system. There are also corresponding generalized velocities, forces, and momenta.

generalized eigenvalue For square matrices $\mathbf{A}$ and $\mathbf{B}$ of the same dimension, the number λ is a *generalized eigenvalue* if $\mathbf{Ax} = \lambda\mathbf{Bx}$ for some nonzero vector $\mathbf{x}$, the corresponding *generalized eigenvector*. The generalized eigenvalues are the roots of the equation $\det(\mathbf{A} - \lambda\mathbf{B}) = 0$. If $\mathbf{B}$ is the *identity matrix, then λ is an *eigenvalue and $\mathbf{x}$ is an eigenvector of $\mathbf{A}$.

generalized function An object that behaves symbolically like a *function. The commonest example is the *Dirac delta function* δ_a, which has the properties that $\delta_a(x) = 0$ if $x \neq a$ but

$$\int_{-\infty}^{\infty} \delta_a(x)\,\mathrm{d}x = 1$$

and, more generally,

$$\int_{-\infty}^{\infty} \delta_a(x)\mathrm{f}(x)\,\mathrm{d}x = \mathrm{f}(a)$$

There are several theories to make the concept of a generalized function more rigorous. The commonest regards one of them as a linear *functional on a suitable space of functions; under this interpretation the Dirac delta function δ_a corresponds to the functional $\mathrm{f} \to \mathrm{f}(a)$.

generalized linear models (J.A. Nelder and R.W.M. Wedderburn, 1972) A model in which some function $g(\mu)$ of the mean μ

of a random variable Y is a linear function of one or more variables called either *explanatory* or *independent variables*. For one explanatory variable x,

$$g(\mu) = \beta_0 + \beta_1 x$$

If Y is normally distributed with mean $\mu = E(Y) = \beta_0 + \beta_1 x$ and variance σ^2, then $g(\mu) = \mu$, and the model reduces to the linear *regression model.

The function $g(\mu)$ is called the *link function*, and its form depends on the distribution of Y, which may be any member of the *exponential family of distributions. For example, if Y is a Bernoulli variable (*see* Bernoulli trials) with mean $\mu = p$, then $g(\mu) = \ln[p/(1-p)]$. This is relevant to studies of the toxicity of insecticides: under fairly commonly occurring conditions, if p_i is the probability that an insect will die if exposed to a dose x_i of an insecticide, then

$$\ln\left(\frac{p_i}{1 - p_i}\right) = \beta_0 + \beta_1 x_i$$

If batches of insects are subjected to doses $x_1, x_2, \ldots, x_i, \ldots, x_n$ of the insecticide, the proportion that dies in batch i provides an estimate of p_i. The parameters β_0 and β_1 may be obtained by *maximum likelihood estimation. Iterative procedures are usually needed because the variance of $\ln[p/(1-p)]$ is a function of p. The model extends readily to more than one explanatory variable.

If Y has a *Poisson distribution with mean λ that varies with x, the link function is $g(\lambda) = \ln \lambda$. *Probit analysis is another special case, although, like least squares regression, its use preceded the formulation of the generalized linear model. *See* deviance; logistic regression.

general linear group The group $GL(n, F)$ of all $n \times n$ *nonsingular matrices with entries from a *field F. The *normal subgroup of $GL(n, F)$ consisting of all the matrices whose determinant equals 1 is called the *special linear group*, $SL(n, F)$.

general solution A solution of a linear *differential equation containing the same number of arbitrary constants as the *order of the equation.

general term *See* sequence; series.

generating angle The angle between the axis and the generators in a circular *cone or conical surface.

generating function A *function $f(t)$ such that if $\{P_i(x)\}$ is a *sequence of functions then

$$f(t) = \sum_{i=0}^{\infty} P_i(x) t^i$$

so that when f is expanded in powers of t, the coefficient of t^i gives the ith function $P_i(x)$. For example, the generating function of the Legendre polynomials is

$$\frac{1}{\sqrt{(1 - 2xt + t^2)}} = \sum_{i=0}^{\infty} P_i(x) t^i$$

When the generating function is expanded it can be seen that $P_0(x) = 1$, $P_1(x) = x$, $P_2(x) = \frac{1}{2}(3x^2 - 1)$, and so on. *See* moment generating function; probability generating function.

generation (of a vector space) *See* vector space.

generator **1. (element)** Any of a set of straight lines or line segments that make up a given surface, The generator can be regarded as a line sweeping out the surface by moving according to some rule. The feminine form *generatrix* is also used.
2. (of a group) A *set of elements in a *group that, with their *inverses, can be combined by the group operation (allowing repetitions) to produce all the other group elements. For example, consider the eight 4×4 matrices

$$I = \begin{pmatrix} 1 & 0 & 0 & 0 \\ 0 & 1 & 0 & 0 \\ 0 & 0 & 1 & 0 \\ 0 & 0 & 0 & 1 \end{pmatrix}, \quad J = \begin{pmatrix} 0 & 0 & -1 & 0 \\ 0 & 0 & 0 & -1 \\ 1 & 0 & 0 & 0 \\ 0 & 1 & 0 & 0 \end{pmatrix}$$

$$K = \begin{pmatrix} 0 & -1 & 0 & 0 \\ 1 & 0 & 0 & 0 \\ 0 & 0 & 0 & 1 \\ 0 & 0 & -1 & 0 \end{pmatrix}, \quad L = \begin{pmatrix} 0 & 0 & 0 & -1 \\ 0 & 0 & 1 & 0 \\ 0 & -1 & 0 & 0 \\ 1 & 0 & 0 & 0 \end{pmatrix}$$

$$-I = \begin{pmatrix} -1 & 0 & 0 & 0 \\ 0 & -1 & 0 & 0 \\ 0 & 0 & -1 & 0 \\ 0 & 0 & 0 & -1 \end{pmatrix}, \quad -J, \ -K, \ -L$$

These form a group G with respect to matrix multiplication, and it is generated by the two elements J and K since each element can be written in terms of them: $I = J^4, J, K, L = JK, -I = J^2, -J = J^3,$ $-K = K^3, \ -L = KJ$. Note that the generators J and K satisfy the identities $J^2 = K^2$ and $J^2 = (JK)^2$; or alternatively $J^2 K^{-2} = I$ and $J^{-1}KJK = 1$. Such expressions of the identity element as products of powers of the generators are called *relations* between the generating elements. The two relations here are actually the *defining relations* for the group G since it is the only group generated by two elements whose squares and the square of whose product are all equal. Equivalently any other identity in powers of J and K, such as $J^4 = I$, is a consequence of $J^2 = K^2 = (JK)^2$.

A group with only one generator a is described as *cyclic*. If the group operation is denoted by juxtaposition, with the identity element denoted by I, then any relation satisfied by a would have to be of the form $a^n = I$ for some natural number n. The group would then be finite, consisting of the n elements $I, a, a^2, \ldots, a^{n-1}$ (with $a^{-1} = a^{n-1}$, etc.). If there is no such relation then the cyclic group is infinite, consisting of all the powers $\ldots, a^{-2}, a^{-1}, I, a,$ $a^2, \ldots$, and they are necessarily all distinct.

In general, a group with no nontrivial relations between its generators is said to be *free* (*see* free group).

generator matrix *See* coding.

generatrix *See* generator.

Gentzen, Gerhard (1909–45)   German mathematician noted for his proof in 1936 of the consistency of elementary number theory. The significance of the proof was, however, muted somewhat by Gentzen's reliance on the principle of transfinite induction, a principle not provable in arithmetic or elementary logic. Earlier, in 1934, he had introduced one of the first systems of natural deduction.

genus 1. The number of handles of a compact closed surface (*see* manifold). For example, the surface of a teacup has genus one.
2. For a Riemann surface, the number of linearly independent holomorphic 1-forms (i.e. expressions that can, locally, be written as $f(z) \, dz$, where f is an *analytic function) that are defined on the surface.
3. For plane *algebraic curves with no *singular points, the genus is $\binom{d-1}{2}$ where d is the degree of the curve. For curves with singular points, the genus is $\binom{d-1}{2} - \Sigma \, \delta$, where each term in the sum corresponds to a singular point of the curve and $\delta = 1$ for a *double point but is a larger integer for more complicated singularities.

In appropriate circumstances it can be shown that the three definitions are equivalent.

geodesic   For a given surface, a geodesic is an arc on the surface between two points that is the shortest curve joining the points. At each point on the geodesic the principal normal to the geodesic coincides with the normal to the surface. On a sphere, for example, a geodesic is part of a great circle of the sphere.

geographical coordinates   Coordinates used to locate position on the earth's surface with respect to the equator and to the prime meridian. The position of a point is specified by its *latitude (angular distance from $0°$ to $90°$ north or south of the equator) and *longitude (angular distance from $0°$ to $180°$ east or west of the prime meridian).

geographical equator A *great circle on the earth's surface that is the intersection of the surface with a plane through the centre perpendicular to the axis through the poles.

geometric distribution A discrete *distribution in which the frequencies decrease in *geometric progression as the variable increases. For example, the distribution of the number of trials, X, up to and including the first success in a series of *Bernoulli trials is geometric. If p is the probability of success and $q (= 1 - p)$ the probability of failure, then X has *frequency function

$$\Pr(X = r) = pq^{r-1}, \text{ where } r = 1, 2, 3, \ldots$$

This distribution has *mean $1/p$ and *variance q/p^2.

The distribution of the number of failures before the first success is also geometric. It has mean q/p and variance q/p^2. *See* negative binomial distribution.

geometric figure *See* figure.

geometric mean *See* mean.

geometric progression (geometric sequence) A *sequence in which the ratio of each term (except the first) to the preceding term is a constant, the *common ratio*. If the first term is a and the common ratio is r, then the sequence takes the form

$$a, ar, ar^2, ar^3, \ldots$$

and the nth term is

$$ar^{n-1}$$

If $r \neq 1$, the sum of the first n terms is

$$\frac{a(1 - r^n)}{1 - r}$$

Compare arithmetic progression.

geometric series A *series of the form

$$a + ar + ar^2 + \cdots + ar^{n-1} + \cdots$$

i.e. a series in which the terms are those of a *geometric progression. If $r \neq 1$ and the number of terms is n, the sum s_n is given by

$$s_n = \frac{a(1 - r^n)}{1 - r}$$

If $|r| < 1$, $s_n \to a/(1 - r)$ as $n \to \infty$, i.e. the infinite series is then *convergent and has sum $a/(1 - r)$.

geometry The branch of mathematics concerned with the properties of space and of figures in space.

Originally, geometry started as a practical subject in ancient Egypt and Babylonia, used in surveying and building. In the time of the ancient Greeks it was realized that properties of figures could be deduced logically from other properties. Around 300 BC, *Euclid drew together a large amount of Greek knowledge in his *Elements*.

The book develops geometry as a formal logical structure based on definitions and axioms, from which propositions (theorems) are proved. The result is the traditional school geometry known as *Euclidean geometry. It is divided into *plane geometry* (in two dimensions) and *solid geometry* (for three-dimensional figures).

Euclidean geometry is mainly concerned with points, lines, circles, polygons, polyhedra, and the conic sections. In 1637, *Descartes published his new *coordinate (or analytic) geometry in which points could be represented by numbers, and lines and curves by equations. The discovery gave mathematicians a new weapon with which to attack geometric problems algebraically, and it also introduced a large number of different types of curve for study. Around the same time, analytic geometry was independently discovered by Fermat. The development of analytic geometry influenced the discovery of the differential calculus, and this in turn led to the study of surfaces by Euler and Monge and, in 1827, to the development of *differential geometry by Gauss.

In 1639, two years after Descartes published his work on coordinate geometry, Desargues invented what is now known as *projective geometry. The subject was neglected at the time, but interest in it was revived in the 19th century with work by Poncelet.

The 19th century saw other major advances in geometry. Cayley developed *algebraic geometry* – i.e. analytic geometry of *n*-dimensional space. Lobachevsky, Bolyai, and Gauss independently developed *non-Euclidean geometries. Finally, Riemann in his lecture *Über die Hypothesen welche der Geometrie zu Grunde liegen* (1854, On the Hypotheses that Lie at the Foundation of Geometry) put forward a view of geometry as the study of any kind of space of any number of dimensions (*see* Riemannian geometry). *See also* topology.

Gergonne, Joseph Diez (1771–1859) French mathematician who from 1810 edited the *Annales de Mathématiques*, the first purely mathematical journal to appear. He was a proponent of analytical geometry; his most important mathematical discovery was the principle of duality, which he formulated in about 1825, about the same time as Poncelet, with whom he disputed priority in the discovery.

Germain, Sophie Marie (1776–1831) French mathematician, mainly self-taught, who initially fell it necessary to adopt in her correspondence with other mathematicians the male pseudonym Louis Le Blanc. As a result of her extensive work on *Fermat's last theorem, other mathematicians were able to show that the theorem held for all $n < 100$. In later life Germain's interests turned to mathematical physics where, following *Euler, she contributed to the mathematical theory of elasticity.

GF(q) Symbol for the *Galois field with q elements.

Gibbs, Josiah Willard (1839–1903) American mathematician and theoretical chemist who in his *Vector Analysis* (1881) introduced into physics the mathematical tools which would eventually replace such competing systems as the quaternions of W.R. Hamilton. In chemistry, he is noted for his development of chemical thermodynamics. He also did important work in the founding of *statistical mechanics.

Gibbs sampler (S. Geman and D. Geman, 1984) A *Markov chain *Monte Carlo technique used to give numerical approximations to Bayesian posterior distributions involving two or more variables. An initial set of values is specified and new values of each variable are successively simulated from their conditional distributions, given the current value of all other variables. If the new value is more in accord with the specified distribution, it replaces the current value; otherwise the current value is retained. The process is continued until equilibrium is reached. The technique is named after Gibbs, who was a pioneer of *statistical mechanics.

giga- *See* SI units.

Giorgi system *See* m.k.s. units.

Girard, Albert (1595–1632) Dutch mathematician who made significant contributions to trigonometry and algebra. He established that an equation of the nth degree has n roots; he also, unlike his contemporaries, allowed for negative and imaginary roots. In trigonometry he introduced the abbreviations sin, tan, and sec.

g.l.b. *Abbreviation for* *greatest lower bound.

glide reflection An *isometry composed of a *reflection in a line and a *translation parallel to the line.

Gödel, Kurt (1906–78) Austrian-American mathematical logician who proved in 1930 the completeness of the first-order functional calculus. This was followed in 1931 by his *Über formal unentscheidbare Sätze der 'Principia Mathematica' und verwandter Systeme* (On Formally Undecidable Propositions in *Principia Mathematica* and Related Systems), in which he proved the first of his two remarkable incompleteness theorems. In 1938 he threw light on Cantor's continuum hypothesis by proving that neither it nor the axiom of choice could ever be disproved within standard set theory (*see* Gödel's proof).

Gödel's proof The proof by Kurt Gödel (1931) that any formal axiomatic system containing arithmetic contains undecidable propositions – i.e. contains sentences S such that neither S nor the negation of S can be proved. This result is known as Gödel's *first incompleteness theorem*.

The method of proof involved giving numbers to the variables and symbols in the *formal system, and using these to assign numbers to expressions so as to give to different expressions different *Gödel numbers*. In this way it was possible to translate the syntax of the system into arithmetic, thereby making the system capable of making statements about its own syntax. It was then possible to show that there is a sentence of the type 'this statement is not provable'.

A corollary, Gödel's *second incompleteness theorem*, states that the consistency of a formal system containing arithmetic cannot be proved by means using the formalization of the system itself – only by using a stronger system. Gödel's work answered the second of *Hilbert's 23 problems and put paid to attempts, like that of Whitehead and Russell, to develop pure mathematics from a few fundamental logical principles. It also damages the scientific ideal of finding a small set of basic axioms in terms of which all natural phenomena can be logically described.

Goldbach's conjecture The conjecture that every even number greater than 2 is the sum of two primes. It was put forward in 1742 by the German mathematician Christian Goldbach (1690–1764), and published in 1770 in Waring's book *Meditationes algebraicae*.

Although Goldbach's conjecture is believed to be true, it has so far resisted all attempts to prove it formally. Similar conjectures, however, have been proved. In 1937 Vinogradov proved that all sufficiently large odd integers are sums of three primes (*see* Vinogradov's theorem); in 1973 Chen Jing-rún proved that every sufficiently large even number is the sum of a prime and a number that is either prime or has two prime factors; and in 1995 O. Ramaré showed that every even number is a sum of at most six primes.

golden section A division of a line into two segments such that the ratio of the larger segment to the smaller segment is equal to the ratio of the whole line to the larger segment. If a line AB is divided at P, then the division is a golden section if

$$AP/PB = AB/AP$$

The ratio AP/PB is $\frac{1}{2}(1 + \sqrt{5})$, or approximately 1.618, a ratio known as the *golden mean* or *golden ratio*. A *golden rectangle* is a rectangle having sides in this ratio.

The golden section has a number of interesting mathematical (and other) properties. It was known to the Pythagoreans, who described it as 'division in mean and extreme ratio'. They discovered it in constructing a pentagram by taking a regular pentagon ABCDE, and drawing the diagonals AC, AD, BE, etc. (*see* diagram (a)). The diagonals intersect at the five points A′, B′, C′, D′, and E′. Each of these points divides a diagonal into two segments in the golden ratio.

(a)

(b)

golden section (a) in the construction of the pentagram; (b) constructed for a line AB.

A golden section can be constructed for a line AB. First, a square ABCD is constructed (*see* diagram (b)). If E is the midpoint of the side DA, DA is produced to F, where EF = EB. The square AFGH is drawn on AF; H then divides AB in the golden section.

The golden ratio is also connected with the *Fibonacci sequence. If u_{n-1} and u_n are two successive terms of the sequence, then the limit of u_n/u_{n-1} as $n \to \infty$ is $\frac{1}{2}(1 + \sqrt{5})$.

The *continued fraction

$$1 + \cfrac{1}{1 + \cfrac{1}{1 + \cfrac{1}{1} \ldots}}$$

is equal to the golden ratio.

The golden section was known simply as 'the section' to the ancient Greeks. Its present name comes from the Renaissance when, around 1500, it was taken up by artists as a 'divine proportion' and used in painting, sculpture, and architecture.

goodness of fit In statistics, the closeness of agreement between a set of observations and a hypothetical model which is suggested as a possible data source. In particular, the term is used when considering the fit of observations to a theoretical distribution such as a *normal distribution with known mean and variance, and to testing whether that fit is acceptable. Closeness of agreement is often measured by a quantity involving the squares of the differences between observed and theoretical values. When the model is so chosen that the quantity has minimum value, the goodness of fit is said to be *best* based on a *least squares criterion. *See also* chi-squared test; Kolmogorov–Smirnov tests.

googol The integer that is written in *decimal notation as 1 followed by a hundred zeroes; in *exponential notation this is 10^{100}.

googolplex The integer that is written in decimal notation as 1 followed by *googol zeroes; in exponential notation this is $10^{(10^{100})}$.

Gordan, Paul Albert (1837–1912) German mathematician noted for his proof in 1868 of his finite base theorem, subsequently known as *Gordan's theorem*. His efforts at generalization to higher-order forms were completed in 1888 by David Hilbert.

grad *See* grade; gradient.

grade 1. (grad) A rarely used unit of angle equal to 1/100 of a right angle. *See* angular measure.
2. *See* gradient.

gradient 1. (grade) In general a slope, i.e. an inclination to the horizontal. A gradient is expressed in various ways:

(1) As the angle the line or path makes with the horizontal, i.e. the slope angle.

(2) As the *tangent of this angle, i.e. the vertical distance travelled per horizontal distance.

(3) As the vertical distance travelled with respect to the actual distance along the path. For example, a gradient of 1 in 4 indicates a vertical distance of 1 unit for 4 units along the slope. This is also indicated as a ratio (1/4, i.e. the sine of the slope angle) or as this ratio expressed as a percentage (a 25 percent gradient).

2. (grad) For a scalar function of position $\phi(\mathbf{r})$, the gradient of ϕ, written as grad ϕ, is given by $\nabla\phi$, where ∇ is the operator *del. Thus

$$\text{grad } \phi = \nabla\phi = \mathbf{i}\frac{\partial\phi}{\partial x} + \mathbf{j}\frac{\partial\phi}{\partial y} + \mathbf{k}\frac{\partial\phi}{\partial z}$$

and

$$\mathbf{r} = x\mathbf{i} + y\mathbf{j} + z\mathbf{k}$$

See curl; divergence; potential.

3. For a *function y of n independent variables x_1, x_2, ..., x_n, the *vector of first *partial derivatives of y with respect to the x_i is called the *gradient vector* of y. *Compare* Hessian.

Graeco-Latin square An extension of a *Latin square allowing classification by four mutually *orthogonal factors usually denoted by rows, columns, Latin letters, and Greek letters. An example of a three-by-three square is

Row	Column		
	1	2	3
1	Aα	Bβ	Cγ
2	Bγ	Cα	Aβ
3	Cβ	Aγ	Bα

Each Latin or Greek letter occurs once in each row or column, and each Latin letter occurs once with each Greek letter. In theory the design may increase precision, but it has technical limitations and, for small squares, insufficient degrees of freedom for the *error mean square. *See* experimental design.

gram Symbol: g. The unit of mass in the *c.g.s. system, equal to 1/1000 of a kilogram.

Gram–Schmidt method A method for converting a set of vectors forming a *basis into an orthonormal basis (*see* orthogonal basis). It is based on the idea of orthogonalizing vectors against each other: if $\mathbf{a}$ and $\mathbf{b}$ are given linearly independent column vectors, then

$$\mathbf{b}' = \mathbf{b} - \frac{(\mathbf{a}^T\mathbf{b})}{(\mathbf{a}^T\mathbf{a})}\mathbf{a}$$

is orthogonal to $\mathbf{a}$. The Gram–Schmidt method effectively forms a *QR factorization of the matrix whose columns comprise the basis vectors. It is named after Jorgen Pedersen Gram (1850–1916) and Erhard Schmidt (1876–1959).

graph 1. A diagram showing a relationship between two *variables. Graphs are most commonly drawn using a Cartesian coordinate system with an x-axis and a y-axis at right angles. In two dimensions, the graph of an equation is a curve for which the coordinates of points on the curve satisfy the equation. A graph of a function $f(x)$ is the graph of the equation $y = f(x)$. A graph of an inequality, in two dimensions, is generally a region in the plane satisfying the inequality.

Graphs of equations (or functions) may be plotted by taking a number of values of x and calculating the values of y from the equation. The points are marked on ruled *graph paper* and a smooth curve is drawn through them. Graphs of observed or measured values of physical quantities are drawn similarly. Although the most common form of graph uses squared graph paper, other types are sometimes employed for special purposes. A *logarithmic graph* is

(a)

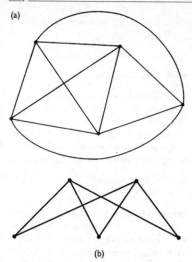

(b)

graph (a) The complete graph with five vertices, K_5. (b) The complete bipartite graph $K_{2,3}$.

one in which both axes are marked with a logarithmic scale. An equation of the form $y = ax^n$ has a straight-line graph when plotted on logarithmic paper. A *semilogarithmic graph* is one with one axis having a logarithmic scale and the other a linear scale. Such graphs are especially useful for plotting relationships of the type $y = a^x$.

2. A set of points (*vertices* or *nodes*) V connected by a set of *edges E. V* and E are the *vertex set* and *edge set* of the graph. A *directed graph* (*digraph*) or *network* is one in which direction is associated with the edges; they are then a set of ordered pairs of vertices, and are called *arcs*. A pair of vertices joined by an edge is *adjacent*. Where two or more edges join a pair of vertices they are called *multiple* or *parallel* edges. A graph may be represented by an *adjacency matrix.

A graph in which there are no edges joining a vertex to itself (*loops*) and no multiple edges is said to be *simple*. (Sometimes the term 'graph' is reserved for a graph of this type; a graph with loops or multiple

edges is then termed a *multigraph*.) A simple graph is *complete* if every pair of vertices is joined by an edge. The complete graph with n vertices is denoted by K_n.

A pair of edges meeting at a vertex is *adjacent*. The *degree* of a vertex is the number of edges meeting at it. A vertex is *odd* or *even* according to whether its degree is odd or even. If every vertex of a graph has the same degree k, it is said to be *regular* or k-*regular*. For example, the graph K_5 (*see* diagram (a)) is 4-regular.

A graph whose vertices fall into two disjoint sets V_1 and V_2 with l and m vertices and such that its edges only join vertices of V_1 to vertices of V_2 is a *bipartite graph*. If the graph is simple and every vertex of V_1 is joined to every vertex of V_2, it is a *complete bipartite graph* and is denoted by $K_{l,m}$.

A *planar* graph is one that can be drawn in the plane without two edges crossing. The graph $K_{2,3}$ can be drawn in such a way and is planar (*see* diagram (b)); the graphs K_5 and $K_{3,3}$ are nonplanar.

See tree; walk; Eulerian graph; Hamiltonian graph; weighted graph; network analysis; Chinese postman problem; travelling salesman problem.

graphical solution A method of solving two *simultaneous equations by plotting the *graphs of each equation. The solutions are given by the coordinates of the points of intersection (since at these points both equations are satisfied by the same values of x and y).

A single equation $f(x) = 0$ may be solved graphically by finding the intersections of $y = f(x)$ and $y = 0$.

Grassmann, Hermann Günther (1809–77) German mathematician noted for his highly original but obscure *Die lineale Ausdehnungslehre, ein neuer Zweig der Mathematik* (1844, The Theory of Linear Extension, a New Branch of Mathematics) in which he tried to develop a calculus of extension to describe and analyse events in physical space. One manifestation of his theory is that *Grassmann coordinates* are used to parametrize sets of spaces of the same dimensions in a Euclidean space.

Graunt, John (1620–74) English merchant who, in his *Natural and Political Observations Mentioned in a Following Index and Made Upon the Bills of Mortality* (1662), introduced the first statistically based ideas of life expectancy, population estimation, and sex ratio. Using the data in the bills, issued weekly since 1603, which gave numbers of christenings, burials, and causes of death, he made the first reasonable estimate of the population of London as 384 000, showed that more boys than girls were born each year, and that women lived longer than men; he also constructed the first *life table.

gravitation The tendency of all material bodies to attract one another. The mutual attraction between bodies is considered as a force – the *gravitational force* – that acts between the bodies and arises because the bodies possess mass. The force decreases as the distance between the bodies increases. This was first expressed in mathematical form as *Newton's law of gravitation*, which gives the magnitude F of the force of attraction between two point masses m_1 and m_2 a distance d apart as

$$F = Gm_1m_2/d^2$$

where G is the *gravitational constant. Gravitation is one of the fundamental forces of nature.

The *gravitational field* due to a material body is the force on a particle of unit mass arising from the mass of the body (*see* field). The *gravitational potential* due to a material body is the potential energy of a particle of unit mass arising from the mass of the body (*see* potential). *See also* relativity (general theory).

gravitational constant Symbol: G. The universal constant appearing in New-

ton's law of *gravitation. Its value is $6.672 \times 10^{-11}\,\mathrm{N\,m^2\,kg^{-2}}$.

gravitational field *See* gravitation; field.

gravitational force *See* gravitation.

gravitational mass The property of a body that determines the gravitational field it can produce. Newton's law of *gravitation is expressed in terms of gravitational mass. The gravitational mass of a body has been found to be equivalent to its *inertial mass.

gravitational potential *See* gravitation; potential.

gravity The tendency for a body to move downwards because it possesses *weight. It is a local manifestation of gravitation on earth or on some other celestial body. *See* acceleration of free fall.

gray Symbol: Gy. The *SI unit of absorbed dose of ionizing radiation, equal to the energy in joules absorbed by 1 kilogram of matter. [After L.H. Gray (1905–65)]

Gray code A *binary code invented by the American physicist Frank Gray in 1947 but a similar code had already been used by Émile Baudot in 1878. The codewords can be listed so that successive words differ in only one place. Gray codes are widely used in transforming analogue data to digital data, and have very useful error-correcting properties.

great circle A circle on a sphere that has its centre at the centre of the sphere; the radius of a great circle therefore equals the radius of the sphere. *Compare* small circle.

greatest common divisor (GCD) *See* common factor.

greatest integer function An alternative name for the floor function. *See* integer part.

greatest lower bound (g.l.b.; infimum) A lower bound l (of a function, sequence, or set) is a greatest lower bound if $l \geqslant m$ for any other lower bound m. *See* bound.

Green, George (1793–1841) English mathematician noted for his *Essay on the Application of Mathematical Analysis to the Theory of Electricity and Magnetism* (1828), in which the fundamental notion of the potential, used earlier by Laplace to determine gravitational attraction, was first used to analyse electrical and magnetic phenomena. It also contained the first formulation of *Green's theorem.

Green's theorem A theorem in *potential theory. For a region R of the x–y plane bounded by a curve C, if functions $P(x, y)$ and $Q(x, y)$ have continuous *partial derivatives then

$$\int_{C} (P \, dx + Q \, dy) = \int_{R} \left(\frac{\partial Q}{\partial x} - \frac{\partial P}{\partial y} \right) dA$$

The analogue in three dimensions is *Gauss's theorem. Both theorems are special cases of the general *Stokes's theorem.

Gregory, David (1661–1708) Scottish mathematician who published many of his uncle James Gregory's results on infinite series in his *Exercitatio geometrica* (1684, Geometrical Essays). He was also the first to publish some of Newton's results in both mathematics and astronomy, and in 1703 he issued the first ever edition of the collected works of Euclid.

Gregory, James (1638–75) Scottish mathematician noted for his expansion of a number of trigonometric functions into infinite series. Gregory was, in fact, one of the first to distinguish between convergent and divergent series. He is, however, known more widely for his description in 1661 of a type of reflecting telescope.

Gregory–Newton interpolation A method of *interpolation which, in its basic form (sometimes called the *forward difference formula*), uses *forward differences. In this form it is especially suited to interpolation between x_0 and x_1, given values y_0, y_1, y_2, $\ldots$, y_n of a function $f(x)$ for equally spaced values $x_0, x_1, x_2, \ldots, x_n$ of the independent variable. If $x_0 < x' < x_1$ and $k = (x' - x_0)/(x_1 - x_0)$, then $y' = f(x')$ is estimated by the formula

$$y' = y_0 + \binom{k}{1} \Delta y_0$$

$$+ \binom{k}{2} \Delta^2 y_0 + \cdots + \binom{k}{n} \Delta^n y_0$$

where $\binom{k}{r}$ is the *binomial coefficient. Note that k is non-integral.

There is a related formula (the *backward difference formula*), useful for interpolation between x_{n-1} and x_n that uses *backward differences.

The *Gauss interpolation formula* uses *central differences and is appropriate for interpolation between x_1 and x_{n-1}.

It is named after James Gregory and Isaac Newton.

Gregory's series (J. Gregory, 1667) The *series expansion for the inverse tangent function:

$$\tan^{-1} x = x - \frac{x^3}{3} + \frac{x^5}{5} - \frac{x^7}{7} + \cdots$$

This is valid for $-1 \leqslant x \leqslant 1$.

Grelling–Nelson paradox A *paradox stated by K. Grelling and L. Nelson in 1908. An adjective is called *autological* if it has the property denoted by itself. Thus the word 'English' is an English word, 'short' is short, and 'polysyllabic' is polysyllabic. If an adjective is not autological, it is *heterological*. Thus 'German' is not German, 'long' is not long, and 'monosyllabic' is not monosyllabic. What of the word 'heterological' itself? Is it heterological? It must be either autological or heterological. Take each alternative in turn:

(1) If it is autological, the adjective 'heterological' has the property denoted by itself,

and it must be heterological. Thus, from the assumption that 'heterological' is autological, it follows that it is heterological.

(2) If it is heterological, the adjective 'heterological' does not have the property denoted by itself, so it is not heterological and must therefore be autological. Thus, from the assumption that 'heterological' is heterological, it follows that it is autological.

The paradox is also known as *Grelling's paradox*.

gross 1. Prior to deductions. *Gross profit*, for instance, is profit before taking away all operating costs.
2. The *gross weight* of an object includes the weight of any wrapper, vessel, vehicle, etc. in which the object is weighed.

Compare net.

group A *set G whose elements can be combined together in a way similar to the addition of integers. If the result of combining the elements a and b of G is denoted by $a \circ b$, then G will be a group if and only if $\circ$ is a *binary operation (so that $a \circ b$ must be in G) and satisfies the following three properties:

(1) the operation is *associative*: given any three members a, b, and c of G then $a \circ (b \circ c) = (a \circ b) \circ c$;

(2) there is a special element I, called the *identity element*, such that for any element a, $a \circ I = I \circ a = a$;

(3) corresponding to each element a there is an element a', called the *inverse* of a (and depending upon a), such that $a \circ a' = a' \circ a = I$.

The set of all integers with the operation of addition is an example of a group. In this case zero is the special element playing the role of I in (2), and the integer which, combined with a, gives zero is $-a$, since $a + (-a) = 0$. The operation of addition of integers has another helpful property, since it does not matter which way round two integers are added: $a + b = b + a$ for

all integers a and b. If the operation $\circ$ in the group G has the analogous property

(4) $a \circ b = b \circ a$ for every pair of elements a, b in G,

then the group is said to be *commutative* or *Abelian* (*see* Abelian group). So the set of integers with the operation of addition forms an Abelian group. An example of a non-Abelian group is the set of all non-singular 2×2 *matrices with matrix multiplication as the group operation. In this case there are many pairs A and B of such matrices with $AB \neq BA$; for example

$$A = \begin{pmatrix} 1 & 1 \\ 0 & 1 \end{pmatrix}, \quad B = \begin{pmatrix} 1 & 1 \\ 1 & 1 \end{pmatrix}$$

In both the above examples the groups concerned are *infinite groups* (i.e. they contain an infinite number of elements). However, *finite groups* are just as common, and for each natural number n there is at least one group having exactly n elements. For example, for each natural number n the set of nth *roots of unity forms a finite group containing n elements (with complex multiplication as the group operation). Indeed, the group concept first arose in finite situations, especially in Galois's work on groups of permutations of the roots of a polynomial equation (*see* Galois theory). Nowadays group theory pervades most of modern algebra and has important applications in several areas of science, such as *crystallography and quantum theory.

grouped data Data consisting of counts of numbers of items with observed values of a variable, such as a measurement of a physical or other distinguishing characteristic, each lying in one of a set of specified non-overlapping intervals. For example, for 200 men the exact height, x cm, of each may be measured and the information reduced to numbers of men within each of the height intervals in the following table:

Height, x (cm)	Number of men
$x < 165$	6
$165 \leqslant x < 170$	39
$170 \leqslant x < 175$	93
$175 \leqslant x < 180$	44
$180 \leqslant x < 185$	15
$x \geqslant 185$	3

With an appropriate choice of interval size, the grouping gives a good 'feel' for the distribution underlying the sample data. The mid-point of each interval other than the unbounded intervals at each end is called the mid-interval value, and these are 167.5, 172.5, 177.5, and 182.5 cm. When, as in this example, it seems that most entries within the end intervals are likely to be in the finite intervals $160 \leqslant x < 165$ and $185 \leqslant x < 190$, it is not unreasonable to assume the mid-interval values 162.5 and 187.5 cm for these. Approximate values for the *mean and *variance of the heights x are computed by assuming that each x for an observation in an interval takes the mid-interval value, e.g. for the above example the mean height is estimated as

$$m = \frac{1}{200}(6 \times 162.5 + 39 \times 167.5$$
$$+ 93 \times 172.5 + 44 \times 177.5$$
$$+ 15 \times 182.5 + 3 \times 187.5)$$
$$= 173.3$$

If it is assumed that the exact measurements x follow a *normal distribution, an improved estimate of the exact variance may be obtained by subtracting $h^2/12$, where h is the interval width, from the estimate of variance based on mid-interval values. The correction, known as *Sheppard's correction*, is not always satisfactory when there are open unbounded end intervals as in the above example.

Approximate *quantiles are also obtainable from grouped data. From the table,

the median is clearly in the interval $170 \leqslant x < 175$, and the estimate may be made more precise by linear interpolation.

A grouped *cumulative frequency table* is obtained by combining counts for groups with the measured variable less than the maximum specified for successive class intervals. With the data above for heights of men, the cumulative frequency table is as follows:

Height (cm)	Cumulative frequency
165	6
170	45
175	138
180	182
185	197
190	200

grouping *See* randomized blocks; matched pairs.

groupoid A *set G, together with a rule or operation, that, given any two elements a and b (in that order) in G, specifies a unique third element of G. An example of a groupoid is the set of all integers together with the operation of subtracting the second of a pair of integers from the first. Here any pair of integers a and b specifies a unique integer $a - b$. However, the set of natural numbers 1, 2, 3, ... with the same operation of subtraction does not form a groupoid. In this case there are pairs (e.g. 4, 7) for which the operation of subtracting the second number from the first does not lead to another natural number.

g-statistics The *statistics g_1 and g_2, which are sample equivalents of γ_1 and γ_2, the population measures of *skewness and *kurtosis.

gyroscope A wheel spinning on a shaft and so mounted that it can rotate freely about any direction. It has two basic properties, either or both of which are used in a variety of instruments. First, the spinning wheel tends to maintain the direction of its rotational axis in space; it is said to have *gyroscopic inertia*. Second, if a twisting force (a *torque) is applied to the shaft so as to try to rotate the shaft about an axis perpendicular to the shaft, the resulting motion will be a *precession, i.e. a rotation of the shaft about an axis that is perpendicular both to the shaft and to the axis of the torque.

H

H_0, H_1 *See* hypothesis testing.

$\mathbb{H}$ Symbol for the set of all *quaternions.

HA *Abbreviation for* *hour angle.

Hadamard, Jacques (1865–1963) French mathematician noted for his proof in 1896 of the *prime number theorem: that the number of primes not greater than n approximately equals $n/\ln n$. The theorem was independently proved at the same time by Vallée-Poussin.

Hadamard matrix An $n \times n$ *matrix H of 1's and –1's whose rows are mutually orthogonal, so that $HH^{\mathrm{T}} = nI$.
  A 2×2 Hadamard matrix is

$$\begin{pmatrix} 1 & 1 \\ -1 & 1 \end{pmatrix}$$

A necessary condition for a Hadamard matrix of dimension n greater than 2 to exist is that n is a multiple of 4, but it is not known whether a Hadamard matrix of dimension n exists for every n that is a multiple of 4.

Hadamard product The element-wise product of two square *matrices of the same dimensions. If A and B are $n \times n$ matrices, their Hadamard product is the $n \times n$ matrix C with $c_{ij} = a_{ij}b_{ij}$.

half-angle formulae 1. Formulae in plane trigonometry that give trigonometric functions of an angle x in terms of the tangent of the half-angle $\frac{1}{2}x$:

$$\sin x = \frac{2t}{1 + t^2}$$

$$\cos x = \frac{1 - t^2}{1 + t^2}$$

$$\tan x = \frac{2t}{1 - t^2}$$

where $t = \tan \frac{1}{2}x$.
2. Formulae in plane trigonometry of the form

$$\tan\frac{1}{2}A = \frac{r}{s - a}$$

$$\tan\frac{1}{2}B = \frac{r}{s - b}$$

$$\tan\frac{1}{2}C = \frac{r}{s - c}$$

where A, B, and C are angles of a triangle and a is the length of the side opposite angle A, b is opposite angle B, and c is opposite angle C; s is the semiperimeter, i.e.

$$s = \frac{1}{2}(a + b + c)$$

and r is the expression

$$\sqrt{\frac{(s - a)(s - b)(s - c)}{s}}$$

Before the use of electronic calculators, these formulae were sometimes used in solving triangles in place of the *cosine rule because they were convenient for use with logarithmic tables.
3. Formulae used in spherical trigonometry to solve *spherical triangles:

$$\tan\frac{1}{2}\alpha = \frac{r}{\sin(s-a)}$$

$$\tan\frac{1}{2}\beta = \frac{r}{\sin(s-b)}$$

$$\tan\frac{1}{2}\gamma = \frac{r}{\sin(s-c)}$$

where α, β, and γ are angles of the spherical triangle, and a is the length of the side opposite angle α, b is opposite angle β, and c is opposite angle γ; s is the semi-perimeter, i.e.

$$s = \frac{1}{2}(a+b+c)$$

and r is the expression

$$\sqrt{\frac{\sin(s-a)\sin(s-b)\sin(s-c)}{\sin s}}$$

half-line (ray) A straight line extending indefinitely in one direction from a fixed point.

half-plane A plane extending indefinitely from a line (the *edge*).

half-range series *See* Fourier's half-range series.

half-side formulae Formulae used in spherical trigonometry to solve *spherical triangles:

$$\tan\frac{1}{2}a = R\cos(S-\alpha)$$

$$\tan\frac{1}{2}b = R\cos(S-\beta)$$

$$\tan\frac{1}{2}c = R\cos(S-\gamma)$$

where α, β, and γ are the angles of the spherical triangle, and a, b, and c are the lengths of the sides, a being opposite α, b opposite β, and c opposite γ; S is half the sum of the angles, i.e.

$$S = \frac{1}{2}(\alpha+\beta+\gamma)$$

and R is the expression

$$\sqrt{\frac{-\cos S}{\cos(S-\alpha)\cos(S-\beta)\cos(S-\gamma)}}$$

half-space A space lying on one side of a given plane.

Halley, Edmond (1656–1742)   English astronomer and mathematician. Although best known for his work on comets, and for his role as editor of Newton's *Principia* (1687), Halley also published a number of mathematical papers. His work ranged over such practical issues as how to use mortality tables to compute annuities, and the computation of logarithms, to more theoretical problems on the nature of infinite quantities. In 1692, on the basis of geometrical arguments, Halley disproved the common assumption that all infinite quantities are equal. In 1710 he produced a Latin translation of the *Conics* of Apollonius.

Halley's method   (E. Halley, 1694) A method of solving an equation in one variable, $f(x) = 0$, by the iteration

$$x_{n+1} = x_n - \frac{f_n}{f'_n - \frac{f_n f''_n}{2f'_n}}, \quad n = 0, 1, 2, \ldots$$

where x_0 is a first approximation to the root, and f_n, f'_n, and f''_n denote the values of f and its first two derivatives at x_n. Halley's method has a faster ultimate speed of convergence than *Newton's method, but requires more computation per iteration.

halting problem   Is there a systematic and mechanical method to prove or disprove any mathematical statement? This fundamental problem was reformulated by Alan Turing in terms of *Turing machines: can a Turing machine A be programmed to determine whether Turing machine B will halt (either solve the problem or determine that the problem cannot be solved in a finite number of steps) when presented with problem X? Turing proved that no such machine A could exist, and thereby demonstrated the existence of undecidable problems in mathematics. *See also* Church's theorem; Gödel's proof.

Hamilton, Sir William Rowan (1805–65) Irish mathematician noted for his introduction in 1843 of *quaternions. Hamilton fully published his results in 1853, while his definitive treatment of the subject, *Elements of Quaternions* (1866), appeared posthumously. He also contributed to dynamics, where the *Hamiltonian function and *Hamilton's principle are still in use.

Hamiltonian A function, H, used to express the rate of change with time of the condition of a dynamic physical system, i.e. one regarded as a set of moving particles. In classical mechanics (as opposed to quantum mechanics) it is a function of the *generalized coordinates q_i and momenta p_i of the system:

$$H = \sum p_i \dot{q}_i - L$$

where L is the *Lagrangian function of the system, $\dot{q}$ is the first derivative of q with respect to time, and p_i, the generalized momenta, are given by

$$p_i = \frac{\partial L}{\partial \dot{q}_i}$$

It follows that

$$\frac{\partial H}{\partial t} = \frac{\partial L}{\partial t}, \quad \frac{\partial H}{\partial p_i} = \dot{q}_i, \quad \frac{\partial H}{\partial q_i} = -\dot{p}_i$$

If H (or L) does not depend explicitly on time, then H is equal to the total energy, kinetic plus potential, of the system.

Hamiltonian graph A connected *graph that contains a closed *path or cycle which includes every vertex (once only). A connected graph that contains a path (not necessarily closed) which includes every vertex is *semi-Hamiltonian*. *Compare* Eulerian graph.

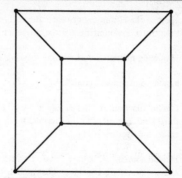
Hamiltonian graph An example

Hamiltonian mechanics The study of mechanics based on *Hamilton's principle.

Hamilton's principle A fundamental principle in dynamics stating that in a *conservative field the motion of a mechanical system can be characterized by requiring that the integral

$$\int_{t_1}^{t_2} (T - V)\, dt$$

be stationary in an actual motion during the time interval t_1 to t_2. T and V are the kinetic and potential energies of the system.

Methods from the *calculus of variations are used to study the circumstances under which the integral is stationary.

Hamming code One of a family of *linear single *error-correcting codes that can also detect (but not correct) two errors. The American mathematician Richard Wesley Hamming (1915–98) published these codes in 1950, and they are in common use in computer memory.

Hamming distance (Hamming metric) The Hamming distance between two *codewords of the same length is the number of places in which they differ. For example, the Hamming distance between 0101 and 1110 is 3. If all pairs of codewords are far apart in the Hamming distance, then multiple transmission errors can be corrected.

Hamming metric *See* Hamming distance.

handle *See* genus; manifold.

Hankel function A *Bessel function of the third kind, named after the German mathematician Hermann Hankel (1839–73).

Hankel matrix A square *matrix whose antidiagonals are constant, illustrated for $n = 3$ by

$$\begin{pmatrix} a & b & c \\ b & c & d \\ c & d & e \end{pmatrix}$$

A Hankel matrix is symmetric.

Hardy, Godfrey Harold (1877–1947) English mathematician noted for his collaboration with J.E. Littlewood in which, between 1910 and 1945, they published nearly 100 papers covering work on number theory, inequalities, and the Riemann hypothesis. On this last topic Hardy proved that there are infinitely many zeroes of the Riemann zeta function on the line $x = \frac{1}{2}$. Hardy also encouraged the Indian mathematician Ramanujan to come to England, and collaborated with him between 1914 and 1917 on a number of topics, of which their work on the partition of integers was the most original.

Hardy–Weinberg law (G.H. Hardy, 1908; W. Weinberg, 1908) In genetics, if two alleles A and a occur in a population in proportions p and $q = 1 - p$, then after one generation of random mating the genotypes AA, Aa, and aa are in proportions p^2, $2pq$, and q^2 and are said to be in *Hardy–Weinberg equilibrium*, because with random mating these proportions are maintained in all future generations.

harmonic **1.** A solution ϕ of the two-dimensional Laplace's differential equation

$$\nabla^2 \phi = 0, \quad \text{i.e.} \quad \frac{\partial^2 \phi}{\partial x^2} + \frac{\partial^2 \phi}{\partial y^2} = 0$$

2. A solution of *Laplace's (differential) equation in *spherical coordinates. A *spherical harmonic* has the form

$$r^n \left[a_n \mathrm{P}_n(\cos\theta) + \sum_{m=1}^{n} \left(a_n^m \cos m\phi \right. \right.$$
$$\left. \left. + b_n^m \sin m\phi \right) \mathrm{P}_n^m(\cos\theta) \right]$$

P_n being a Legendre polynomial and P_n^m an associated Legendre function (*see* Legendre's differential equation). If $r = 1$, the expression is a *surface harmonic*. A surface harmonic of the form $\cos m\phi \, \mathrm{P}_n^m(\cos\theta)$ or $\sin m\phi \, \mathrm{P}_n^m(\cos\theta)$ may be of two types: a *tesseral harmonic* $(m < n)$ or a *sectoral harmonic* $(m = n)$. The function $\mathrm{P}_n(\cos\theta)$ is a *zonal harmonic*.

harmonic analysis The study of functions by expressing them as the sum of a series of a family of functions such as sine and cosine. *See* Fourier series; wavelets.

harmonic mean *See* mean.

harmonic motion A form of *periodic motion, characteristic of elastic bodies (*see* elasticity), in which there is a linear restoring *force acting on the moving particle, point, etc. There may also be additional disturbing forces. In the simplest case, known as *simple harmonic motion* (SHM), there is periodic motion in a straight line: a particle moves to and fro about an equilibrium position such that the restoring force is proportional to the particle's displacement x from this point. The equation of motion is

$$\frac{\mathrm{d}^2 x}{\mathrm{d}t^2} = -\omega^2 x$$

This gives the displacement as

$$x = a \cos(\omega t + \alpha)$$

where a is the maximum displacement, or amplitude, of the motion from the equilibrium position, ω is the angular frequency, $\omega t + \alpha$ is the *phase*, and α is the *initial phase* or *phase angle*. At $t = 0$ the displacement is $a \cos \alpha$. The motion repeats itself in a time $2\pi/\omega$, which is the *period* of the motion. Simple harmonic motion is thus a pure sinusoidal displacement in time with a single amplitude and frequency. Two sinusoidal quantities

$$x_1 = a_1 \cos(\omega t + \alpha_1)$$
$$x_2 = a_2 \cos(\omega t + \alpha_2)$$

where $\alpha_1 \neq \alpha_2$ are said to be *out of phase*, with a phase difference of $|\alpha_1 - \alpha_2|$; if $\alpha_1 = \alpha_2$, the two are said to be *in phase*.

More complex harmonic motion is made up of two or more simple components. For example, uniform circular motion has two simple harmonic components of the same period and amplitude moving at right angles and out of phase by a quarter of the period (i.e. $\frac{1}{2}\pi$). If the phase difference is not a quarter of the period then the motion is elliptical: the resulting motion follows an elliptical path.

See also damped harmonic motion.

harmonic pencil A set of four coplanar concurrent lines such that the four points of intersection with a fifth line form a harmonic range (*see* cross-ratio).

harmonic range *or* **ratio** *See* cross-ratio.

harmonic sequence (harmonic progression) A sequence a_1, a_2, a_3, ... for which the reciprocals of the terms, $1/a_1$, $1/a_2$, $1/a_3$, ... form an *arithmetic sequence.

harmonic series A *series whose terms form a *harmonic sequence. The name is sometimes confined to the divergent series

$$\sum \frac{1}{n} = 1 + \frac{1}{2} + \frac{1}{3} + \cdots$$

harmonic set *See* cross-ratio.

Harriot, Thomas (1560–1621) English mathematician, physicist, and astronomer who, in his posthumous *Artis analyticae praxis* (1631, Applied Analytical Arts), dealt with equations up to the fourth degree and introduced into mathematics the signs > for 'greater than' and < for 'less than'.

Hausdorff, Felix (1868–1942) German mathematician who worked in set theory and was one of the main creators of modern topology. He also introduced the concept of a partially ordered set. He was the first to define a fractional dimension (usually called the *Hausdorff dimension*) of a fractal set.

Hausdorff dimension *See* fractal.

Hausdorff metric A distance defined between two *compact subsets A and B in a *metric space X and which makes the set of all compact subsets of X into a metric space. The distance $\mathrm{d}(A, B)$ is the smallest r such that A and B are contained in r-neighbourhoods of each other, where the r-neighbourhood of a compact subset K of X is

$$\{x : \text{there is a } y \in K \text{ with } \mathrm{d}(x, y) \leqslant r\}$$

Hausdorff space A *topological space X in which distinct points have disjoint *neighbourhoods. That is, if $x \neq y \in X$ then there are open sets U and V with $U \cap V = \varnothing$ and $x \in U$, $y \in V$. Every *metric space is a Hausdorff space.

HCF *Abbreviation for* highest common factor. *See* common factor.

heat equation The *partial differential equation that models the flow of heat in a body. For a three-dimensional body, the equation for the temperature u at time t and position (x, y, z) is

$$\frac{\partial u}{\partial t} = \frac{\partial^2 u}{\partial x^2} + \frac{\partial^2 u}{\partial y^2} + \frac{\partial^2 u}{\partial z^2}$$

The equation can often be solved when the temperature on the boundary of the body is known as a function of time.

hectare An *SI unit of area, equal to $10\,000\,\mathrm{m}^2$.

hecto- *See* SI units.

Heine–Borel theorem *See* compact.

helix A twisted *curve whose tangent always makes a constant angle with a fixed line, called the *axis* of the helix.

A *circular helix* lies entirely on the curved surface of a right circular cylinder, and the axis of the helix is the axis of the cylinder. If the axes are as shown in the diagram, the *parametric equations of the helix are $x = a\cos t$, $y = a\sin t$, and $z = bt$, where a is the radius of the cylinder, b a constant, and t the parameter.

The *pitch* of a helix is the amount by which a point on the helix is displaced, in a direction parallel to the axis, in making one revolution about the axis. For the above circular helix the pitch is $2\pi/b$.

helix A circular helix.

Helly's theorem (E. Helly, 1923) If A_1, A_2, $\ldots$, $A_r \subset \mathbb{R}^n$ are *convex sets with $r > n$ and with the property that every collec-

tion of $n + 1$ of the sets A_i have a point in common, then they all have a point in common.

hemisphere A half-sphere: part of a sphere cut off by a plane through its centre. A hemisphere is a *zone of one base with an altitude equal to the sphere's radius.

Hénon attractor *See* chaos.

henry Symbol: H. The *SI unit of inductance, equal to the inductance of a closed circuit that produces 1 weber of magnetic flux per ampere. [After J. Henry (1797–1878)]

heptagon A *polygon that has seven interior angles (and seven sides).

heptahedron (*plural* **heptahedra**) A *polyhedron that has seven faces, for example a pentagonal prism or a hexagonal pyramid.

Hermite, Charles (1822–1901) French mathematician who in 1873 demonstrated the transcendence of e. He also, using elliptic functions, solved in 1858 the general quintic equation in one variable. Other work on complex numbers led to the definition of *Hermite polynomials, which have since found wide application in modern quantum theory.

Hermite polynomial A *polynomial

$$(-1)^n \exp(x^2)\frac{\mathrm{d}^n}{\mathrm{d}x^n}[\exp(-x^2)]$$

that satisfies the differential equation

$$\frac{\mathrm{d}^2 y}{\mathrm{d}x^2} - 2x\frac{\mathrm{d}y}{\mathrm{d}x} + 2ny = 0$$

Hermitian conjugate A matrix that is the *transpose of the *complex conjugate of a given matrix. The Hermitian conjugate of a matrix A is denoted commonly by A^*, sometimes by $A^\dagger$. It is also called the *associate matrix* or, sometimes, as in quantum mechanics, the *adjoint. If a matrix has a Hermitian conjugate that is equal to the matrix itself, i.e. if $A = A^*$, then the matrix is said to be *Hermitian*. In a Hermitian matrix each element a_{ij} is equal to the complex conjugate of the element a_{ji}. If $A = -A^*$, then the matrix A is said to be *skew-Hermitian* or *anti-Hermitian*. Every real symmetric matrix is Hermitian. For example, the matrices

$$\begin{pmatrix} 1 & i \\ -i & 1 \end{pmatrix} \quad \text{and} \quad \begin{pmatrix} i & i \\ i & 0 \end{pmatrix}$$

are Hermitian and skew-Hermitian, respectively. *See* unitary matrix.

Hermitian matrix *See* Hermitian conjugate.

Hero (*or* **Heron**) of Alexandria (*fl.* AD 62) Greek mathematician and engineer, author of a number of works on mensuration of which the *Metrica* is the most important. In addition to showing how to work out the volume of cones, prisms, pyramids, spherical segments, the five regular polyhedra, and other figures, Hero described a method of approximating square roots and the formula for the area of a triangle that bears his name. He also worked on optics. His best-known book is the *Pneumatica*, in which he describes about a hundred mechanical devices.

Hero's formula (Heron's formula) A formula for the area A of a triangle:

$$A = \sqrt{[s(s - a)(s - b)(s - c)]}$$

where a, b, and c are the lengths of the sides and s is the semiperimeter, i.e.

$$s = \tfrac{1}{2}(a + b + c)$$

Compare Brahmagupta's formula.

Hero's method (Heron's method) An *iterative method of approximating the square *root of a number by estimating the value, dividing this into the number, and taking the average of the result and the initial estimate. For example, the square root of 92 can be estimated as 8.5.

Dividing 92 by 8.5 gives 10.8235. Taking the average of this and 8.5 gives 9.6618. This value can then be used similarly to give a better estimate, of 9.5919 (the 'true' value is 9.5917).

hertz Symbol: Hz. The *SI unit of frequency, equal to one cycle per second. [After H.R. Hertz (1857–94)]

Hessian For a *function y of n independent variables $x_1, x_2, \ldots, x_n$, the *matrix of second-order *partial derivatives of y with respect to the x_i is called the *Hessian matrix* of y. Its *determinant is the *Hessian* of y. The element in the ith row and jth column of the matrix (or determinant) is $\partial^2 y / \partial x_i \partial x_j$. The function is named after the German mathematician Ludwig Otto Hesse (1811–74). *See also* Jacobian; Wronskian.

hexadecimal system A *number system using the base sixteen, in which the digits A to F are used for the numbers represented as 10 to 15, respectively, in the decimal system. For example, in hexadecimal 4B is the number represented in decimal by 75 ($= 4 \times 16 + 11$). *See also* duodecimal system.

hexagon A *polygon that has six interior angles (and six sides).

hexahedron (*plural* **hexahedra**) A *polyhedron that has six faces. A cube is a regular hexahedron.

highest common factor (HCF) *See* common factor.

Hilbert, David (1862–1943) German mathematician who made major contributions to several branches of mathematics. In 1888 he generalized an important theorem of *Gordan's to higher-order systems, while in 1899 he published his famous *Grundlagen der Geometrie* (Foundations of Geometry) in which he provided a rigorous axiomatic foundation for the subject. He also demonstrated that geometry was as consistent as the arithmetic of the real numbers. In 1900 Hilbert posed 23 problems as a challenge to the mathematicians of the 20th century; solutions have been found or substantial advances made for about three-quarters of them.

In later life, Hilbert devoted himself increasingly to work in theoretical physics and the foundations of mathematics. In the latter he developed a strictly formalist position (*see* formalism) which culminated in the two-volume *Grundlagen der Mathematik* (1934, 1939; Foundations of Mathematics), co-written with Paul Bernays. Other work of Hilbert's included his proof of Waring's conjecture, his development of the notion of a *Hilbert space, and contributions to the study of integral equations and algebraic number theory.

Hilbert matrix The $n \times n$ *matrix H_n whose (i,j)th entry is $1/(i+j-1)$. It was introduced in 1894 by Hilbert, who obtained an explicit formula for the determinant of H_n. The matrix is *symmetric positive definite, and is extremely *ill-conditioned, even for moderate values of n. For $n = 3$,

$$H_3 = \begin{pmatrix} 1 & \frac{1}{2} & \frac{1}{3} \\ \frac{1}{2} & \frac{1}{3} & \frac{1}{4} \\ \frac{1}{3} & \frac{1}{4} & \frac{1}{5} \end{pmatrix}$$

and $\det(H_3) = 1/12\,160$.

Hilbert's axioms *See* Euclidean geometry.

Hilbert space A *complete normed *inner product space. Usually the space is considered to have infinite dimensions. *See also* quantum mechanics.

Hindu–Arabic numerals Arabic numerals. *See* number system.

Hipparchus (*c*.190–*c*.126 BC) Greek mathematician and astronomer noted as the author of the first chord table – the equivalent of a modern table of sines – and also for his discovery of the precession of the equinoxes.

Hippasus of Metapontum (*c*.470 BC) Greek mathematician, a Pythagorean, who was said to have revealed the secret that $\sqrt{2}$ was irrational. For this, and for his further discovery of the regular dodecahedron, he was allegedly punished by death by drowning.

Hippias of Elis (*c*.420 BC) Greek mathematician best known for his description of the *quadratrix, a curve which allows an angle to be divided into any given ratio and thus provides a method for the *trisection of an angle.

Hippocrates of Chios (*fl.* 440 BC) Greek mathematician noted as the first geometer to determine the area of a curvilinear figure, namely the lune. He is also supposed to have contributed to the problem of duplication of the cube.

histogram A graphical representation of *grouped data. The x-axis is divided into segments with lengths proportional to each class interval; on these segments as bases, rectangles are drawn with areas proportional to the numbers in the classes. If all class intervals are equal the heights are also proportional to the numbers.

hodograph A curve used to determine the acceleration of a point P moving with known velocity along a curved path. The hodograph is drawn through the ends of lines drawn from a reference position, O, whose length and direction represent the velocity of P at successive positions, i.e. at successive instants, along its path; these lines are thus vectors. If $\overrightarrow{OH}$ and $\overrightarrow{OH'}$ represent the velocity of P at times t and $t + \delta t$, then $\overrightarrow{HH'}$ represents the change in velocity during a time δt. Thus $\overrightarrow{HH'}/\delta t$ represents the average change in velocity (i.e. acceleration) of P during the interval, and also the average velocity of the point H in the hodograph. It follows that, after letting $\delta t \rightarrow 0$, the velocity of the point H represents in magnitude and direction the acceleration of P.

hodograph

Hölder's inequality An inequality used in the study of both Euclidean and function spaces. For the Euclidean plane, the inequality is

$$x_1y_1 + x_2y_2 \leqslant \left(x_1^a + x_2^a\right)^{1/a}\left(y_1^b + y_2^b\right)^{1/b}$$

where $a, b > 1$ and $\dfrac{1}{a} + \dfrac{1}{b} = 1$

It is named after the German mathematician Otto Ludwig Hölder (1859–1937).

holomorphic function *See* analytic function.

homeomorphism Given *topological spaces X and Y, a continuous map $f: X \to Y$ is a homeomorphism if there exists another continuous map $g: Y \to X$ such that $gf(x) = x$ for all $x \in X$ and $fg(y) = y$ for all $y \in Y$. If such f and g exist, X and Y are said to be *homeomorphic*.

For example, the open interval $(-\frac{1}{2}\pi, \frac{1}{2}\pi)$ and the real line $\mathbb{R}^1$ are homeomorphic, a suitable homeomorphism f being defined by $f(x) = \tan x$ (since both the function and its inverse are continuous).

homogeneity of variance Statistical techniques such as those for comparing population means on the basis of observed samples often require it to be assumed that all population variances are equal. A number of tests have been proposed for checking whether this is an acceptable assumption, given some relevant data. Some of these tests, such as *Bartlett's test* (M.S. Bartlett,

1934), work well under an assumption of normality, but sometimes less well if that assumption is violated. If there is doubt about the assumption of normality, then *Levene's test* (H. Levene, 1960) is sometimes preferred to Bartlett's test. There are also a number of nonparametric tests, including the *Siegel–Tukey test* (S. Siegel and J.W. Tukey, 1960), the *Conover squared rank test* (W.J. Conover, 1980), and the *Ansari–Bradley test* (A.R. Ansari and R.A. Bradley, 1960) and others similar to it. *See also* variance ratio.

homogeneous Describing an expression in which the *variables can be replaced by the product of a (nonzero) constant and the variable, and the constant can then be taken out as a factor of the expression. A *homogeneous polynomial* is one in which all the terms have the same total degree. An example is

$$x^2 + 3xy + y^2$$

in which the degree of each term is 2. If x is replaced by kx and y by ky, then the polynomial becomes

$$k^2x^2 + 3k^2xy + k^2y^2$$

i.e.

$$k^2(x^2 + 3xy + y^2)$$

Similarly, a function such as

$$x^2 \sin(x/y) + y^2 \cos(x/y)$$

is homogeneous because if x and y are replaced by kx and ky, respectively, the result is

$$k^2[x^2 \sin(x/y) + y^2 \cos(x/y)]$$

A *homogeneous equation* is one formed by putting a homogeneous polynomial (or other function) equal to zero. For instance,

$$x^2 + y^2 = 0$$

is a homogeneous equation, whereas

$$x^2 + y^2 = 3$$

is not homogeneous.

homogeneous coordinates Numbers a_1, a_2, and a_3 associated with a point (x, y) in Cartesian coordinates, such that

$$x = a_1/a_3 \quad \text{and} \quad y = a_2/a_3$$

A polynomial equation in Cartesian coordinates becomes a homogeneous equation when changed to homogeneous coordinates. For instance,

$$2x^2 + x + 7 = y$$

becomes

$$2a_1^2 + a_1a_3 + 7a_3^2 = a_2a_3$$

homogeneous differential equation A *differential equation of the form

$$\frac{dy}{dx} = f\left(\frac{y}{x}\right)$$

homology group A basic tool in *algebraic topology, first defined by Poincaré (1895). The nth homology group $H_n(X)$ of a topological space X gives an idea of the number of independent 'holes' in X. For example, if Γ is a finite graph, the number of independent generators for the group $H_1(\Gamma)$ equals the number of independent cycles in Γ; and if X denotes 3-dimensional space with 28 (disjoint) balls removed, then $H_2(X)$ has 28 independent generators.

There are a number of approaches to define the homology groups of a space X. All of them involve constructing, from X, a *chain complex* – a sequence of groups C_k with differentials d_{k+1}: $C_{k+1} \to C_k$ satisfying $d_k d_{k+1} = 0$. $H_k(X)$ is then defined to be the quotient $\text{Ker}(d_k)/\text{Im}(d_{k+1})$ of the *kernel $\text{Ker}(d_k)$ of one differential by the image $\text{Im}(d_{k+1})$ of another. For example, if $X = S^n$, the sphere of radius 1 in $(n + 1)$-dimensional space, then $H_k(X) = 0$ unless $k = 0$ or n, and equals the integers $\mathbb{Z}$ when

$k = 0$ or n. Brouwer's *fixed-point theorem follows from this calculation.

homomorphism A *map from one *algebraic structure to another like structure, linked to the algebraic operations in the two structures. Thus, suppose that G and G' are *groups whose operations are written as $\circ$ and $\cdot$ respectively. A group homomorphism from G to G' is a map ϕ whose *domain is G, whose *range is contained in G', and that satisfies

$$\phi(x \circ y) = \phi(x) \cdot \phi(y)$$

for every x and y in G.

For example, if G_1 is the group of all nonzero real numbers with multiplication as the operation, and G_2 is the group of all positive real numbers with the same operation, then the map ϕ_1, given by $\phi_1(x) = x^2$ for each x in G_1, is a homomorphism from G_1 to G_2. This is because

$$\phi_1(xy) = (xy)^2 = x^2y^2 = \phi_1(x) \cdot \phi_1(y)$$

Similarly, a homomorphism from a *ring or *field R with operations $+_R$ and $\times_R$ to another ring Q with operations $+_Q$ and $\times_Q$ is a map f whose domain is R, whose range is contained in Q, and that satisfies

$$f(a +_R b) = f(a) +_Q f(b)$$

and

$$f(a \times_R b) = f(a) \times_Q f(b)$$

for every a and b in R. Likewise, a homomorphism from a (left) *module M to a (left) module N, both over the ring R, is a map g from M to N satisfying

$$g(x +_M y) = g(x) +_N g(y)$$

and

$$g(ax) = ag(x)$$

for every a in R and x, y in M.

homoscedastic Having the same *variance, used in particular when referring to observational error distribution.

homothetic Describing two figures, one of which is mapped onto the other by an *enlargement (homothety). *Homothetic triangles* have corresponding sides parallel and proportional in length.

homothety See enlargement.

homotopy Two continuous maps $f, g: X \to Y$ between *topological spaces are *homotopic* (written as $f \simeq g$) if one can be 'continuously deformed' into the other; that is, if there exists a continuous map $F: X \times I \to Y$ (where I denotes the closed unit interval $[0, 1]$ in $\mathbb{R}^1$) such that $F(x, 0) = f(x)$ and $F(x, 1) = g(x)$ for all $x \in X$. Such an F is called a *homotopy* between f and g.

For example, if Y is a subspace of some Euclidean space $\mathbb{R}^n$, and for all $x \in X$ the line segment in $\mathbb{R}^n$ from $f(x)$ to $g(x)$ is contained in Y, then $f \simeq g$ by reason of $F: X \times I \to Y$, where F is defined by

$$F(x, t) = (1 - t)f(x) + tg(x)$$

(this is called a *linear homotopy*).

A continuous map $f: X \to Y$ is a *homotopy equivalence* if there exists a continuous map $g: Y \to X$ such that gf is homotopic to the identity map of X and fg is homotopic to the identity map of Y (*compare* homeomorphism). If such f and g exist, X and Y are said to be *homotopy-equivalent*. A space X homotopy-equivalent to a single point is said to be *contractible*: $\mathbb{R}^n$, for example, is contractible for all n.

A continuous map $f: X \to Y$ is an *inessential map* if it is homotopic to a continuous map that sends all of X to a single point; otherwise, f is an *essential map*.

See also homotopy group.

homotopy equivalence See homotopy.

homotopy group A tool in *homotopy theory. Given a *topological space X, a point $x_0 \in X$, and an integer $n \geqslant 1$, the nth

homotopy group of X, $\pi_n(X, x_0)$, consists of the *equivalence classes of continuous maps $f: S^n \to X$ (where S^n is the n-sphere) that send $(1, 0, \ldots, 0)$ to x_0, two such maps being defined to be equivalent if they are homotopic, keeping the point $(1, 0, \ldots, 0)$ fixed. A continuous map $g: Y \to X$ gives rise to homomorphisms

$$g_*: \pi_n(X, x_0) \to \pi_n(Y, y_0), \text{ where } y_0 = g(x_0)$$

and g_* is an isomorphism for all n if g is a homotopy equivalence.

For $n = 1$, $\pi_1(X, x_0)$ is sometimes called the *fundamental group* or *Poincaré group* of X; it can readily be calculated if X is a polyhedron, although it is usually a non-Abelian group.

The fundamental group was defined by Poincaré in 1895. Poincaré's definition was extended to the case $n > 1$ by E. Čech (1932) and W. Hurewicz (1935).

homotopy theory A branch of *algebraic topology concerned with the study of those properties of *topological spaces that are invariant under homotopy equivalence. Many problems in homotopy theory are attacked by calculating *homotopy groups.

Hooke, Robert (1635–1703) English mathematical physicist. Hooke's work ranged widely over much of the science of his day and included major contributions to optics and mechanics. In correspondence with Newton in 1679, Hooke made the important proposal that planetary motion was compounded out of 'an attractive motion towards the central body' and direct tangential motion. The proposal turned out to be an essential ingredient in Newton's definitive analysis of curvilinear motion. He is remembered for *Hooke's law of elasticity: within certain limits, strain is proportional to stress. He also invented the conical pendulum.

Hooke's law A law that is the basis of the theory of *elasticity, stating in its most

stress

Hooke's law Stress–strain diagram of a material under tension.

general form that, up to a certain *stress, the *strain produced in a body is proportional to the stress and disappears when the stress is removed. The diagram shows a typical graph of stress versus strain. The segment OA is linear and corresponds to the conditions under which Hooke's law holds. The slope of OA is the *modulus of elasticity for the material under study; different moduli apply to different types of strain. A body obeying Hooke's law is said to be *perfectly elastic*. A body can still be considered *elastic* if it returns to its original shape once the stress is removed. This occurs for small stresses and is represented in the diagram by the *elastic* region OB. Above what is known as the *yield stress*, brittle materials tend to crack while others become *plastic*; BD represents the plastic region on the graph. A material stressed into the plastic state cannot resume its original shape but takes on a permanent deformation or *set*; OF represents the permanent set for a stress (load) removed at point C. (The diagram may also be regarded as a plot of load versus extension – the deformation of the body in the direction of the applied load.) As the stress is increased the material will eventually fracture; this is represented by point E, where the stress is known as the *breaking stress*.

Hopf bifurcation *See* bifurcation.

horizon The circle that is the intersection of a horizontal plane through the position of an observer with the *celestial sphere. The zenith and nadir are poles of the horizon. *See* horizontal coordinate system.

horizontal coordinate system An *astronomical coordinate system in which measurements are based on the horizon. A point on the *celestial sphere is located by two angular measurements. The *azimuth (A) is the angular distance measured eastwards from the north point. The altitude (h) is the angular distance north or south of the horizon. Sometimes *zenith distance (ζ), which is the complement of altitude (i.e. $\zeta = 90° - h$), is used.

Horner's method 1. A method for evaluating a polynomial, also known as *nested multiplication*. A polynomial

$$p(x) = a_n x^n + a_{n-1} x^{n-1} + \cdots + a_1 x + a_0$$

can be written in the nested form

$$p(x) = (\cdots ((a_n x + a_{n-1})x + a_{n-2})x + \cdots + a_1)x + a_0$$

For example, the cubic $4x^3 - 2x^2 + 3x - 1$ may be written as

$$((4x - 2)x + 3)x - 1$$

Horner's method corresponds to evaluating the polynomial in the order suggested by the parentheses. It requires fewer multiplications than the more obvious method that involves forming the powers x^i, multiplying each by the corresponding a_i, and summing.

Horner's method is closely related to the method of *synthetic division* for dividing a polynomial by a linear factor. If we divide $p(x)$ by $x - a$, we obtain

$$p(x) = (x - a)q(x) + r$$

where $q(x)$ is a polynomial of degree $n - 1$ and r is a constant. When we apply Horner's method to evaluate $p(a)$, the coefficient a_n together with the intermediate quantities produced as each pair of parentheses is removed are the coefficients of q, and the final result is $r = \mathrm{p}(a)$. Thus, when $a = 2$ the coefficient and the intermediate values for the above cubic are 4, 6, and 15, and the final result is 29. Moreover,

$$4x^3 - 2x^2 + 3x - 1 =$$
$$(x - 2)(4x^2 + 6x + 15) + 29$$

2. An *iterative method for finding real roots of *polynomial equations. If, for example, the root required is the positive decimal $abc.def\ldots$, the process begins by finding the leading digit a by inspection. A new equation is formed whose roots are $100a$ less than those of the given equation. This will have a root $bc.def\ldots$ (in decimal form). The digit b is then found by inspection, a new equation is formed with roots $10b$ less than before, and so on.

The method was given in 1819 by the English mathematician William Horner (1786–1837), but also independently in 1804 by the Italian algebraist Paolo Ruffini (1765–1822), and is more properly known as the *Ruffini–Horner method*. Convergence to a root is certain, but slow.

See also Qin Jiushao.

horsepower Symbol: hp. An *f.p.s. unit of power, equal to a rate of doing work of 550 foot-pounds per second. 1 hp = 745.7 watts.

Hotelling's T-test (H. Hotelling, 1931) A generalization of the *t-test to hypothesis tests about multivariate *normal distribution means.

Hough transform A general technique now used to identify the locations of features in digital images. The transform is a development of the *Radon transform suitable for numerical calculation and was invented in its original form by the American physicist Paul Hough in 1959.

hour Symbol: h. A unit of time equal to 60 minutes. It was formerly defined as 1/24 of a mean solar *day.

hour angle (HA) Symbol: t. The angle on the *celestial sphere between an observer's *meridian and the *hour circle of a given point. It is measured westwards along the celestial equator and expressed in units of hours, minutes, and seconds. The hour angle of a star changes daily from 0 to 24 hours because of the rotation of the earth. It is sometimes used in place of *right ascension. *See* equatorial coordinate system.

hour circle A great circle on the *celestial sphere passing through the celestial poles.

hull, convex *See* convex hull.

hundredweight Symbol: cwt. An *avoirdupois unit of mass, equal to 112 pounds in the UK. In the USA it is equal to 100 pounds and is sometimes known as the *short hundredweight*.

Huygens, Christiaan (1629–95) Dutch mathematical physicist and astronomer known for his *Horologium oscillatorium* (1673, The Pendulum Clock) in which he dealt with the problem of accelerated bodies falling freely. He demonstrated that the *cycloid was the tautochronous curve and introduced his theory of evolutes and centrifugal force. Other mathematical work by Huygens was concerned with the cissoid, the catenary, the logarithmic curve, and probability theory.

hydrodynamics *See* hydrostatics.

hydrostatics The study of the mechanical properties and behaviour of fluids, particularly liquids, that are not in motion. It is concerned mainly with the forces that arise from the presence of fluids. The study of fluid motion is more complex and is known as *hydrodynamics*.

Hypatia ($c.$370–415) The first woman to be named in the history of mathematics. She was the daughter of a mathematician and astronomer, Theon of Alexandria, and is credited with commentaries on Diophantus and Apollonius.

hyperbola A type of *conic that has an *eccentricity (e) greater than 1. It is an open curve with two symmetrical branches. In a Cartesian coordinate system the standard equation of the hyperbola is

$$\frac{x^2}{a^2} - \frac{y^2}{b^2} = 1$$

In this form of the equation each branch of the curve cuts the x-axis, one on each side of the origin. The x- and y-axes are two axes of symmetry for the curve. The one along the x-axis is the *transverse axis*; the one along the y-axis is the *conjugate axis*. These terms for the axes of symmetry are applied to any hyperbola (not necessarily having axes that coincide with the coordinate axes). The terms are also used for line segments on these axes. The transverse axis is the segment between the two branches of the curve (length $2a$). If two ordinates are drawn at each of the points (*vertices*) at which the branches of the curve meet the transverse axis, then these cut the *asymptotes at four points: A, B, C, and D. The line segment on the conjugate axis cut off by two parallel sides of the rectangle ABCD is also called the *conjugate axis*, and its length is $2b$. A line segment of length a from the centre of the hyperbola along the transverse axis is a *semitransverse axis*. One from the centre of length b along the conjugate axis is a *semiconjugate axis*.

The hyperbola has two directrices and two foci, one each side of the centre (*see* diagram (a)). The eccentricity of a hyperbola is given by $c/2a$, where c is the distance between the two foci. Alternatively, it can be given by

$$e^2 = 1 + \frac{b^2}{a^2}$$

hyperbola (a) F_1 and F_2 are foci.

Either of the two chords through a focus and perpendicular to the transverse axis is a *latus rectum*. The length of the latus rectum is $2b^2/a$.

With the equation in the form

$$\frac{x^2}{a^2} - \frac{y^2}{b^2} = 1$$

the two asymptotes of the hyperbola are the lines $y = bx/a$ and $y = -bx/a$. If the transverse and conjugate axes are equal (i.e. $a = b$), the equation becomes

$$x^2 - y^2 = a^2$$

In this case, the asymptotes are mutually perpendicular (the lines $y = x$ and $y = -x$) and the hyperbola is said to be a *rectangular* (or an *equilateral* or *equiangular*) *hyperbola*. If the coordinate axes are rotated clockwise through $45°$, the equation of the hyperbola becomes (with respect to new axes x and y) $xy = c^2$, where $c = a/\sqrt{2}$. In this form, the x- and y-axes are the asymptotes; the transverse and conjugate axes are the lines $y = x$ and $y = -x$, respectively, and the hyperbola has the *parametric equations $x = ct$ and $y = c/t$.

A circle with its centre at the centre of any hyperbola and passing through the vertices (i.e. with radius a) is an *eccentric circle* of the hyperbola (*see* diagram (b)). The similar circle with radius b is also an eccentric circle. The one with radius a is called the *auxiliary circle* of the hyperbola.

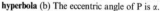

hyperbola (b) The eccentric angle of P is α.

If the hyperbola has its centre at the origin and its transverse axis along the x-axis, the *eccentric angle*, α, is defined as follows. For any point P on the hyperbola an ordinate is drawn to the x-axis. From the point at which this meets the axis a tangent is drawn to the auxiliary circle at P' (on the same side of the x-axis); α is the positive angle between the x-axis and the radius OP'. The parametric equations of the hyperbola are $x = a \sec \alpha$ and $y = b \tan \alpha$.

The hyperbola has two properties connected with its foci, F_1 and F_2. For any point P on the hyperbola, the difference $|PF_1 - PF_2|$ is constant (equal to $2a$). The *focal property* of the hyperbola is that the tangent at any point P, APB, makes equal angles with straight lines from the foci to the point; i.e. $\angle F_1 PA = \angle F_2 PA$ (*see* diagram (c)). This is also called the *reflection property*, since a reflector shaped like a hyperbola would reflect

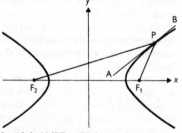

hyperbola (c) $|PF_1 - PF_2| = 2a$, and angle $F_1 PA$ = angle $F_2 PA$.

rays of light from a source at one focus so as to appear to come from the other focus (called the *optical property*). The analogous reflection of sound leads to the alternative term *acoustical property*.

hyperbolic functions Functions analogous to *trigonometric functions, related to the *hyperbola, $x^2 - y^2 = r^2$, in a similar way to that in which the trigonometric functions are related to the circle. They are termed *hyperbolic sine*, *hyperbolic cosine*, etc. The abbreviations of hyperbolic functions are those for the corresponding trigonometric functions with 'h' on the end: sinh (pronounced 'shine'), cosh, tanh (pronounced 'than' or 'tansh'), csch (pronounced 'cosesh'), sech, and coth. The functions are defined as follows:

Hyperbolic sine: $\quad \sinh x = \frac{1}{2}(e^x - e^{-x})$

Hyperbolic cosine: $\cosh x = \frac{1}{2}(e^x + e^{-x})$

These two equations, together with the relationship

$$e^x = \cosh x + \sinh x$$

are the hyperbolic analogues of *Euler's identities. The other hyperbolic functions are given by relationships analogous to those for the trigonometric functions:

Hyperbolic tangent: $\quad \tanh x = \dfrac{\sinh x}{\cosh x}$

Hyperbolic cosecant: $\quad \operatorname{csch} x = \dfrac{1}{\sinh x}$

Hyperbolic secant: $\quad \operatorname{sech} x = \dfrac{1}{\cosh x}$

Hyperbolic cotangent: $\quad \coth x = \dfrac{1}{\tanh x}$

There are various identities between the hyperbolic functions:

$$\sinh(-x) = -\sinh x$$

$$\cosh(-x) = \cosh x$$

$$\cosh^2 x - \sinh^2 x = 1$$

$$\coth^2 x - \operatorname{csch}^2 x = 1$$

$$\operatorname{sech}^2 x + \tanh^2 x = 1$$

The hyperbolic functions can be written as series:

$$\sinh x = x + \frac{x^3}{3!} + \frac{x^5}{5!} + \cdots$$

$$\cosh x = 1 + \frac{x^2}{2!} + \frac{x^4}{4!} + \cdots$$

A hyperbolic function is related to the corresponding trigonometric function:

$$\sinh \mathrm{i}x = \mathrm{i}\sin x$$

$$\cosh \mathrm{i}x = \cos x$$

$$\tanh \mathrm{i}x = \mathrm{i}\tan x$$

See also inverse hyperbolic functions.

hyperbolic geometry *See* non-Euclidean geometry.

hyperbolic logarithm A natural *loga-rithm.

hyperbolic paraboloid *See* paraboloid.

hyperbolic spiral *See* spiral.

hyperboloid A surface that has plane sections which are either *hyperbolas or *ellipses. A special case is a *hyperboloid of revolution*, generated by revolving a hyperbola about one of its axes. If the hyperbola is revolved about its conjugate axis (which lies between the two branches), a *hyperboloid of one sheet* is formed. If revolved about the transverse axis (passing through the vertices), the surface is a *hyperboloid of two sheets*. In each case, plane sections perpendicular to the axis of revolution are circles; sections parallel to the axis are hyperbolas (*see* diagram).

The general forms of the equation for a hyperboloid are

$$\frac{x^2}{a^2} + \frac{y^2}{b^2} - \frac{z^2}{c^2} = 1$$

for a hyperboloid of one sheet, and

$$\frac{x^2}{a^2} - \frac{y^2}{b^2} - \frac{z^2}{c^2} = 1$$

for a hyperboloid of two sheets.

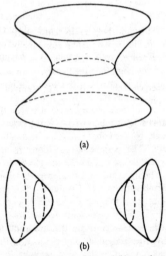

(a)

(b)

hyperboloid of (a) one sheet and (b) two sheets.

hypercube A regular *polytope in a space of four or more dimensions which is the analogue of the cube in three-dimensional space. A hypercube in four-dimensional space is called a *tesseract*.

hypergeometric differential equation A second-order *differential equation of the form

$$x(1-x)\frac{\mathrm{d}^2 \phi}{\mathrm{d}x^2} +$$

$$[c - (a+b-1)x]\frac{\mathrm{d}\phi}{\mathrm{d}x} - ab\phi = 0$$

where a, b, and c are constants.

hypergeometric distribution If a sample of n units is taken without replacement from a *population with M satisfactory units and N unsatisfactory units, $M, N > n$, then the number of satisfactory units in the sample has a hypergeometric distribution. If X is the number of satisfactory units, then

$$\Pr(X = r)$$

$$= \binom{M}{r}\binom{N}{n-r} \bigg/ \binom{M+N}{n}$$

where $\binom{M}{r}$, etc. are *binomial coefficients. When M and N are large compared with n, the distribution approaches the *binomial distribution with parameters n and $p = M/(M+N)$.

hypergeometric function The sum of a *convergent *hypergeometric series.

hypergeometric series The power series

$$\sum_{n=0}^{\infty} \alpha_n x^n, \quad \text{where } \alpha_n = \frac{(a)_n (b)_n}{(c)_n n!}$$

and $(a)_n$ denotes the *rising factorial*

$$(a)_n = a(a+1)(a+2)\cdots(a+n-1)$$

with n terms. It defines a solution to the *hypergeometric differential equation.

hypocycloid A plane curve that is the *locus of a point on a circle which rolls on the inside of a fixed circle. The hypocycloid is thus similar to the *epicycloid (in which the generating circle rolls on the outside of the fixed circle). The *hypotrochoid* is analogous to the *epitrochoid. *See also* astroid; deltoid; cycloid.

hypotenuse The side opposite the right angle in a right-angled triangle.

hypothesis *See* conjecture.

hypothesis testing A procedure for deciding whether a hypothesis H_0, known as the *null hypothesis*, should be accepted or rejected in favour of an *alternative hypoth-*

esis, H_1. To perform the test, an appropriate *statistic is calculated from observed sample values. The sample space for this statistic is divided into an *acceptance region* and a *critical* (or *rejection*) *region*, the latter being chosen such that the probability that the statistic takes a value in this region when H_0 is true takes a pre-specified value α. Traditional choices for α are 0.05, 0.01, and 0.001. If the probability that the statistic takes a value in the critical region when H_1 is true is always greater than α, the test is said to be unbiased (*see* unbiased hypothesis test). If the statistic takes a value in the acceptance region, then H_0 is accepted; if it takes a value in the critical region, H_0 is rejected at significance level α (or the 100α percent significance level). The value of the statistic marking the boundary between the acceptance and critical regions is called the *critical value*.

Critical values corresponding to $\alpha = 0.05$, 0.01, and 0.001 are tabulated for many commonly used statistics, such as those for the *t-test, F-test (*see* F-distribution), and *chi-squared test, and this has made the use of these levels traditional. However, modern statistical software makes it easy to calculate the probability p, called the *p-value, that a statistic will take a value greater than or equal to (or sometimes less than or equal to) a calculated value when H_0 is true. If p is less than a pre-assigned significance level α, the term *exact significance level* is sometimes used for p. A statement that, for example, $p = 0.023$ is more informative than simply saying that a result is significant at the 5 percent but not at the 1 percent level. While the ready availability of p-values gives us freedom to set critical values at less conventional levels than $\alpha = 0.05$ or 0.01, these conventional values should still be regarded as useful criteria for declaring significance.

Rejecting a hypothesis when it is true (and for a large number of independent tests with fixed α, this will happen in 100α percent of the tests) is called an *error of the first kind* or a *Type I error*. Accepting H_0 when it is not true is called an *error of the second kind* or a *Type II error*. The probability of a Type I error is controlled by fixing α, but in general the probability of a Type II error depends on the alternative hypothesis H_1, and for a particular H_1 it is often denoted by β. The probability of rejecting the null hypothesis when it is false is called the *power* of a test. Thus the power is $1 - \beta$, and so it also depends on H_1. While H_0 often, but not always, states the specific value of a parameter θ, say 2 (written as $H_0: \theta = 2$), the alternative hypothesis is often more general, for example $H_1: \theta > 2$ or $H_1: \theta \neq 2$. The former is called a *one-sided alternative* and the corresponding test is a *one-tail test*; the latter is a *two-sided alternative* (θ may be either greater or less than 2), and the corresponding test is a *two-tail test*.

A hypothesis is *simple* if the distribution function of the population random variable is completely specified, otherwise it is *composite*. For example, the hypothesis 'X is $N(2, 9)$' is simple, but the hypothesis 'X is normally distributed with mean 2 but unknown variance' is composite.

See also estimation.

hypotrochoid *See* hypocycloid.

I

i The symbol for $\sqrt{-1}$; the *complex number with unit modulus and an argument of $\frac{1}{2}\pi$.

icosagon A *polygon that has 20 interior angles (and 20 sides).

icosahedron (*plural* **icosahedra**) A *polyhedron that has 20 faces. A regular icosahedron, in which all the faces are equilateral triangles, is one of the five regular polyhedra.

icosidodecahedron (*plural* **icosidodecahedra**) A type of *polyhedron with 32 faces.

ideal A *set I of elements of a *ring R such that I is a *subring of R, and for every a in R and x in I

(1) ax is in I; and

(2) xa is in I.

If just (1) is satisfied, then I is a *left ideal* of R; if just (2) holds, it is a *right ideal*. An ideal I is a *principal ideal* if it contains an element g such that I is the smallest ideal containing g. It is the *intersection of all the ideals containing g and is said to be generated by g.

For a fixed a in R, the *coset $(a + I)$ is the set of all elements of the form $(a + y)$, where y can be any element of I. If the subring I is an ideal, then (and only then) the set of all cosets forms a ring, with addition and multiplication of the typical cosets $(a + I)$ and $(b + I)$ defined by

$$(a + I) + (b + I) = (a + b) + I$$

$$(a + I)(b + I) = ab + I$$

The resulting ring is called the *quotient ring R/I, and the ideal I is its zero element.

In the ring Z of all integers, with the usual addition and multiplication, the set $3Z$ of all multiples of 3 is a principal ideal generated by 3. The quotient ring $Z/3Z$ consists of just three elements:

$$0 + 3Z = 3Z = \{0, \pm 3, \pm 6, \ldots\}$$

$$1 + 3Z = \{\ldots, -5, -2, 1, 4, 7, \ldots\}$$

$$2 + 3Z = \{\ldots, -4, -1, 2, 5, 8, \ldots\}$$

ideal point A point at infinity in *projective geometry.

idempotent Describing a quantity that is unchanged by multiplication by itself. Unity is idempotent, as is any *identity matrix.

identical (of geometric figures) *See* congruent.

identity A statement that two mathematical expressions are equal for all values of their *variables.

Sometimes the sign $\equiv$ is used to indicate that the statement is an identity. For example,

$$a^2 - b^2 \equiv (a + b)(a - b)$$

See equation.

identity element For a given *binary operation, an element I of a set for which

$$I \circ x = x \circ I = x$$

for all members x of the set. For example, in multiplication of numbers, the identity element is 1 ($x.1 = x$); in addition of numbers the identity element is 0 ($x + 0 = x$).

identity mapping (identity function) For any *domain X, the mapping f: $x \mapsto x$ that maps each element x of X into itself. *See* monoid.

identity matrix (unit matrix) A *diagonal matrix in which all the elements on the leading diagonal are unity, and all other elements are zero.

IEEE arithmetic A standard for floating-point arithmetic published in the 1980s by the Institute of Electrical and Electronic Engineers' Computer Society Computer Standards Committee. The standard specifies, among other things, floating-point number formats and the results of the basic *floating-point operations. IEEE standard arithmetic is used on most modern computers.

iff *Abbreviation for* if and only if. *See* biconditional; equivalence.

ill-conditioned A problem is ill-conditioned if its solution is sensitive to small changes in the input data. This sensitivity affects any method for solving the problem, so ill-conditioned problems are hard to solve accurately. Consider the zeroes of the polynomials p(x) $= x^8$ and q(x) $= x^8 - 10^{-8}$. The zeroes of p are all 0, whereas the zeroes of q, a tiny perturbation of p, are all of modulus 0.1. Therefore the zeroes of p are ill-conditioned functions of the constant term (0 for p and -10^{-8} for q) in the polynomial. For another example, the solution of the equations

$$x - y = 1, \quad x - 1.0001y = 0$$

is $x = 10\,001$, $y = 10\,000$, while the solution of the equations

$$x - y = 1, \quad x - 0.9999y = 0$$

is $x = -9999$, $y = -10\,000$. Here, a change in the fourth decimal place of one coefficient has changed the solution completely. This can be explained by the fact that the *coefficient matrix is nearly *singular; such a matrix is said to be ill-conditioned.

image If f is a *function that assigns to each element of its *domain A a unique element of its *codomain B, then the element of B assigned to a particular element of A is known as the image of that element.

For example, if $A = \{a, b, c, d\}$ and $B = \{w, x, y, z\}$, and the function f assigns to a the unique element z, then z is the image of a. If A' is a subset of the domain A, then the image of A' is the subset of B containing the images of all the members of A'. The image of a *function or mapping f: $A \rightarrow B$ is the subset $\text{Im(f)} = \{f(a): a \in A\}$.

imaginary axis *See* Argand diagram.

imaginary circle The set of (imaginary) points that satisfy an equation of the form

$$(x - a)^2 + (y - b)^2 = -r^2$$

imaginary number *See* complex number.

imaginary part *See* complex number.

imperial units A system of units that developed in the UK and was formally defined in 1824. It has been widely used in various derivative forms in most English-speaking countries. Imperial units are usually *f.p.s. units. Factors of 12 and 60 frequently feature in their submultiples. The imperial system, with its arbitrary and illogical structure, has been replaced for many purposes by the *metric system, and for scientific purposes by *SI units. *See also* avoirdupois; British units of length; United States customary units.

implication **1. (material implication)** A truth-functional connective (*see* truth function), often symbolized in a formal system as $\supset$ or $\rightarrow$, whose meaning is given by the following *truth table:

A	B	$A \supset B$
T	T	T
T	F	F
F	T	T
F	F	T

Material implication is the relation between two statements A and B when the conditional $A \supset B$ (if A then B) is true. Material implication does not necessarily represent the logical force of conditional statements. Thus, if A is false, $A \supset B$ is true, and if A is true, $B \supset A$ is true, no matter what the statement B is. On the basis of material implication, the statement 'if elephants have two heads, cats can walk on water' is true. These are the so-called *paradoxes of material implication*, which have led to the search for definitions of strict implication (*see* below).

2. (strict implication) A *connective of *modal logic, often symbolized as $\Rightarrow$ and usually defined as $\Box (A \supset B)$. As the truth value of $\Box (A \supset B)$ does not depend wholly on the truth values of A and B, $\Rightarrow$ is not a truth-functional connective. For example, the truth of 'snow is white $\supset$ Aristotle was a philosopher' does not determine whether or not it is necessarily true.

The use of strict implication avoids the paradoxes of material implication, while allowing the derivation of the following two comparable paradoxes of strict implication:

(1) an impossible proposition strictly implies any proposition;
(2) a necessary proposition is strictly implied by any proposition.

Thus, the proposition '$2 + 2 = 5$' strictly implies 'grass is blue', while 'grass is blue' strictly implies '$2 + 2 = 4$'.

See also conditional.

3. (entailment) Attempts have been made to define a connective which avoids the paradoxes of material and strict implication by insisting that, before p can imply q, it must be both relevant to and actually used in the derivation of q; it is then said that p *entails* q (written as $p \Rightarrow q$ or $p \prec q$). This approach to implication rejects, on the grounds of relevance, the classical inference known as the *disjunctive syllogism* that, from $\sim p$ and $p \lor q$, we can derive q, and consequently that $p \rightarrow q$ can be defined as $\sim p \lor q$.

implicit differentiation The *differentiation of an *implicit function with respect to the *independent variable to find the derivative. For example, the function

$$y^3 + 2x^2 y + 8 = 0$$

can be differentiated with respect to x to give

$$3y^2 \frac{\mathrm{d}y}{\mathrm{d}x} + 4xy + 2x^2 \frac{\mathrm{d}y}{\mathrm{d}x} = 0$$

which can then be rearranged to give $\mathrm{d}y/\mathrm{d}x$ in terms of y and x.

implicit function A *function defined by $F(x_1, x_2, \ldots, x_n, y) = 0$ where y is the *dependent variable. An example is

$$yx_1 + y^2 x_2 + x_3^2 = 0$$

It is sometimes possible to derive *explicit functions exactly with the form $y = f(x_1, x_2, \ldots, x_n)$ from an implicit function. For example, the implicit function $x^3 + y^2 = 1$ gives explicit functions

$$y = +\sqrt{(1 - x^3)}, \quad y = -\sqrt{(1 - x^3)}$$

In other cases approximate explicit functions can be obtained. *Compare* explicit function.

implicit function theorem The theorem that gives conditions on derivatives which ensure that an implicitly defined set is (locally) the *graph of a *function. For example, the circle $x^2 + y^2 = 1$ is locally the graph of a function near all points where $y \neq 0$. The theorem is often used to check that a subset of a Euclidean space is a *manifold.

improper fraction A fraction in which the numerator is greater than the denominator. For example, 4/3 is an improper fraction (3/4 is a *proper fraction*).

improper integral See infinite integral.

impulse 1. The time integral of a *force **F** acting on a particle over a finite time, say from t_1 to t_2:

$$\int_{t_1}^{t_2} \mathbf{F}\,dt$$

For a constant force this reduces to the product $\mathbf{F}(t_2 - t_1)$. Newton's second law of motion,

$$\mathbf{F} = m\mathbf{a} = \frac{d(m\mathbf{v})}{dt}$$

where m is the mass of a particle whose velocity and acceleration at time t are **v** and **a**, indicates that the impulse of a force is equal to the change of *momentum, $m\mathbf{v}$, experienced by the particle in this time.
2. (impulsive force) A large *force acting for a very short time, such as the blow of a hammer.

incentre The centre of the *incircle of a polygon. In the case of a triangle, the incentre is the point of intersection of the bisectors of the interior angles of the triangle. Compare excentre.

inch Symbol: in. A *British unit of length equal to 1/12 foot. 1 inch = 0.0254 m.

incircle (inscribed circle) A circle *inscribed in a given *polygon. Compare excircle.

inclined plane A plane that is not horizontal.

inclusion A *set A is included in a set B, denoted by $A \subseteq B$, if and only if A is a *subset of B. See also proper inclusion.

incommensurable Not *commensurable. Two numbers are incommensurable if they cannot be expressed as integral multiples of the same number. Thus, 6 and $\sqrt{3}$ are incommensurable because 6 is rational and $\sqrt{3}$ is irrational.

incomplete decoding A *decoding that has detected some errors but has been unable to correct them.

incompleteness theorems See Gödel's proof.

inconsistent equations See consistent.

increasing function See monotonic increasing function.

increasing sequence A *sequence $a_1, a_2, \dots$ for which $a_n < a_{n+1}$ for all n is said to be *strictly increasing*. The sequence is described as *monotonic increasing* if $a_n \leqslant a_{n+1}$ for all n.

If a monotonic increasing sequence $\{a_n\}$ has an upper bound (*see* bounded sequence) then it tends to a finite limit; without an upper bound, $a_n \to \infty$ as $n \to \infty$.

Compare decreasing sequence.

increment A positive or negative change in a *variable. The term is generally used to mean a small change.

indefinite integral An integral without any specified *limits, whose solution includes an undetermined constant C (the constant of integration). Compare definite integral; see integration.

independence 1. Events A and B are said to be *independent events* in the probabilistic sense if the probability that both occur is the product of the probability of either occurring: $\Pr(A \& B) = \Pr(A) \cdot \Pr(B)$. See probability.
2. *Random variables X and Y are said to be independent if the *distribution function $F(x, y)$ factorizes into the product of the two marginal distribution functions $F_1(x)$ and $F_2(y)$. Corresponding properties hold for the *frequency function. See random variable; bivariate distribution.

independent 1. Describing an *axiom of a *formal (logical) system that is not a formal consequence of any other axioms.
2. Describing a rule of *inference of a *formal system that cannot be derived from the axioms and the remaining rules of inference. If an axiom or a rule of inference fails to be independent then the formal system may still be acceptable, even though it may not be economical. *See also* proof theory.

independent equations *See* dependent equations.

independent events *See* independence.

independent variable *See* function; regression; variable.

indeterminate equation An equation in two or more variables with an infinite set of solutions. For example, the equation

$$3x + 4y = 50$$

is indeterminate.

A system of *simultaneous *linear equations with an infinite set of solutions is also said to be indeterminate. For instance, the system

$$x + y = 5, \quad x + z = 6$$

with three variables, is indeterminate.

indeterminate expression (indeterminate form) An undefined expression, such as $0/0$, ∞/∞, $0 \times \infty$, $\infty - \infty$, etc.

index (*plural* **indices**) **1. (index number)** In statistics, a measure of change in magnitude of business activity, wages, cost of living, share prices, imports, etc. with time. Constructing a meaningful index requires reliable information on relevant components, usually weighted according to importance. For the *base* year the index value is usually 100 (sometimes 1000). If the price of an item is 50 in the base year, and 55 and 70 in the next two years, the price indices for that item would be 100, 110, and 140, respectively. Most indices are

weighted *means of a number of such simple indices, often called *relatives*.

If the prices of a set of k commodities in the base year are p_{01}, p_{02}, ..., p_{0k}, the quantities of each sold are q_{01}, q_{02}, ..., q_{0k}, and the corresponding prices and quantities sold in year n are p_{n1}, p_{n2}, ..., p_{nk} and q_{n1}, q_{n2}, ..., q_{nk}, then the index

$$L_{0n} = \sum_j p_{nj} q_{0j} \Big/ \sum_j p_{0j} q_{0j}$$

is the *Laspeyres index*, and the index

$$P_{0n} = \sum_j p_{nj} q_{nj} \Big/ \sum_j p_{0j} q_{nj}$$

is the *Paasche index*. The weights in the Laspeyres index are base year quantities, and those in the Paasche index are current year quantities (É. Laspeyres, 1871; H. Paasche, 1874).
2. *See* exponent.
3. *See* radical.
4. (of a subgroup) If G is a *group and H is a *subgroup of G, then the index of H in G is the number of left *cosets of H in G. This is also the number of right cosets of H in G. The index of H in G is denoted by $[G:H]$; and if K is a subgroup of G which is also a subgroup of H, then $[G:K] = [G:H][H:K]$.

index notation *See* exponential notation.

index of determination *See* coefficient of determination.

index set A *set whose elements are being used to label (or index) a family of mathematical objects. For example, the index set for the set of functions f_1, f_2, and f_3 is $\{1, 2, 3\}$.

indicator diagram A curve in which the y-coordinates represent a varying *force or pressure in a system and the x-coordinates represent the corresponding distances through which a component of the system has moved. The area under the curve, or enclosed by the curve when a cycle is involved, indicates the *work done.

indicator function *See* characteristic function.

indices *Plural of* index.

indirect proof (proof by contradiction) A proof used in circumstances when it is more convenient, in setting out to prove P, to begin with the negation of P, $\sim P$, and to show that this assumption leads to a contradiction. Thus, if the assumption of $\sim P$ leads to an absurdity, then $\sim P$ must be false and P must be true. For example, to prove that there are an infinite number of primes, begin by assuming that there is in fact a greatest prime N. If from this we can deduce that there is a prime greater than N, it will follow indirectly that there can be no N such that N is the greatest prime, and consequently that there must be an infinite number of primes. Such a method of proof was known traditionally as *reductio ad absurdum* [Latin: reduction to absurdity].

Mathematicians sometimes adopt a double *reductio ad absurdum*. Thus, to show that the area A of a circle is the same as the area T of a certain triangle, Archimedes argued as follows:

(1) the assumption that $A > T$ leads to a contradiction, therefore $A \ngtr T$;
(2) the assumption that $A < T$ leads to a contradiction, therefore $A \nless T$;
(3) it follows that, since $A \ngtr T$ and $A \nless T$, then $A = T$.

individual constant *See* constant.

indivisibles (method of) *See* calculus.

induction 1. (in mathematics) A common method of proving that each of an infinite *sequence of mathematical statements is true by proving that

(1) the first statement is true;
(2) the truth of any one of the statements always implies the truth of the next one.

For if (1) and (2) hold, then the truth of the first statement will imply the truth of the second statement, which in turn will imply the truth of the third statement, and so on.

As an example, consider the theorem that the sum of the first n natural numbers is $\frac{1}{2}n(n+1)$. This is really an infinite sequence of statements: one for each $n = 1, 2, \ldots$. The first statement is true, as the first sum is $1 = \frac{1}{2} \times 1 \times 2$. The requirement (2) amounts to showing, for any n, that if

$$1 + 2 + \cdots + n = \frac{1}{2}n(n+1)$$

then

$$1 + 2 + \cdots + n + (n+1)$$
$$= \frac{1}{2}(n+1)[(n+1)+1]$$

But if

$$1 + 2 + \cdots + n = \frac{1}{2}n(n+1)$$

then

$$1 + 2 + \cdots + n + (n+1)$$
$$= \frac{1}{2}n(n+1) + (n+1)$$

This equals $\frac{1}{2}(n+1)(n+2)$, so in this case the truth of the general nth statement does imply the truth of the $(n+1)$th statement; and as the first statement is true, then, by the method of induction, they must all be true.

It is often convenient just to refer to the numbers that label the various statements (as statement 1, statement 2, etc.) and to concentrate attention on the set, or collection, of number labels of the true statements. So the inductive method can be formulated equivalently in the language of set theory as the *principle of induction*: if a set of natural numbers contains 1, and if it contains $n + 1$ whenever it contains a number n, then it must contain every natural number.

The above method of induction can still be used in some cases where the first statement is not true, or does not make sense. If there is a certain natural number k such that

(1) the kth statement is true; and
(2) the truth of each statement, from the kth one onwards, implies the truth of the next;

then every statement, from the kth one onwards, is true. An example of such a situation is given by the sequence of statements

$$n^3 - 23 > (4n - 7)^2, \quad n = 1, 2, \ldots$$

which are true only for $n \geqslant 12$.

There is another form of induction, called the method of *complete induction*, which can sometimes be applied to sequences of statements where the original method is hard to use directly. If it can be proved that

(1) the first statement is true; and
(2) for each n the truth of every statement, from the first to the nth inclusive, would imply the truth of the $(n + 1)$th;

then each of the statements must be true.

Again, this can be rephrased in set theory terms as the *principle of complete induction*: if a set of natural numbers contains 1 and, for each n, it contains $n + 1$ whenever it contains all numbers less than $n + 1$, then it must contain every natural number. Complete induction is used for example in proving that every natural number is a product of primes.
2. (in logic) A method of reasoning in which general laws are inferred from a number of particular observations. An example of inductive reasoning is the observation of a large number of crows, all of which are black, leading to the formulation of a general law that all crows are black. The conclusion 'all crows are black' does not follow logically from the premise 'all crows observed so far have been black', and the observation of one white crow at any time would disprove the law. Although inductive thinking is not, then, rational in the logical sense, it is the basis on which people often come to conclusions.

inductive *See* recursive.

inelastic collision *See* collision.

inequality **1.** A mathematical statement that two expressions are not equal to one another in value. The symbol $\neq$ is used; for example, $x \neq y$ means 'x is not equal to y'. **2.** A mathematical statement that one expression is greater than or less than another in value. The following symbols are used:

$x > y$ for 'x is greater than y'

$x < y$ for 'x is less than y'

The two expressions above are said to have opposite *senses*. Obviously $x > y$ is the same as $y < x$. The symbols $\geqslant$ for 'greater than or equal to' and $\leqslant$ for 'less than or equal to' are also used. To distinguish an inequality such as $x > y$ from $x \geqslant y$, the former is called a *strict inequality*.

As with equations, inequalities can be unconditional or conditional. An *unconditional inequality* is one that holds for all values of the variables (i.e. it is the analogue of an identity in equations). An example would be

$$2x^2 + 1 > x - 1$$

which is true for all values of x. A *conditional inequality* is true only for certain values of the variables: for instance,

$$2x + 1 > 11$$

is true only for $x > 5$, i.e. the inequality is satisfied by all values of x greater than 5. Such a set of values satisfying an inequality is a *solution* (or *solution set*) of the inequality.

Inequalities involve *transitive relationships. Thus, if $a > b$ and $b > c$, it follows that $a > c$. They can also be manipulated, but not always in exactly the same way as equations. Thus, in algebraic addition if $x > y$ then $x + a > y + a$ for all a, whereas for multiplication if $x > y$ then $ax > ay$ if $a > 0$ and $ax < ay$ if $a < 0$.

inertia **1.** A property of all forms of matter, manifest in a body by its resistance to *acceleration, i.e. by a tendency to remain at rest or to resist any change in motion. The inertia of a body requires a force to be exerted if the body is to be accelerated. The *mass of a body can be considered as a consequence of its inertia, and thus as a measure of the body's inertia: the greater the inertia, the greater the mass. *See also* inertial mass.
2. The inertia of a *Hermitian matrix is an ordered triple of integers comprising the number of positive, negative, and zero *eigenvalues, respectively.

inertial coordinate system Any set of coordinate axes moving at constant velocity with respect to a set of axes that are fixed in space relative to the positions of distant stars. These are axes of an inertial *frame of reference.

inertial force A force, such as a *centrifugal force or *Coriolis force, that is introduced in order to treat a noninertial *frame of reference as though it were a Newtonian frame. Thus a particle at rest in a frame rotating with constant angular speed ω can be treated as a particle in a fixed Newtonian frame experiencing a radial force $mr\omega^2$. Inertial forces are sometimes referred to as 'fictitious forces'.

inertial frame of reference *See* frame of reference.

inertial mass The property of a body that determines its resistance to acceleration, and is thus a measure of the body's *inertia. Newton's second law of motion is expressed in terms of inertial mass. The *mass of a body is usually considered in terms of inertial mass; this has, however, been found to be equivalent to the body's *gravitational mass.

inessential map *See* homotopy.

inf Infimum. *See* greatest lower bound.

inference **1.** The drawing of a conclusion from a set of premises.
2. (rule of) A rule in *logic that allows us to pass from a set of sentences (premises) to another sentence (conclusion). When a formal language is interpreted, these rules should be such as to guarantee that if the premises are true then the conclusion is also true. *See also* consequence; logic; sound.
3. In *statistics, the process of drawing conclusions about a *population or making predictions using *random samples. *See* Bayesian inference; confidence intervals; decision theory; estimation; fiducial inference; hypothesis testing; random sample.

infimum (inf) *See* greatest lower bound.

infinite decimal (nonterminating decimal) *See* decimal.

infinite discontinuity *See* discontinuity.

infinite group A *group with an infinite number of elements.

infinite integral (improper integral) An integral in which one or both of the *limits is infinite or in which the integrand is infinite at some point in the range or region of *integration. An example of the first type is

$$\int\limits_{a}^{\infty} f(x)\,dx$$

which is short for

$$\lim_{b \to \infty} \int\limits_{a}^{b} f(x)\,dx$$

If the limit exists, the integral is said to be *convergent*; if not it is *divergent*.
  An integral of the second type, whose integrand is a function $f(x)$ that is finite for $a \leqslant x < b$, but infinite for $x = b$, is

$$\int\limits_{a}^{b} f(x)\,dx$$

which is short for

$$\lim_{\delta \to \infty} \int_a^{b-\delta} f(x)\,dx$$

where $\delta > 0$. If the limit exists the integral is again said to be *convergent*.

infinite product A *continued product of an infinite number of terms. An infinite product of terms

$$T_1 \times T_2 \times \cdots \times T_n \times \cdots$$

is written using the notation

$$\prod_1^\infty T_n$$

Such a product might have a value of zero (e.g. $1 \times \frac{1}{2} \times \frac{1}{3} \times \frac{1}{4} \times \cdots$) or might be infinite (e.g. $1 \times 2 \times 3 \times 4 \times \cdots$). In either case, the product is said to be *divergent*. If the product has a nonzero value, it is *convergent*. In this case the value of the infinite product is the limit of the sequence

$$T_1, \; T_1 \times T_2, T_1 \times T_2 \times T_3, \ldots$$

If the product is neither convergent nor divergent, it is an oscillating product, for example

$$\prod (-1)^n$$

oscillates about the values 1 and -1. *See* Wallis's product; Viète's product.

infinite sequence A *sequence that has an unlimited number of terms. *See* convergent sequence; divergent sequence.

infinite series A *series with an unlimited number of terms. *See* convergent series; divergent series.

infinite set A *set that is not finite, i.e. one that can be put into a *one-to-one correspondence with a proper *subset of itself. The set of natural numbers is infinite because it can be put into a one-to-one correspondence with a proper subset of

itself, e.g. the set of even numbers. Infinite sets are either *countable (like the set of natural numbers) or uncountable (like the set of irrational numbers). *See* cardinal number.

infinitesimal A variable whose *limit is zero. Two variables x and y, each of which tend to zero, are infinitesimals of the same *order* if the ratio x/y is finite, and does not tend to zero. *See also* order (of infinitesimals).

infinitesimal calculus *See* calculus.

infinity Symbol: ∞. The idea of something that is unlimited, in the sense of being greater than any fixed bound. It arises in mathematics in various ways:

(1) In limits. For example, the function $y = 1/x$, for positive values of x, becomes larger as x decreases. In the limit as x tends to zero, y tends to infinity ($y \to \infty$). This means that for any number C greater than zero, there is a number $a > 0$ such that $y > C$ when $0 < x < a$. Similarly, for negative values of x it can be said that $y < -C$ when $-a < x < 0$, in which case y approaches $-\infty$ as $x \to 0$. When $y \to +\infty$ it is said to become *positively infinite* and when $y \to -\infty$ it becomes *negatively infinite*. Ideas of infinity in limits date back to Zeno of Elea (5th century BC) and Eudoxus of Cnidus (4th century BC). The symbol ∞ for infinity was introduced by John Wallis in 1655.

(2) In geometry. Infinity is regarded as a 'location': for example, parallel lines can be said to intersect at a point at infinity; parallel planes at a line at infinity. The asymptote to a curve can be regarded as intersecting the curve at infinity. The idea of infinity as a location was introduced by Johann Kepler, who pointed out that a parabola could be regarded as an ellipse or a hyperbola with one focus at infinity. The idea was developed by Girard Desargues in his formulation of *projective geometry, which assumed the existence of an *ideal point* at infinity.

(3) In set theory. *See* infinite set; Cantor's theory of sets.

infix notation The traditional notation for *binary operations in which the operator is written between its arguments, as in $2 + 3$. In more complicated expressions this notation necessitates the use of parentheses to avoid ambiguity: $2 + 3 \times 4$ is ambiguous, whereas $2 + (3 \times 4)$ or $(2 + 3) \times 4$ makes the intended meaning clear. Conventionally, however, multiplication takes precedence over addition in mathematical notation, so that $2 + 3 \times 4$ would usually be interpreted as $2 + (3 \times 4)$. In computer programming languages rules of precedence are specified for all binary operations. *Compare* prefix notation; postfix notation.

inflection In general, change from concavity to convexity or vice versa. A *point of inflection* is a point on a curve at which the tangent changes from rotating in one sense to rotating in the opposite sense. A horizontal point of inflection is an example of a *stationary point. At a point of inflection the second derivative is zero. Note that this is a *necessary but not sufficient condition for a point of inflection. *See* turning point.

inflectional tangent A *tangent to a curve at a point of *inflection.

information 1. *See* information theory.
2. In *estimation theory, if L is the logarithm of the *likelihood function for a parameter θ, the amount of information is given by $E((\partial L/\partial \theta)^2)$. For a sample of n independent observations from a distribution with *frequency function $f(x, \theta)$, the information I is given by

$$I = n E\left[\left(\frac{\partial \ln f}{\partial \theta} \right)^2 \right]$$

Under certain regularity conditions, and if the extremes do not depend on θ,

$$I = -n E\left(\frac{\partial^2 \ln f}{\partial \theta^2} \right)$$

and $1/I$ gives a lower bound (the *Cramér–Rao lower bound*) to the variance of any *unbiased estimator of θ. An unbiased estimator T such that $\text{Var}(T) = 1/I$ is called a *minimum variance unbiased estimator*. An implication of this result is that, the smaller the variance of an unbiased estimator, the greater its information content. The concept may be extended to $p \geqslant 2$ parameters θ_i, $i = 1, 2, \ldots, p$, where I is now replaced by the $p \times p$ *information matrix* with the element in row i, column j given by

$$n E\left[\left(\frac{\partial \ln f}{\partial \theta_i} \right) \left(\frac{\partial \ln f}{\partial \theta_j} \right) \right]$$

See Cramér–Rao inequality.

information theory A branch of mathematics concerned with the transmission and processing of information. A general theory of the subject was propounded in 1948 by Claude E. Shannon, in his article 'A Mathematical Theory of Communication'. The subject is based on the idea that it is possible to give a quantitative measure of *information*. The usual method of assigning such a measure can be illustrated by the example of transmitting and receiving a single letter of the alphabet (i.e. any one of the 26 letters). The amount of information in such a message (if correctly received) is measured with reference to the situation in which there are only two letters, and is given by $\log_2 26 / \log_2 2 = 4.7$, i.e. there is 4.7 times as much information in receiving a single letter of the 26-letter alphabet as in receiving a single *bit. The information content is said to be 4.7 bits. In fact, this applies only if the letters in the alphabet are equally likely to occur. In practice, this is not the case and information content is measured by a quantity known as *entropy*, given by

$$-p_1 \log_2 p_1 - p_2 \log_2 p_2 - p_3 \log_2 p_3 - \cdots$$

where $p_1, p_2, p_3, \ldots$ are the probabilities of

different values of the variable (in the example, the letter sent). This idea of entropy is similar to the concept originally developed in thermodynamics and statistical dynamics.

In considering information, it is usual to have a model consisting of:

(1) a source of information;
(2) an encoder, which changes this into a form suitable for transmission;
(3) a channel along which the information is transmitted;
(4) a decoder, which converts the information back into a useful form; and
(5) a destination or user, which receives the information.

The signal transmitted via the channel may be subject to extraneous *noise*. In its most restricted sense, information theory deals with the entropies of sources and channels. More generally, the term is also used to encompass *coding theory (ways of encoding information to ensure effective transmission). The term *communication theory* is often used to include both information theory and coding theory.

Information theory is essentially an application of probability theory. It has obvious uses in telegraphy, radio transmission, and the like, but has also been applied to language studies and cybernetics.

initial conditions *See* boundary conditions.

initial meridian plane *See* spherical coordinate system.

injection An injection from a *set A to a set B is a *one-to-one function whose *domain is A and whose *range is part of B. For example, if $A = \{3, 6\}$ and $B = \{9, 36, 150\}$ then the function $f: x \mapsto x^2$ is an injection (or *injective function*). *See also* surjection; bijection.

inner product *See* scalar product.

inradius The radius of the *incircle of a polygon. *Compare* exradius.

inscribed Describes a figure that is *circumscribed by another figure. For example, a polygon lying inside a circle with all its vertices on the circumference is said to be *inscribed in* the circle. A circle inside a polygon with all the sides of the polygon tangent to the circle is inscribed in the polygon (it is the *incircle* of the polygon).

instantaneous Occurring at or associated with a particular instant. The *instantaneous acceleration* or *velocity* is strictly the limit of the average value as the time interval over which the acceleration or velocity is considered approaches zero. The *instantaneous centre of rotation* is a point about which a moving body may be considered to be rotating at a particular instant.

instanton *See* integrable system.

integer Any of the positive and negative whole numbers $0, \pm 1, \pm 2, \pm 3, \ldots$. The positive integers $1, 2, 3, \ldots$ are called the *natural numbers* or *counting numbers*.

The set of all integers is usually denoted by $\mathbb{Z}$; and the set of all positive integers by $\mathbb{Z}^+$ or $\mathbb{N}$.

integer lattice *See* lattice (**2**).

integer part The integer part of a real number x is the largest integer n such that $n \leqslant x$; it is denoted by $[x]$, or sometimes $\lfloor x \rfloor$. The difference $x - n$ is called the *fractional part* of x. An alternative term is that the function $\lfloor x \rfloor$ is called the *floor function* or *greatest integer function*. Analogously, the *ceiling function* or *least integer function* is denoted by $\lceil x \rceil$, and defined as the smallest integer m such that $m \geqslant x$. If x is not itself an integer, then $\lceil x \rceil = \lfloor x \rfloor + 1$. For example, $\lceil 4.35 \rceil = \lfloor 4.35 \rfloor = 4$, and $\lceil 4.35 \rceil = 5$.

integrability The property of having an integral (*see* integration). The question of whether a function is integrable depends on the sense in which the integral is defined. *Darboux's theorem gives the necessary and sufficient condition for a function to have a Riemann integral. A function that has a Riemann integral also has a *Lebesgue integral, although the converse is not necessarily true.

integrable Describing a *function that has an integral (*see* integration).

integrable system A system of one or more *partial differential equations which are nonlinear and whose solutions have very special properties. An important example is the Korteweg–de Vries equation of 1895:

$$\frac{\partial y}{\partial t} + \frac{\partial^3 y}{\partial x^3} + 6y\frac{\partial y}{\partial x} = 0$$

where t denotes time and x denotes a space variable. It is named after the Dutch mathematicians Diederik Johannes Korteweg (1848–1941) and Gustav de Vries (1866–1934). A solution of this equation describes certain solitary water waves of the type that had already been observed by the Scottish engineer John Scott Russell in 1834. These waves have remarkable stability properties and are called *solitons*. They also appear in optical fibres. Closely related are *instantons*, which arise as solutions of equations in *gauge theories.

integral 1. *See* integration.
2. Describing or denoting an integer.

integral calculus *See* calculus.

integral domain A commutative *ring that has an *identity element, and in which there are no *proper divisors of zero*, i.e. there are no nonzero elements a and b with $ab = 0$. The absence of proper divisors of zero is equivalent to the existence of the cancellation laws: namely that if $a \neq 0$ and $ax = ay$ then $x = y$, and similarly, if $b \neq 0$ and $wb = zb$, then $w = z$. The ring of all integers is a typical integral domain.

integral equation An equation that involves an integral of an unknown *function (*see* integration). A general integral equation of the third kind has the form

$$u(x)g(x) = f(x) + \lambda \int_a^b K(x,y)g(y)\,dy$$

where the functions $u(x)$, $f(x)$, and $K(x,y)$ are known and g is the unknown function. The function K is the *kernel* of the integral equation and λ is the *parameter*. The limits of integration may be constants or may be functions of x. If $u(x)$ is zero, the equation becomes an integral equation of the first kind, i.e. it can be put in the form

$$f(x) = \lambda \int_a^b K(x,y)\,g(y)\,dy$$

If $u(x) = 1$, the equation becomes an integral equation of the second kind:

$$g(x) = f(x) + \lambda \int_a^b K(x,y)g(y)\,dy$$

An equation of the second kind is said to be *homogeneous* if $f(x)$ is zero.

If the limits of integration, a and b, are constants then the integral equation is a *Fredholm integral equation. If a is a constant and b is the variable x, the equation is a *Volterra integral equation.

integral function *See* entire function.

integral sign *See* integration.

integral test *See* Cauchy integral test.

integral transform A relationship between two *functions that can be expressed by a homogeneous *integral equation, as in

$$f(t) = \int K(x,t)F(x)\,dx$$

Here $f(t)$ is an integral transform of $F(x)$; $K(x,t)$ is the *kernel* of the transform.

Inversion of the transform is the process of finding $F(x)$, i.e. of solving the integral equation. If this can be done there is a

reciprocal relationship

$$F(x) = \int K'(x, t) f(t)\, dt$$

Integral transforms are useful for simplifying problems, as in the transformation of certain types of differential equations into linear equations. Many special cases have been studied, differing in the kernel and the limits of integration. *See* Fourier transform; Laplace transform.

integrand An expression that is to be integrated. *See* integration.

integrating factor A quantity by which each term of a differential equation is multiplied to enable integration to be performed. *See* differential equation.

integration The inverse process to *differentiation, i.e. the process of finding a *function with a *derivative that is a given function. It is sometimes called *antidifferentiation*. If $F(x)$ is a function of x which, when differentiated, gives $f(x)$, then $F(x)$ is said to be an *integral* (or *antiderivative*) of $f(x)$, written as

$$F(x) = \int f(x)\, dx$$

which is equivalent to

$$\frac{dF(x)}{dx} = f(x)$$

The symbol $\int$ is called the *integral sign*.

If $F(x)$ is an integral of $f(x)$, then $F(x) + C$ will also be an integral (since $dC/dx = 0$). C is an arbitrary constant called the *constant of integration*; $f(x)$ is the *integrand*. An integral of this type is called an *indefinite integral*. Methods of integration include *change of variable, *integration by parts, and *integration by partial fractions. A table of integrals is given in the Appendix.

The difference between the values of an integral for two values of the independent variable is a *definite integral. The two

values of the variable are the *limits* of the integral, and the notation is

$$F(b) - F(a) = \int_a^b f(x)\, dx$$

Note that here the constants of integration cancel out. The values $x = b$ and $x = a$ are called the *upper* and *lower limits of integration*.

An integral can also be regarded as the limit of a sum, as in finding the area under a curve between two points $x = a$ and $x = b$ (*see* diagram). The area is divided into a number of narrow strips parallel to the y-axis, each of width δx. For the curve $y = f(x)$ the area of each strip δA is given approximately by $f(x)\,\delta x$ (i.e. regarding each as a rectangle). The approximate value of the total area is given by the sum

$$A \approx \sum f(x)\,\delta x$$

This method of forming an integral was first put forward by A.-L. Cauchy; an integral so formed is sometimes called a *Cauchy integral.*

Formally, it is possible to define a definite integral as the limit of a sum in the following way. For a function $f(x)$ with $a \leqslant x \leqslant b$, the interval $[a, b]$ is subdivided into n parts by points $a = x_0 < x_1 \cdots < x_n = b$. The lengths of these subintervals are $x_1 - x_0, x_2 - x_1, \ldots, x_n - x_{n-1}$. In the

integration

n subintervals, n intermediate points are taken: t_0 in $[x_0, x_1]$, t_1 in $[x_1, x_2]$, ... Then a sum, called the *Riemann sum*, is defined by

$$R = \sum_{0}^{n-1} (x_{k+1} - x_k)\mathrm{f}(t_k)$$

If the largest subinterval $[x_k, x_{k+1}]$ is of length δ, the definite integral of $\mathrm{f}(x)$ on the interval $[a, b]$ is defined by

$$\int_{a}^{b} \mathrm{f}(x)\,\mathrm{d}x = \mathop{\mathrm{Lim}}_{\delta \to 0} R$$

This integral is called the *Riemann integral*. The definition is in fact a generalization of the 'area under a curve' idea above, in which the strips have different widths and the height of a strip is taken at any point on the strip's base. It can be shown that a function has a Riemann integral if it is a continuous function. Note that this definition of an integral is different from that of an antiderivative. Integrals and derivatives are connected by the *fundamental theorem of calculus. In the 19th century the idea of an integral was extended using the concept of *measure.

See also multiple integral; Darboux's theorem; Lebesgue integral; numerical integration; contour integral.

integration by partial fractions A method of integrating *rational functions that are fractions in which the denominator has a higher degree than the numerator. For example, in the integral

$$\int \frac{x + 3}{x^2 + 3x + 2}\,\mathrm{d}x$$

the integrand can be split into two *partial fractions, to give

$$\int \frac{2}{x + 1}\,\mathrm{d}x - \int \frac{1}{x + 2}\,\mathrm{d}x$$

integration by parts A method of integrating a product using the formula

$$\int u\,\frac{\mathrm{d}v}{\mathrm{d}x}\,\mathrm{d}x = uv - \int v\,\frac{\mathrm{d}u}{\mathrm{d}x}\,\mathrm{d}x$$

For example, it is possible to integrate $x\cos x$ using $x = u$ and $\cos x = \mathrm{d}v/\mathrm{d}x$, so that $\mathrm{d}u/\mathrm{d}x = 1$ and $v = \sin x$. Then the formula gives

$$\int x\cos x\,\mathrm{d}x = x\sin x - \int \sin x\,\mathrm{d}x$$

$$= x\sin x + \cos x$$

The formula for integration by parts can be derived from the formula for differentiating a product:

$$\frac{\mathrm{d}}{\mathrm{d}x}(uv) = u\,\frac{\mathrm{d}v}{\mathrm{d}x} + v\,\frac{\mathrm{d}u}{\mathrm{d}x}$$

$$u\,\frac{\mathrm{d}v}{\mathrm{d}x} = \frac{\mathrm{d}}{\mathrm{d}x}(uv) - v\,\frac{\mathrm{d}u}{\mathrm{d}x}$$

Integrating both sides gives the formula.

integration by substitution *See* change of variable.

intensive definition An attempt to define a term by expressing its meaning. Thus the set of regular convex polyhedra is {convex figures that have regular polygons forming their faces and all their polyhedral angles congruent}. *Compare* extensive definition.

interaction *See* factorial experiments.

intercept A cutting of a line, curve, or surface by another line, curve, or surface. In a Cartesian coordinate system, it is the distance from the origin to the point at which a line, curve, or surface cuts a given axis.

intercept form *See* line; plane.

intercept theorem (parallel transversal theorem) In a triangle ABC, if a *transversal is parallel to BC and meets AB and AC at D and E respectively, then AD/DB = AE/EC (*see* diagram). The *converse is also true, i.e. if the points D and E on AB and AC are such that AD/DB = AE/EC, then DE is parallel to BC. *See also* mid-point theorem.

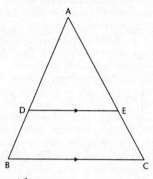

intercept theorem

interest Money paid by a borrower or to an investor for the use of money. The amount borrowed (or invested) is the *principal*. *Simple interest* is calculated on the principal only. For example, the interest on £1000 borrowed at 8 percent simple interest per annum is £80 per annum. *Compound interest* is calculated by adding the interest to the principal and calculating the interest at the end of agreed *conversion periods.*

For example, suppose that £1000 is invested for 2 years at 8 percent per annum and it is agreed that the interest is *compounded* half-yearly. At the end of the first six months the interest will be $8/100 \times \frac{1}{2} \times £1000 = £40$. At the end of the next six months the interest will be 4 percent of £1040 = £41.60. After eighteen months it will be 4 percent of £1081.60 = £43.26; and after two years the interest on the half-yearly period will be 4 percent of £1124.86 = £44.99. The total interest earned over the two-year period is £169.85.

The formula for compound interest is

$$I = P[(1 + r)^n - 1]$$

where P is the principal, r the rate for each conversion period, and n the total number of conversion periods. In the second example above, r is 0.04 (half of 8 percent) and n is 4.

The *nominal rate* of interest is the rate stated for a year when the interest is calculated over periods of less than a year. The *effective rate* of interest is the annual rate that would give the same yield as the nominal rate calculated over conversion periods of less than a year. In the second example above the nominal rate is 8 percent per annum; the effective rate is 8.16 percent. Tables of compound interest are used to help calculations. These generally give four values based on 1 unit of money:

(1) The *accumulation factor* $(1 + r)^n$, which gives the amount to which 1 unit will increase after n conversion periods at rate r.

(2) The *discount factor* $(1 + r)^{-n}$, which gives the amount that will give 1 unit after n periods at a rate r. It is often written as γ^n.

(3) The *amount* of an annuity, which is the value after n periods of 1 unit invested per period after addition of compound interest at rate r. It is also called the *accumulated value* and given the symbol S_n.

(4) The *present value* of an annuity, which is the amount necessary to provide one unit payment at the end of each of n payment periods.

interior (of a set) *See* frontier.

interior angle 1. An angle between two sides of a *polygon lying within the polygon. An interior angle greater than 180° is a *re-entrant angle*; one less than 180° is a *salient angle*.
2. *See* transversal.

internal force A *force that is exerted by one particle of a body (considered as a system of particles) on another particle of that body, and to which there is an equal but opposite reaction by this other particle. Internal forces thus occur in pairs whose individual resultants are zero. Hence the resultant of all forces internal to a body is zero. Only an *external force can affect a body considered as a whole.

internal tangent *See* common tangent.

interpolation For known values $y_1, y_2, \ldots,$ y_n of a *function $f(x)$ corresponding to values $x_1, x_2, \ldots, x_n$ of the independent variable, interpolation is the process of estimating a value y' of the function for a value x' lying between two of the values of x, e.g. x_1 and x_2.

Linear interpolation assumes that (x_1, y_1), (x', y'), and (x_2, y_2) all lie on a straight-line segment (*see* diagram). This implies that

$$\frac{y' - y_1}{x' - x_1} = \frac{y_2 - y_1}{x_2 - x_1}$$

whence

$$y' = y_1 + \frac{(x' - x_1)(y_2 - y_1)}{x_2 - x_1}$$

Only if $f(x)$ is a straight line (for $x_1 < x < x_2$) is linear interpolation certain to yield the correct value of $f(x')$.

Improved methods of interpolation take into account other data values; examples are *Lagrange interpolation and *Gregory–Newton interpolation. *Extrapolation is the process of estimating $f(x)$ when x' lies outside the range of observed x_i.

interpretation In *logic, a set of entities (the *domain) together with a *function that assigns to suitable expressions of a *formal language entities in the domain.

By interpreting a formal language we confer meaning on its expressions; for example, the sign 'Aristotle' has no meaning in itself, but acquires meaning when interpreted as standing for the person Aristotle. For a given expression, the function assigning it an entity in the domain is called a *semantic rule*, and the entity so assigned is the *semantic value* of the expression.

In the *propositional calculus, the domain consists of a set of *truth values, usually 'True' and 'False', and the semantic rules assign to each *wff of the propositional calculus one or other of these truth values. The truth-functional connectives are assumed to have some fixed meaning.

See model; logic; valid.

interquartile range In *statistics, a measure of *dispersion represented by the difference between the first and third *quartiles of a sample. Half this difference is called the *semi-interquartile range*. *See* box-and-whisker diagram; quantiles.

intersecting chords theorem *See* circle.

intersection 1. (meet, product) The intersection of two *sets A and B, denoted by $A \cap B$, consists of those elements that belong to both A and B:

$$A \cap B = \{x : (x \in A) \ \& \ (x \in B)\}$$

For example, if A is $\{1, 2, 3, 4, 5, 6\}$ and B is $\{1, 4, 5, 6, 7, 8\}$ then $A \cap B$ is $\{1, 4, 5, 6\}$. *Compare* union.
2. The point, line, etc. that is common to two or more geometric figures. Two curves, for instance, may intersect at one or more points. Two surfaces generally intersect in one or more curves.

interval A *set of numbers containing all *real numbers between two given numbers. The given numbers are called the *end points*; the interval can be represented as a segment of a number line. If the interval contains the end points (a and b) it is a *closed interval*, written as $[a, b]$. In this case the set is the set of numbers x for which

$a \leqslant x \leqslant b$. If it does not contain the end points, it is an *open interval*, written as (a, b). Here, $a < x < b$. An interval can also be partly open (and partly closed). The convention is to use a combination of round and square brackets:

$(a, b]$ contains b but not a;
$[a, b)$ contains a but not b.

The idea of an interval can be generalized to n dimensions by defining a closed interval as a set of *n-tuples for which $a_1 \leqslant x_1 \leqslant b_1, a_2 \leqslant x_2 \leqslant b_2, \ldots, a_n \leqslant x_n \leqslant b_n$. Open intervals are similarly defined using $<$ rather than $\leqslant$. *See also* bound.

interval estimate *See* estimation.

interval of convergence *See* power series.

interval scale *See* scales of measurement.

intraclass correlation A concept now superseded by the related idea of a variance ratio in the *analysis of variance.

intransitive relation *See* transitive relation.

intrinsic equation A method of defining a curve without reference to a set of coordinate axes. For a plane *curve this can be done by relating *arc length s to the *curvature κ or the radius of curvature ρ at the locus point. For example, an intrinsic equation for the *catenary is

$$c\rho = c^2 + s^2$$

where c is a constant.

The name is also given to equations relating s to ψ, the inclination of the tangent at the locus point to a fixed line. The equation of the catenary can then be put in the form

$$s = c \tan \psi$$

intuitionism The view, originated by Brouwer, that mathematical objects are mental entities that do not exist independently of our ability to provide a proof of their existence in a finite number of steps,

intrinsic equation of the catenary: $c\rho = c^2 + s^2$ or $s = c \tan \psi$.

and that a mathematical statement is true only if it is proved to be so in a finite number of steps. This is in contrast to a classical conception of mathematics and logic according to which mathematics, like natural science, is concerned with discovering truths about a world independent of human mentality.

According to the intuitionist, the sequence of natural numbers is to be taken as primitive, as are the familiar operations of arithmetic. But any mathematical proof is unacceptable to the intuitionist if it requires an infinite number of steps to complete, and is thus *nonconstructive* since no person would have time to carry it out.

Acceptance of intuitionism is incompatible with classical logic. The principle of *bivalence, according to which every sentence is either true or false, is rejected on the grounds that we may not be able to prove the truth or falsity of the sentence. Thus, the claim that *Goldbach's conjecture is either true or false is denied by the intuitionist, since we have neither a proof nor a disproof of the conjecture. As a result, intuitionists reject the law of the excluded middle, '$A \lor \sim A$', and, consequently, many other laws of classical logic, such as that of *double negation.

Intuitionists also reject impredicative definitions, those in which a particular member of a set is defined by reference to the totality of members of the set.

See formalism; logicism.

invariance **1.** The property of being *invariant.

2. In statistics, either:

(1) a quantity that is unchanged by a transformation; e.g. the statistic t used in the *t-test is unchanged by a *linear transformation of the sample values such as replacing x_i by $3x_i + 7$; or

(2) a property that is not changed by a transformation, e.g. the property of independence and normality of a set of independent normal variables is invariant under an *orthogonal transformation.

3. An essential property of *tensors under admissible transformations.

invariant Describing a property or quantity that is unchanged by a given *transformation. For example, the *discriminant of a conic is an invariant under translation or rotation of axes.

Hilbert proved important general results about the set of all algebraic invariants of certain groups of transformations. However, the calculation and full description of all possible invariants can be rather an intractable problem. The algebraic invariants for the group of all permutations of the variables $x_1, x_2, \ldots, x_n$ are those generated by the *elementary symmetric polynomials*

$$p_1 = x_1 + x_2 + \cdots + x_n$$
$$p_2 = x_1x_2 + x_1x_3 + \cdots + x_{n-1}x_n$$
$$\vdots$$
$$p_n = x_1x_2 \cdots x_n$$

See form.

invariant measure Given a map T of a space X to itself, a *measure μ that associates to each set A the same measure as its pre-image (*see* function); i.e if $T^{-1}(A) =$ $\{x: T(x) \in A\}$, then $\mu(T^{-1}(A)) = \mu(A)$. For example, the standard measure given to arcs on the unit circle $\{z: |z| = 1\}$ is invariant under the map $T(z) = z^2$. If $T: X \to X$ is a *bijection, then μ is such that $\mu(T(A)) = \mu(A)$.

Similarly, given a *flow $x(t)$, this is a measure μ for which every set A has the same measure at all times.

inverse **1.** (of a function) A *function that assigns to every element y of a *set Y a unique element $x = g(y)$ of a set X, where X is the *domain of the given (single-valued) function f and Y is the *range of the function. $y = f(x)$ is equivalent to $x = g(y)$ and g is said to be the inverse of f, written as f^{-1}. Also, $f(f^{-1}(y)) = y$ for all y in Y and $f^{-1}(f(x)) = x$ for all x in X, the domain of f being the range of g and vice versa.

If f is continuous, monotonic, and defined on a real interval $[a, b]$ then a continuous monotonic inverse f^{-1} exists. For instance,

$$f(x) = y = 2x + 3$$

where $0 \leqslant x \leqslant 1$, has inverse

$$f^{-1}(y) = x = \tfrac{1}{2}(y - 3)$$

where $3 \leqslant y \leqslant 5$. The variables x and y are often interchanged in the inverse function, so that in this instance

$$f(x) = y = 2x + 3$$

is said to have inverse

$$f^{-1}(x) = y = \tfrac{1}{2}(x - 3)$$

This can be written as

$$f: x \mapsto 2x + 3 \text{ on } [0, 1]$$

$$f^{-1}: x \mapsto \tfrac{1}{2}(x - 3) \text{ on } [3, 5]$$

2. In a *group (or, more generally, a *groupoid) with an *identity 1 (and operation $\circ$) an inverse for the element u is an element v such that $u \circ v = v \circ u = 1$. If the operation is addition (multiplication), the element v is said to be an *additive* (*multiplicative*) *inverse* of the element u. The element v is a *right inverse* for u if $u \circ v = 1$; it is a *left inverse* if $v \circ u = 1$.

In a group, such as the positive rational numbers under multiplication, every element has an inverse. Thus $4/23$ and $5\frac{3}{4}$ are mutual inverses since $(4/23)(23/4) = (23/4)(4/23) = 1$. For another example consider the *monoid of real continuous functions with domain and codomain the interval $[0, 1]$ and function composition as operation. Some functions here do not have inverses, but, for instance, the function that maps each number to its positive square root is the inverse of the function that squares each number.

3. (of a matrix) (**reciprocal**) A square matrix constructed from a given *nonsingular matrix A by taking the *cofactors of the elements of A, dividing each by the *determinant of A, and taking the *transpose. The inverse is denoted by A^{-1}, and $AA^{-1} = I$, where I is the *identity matrix.

For example, if $ad - bc \neq 0$ then

$$\begin{pmatrix} a & b \\ c & d \end{pmatrix}^{-1} = \frac{1}{ad - bc} \begin{pmatrix} d & -b \\ -c & a \end{pmatrix}$$

See also pseudoinverse.
4. (of a relation) *See* relation.
5. (of a point, curve, or surface) *See* inversion.

inverse function theorem The theorem that guarantees the existence of a (local, single-valued) differentiable inverse for a function $f : \mathbb{R}^n \to \mathbb{R}^n$ if the *Jacobian matrix of f is nonsingular at the locality. The function $f(x) = x^2$ does not have a local inverse around $x = 0$ (because $f'(0) = 0$), but it does have a local inverse around every $a \neq 0$ (because $f'(a) \neq 0$). The theorem is closely related to the *implicit function theorem.

inverse hyperbolic functions The *inverses of the *hyperbolic functions, written as $\cosh^{-1} x$, $\sinh^{-1} x$, etc. They are also called *arccosh*, *arcsinh*, etc. It can be shown that

$$\sinh^{-1} x = \ln [x + \sqrt{(x^2 + 1)}]$$

$$\cosh^{-1} x = \ln [x \pm \sqrt{(x^2 - 1)}]$$

where $x \geqslant 1$, and

$$\tanh^{-1} x = \frac{1}{2} \ln \left(\frac{1+x}{1-x} \right)$$

where $-1 < x < 1$.

inversely proportional *See* variation.

inverse probability A term used in *Bayesian inference when applying a probability approach that reasons from observed events to hypotheses which may explain them. In a more general context, the term is sometimes used in a problem that attempts to find the probability of an event (a cause) conditional on another event (a consequence or effect) having taken place. For instance, finding the probability that a person has a disease (a cause) given that they have tested positive for it (an effect) is a problem in inverse probability that may be solved using Bayes' theorem if other relevant information for the application of this theorem is available (*see* Bayes' theorem for an example).

inverse ratio (**reciprocal ratio**) The reciprocal of the ratio of two quantities.

inverse sine series the *series expansion for the inverse sine function:

$$\sin^{-1} x = x + \frac{1}{2} \frac{x^3}{3} + \frac{1.3}{2.4} \frac{x^5}{5}$$
$$+ \frac{1.3.5}{2.4.6} \frac{x^7}{7} + \cdots$$

This is valid for $-1 \leqslant x \leqslant 1$.

inverse square law A relation between two variables, one of which is proportional to the reciprocal of the square of the other. Examples include laws relating the *force of interaction between two particles to the reciprocal of the square of the distance between them, as in Newton's law of *gravitation, or relating the intensity of an effect to the reciprocal of the square of the distance from the cause, as with the illumination provided by a source of light.

inverse tangent series *See* Gregory's series.

inverse trigonometric functions (antitrigonometric functions) Functions that are the *inverses of trigonometric functions. For example, if

$$y = \tan x$$

then the inverse is written as

$$x = \tan^{-1} y$$

i.e. x is the angle whose tangent is y. The inverse trigonometric functions $\sin^{-1}$, $\cos^{-1}$, $\tan^{-1}$, etc. are sometimes written as *arcsin*, *arccos*, *arctan*, etc. Graphs of the inverse functions are like graphs of the original functions with axes interchanged. The inverse trigonometric functions are regarded as single-valued functions, having values (*principal values*) lying within a restricted range:

inverse sine $[-\frac{1}{2}\pi, \frac{1}{2}\pi]$

inverse cosine $[0, \pi]$

inverse tangent $[-\frac{1}{2}\pi, \frac{1}{2}\pi]$

See Gregory's series; inverse sine series.

inversion 1. For a circle of radius r with centre at O, and a point P outside the circle, inversion is the process of finding another point P′ on OP for which OP. OP′ $= r^2$. It is said that P′ is the *inverse* of P (it follows that P is the inverse of P′). O is the *centre of inversion* and r the *radius of inversion*. The inverse of a given curve is the curve produced by the inverses of the points on the given curve. A curve $f(x, y) = 0$ has an

inverse $f(x', y') = 0$, where

$$x' = \frac{r^2 x}{x^2 + y^2}, \quad y' = \frac{r^2 y}{x^2 + y^2}$$

The inverse of a circle is a circle unless the circle passes through the centre of inversion, in which case the inverse is a straight line. Two curves intersect at the same angle as their inverses, i.e. inversion is a *conformal transformation. Inversion can also be performed on surfaces with respect to a sphere.

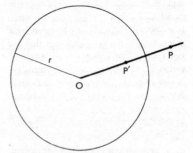

inversion Inverse points: OP. OP′ $= r^2$.
2. *See* integral transform.

invertible Possessing an *inverse.

involute A curve that is the *locus of a fixed point on a *tangent line to a given curve as this tangent line rolls on the given curve. The involute is the path that would be followed by a point on a string 'unwound' under tension from the curve. In the case of a circle, the parametric equations of the involute are

$$x = r(\cos \theta + \theta \sin \theta)$$

$$y = r(\sin \theta - \theta \cos \theta)$$

where r is the radius of the circle and θ the angle between the x-axis and a radius to the point of contact. *See also* evolute.

involution The process of finding a *power of a number or expression. *Compare* evolution.

involutory Describing a *matrix A such that $A^2 = I$, where I is the *identity matrix. *Compare* nilpotent.

irrational number A number that cannot be written as an *integer or as a quotient of two integers. The real irrational numbers are infinite, nonrepeating decimals. Every *complex number with a nonzero imaginary part is irrational.

There are two types of irrational number. *Algebraic irrational numbers* are irrational numbers that are roots of polynomial equations with rational coefficients; an example is $\sqrt{5}(2.2360\ldots)$, which is a root of $x^2 - 5 = 0$. *Transcendental numbers* are irrational numbers that are not roots of polynomial equations with rational coefficients; π and e are transcendental numbers. *Compare* rational number; *see also* Dedekind cut; real number.

irreducible equation *See* reducible polynomial.

irreducible fraction A common fraction such as 2/7 in which the numerator and denominator are *relatively prime. *Compare* reducible fraction.

irreducible polynomial *See* reducible polynomial.

irreducible radical A *radical that cannot be written in a rationalized form, i.e. a form not containing radicals. For example, $\sqrt{3}$ and $\sqrt{7}$ are irreducible. *Compare* reducible radical.

irreflexive relation *See* reflexive relation.

irrotational vector (in a region) A *vector function $\mathbf{V}$ such that curl $\mathbf{V} = 0$ at every point in a given region. *See* curl.

ISBN *Abbreviation for* International Standard Book Number. A *codeword assigned to every book by its publisher. One of the digits represents the language, two represent the publisher, and there is a *check digit to detect errors in transcription.

isochrone (tautochrone) A curve with the property that a particle sliding freely down the curve will reach the lowest point in the same time, irrespective of its starting point on the curve. *See* cycloid.

isogonal Having equal angles.

isogonal transformation *See* conformal transformation.

isolate (a root) To find two numbers between which a *root of an equation lies.

isolated point (acnode) A *singular point that does not lie on a given curve but does have coordinates that satisfy the equation of the curve. For instance, the curve $y^2 = x^3 - x^2$ has an isolated point at the origin $(0, 0)$.

isolated singularity *See* singular point.

isometry (isometric map) A *transformation that preserves distances between points. Thus any two points P and Q will have images P' and Q' such that $P'Q' = PQ$. Translation, rotation, and reflection are isometrics. An isometry transforms a geometric figure into a directly or oppositely *congruent figure.

isomorphism If A and B are two *sets in each of which a *binary operation is defined, a one-to-one mapping f of A onto B (*see* one-to-one function) that preserves the binary operations is known as an *isomorphism*. For example, the natural numbers can be mapped onto the even natural numbers by the one-to-one mapping that assigns, to each n of the set of natural numbers, the number $2n$ of the set of even natural numbers. The binary operation $+$ defined on the natural numbers is preserved by the mapping, and the two sets are consequently isomorphic for addition.

isoperimetric Describing figures that have equal *perimeters.

isoperimetric inequality If C is a simple, closed, plane curve of length l enclosing an area A, then $l^2 \geqslant 4\pi A$. If equality holds, i.e. if $l^2 = 4\pi A$, then C is a circle. In other words, amongst all curves of a given length, the circle encloses the greatest area. There are numerous generalizations of this result to surfaces and higher-dimensional manifolds. Closely related results are also used in applications of mathematics to other subjects.

isosceles trapezium A *trapezium in which two sides (the nonparallel sides) are of equal length.

isosceles triangle A triangle that has two sides equal (and unequal to the third). The angles opposite the equal sides are also equal.

iterated integral *See* multiple integral.

iterated map A *map $T: x \mapsto T(x)$ of a space X to itself with an associated *iteration $x_{n+1} = T(x_n)$, $n = 0, 1, 2, \ldots$. *See* Julia set; logistic map; Mandelbrot set.

iteration Successive repetition of a mathematical process, using the result of one stage as input for the next. Iteration is the basis of many approximation methods in numerical analysis. To solve an equation of the form $x = \phi(x)$, *direct* or *fixed-point iteration* is often appropriate, in which we iterate

$$x_{n+1} = \phi(x_n), \quad n = 0, 1, 2, \ldots$$

where x_0 is chosen as a first approximation to the desired root. The given equation can be rearranged in other ways, for example $x = \phi^{-1}(x)$, and the same technique applied. For example, the iteration $x_{n+1} = \cos x_n$ with $x_0 = 1$ may be used to find the root of the equation $x = \cos x$ (*see* diagram). Perhaps the most widely used iterative method is *Newton's method. *See also* bisection method; Halley's method; secant method.

iteration Direct iteration: $x_{n+1} = \cos x_n$.

J

jack-knife (M.L. Quenouille, 1949) A statistical procedure used mainly for estimating *bias in a sample *estimator of a population parameter. It is particularly useful when there is no analytic theory to estimate bias, as in the case when the sample correlation coefficient r is used as an estimator of the population coefficient ρ. The method involves computation of the sample estimator with the omission of one observation at a time over the entire set, and at that stage computation is equivalent to that for leave-one-out *cross-validation.

The computations are computer intensive, and the method has largely given way to use of the more versatile *bootstrap.

Jacobi, Carl Gustav Jacob (1804–51) German mathematician noted for his *Fundamenta nova theoriae functionum ellipticarum* (1829, New Elements in the Theory of Elliptic Functions) in which, starting from Legendre's work on elliptic integrals, he defined and explored the properties of elliptic functions obtained by inverting the integrals. Abel and Gauss had independently discovered their double periodicity earlier; Jacobi applied them to the theory of numbers and was able to prove with them Fermat's conjecture that every integer is the sum of four squares. He also contributed to the theory of determinants, to the theory of Abelian functions, and to the discipline of dynamics.

Jacobian For n *functions, $f_1, f_2, \ldots, f_n$ in n variables $x_1, x_2, \ldots, x_n$, the Jacobian is the *determinant

$$\begin{vmatrix} \dfrac{\partial f_1}{\partial x_1} & \dfrac{\partial f_1}{\partial x_2} & \cdots & \dfrac{\partial f_1}{\partial x_n} \\[2mm] \dfrac{\partial f_2}{\partial x_1} & \dfrac{\partial f_2}{\partial x_2} & \cdots & \dfrac{\partial f_2}{\partial x_n} \\[2mm] \vdots & \vdots & & \vdots \\[2mm] \dfrac{\partial f_n}{\partial x_1} & \dfrac{\partial f_n}{\partial x_2} & \cdots & \dfrac{\partial f_n}{\partial x_n} \end{vmatrix}$$

It is often written as

$$\frac{\partial(f_1, f_2, \ldots, f_n)}{\partial(x_1, x_2, \ldots, x_n)}$$

For m functions $f_1, f_2, \ldots, f_m$ in n variables $x_1, x_2, \ldots, x_n$, the *Jacobian matrix* is the $m \times n$ matrix whose element in the ith row and jth column is $\partial f_i / \partial x_j$. *See also* chain rule; Hessian; Wronskian.

Jacobi method An *iterative method for solving a system of linear equations $\mathbf{Ax} = \mathbf{b}$ published by Jacobi in 1845. Let $\mathbf{A} = \mathbf{D} + \mathbf{B}$, where $\mathbf{D}$ denotes the diagonal matrix with (i, i) entry a_{ii} and the a_{ii} are assumed to be nonzero. Let $\mathbf{x}_0$ be a first approximation for the vector $\mathbf{x}$. The Jacobi method generates a sequence of vectors $\mathbf{x}_1$, $\mathbf{x}_2, \ldots$ from the formula

$$\mathbf{x}_{n+1} = \mathbf{D}^{-1}(\mathbf{b} - \mathbf{Bx}_n)$$

The iteration is intended for matrices with relatively large diagonal elements, such as matrices that are *diagonally dominant.

Jeans, Sir James Hopwood (1877–1946) English mathematician and astronomer who, before he devoted himself to the study of astrophysics and cosmology, published a number of influential works in mathematical physics. They include *Dynamical Theory of Gases* (1904), *Theoretical Mechanics* (1906), and *The Mathematical Theory of Electricity and Magnetism* (1908).

Jiuzhang suanshu (Chiu-chang Suan-shu) The 'Nine Chapters on the Mathematical Art', the classic text of ancient Chinese mathematics; its authorship is unknown. Commentaries and extensions appear from the 2nd to the 15th century AD. It contains applications of Pythagoras' theorem, rules for extracting square and cube roots, and methods for the solution of simultaneous equations which foreshadow *Gaussian elimination.

join *See* union.

joint distribution *See* bivariate distribution; multivariate distribution.

Jonckheere–Terpstra test (A.R. Jonckheere, 1954; T.J. Terpstra, 1952) A *nonparametric test where the null hypothesis (H_0) is that three or more independent samples all come from the same population, against an alternative hypothesis (H_1) that the means, μ_i, taken in order, show a monotonic trend. For k samples the test is one of H_0: $\mu_1 = \mu_2 = \cdots = \mu_k$ against either H_1: $\mu_1 \leqslant \mu_2 \leqslant \cdots \leqslant \mu_k$ or H_1: $\mu_1 \geqslant \mu_2 \geqslant \cdots \geqslant \mu_k$, where in either version of H_1 at least one inequality is strict (i.e. of the form < or >).

Jones polynomial *See* knot polynomial.

Jordan, Camille (1838–1922) French mathematician who, in his *Traité des substitutions et des équations algébriques* (1870, Treatise on Substitutions and Algebraic Equations), revived interest in the work of Galois and established several fundamental results in group theory. His influential *Cours d'analyse de l'École Polytechnique* (1882) describes his research on

analysis and (in a later edition) the *Jordan curve theorem.

(a)

(b)

Julia set (a) Julia set for $c = -0.13 + 0.75i$. (b) Filled Julia set for $c = -0.75$.

Jordan canonical form A certain *block diagonal *canonical form to which any square matrix can be reduced by a similarity transformation. The diagonal blocks are *Jordan matrices and contain the eigenvalues on the principal diagonal.

Jordan curve theorem The theorem, first stated by C. Jordan (1893) and proved by O. Veblen (1905), to the effect that a *simple *closed curve (a *Jordan curve*) divides the plane into two connected regions, an 'inside' and an 'outside'.

Similarly, any *connected *manifold M of dimension $n - 1$ and without *boundary which is embedded in Euclidean space $\mathbb{R}^n$ divides the space into an inside and an outside.

Jordan–Hölder theorem (C. Jordan 1869, O. Hölder 1889) Suppose that a *group G with *identity element e has two *composition series, say $\{e\} = H_0, H_1, \ldots, H_n = G$ and $\{e\} = F_0, F_1, \ldots, F_m = G$. Then

$n = m$, and the two sets of *factor groups, $H_1/H_0, H_2/H_1, \ldots, H_n/H_{n-1}$ and $F_1/F_0, F_2/F_1, \ldots, F_m/F_{m-1}$, consist of exactly the same set of groups, apart possibly from their order of occurrence.

Jordan matrix A square *matrix in which the elements are equal on the main diagonal, unity on the first superdiagonal, and zero elsewhere: for example,

$$\begin{pmatrix} \lambda & 1 & 0 & 0 \\ 0 & \lambda & 1 & 0 \\ 0 & 0 & \lambda & 1 \\ 0 & 0 & 0 & \lambda \end{pmatrix}$$

An $n \times n$ Jordan matrix is not *diagonalizable when $n > 1$.

joule Symbol: J. The *SI unit of work or energy, equal to the work done when the point of application of a force of 1 newton moves through a distance of 1 metre in the direction of the force. 1 joule $= 10^7$ erg. [After J.P. Joule (1818–89)]

Julia set A closed subset of the *extended complex plane which is invariant under a *rational map. Julia sets were introduced by the French mathematician Gaston Maurice Julia (1893–1978) in 1918. The Julia set J of an *iterated rational map is the *boundary of the set of those points whose *orbits are bounded (*see* diagram (a)); the *filled Julia set* is the set itself of those points whose orbits are bounded (*see* diagram (b)). The Julia set for the map $z \mapsto z^2$ is the unit circle. However, most Julia sets are *fractal. In the case of the quadratic maps $z \mapsto z^2 + c$, the Julia set is a *connected set precisely when c is in the *Mandelbrot set.

jump discontinuity *See* discontinuity.

K

Kakeya's problem (S. Kakeya, 1917) The problem of determining the smallest possible area of a *set in the plane inside which a needle of length 1 can be moved continuously to reverse its direction. It was shown by A.S. Besicovitch in 1928 that, for any $\varepsilon > 0$, however small, there is a set of area ε in which this can be done.

kappa curve A plane *curve with the equation

$$x^4 + x^2y^2 = a^2y^2$$

in rectangular Cartesian coordinates. The curve is symmetrical about the axes and the origin, and has asymptotes $x = \pm a$. There is a double *cusp at the origin.

kappa statistic *See* Cohen's kappa statistic.

Karmarkar's algorithm (N.K. Karmarkar, 1984) An *algorithm for solving *linear programming and some more general constrained optimization problems that for large problems is often faster than the *simplex method. The latter searches the boundaries of the feasible region for the optimum solution, whereas Karmarkar's method, and some refinements thereof, commence the search from an interior point.

Kelvin, Sir William Thomson, Baron (1824–1907) Scottish mathematical physicist responsible for numerous innovations in both the theory and formalism of electromagnetism and thermodynamics. In the latter field he introduced the concept of absolute zero and the absolute scale of temperature since known as the kelvin scale, and published one of the first formulations of the second law of thermodynamics.

kelvin Symbol: K. The *SI unit of thermodynamic temperature, equal to 1/273.16 of the thermodynamic temperature of the triple point of water. The kelvin is equal in magnitude to the degree *Celsius. A temperature in kelvin is equal to the temperature on the Celsius scale plus 273.15. (The freezing point of water is 273.15 K.) [After Lord Kelvin]

Kendall, Sir Maurice George (1907–83) English statistician noted for his work on time series, multivariate analysis, and nonparametric methods, especially those involving correlation and concordance. His *The Advanced Theory of Statistics* (2 vols, 1943, 1946) was the first all-embracing treatment of the subject.

Kendall's coefficient of concordance *See* coefficient of concordance.

Kendall's rank correlation coefficient *See* correlation coefficient.

Kepler, Johann (1571–1630) German astronomer and mathematician who in his *Stereometria doliorum* (1615, Measurement of the Volume of Barrels) made one of the first ever attempts to determine the areas and volumes of figures generated by curves with the aid of infinitesimals. He is best known for his exposition of *Kepler's laws of planetary motion.

Kepler–Poinsot solid *See* Poinsot; polyhedron.

Kepler's conjecture (J. Kepler, 1611) The packing of three-dimensional space by balls of radius 1 has the greatest possible density when the balls are centred at the points of a face centred *cubic lattice; the density is $\pi/\sqrt{18}$. The conjecture was proved in 2006 by T.C. Hales. This packing can be described as made up of layers of balls placed on top of each other, the centres of the balls in each layer forming a planar equilateral triangular lattice as depicted in diagram (b) of *lattice.

Kepler's laws Three laws of planetary motion established empirically by Johann Kepler and based on detailed observations made by Tycho Brahe. The first two laws were published in 1609, the third in 1619.

They are as follows:

(1) Each planet moves in a path that is an *ellipse with the sun at one focus.
(2) The line joining a planet to the sun sweeps out equal areas in equal times during orbital motion.
(3) The squares of the periods of revolution of any two planets are proportional to the cubes of the major axes of their elliptical orbits.

Kepler realized that the sun was a controlling factor in planetary motion but he was unable to explain how the control was exercised. The explanation was to be provided by Newton when he formulated his law of *gravitation, which can be universally applied and from which Kepler's laws can be derived.

kernel If $f: G_1 \rightarrow G_2$ is a *homomorphism between two groups G_1 and G_2, its kernel is the subset consisting of those elements of G_1 that are mapped to the identity of G_2 by f, i.e. $\mathrm{Ker}(f) = \{g \in G_1 : f(g) = e \in G_2\}$. $\mathrm{Ker}(f)$ is always a *normal subgroup of G_1. A special case is when $f: V \rightarrow W$ is a linear transformation between vector spaces; then $\mathrm{Ker}(f)$ is a *vector subspace

of V called the *null space* of f. *See* integral equation; integral transform

kernel density estimation A *nonparametric method for estimating a *frequency function $f(x)$ of a continuous *random variable X based on data forming a (usually large) sample. The 'kernel' is a probability function $k(u)$ symmetric about $u = 0$. Suppose that there are n data points, $x_1, x_2, \ldots, x_n$. For each data point x_i, substituting $u = (x - x_i)/h$, where h is a constant called the *bandwidth*, produces a function of x. The estimated density function is then given by

$$f(x) = \frac{1}{nh} \sum_{i=1}^{n} k\left(\frac{x - x_i}{h}\right)$$

There are several widely used choices for $k(u)$, a common one being the Gaussian or normal kernel

$$k(u) = \frac{1}{\sqrt{(2\pi)}} e^{-u^2/2}$$

Choice of bandwidth strongly influences the estimate, and rules exist to select a width that avoids either overemphasizing local irregularities or obtaining an oversmoothed estimate. In practice the method is highly computer intensive.

key A piece of information held by the recipient that allows a *ciphertext to be decoded. A (possibly different) key is used by the sender to encode a message.

kilo- *See* SI units.

kilogram Symbol: kg. The *SI unit of mass, equal to the mass of the international prototype maintained by the Bureau International des Poids et Mesures at Sèvres, near Paris. 1 kilogram $= 2.204\,62$ pounds.

kilogram-force Symbol: kgf. A unit of force, equal to the force required to impart to a mass of 1 kilogram an acceleration equal to the standard acceleration of free fall. 1 kilogram-force $= 9.806\,65$ newtons.

kilowatt-hour Symbol: kWh. A unit of energy, widely used in charging for electrical energy, equal to the energy expended when a power of 1000 watts is applied for 1 hour. It is equal to 3.6×10^6 joules.

kinematics The study of the motion of objects without regard to the mechanisms that cause motion. Kinematics is thus concerned with the position of an object at different times, and hence with its velocity and acceleration. The object is usually considered as a *particle or system of particles. The motion can be along a straight line or a curve, and can therefore be considered in one, two, or three dimensions with respect to some coordinate system. *See also* dynamics; kinetics.

kinetic energy *Energy possessed by virtue of motion. It is equivalent to the work that would be required to bring a moving body to rest. Kinetic energy is a scalar quantity, usually denoted by T. A body with speed v has kinetic energy $\frac{1}{2}mv^2$, where m is the body's mass; this holds only when v is considerably less than the speed of light (*see* rest mass). A body with rotational motion, with angular speed ω, has kinetic energy $\frac{1}{2}I\omega^2$, where I is the body's *moment of inertia about the rotational axis. Kinetic energy can be converted to *potential energy and vice versa. For motion under a conservative force, such as gravitation, the total energy (kinetic plus potential) is conserved in an isolated system.

kinetic potential *See* Lagrangian function.

kinetics The study of the effects of *forces and *torques on the motion of material bodies. The word is used in classical mechanics in several ways. It can be considered as synonymous with *dynamics, the two fields being effectively concerned with the same subject matter; alternatively it can be used to denote a subsection of dynamics, usually with *kinematics forming the other subsection. Some prefer not to use the word kinetics: they divide classical mechanics into dynamics and kinematics, and consider *statics as a part of dynamics.

kite A convex *quadrilateral that has two pairs of equal adjacent sides. *See* convex polygon; *compare* deltoid; *see also* Penrose tiles.

Klein, Christian Felix (1849–1925) German mathematician who in 1871 proved the relative consistency of the various geometries by providing projective models of hyperbolic, elliptic, and Euclidean geometry. In the following year Klein announced his *Erlangen Programm* in which he sought to set up invariants of groups on which various geometries were based. Other work by Klein was concerned with group theory, the theory of functions, and topology.

Klein bottle An example of a one-sided closed surface. Formally, it is the 2-manifold obtained from the square

$$\{(x_1, x_2) \in \mathbb{R}^2 : |x_1|, |x_2| \leqslant 1\}$$

by identifying the edges $x_1 = \pm 1$ 'with a twist' and the edges $x_2 = \pm 1$ 'without a twist'; that is, by identifying $(-1, x_2)$ with $(1, -x_2)$ for all x_2 and $(x_1, -1)$ with $(x_1, 1)$ for all x_1. *See* manifold.

Klein bottle

Klein's four group A *group with four elements, say e, a, b, and c, that are combined according to the following table:

	e	a	b	c
e	e	a	b	c
a	a	e	c	b
b	b	c	e	a
c	c	b	a	e

It is an *Abelian group with e as its *identity element. A geometrical instance of this group is the group of all *symmetries of a non-square rectangle, where e is the identity map, a is rotation in its plane through $180°$ about its centre, and b and c are reflections in a line that bisects a pair of opposite sides.

klothoid *See* spiral.

knot 1. *See* knot theory.
2. A unit of speed or velocity equal to 1 *nautical mile per hour. Because the international nautical mile (defined as 1852 metres) differs from the UK nautical mile (6080 feet) the unit is not suitable for accurate measurements. 1 knot is approximately equal to 1.15 (land) miles per hour.

knot polynomial A polynomial that is associated with a knot. If two knots are equivalent, then their knot polynomials are equal; if the associated polynomials are different, the knots are not equivalent. The *Alexander polynomial* was discovered in 1928 by the American mathematician J.W. Alexander (1888–1971) using homology theory. In the late 1960s the English mathematician J.H. Conway discovered a simpler and explicit method of finding the Alexander polynomial, by using transformations applied to the planar representation of the knot. In 1984 the New Zealand born mathematician V.F.R. Jones, using ideas from mathematical physics, discovered another polynomial associated with a knot. The *Jones polynomial* can also be calculated using the approach discovered by Conway to calculate the *Alexander–Conway polynomial*. More knot polynomials have been found, but a common framework within which to understand them has yet to be discovered. One was suggested in 1990 by the Russian mathematician V.A. Vassiliev who introduced *Vassiliev invariants* based on his work in singularity theory.

knot theory The branch of geometry that studies entwined circles in Euclidean space $\mathbb{R}^3$. It was first studied by Gauss and his student J.B. Listing, and then in more detail by the Scottish physicists Kelvin, Maxwell, and P.G. Tait in the late 19th century. Tait constructed tables of knots whose planar representations had few crossings, and made a particular study of those with *alternating diagrams* – those in which the string crosses over and under alternately. He conjectured that such a knot could not be 'unknotted', and this was eventually proved to be correct using the Jones polynomial (*see* knot polynomial). Much of knot theory is studied using the planar diagrams representing the knots; an example is shown in diagram (a).

There are three basic *Reidemeister moves* that can be applied to a small portion of the diagram of a knot without changing any other part (*see* diagram (b)). Such moves applied successively reflect all the possible equivalences of knots in space. Each of the Reidemeister moves corresponds to moving a piece of string in space. They are named after the German mathematician Kurt Werner Friedrich Reidemeister (1893–1971).

Knot theory is used in the study of complicated molecules such as DNA.

(a)

knot theory (a) Trefoil knot.

knot theory (b) The three types of Reidemeister move.

Koch curve A *fractal curve in the plane constructed by iterating a procedure. The procedure starts with a line segment and replaces the middle third by two sides of an equilateral triangle erected upon it. The same process is then applied to each of the four new segments (*see* diagram). Continuing the process indefinitely produces the Koch curve, introduced in 1904 by the Swedish mathematician Helge von Koch (1870–1924) in demonstrating a curve of infinite length joining the ends of a finite segment and enclosing a finite area. *See* snowflake curve.

Kolmogorov, Andrei Nikolaevich (1903–87) Russian mathematician best known for his work in probability theory. In 1933 he presented the first general axiomatic treatment of probability theory, later translated into English under the title *Foundations of the Theory of Probability* (1950). He also made important contributions to the study of Markov processes, Fourier analysis, and topology. In 1925, following the work of Heyting, Kolmogorov succeeded in establishing new foundations for intuitionistic logic. In later work, based on topological analysis, he demonstrated the stability of the solar system.

Kolmogorov–Smirnov tests A.N. Kolmogorov (1933) proposed a nonparametric test to determine whether sample data are consistent with a specified distribution function; it was extended by N.V. Smirnov (1939) to test whether two samples may reasonably be supposed to come from the same unspecified distribution. The tests require the calculation of the sample cumulative distribution functions. *See* nonparametric methods.

Koch curve A stage in its construction.

Königsberg bridge problem A famous problem solved by Euler in 1736. The problem was to plan a walk in which each of the seven river-bridges of Königsberg (*see* diagram) would be crossed once and once only. Euler showed that such a walk was impossible, since each of the four areas of land had an odd number of bridges connecting it to the other areas, and so would have to contain either the starting point or the end point of such a walk. *See* Eulerian graph.

Korteweg–de Vries equation *See* integrable system.

Kovalevsky, Sonya (1850–91) Russian mathematician who in 1875 improved and generalized a result of Cauchy's on partial differential equations, and thus established the *Cauchy–Kovalevsky theorem*. She also worked on elliptic integrals. Other work was concerned with the rings of Saturn, the propagation of light in a crystalline medium, and the rotation of bodies.

Königsberg bridge problem The seven bridges of Königsberg.

kriging (D.G. Krige, 1951) A generalized *least-squares technique for *interpolation widely used in geostatistics for problems such as establishing contour lines when given a set of spot values of a relevant random variable, e.g. spot heights, or rainfall at a number of weather stations.

Kronecker, Leopold (1823–91) German mathematician noted for his work on algebraic numbers, beginning with his *De unitatibus complexis* (1845, On Complex Units) and continuing through much of his career. Kronecker was also well known for his opposition to the proposed transfinite cardinals of Cantor, declaring that only the whole numbers came from God, all else is *Menschenwerk* ('the work of Man'). He consequently rejected the treatment of irrationals put forward by Weierstrass and at one time went so far as to deny the existence of such numbers.

Kronecker delta The function δ_{ij} defined by the equations

$$\delta_{ij} = 1 \quad \text{when} \quad i = j$$

$$\delta_{ij} = 0 \quad \text{when} \quad i \neq j$$

The *tensor notation for the Kronecker delta is δ_j^i.

Kruskal's algorithm *See* tree.

Kruskal–Wallis test (W.H. Kruskal and W.A. Wallis, 1952) An extension of the *Wilcoxon rank sum test to three or more independent samples.

K-theory A method introduced by Alexandre Grothendieck in 1956 to enable a loose classification of vector *bundles in algebraic geometry. The method has been very successfully adapted to enable the study of a wide range of objects in algebra and geometry and hence the solution of a number of difficult problems. It takes its name from *Klasse*, German for 'class'.

k-tuple point *See* multiple point.

Kummer, Ernst Eduard (1810–93) German mathematician noted for his creation in 1845 of the theory of ideals. In 1850 he demonstrated that Fermat's last theorem holds for every exponent that is a regular prime.

kurtosis The degree of peakedness of a *frequency function near the mode. The normal distribution is said to be *mesokurtic*, one less peaked is said to be *platykurtic*, and one more peaked is said to be *leptokurtic*. If μ_i is the ith *moment about the mean the *coefficient of kurtosis* is

$$\gamma_2 = \mu_4/\mu_2^2 - 3$$

It has the value zero for the normal distribution; it is positive for leptokurtosis and negative for platykurtosis. *See also* g-statistics.

L

labelled tree *See* tree.

lag *See* autocorrelation.

Lagrange, Joseph Louis, Comte (1736–1813) Italian-French mathematician noted for his *Mécanique analytique* (1788), the definitive text on the post-Newtonian mechanics of the 18th century, written in a purely formal rigorous manner and lacking any diagrams. As a pure mathematician, Lagrange published two important memoirs on the theory of equations in 1770 and 1771, advancing a uniform principle for the solution of all equations up to the quintic. In the course of this work the result known as *Lagrange's theorem (on groups) was first formulated.

Other mathematical work was on the foundations of the calculus, the theory of differential equations, and number theory. Lagrange also contributed to astronomy, publishing a special solution of the three-body problem. In addition he played a leading role in the introduction of the metric system into revolutionary France.

Lagrange multipliers A means of evaluating maxima or minima of a *function $f(x_1, x_2, \ldots, x_n)$, subject to one or more equality constraints $g_i(x_1, x_2, \ldots, x_n) = 0$. The solution is found by minimizing $L = f + \lambda_1 g_1 + \lambda_2 g_2 + \cdots$ with respect to the x_i and the λ_i, where the λ_i are Lagrange multipliers (sometimes called *undetermined multipliers*).

For example, to find the maximum of $u = xy$ subject to the constraint $x + y = 1$, we write $L = xy + \lambda(x + y - 1)$. Differen-tiating with respect to x, y, and λ, and equating derivatives to zero, then gives $y + \lambda = 0$, $x + \lambda = 0$, and $x + y - 1 = 0$. The solutions are easily found to be $\lambda = -\frac{1}{2}$, $x = y = \frac{1}{2}$, giving $u = \frac{1}{4}$. It may be verified that this is a maximum.

Lagrange's equations *See* Lagrangian function.

Lagrange's interpolation formula A formula for *interpolation. If a function $y = f(x)$ has known values $y_1, y_2, y_3, \ldots, y_n$ at points $x_1, x_2, x_3, \ldots, x_n$, and a value y' is to be estimated at x', the formula is

$$y' = \frac{y_1(x' - x_2) \cdots (x' - x_n)}{(x_1 - x_2) \cdots (x_1 - x_n)}$$
$$+ \frac{y_2(x' - x_1)(x' - x_3) \cdots (x' - x_n)}{(x_2 - x_1)(x_2 - x_3) \cdots (x_2 - x_n)}$$
$$+ \cdots$$

and so on for n terms.

It is equivalent to interpolating by the polynomial in x of degree at most $n - 1$ whose graph passes through all the n points (x_1, y_1), (x_2, y_2), $\ldots$, (x_n, y_n).

Lagrange's theorem 1. (J.L. Lagrange, 1772) The theorem that every *natural number can be written as a sum of four squares of integers. It is sometimes called the *four squares theorem*. For example,

$$1 = 1^2 + 0^2 + 0^2 + 0^2$$
$$23 = 3^2 + 3^2 + 2^2 + 1^2$$

$$59 = 7^2 + 3^2 + 1^2 + 0^2$$
$$= 5^2 + 5^2 + 3^2 + 0^2$$
$$= 5^2 + 4^2 + 3^2 + 3^2$$

Every natural number that, like 23, is of the form $8k + 7$, or a power of four times such a number, needs four nonzero summands.

2. The theorem that if G is a finite *group and H is a *subgroup of G then the number of elements in H (called the *order* of H) must divide the number of elements of G (the order of G). It is not always true that a given divisor d of the order of G must be the order of some subgroup, but this is so if d is a power of a prime.

Lagrangian function Symbol: L. A *function of the *generalized coordinates, q_i, and generalized velocities, v_i, of a dynamical system. In a conservative system, in which both a *potential energy V and a *kinetic energy T can be defined, the Lagrangian function is given by

$$L = T - V$$

The function is then also known as the *kinetic potential*. The equations of motion for a conservative system are

$$\frac{d}{dt}\left(\frac{\partial L}{\partial \dot{q}_j}\right) - \frac{\partial L}{\partial q_j} = 0$$

This is the simplest form of what are known as *Lagrange's equations*. They are derived using the *calculus of variations from the stationary nature of L. *See also* Hamiltonian.

Lagrangian mechanics The development of mechanics through the application of Lagrange's equations.

Laguerre's differential equation The *differential equation

$$x\frac{d^2 y}{dx^2} + (1 - x)\frac{dy}{dx} + \alpha y = 0$$

It is satisfied for $\alpha = n$ by a *Laguerre polynomial* $L_n(x)$, given by

$$L_n(x) = e^x \frac{d^n}{dx^n}\left(x^n e^{-x}\right)$$

The equation is named after the French mathematician Edmond Nicolas Laguerre (1834–86).

Lambert, Johann Heinrich (1728–77) German mathematician, physicist, and philosopher, who in 1767 was the first to prove that π is irrational. He also worked on Euclid's parallel postulate, coming close to the discovery of non-Euclidean geometry. In this work he suggested that a surface might exist on which triangles had an angular sum of less than two right angles (a surface later discovered and named the pseudosphere). He also developed the notation and theory of hyperbolic functions.

Lamé's theorem *See* Euclidean algorithm.

lamina An idealized plane material object having area and density, but no thickness. If the density is constant, the lamina is said to be *uniform*.

Lamy's theorem (B. Lamy, 1679) The theorem that if three forces acting at a point have zero *resultant, then the magnitude of each force is proportional to the sine of the angle between the directions of the other two forces.

Lanczos method An iterative method for computing the eigenvalues of a *symmetric matrix. It is most often used for large, *sparse matrices, for which it is particularly appropriate because each iteration involves just a single product between the matrix and a vector. It is closely related to the *conjugate gradient method for solving linear systems. Named after the Hungarian-Irish physicist Cornelius Lanczos (1893–1974).

language *See* formal language.

Laplace, Pierre-Simon, Marquis de (1749–1827) French mathematician and physicist noted for his *Traité de mécanique céleste* (1799–1825, 5 vols, Celestial Mechanics) in which he tried to develop a rigorous mechanics capable of describing all motions of heavenly bodies including the various anomalies and inequalities that had emerged since the time of Newton. Equally notable was his *Théorie analytique des probabilités* (1812, Analytic Theory of Probability) which advanced the subject considerably. Specific contributions of Laplace's include the development of the concept of potential and the related *Laplace's equation, the *Laplace transform, and, in astronomy, the nebular hypothesis.

Laplace's equation The partial *differential equation

$$\frac{\partial^2 V}{\partial x^2} + \frac{\partial^2 V}{\partial y^2} + \frac{\partial^2 V}{\partial z^2} = 0$$

or $\nabla^2 V = 0$, where ∇ is the differential operator *del.

It is important in potential theory, and when expressed in spherical coordinate form becomes

$$\frac{1}{r^2}\frac{\partial}{\partial r}\left(r^2\frac{\partial V}{\partial r}\right) + \frac{1}{r^2\sin^2\theta}\frac{\partial^2 V}{\partial \phi^2}$$

$$+ \frac{1}{r^2\sin\theta}\frac{\partial}{\partial \theta}\left(\sin\theta\frac{\partial V}{\partial \theta}\right) = 0$$

See also harmonic.

Laplace transform The *transformation of a *function into another function of a different variable by multiplying by e^{-pt} and integrating with respect to t between the *limits 0 and ∞. If $f(t)$ is the original function, integration will give a function in p, say $F(p)$; this is the Laplace transform of the original function, written as $L(f(t))$:

$$L(f(t)) = \int_0^\infty e^{-pt} f(t)\, dt = F(p)$$

See also differential equation.

Laplacian (Laplace operator) The differential operator

$$\frac{\partial^2}{\partial x^2} + \frac{\partial^2}{\partial y^2} + \frac{\partial^2}{\partial z^2}$$

or ∇^2, where ∇ is the differential operator *del. *See also* Laplace's equation.

Laspeyres index *See* index number.

latent root Alternative name for an *eigenvalue.

lateral Denoting or concerned with a surface, edge, etc. that is regarded as on the side of a geometric figure, as opposed to the base. *See* cone; cylinder; prism; pyramid.

Latin square An *experimental design allowing classification by three mutually *orthogonal factors, usually denoted by rows, columns, and Latin letters. Treatments are designated by Latin letters and allocated to units under restricted randomization, each treatment occurring exactly once in each row or column. An example of a three-by-three Latin square is

A	B	C
C	A	B
B	C	A

The Latin square provides a useful double-blocking system to increase precision by reducing two potential sources of variation not related to treatments. In the *analysis of variance, the degrees of freedom for the error mean square are low for Latin squares smaller than six by six; this difficulty may be overcome by using more than one Latin square. The restriction that the number of treatments equals the number of rows or columns sometimes leads to practical difficulties.

See also Graeco-Latin square.

latitude 1. The angular distance of a point on the earth's surface, measured from the equator along the *meridian passing through the point. Latitude is measured from the equator, from $0°$ to $90°$ north and from $0°$ to $90°$ south.
2. *See* celestial latitude.
3. *See* galactic latitude.

lattice 1. (in algebra; R. Dedekind, 1894) A partially ordered set in which any two elements have a *least upper bound and a *greatest lower bound. *See* partial order.
2. (in geometry; C.F. Gauss, 1831) The *set of all expressions of the form

$$a_1\mathbf{v}_1 + \cdots + a_n\mathbf{v}_n$$

where $\mathbf{v}_1, \ldots, \mathbf{v}_n$ are n fixed linearly independent vectors or points, in a Euclidean space, and $a_1, \ldots, a_n$ are any integers. Each expression $a_1\mathbf{v}_1 + \cdots + a_n\mathbf{v}_n$ is called a *lattice point*. Equivalently a lattice is a *group with respect to the operation of vector addition and has a finite number of *generators ($\mathbf{v}_1, \ldots, \mathbf{v}_n$ here).

The set $\{\mathbf{v}_1, \ldots, \mathbf{v}_n\}$ is a *basis* for the lattice and n (the number of vectors in a basis) is its *dimension*. The lattice may have several bases, but the quantity $|\det(\mathbf{v}_1, \ldots, \mathbf{v}_n)|$ (the absolute value of the *determinant of the matrix formed by writing the coordinates of $\mathbf{v}_1, \ldots, \mathbf{v}_n$ in rows) is independent of the choice of basis and is called the *determinant* of the lattice.

An example of a lattice in n dimensions is the *integer lattice* formed by the collection of all n-dimensional vectors with integer coordinates. A basis for it is the set of n vectors $(1, 0, 0, \ldots, 0)$, $(0, 1, 0, 0, \ldots, 0)$, $\ldots$, $(0, \ldots, 0, 0, 1)$, and it has determinant equal to 1. The lattice points of the two-dimensional integer lattice are shown in diagram (a). As well as the basis $\{(1, 0), (0, 1)\}$, this lattice has the basis $\{(0, -1), (1, 1)\}$, for instance. Another two-dimensional lattice is that in diagram

(b), which is generated by the vectors $(2, 0)$ and $(1, \sqrt{3})$.

The most efficient way to pack circles of unit radius in the plane is to place their centres at the lattice points in this example. *See* crystallography; cubic lattice; Kepler's conjecture.

(a)

(b)

lattice (a) The points of the two-dimensional integer lattice. (b) The points of the lattice generated by vectors $(2, 0)$ and $(1, \sqrt{3})$.

latus rectum A focal chord of a *conic that is perpendicular to an axis through the vertex or vertices. [Latin: right side]. *See* ellipse; hyperbola; parabola.

Laurent expansion (of an analytic function) If a function f is an *analytic function in

$$r_1 \leqslant |z - z_0| \leqslant r_2$$

then the Laurent expansion is the *series

$$f(z) = \sum_{-\infty}^{\infty} a_n(z - z_0)^n$$

for $r_1 < |z - z_0| < r_2$, where

$$a_n = \frac{1}{2\pi i} \int_C f(z)(z - z_0)^{n+1} \, dz$$

and C is a circle with centre z_0 and radius r, $r_1 < r < r_2$ (*see* contour integral). $f(z)$ is the sum of two expressions:

$$\sum_{-\infty}^{-1} a_n(z - z_0)^n$$

called the *principal part*, and

$$\sum_{0}^{\infty} a_n(z - z_0)^n$$

called the *analytic part*. The expansion is named after the French mathematician and physicist Pierre Alphonse Laurent (1813–54). *See also* singular point.

law of cosines *See* cosine rule.

law of sines *See* sine rule.

law of species (law of quadrants) *See* species.

laws of indices *See* exponent.

laws of large numbers If $\{X_i\}$, $i = 1, 2, \ldots,$ n is a *sequence of *random variables with *expectations μ_i, the *weak law of large numbers* gives conditions under which

$$\Pr\left\{ \left| \frac{1}{n} \sum_{i=1}^{n} (X_i - \mu_i) \right| > \varepsilon \right\} \to 0 \quad \text{as } n \to \infty$$

for any given $\varepsilon > 0$. If the X_i are independently and identically distributed random variables, the weak law holds if and only if the means $E(X_i)$ exist and equal μ. This implies that

$$\frac{1}{n} \sum_{i=1}^{n} X_i$$

converges to μ with probability tending to 1.

There is also a *strong law of large numbers* which gives conditions under which

$$\Pr\left\{ \lim_{n \to \infty} \frac{1}{n} \sum_{i=1}^{n} (X_i - \mu_i) = 0 \right\} = 1$$

laws of motion *See* Newton's laws of motion.

LCD *Abbreviation for* least common denominator. *See* common denominator.

LCM *Abbreviation for* least common multiple. *See* common multiple.

leading coefficient The coefficient of the highest-degree term in a *polynomial.

leading diagonal An alternative name for the main *diagonal of a square array.

least action, principle of A principle first put forward by Pierre de Maupertuis in 1744 and since modified. It states that for a dynamical system moving under *conservative forces, the actual motion of the system from point A to point B takes place in such a way that the *action has a stationary value with respect to all other possible paths from A to B with the same kinetic plus potential energy.

least common denominator *See* common denominator.

least common multiple *See* common multiple.

least integer function An alternative name for the ceiling function. *See* integer part.

least squares **1.** A method in *approximation theory for estimating true values of quantities from observed values subject to error. The criterion used is to estimate the true values so as to minimize the sums of squares of deviations of the observed values from these estimates. For example, if two items are weighed, first individually and then together, on a faulty balance and the recorded weights are 17 g and 25 g for the separate items and 40 g for the combined weight, then the least-squares estimates of the true weights are the values of $\hat{w}_1$ and $\hat{w}_2$ that minimize

$$L = (w_1 - 25)^2 + (w_2 - 17)^2$$
$$+ (w_1 + w_2 - 40)^2$$

Differentiating with respect to w_1 and w_2, equating derivatives to zero, and solving gives $\hat{w}_1 = 16.33$ and $\hat{w}_2 = 24.33$.

2. A method used in statistics for estimation of parameters, especially in regression models (see regression). For example, if the expected value of a response y is of the form

$$E(y) = \alpha + \beta x$$

and a set of n pairs (x_i, y_i) is given, the least-squares estimators of the unknown parameters α and β are a and b, chosen to minimize

$$\sum_i (y_i - a - bx_i)^2$$

If the x-variable is error-free and errors in y are assumed to be identically and independently normally distributed with mean zero, the method is equivalent to *maximum likelihood estimation. The method extends to nonlinear models; related procedures known as *weighted least squares* and *generalized least squares* may have optimum properties when the assumption of identically distributed and independent errors is relaxed, or the x-variables are not error-free. See Gauss–Markov theorem.

least upper bound (l.u.b.; supremum) An upper bound u (of a function, sequence, or set) is a least upper bound if $u \leqslant v$ for any other upper bound v. See bound.

leave-one-out See cross-validation.

Lebesgue, Henri Léon (1875–1941) French mathematician noted for his work on measure theory and the theory of integration. He developed, around the end of the

19th century, a concept of integration more general than that of the Riemann integral, based on the Lebesgue measure of the set. This work was stimulated by Borel's work on sets. He also worked on point-set theory and on the calculus of variations.

Lebesgue integral For a bounded *measurable function $f(x)$ over a measurable *set E having finite measure, the Lebesgue integral is defined as follows. U is the upper bound of $f(x)$ over E, and L is the lower bound. The interval $[L, U]$ is divided into n subintervals by numbers $L = t_0 < t_1 < t_2 < \cdots < t_n = U$. The set E is divided into sets $e_1, e_2, \ldots$. Here, e_1 is the set of points of E for which $t_0 \leqslant f(x) \leqslant t_1$; e_2 is the set for which $t_1 \leqslant f(x) \leqslant t_2$; and in general e_i is the set for which $t_{i-1} \leqslant f(x) \leqslant t_i$. The Lebesgue *measure of the set e_i is written as $m(e_i)$. Two sums can be formed:

$$\sum_1^n t_{i-1} m(e_i) \quad \text{and} \quad \sum_1^n t_i \, m(e_i)$$

If δ is the greatest of the numbers $t_i - t_{i-1}$, then the Lebesgue integral is defined as the limit of either of the above sums as $\delta \to 0$. A function that has a Riemann integral necessarily has a Lebesgue integral, although the converse is not necessarily the case. See also calculus; integration.

Lefschetz, Solomon (1884–1972) Russian-American who trained as an engineer but, following a serious accident in which he lost both hands, became a mathematician. He pioneered much of algebraic topology, including his *fixed-point theorem, and made notable discoveries in algebraic geometry. He also made many contributions to other parts of mathematics, particularly in the theory of nonlinear ordinary differential equations.

Lefschetz number See fixed-point theorem; Euler–Poincaré characteristic.

Lefschetz theorem See fixed-point theorem.

left coset *See* coset.

left-handed triad *See* Cartesian coordinate system.

leg One of the sides containing the right angle in a right-angled triangle.

Legendre, Adrien Marie (1752–1833) French mathematician who spent many years studying elliptic integrals. He also worked on problems in number theory, collecting his results in his *Théorie des nombres* (1830). Legendre wrote a popular and influential geometry textbook, *Éléments de géométrie* (1794), and contributed to the development of the calculus and mechanics.

Legendre's differential equation The *differential equation

$$(1 - x^2)\frac{d^2y}{dx^2} - 2x\frac{dy}{dx} + n(n-1)y = 0$$

Its solutions are a set of polynomials $P_n(x)$ (*Legendre polynomials*). These are obtained by expanding

$$\frac{1}{\sqrt{(1 - 2xy + y^2)}}$$

in ascending powers of y and taking the coefficients in the resulting series, $P_0(x) = 1$, $P_1(x) = x$, etc. The *associated Legendre functions* are functions $P_n{}^m(x)$ defined by

$$P_n{}^m(x) = (1 - x^2)^{m/2}\frac{d^m}{dx^m}P_n(x)$$

where $P_n(x)$ are Legendre polynomials. *See* harmonic.

Legendre symbol A symbol that concisely represents whether or not an *integer a is a *quadratic residue modulo an odd *prime p. It is actually a *function of two variables, an integer a and an odd prime p, and is traditionally written as

$\left(\frac{a}{p}\right)$. The function is defined as follows:

$$\left(\frac{a}{p}\right) = \begin{cases} +1 & \text{if } a \text{ is a quadratic} \\ & \text{residue modulo } p \\ -1 & \text{if } a \not\equiv 0 \ (\text{mod } p) \text{ is} \\ & \text{not a quadratic residue} \\ 0 & \text{if } a \equiv 0 \ (\text{mod } p) \end{cases}$$

For instance, $\left(\frac{2}{7}\right) = +1$, since $3^2 \equiv 2$ (mod 7); but $\left(\frac{3}{7}\right) = -1$, since $x^2 \equiv 3$ (mod 7) has no solutions. There are rules for manipulating and evaluating Legendre symbols, and so determining whether any integer a is a quadratic residue modulo a given odd prime p. *See also* quadratic reciprocity.

Leibniz, Gottfried Wilhelm (1646–1716) German mathematician, physicist, and philosopher noted for his discovery of the differential *calculus which he first made public in his *Nova methodus pro maximis et minimis* (1684, A New Method for Determining Maxima and Minima). In subsequent works Leibniz also developed the integral calculus (the now-familiar symbols are in fact his innovations). Much of Leibniz's time was spent on his attempts to develop a *characteristica generalis*, a universal language, work which can be seen now as one of the earliest attempts to advance beyond the traditional logic of Aristotle to the mathematical logic later formulated by Boole.

Leibniz theorem The formula for finding the nth *derivative of the product of two *functions. If u and v are functions of x, and their first, second, etc. derivatives are $u_1, u_2, \ldots, v_1, v_2, \ldots$, then the nth derivative $(uv)_n$ is given by

$$(uv)_n = u_nv_0 + nu_{n-1}v_1$$
$$+ [n(n-1)]u_{n-2}v_2/2!$$
$$+ \cdots + nu_1v_{n-1} + uv_n$$

For example,

$$\frac{d}{dx}(uv) = v\frac{du}{dx} + u\frac{dv}{dx}$$

$$\frac{d^2}{dx^2}(uv) = v\frac{d^2u}{dx^2} + 2\frac{du}{dx}\frac{dv}{dx} + u\frac{d^2v}{dx^2}$$

lemma *See* theorem.

lemniscate A type of plane curve, with the equation in Cartesian coordinates

$$(x^2 + y^2)^2 = a^2(x^2 - y^2)$$

It is the *locus of a point that is the foot of the perpendicular from the origin to a variable tangent on a rectangular hyperbola. The curve is also known as the *lemniscate of Bernoulli*. *See also* Cassini's ovals.

length For a line segment, the length is taken as the *absolute value $|\mathbf{a} - \mathbf{b}|$ where $\mathbf{a}$ and $\mathbf{b}$ are the *position vectors of the end points. For a curve, *arc length* is obtained by integration. In a Cartesian coordinate system a curve $y = f(x)$ has an element of length ds given by $\sqrt{(dx^2 + dy^2)}$. The length of the curve between points $x = a$ and $x = b$ is given by the integral

$$\int_a^b \sqrt{\left[1 + \left(\frac{dy}{dx}\right)^2\right]} dx$$

In polar coordinates, the length between $r = u$ and $r = v$ is

$$\int_u^v \sqrt{\left[1 + r^2\left(\frac{d\theta}{dr}\right)^2\right]} dr$$

Alternatively, in polar coordinates the length between $\theta = \alpha$ and $\theta = \beta$ is

$$\int_\alpha^\beta \sqrt{\left[\left(\frac{dr}{d\theta}\right)^2 + r^2\right]} d\theta$$

For a curve given in terms of a parameter t,

the length between $t = t_1$ and $t = t_2$ is

$$\int_{t_1}^{t_2} \sqrt{\left[\left(\frac{dx}{dt}\right)^2 + \left(\frac{dy}{dt}\right)^2\right]} dt$$

Leonardo of Pisa *See* Fibonacci.

leptokurtic *See* kurtosis.

Leslie matrix (P.H. Leslie, 1945) An $n \times n$ matrix of the form

$$\begin{pmatrix} a_1 & a_2 & \cdots & a_{n-1} & a_n \\ b_1 & 0 & \cdots & 0 & 0 \\ 0 & b_2 & \cdots & 0 & 0 \\ \vdots & \vdots & & \vdots & \vdots \\ 0 & 0 & \cdots & b_{n-1} & 0 \end{pmatrix}$$

This type of matrix can be used to model population growth, with the a_i describing birth rates and the b_i survival rates associated with different age groups in a population.

Levene's test *See* homogeneity of variance.

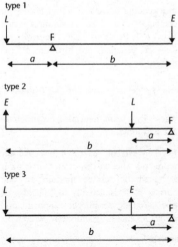

lever The three types of lever.

lever A simple *machine composed essentially of a rigid bar pivoted in such a way that a *force can be transferred to a load, usually with a mechanical advantage. The lever pivots about a point known as the *fulcrum*. The position of the fulcrum, F, relative to that of the load, L, and applied force, or effort, E, determines the type of lever (*see* diagram). In equilibrium the algebraic sum of the *moments of all forces about the fulcrum is zero. Thus in all three types (assuming the system to be frictionless)

$$La = Eb$$

The mechanical advantage, L/E, is then b/a.

Many everyday mechanical devices employ the principle of the lever: pliers and scissors are type 1 levers; wheelbarrows and traditional nutcrackers are type 2 levers. Type 3 levers amplify movement rather than force, working at a mechanical advantage less than unity; foot treadles are type 3 levers. The skeletal elements to which muscles are attached are often lever systems, mainly type 3, where a joint acts as the fulcrum.

Levi ben Gerson (1288–1344) French-born Jewish mathematician and astronomer who produced in his *Sefer ha mispar* (1321, Book of Number) one of the first texts to establish simple rules for calculating permutations and combinations, and use the principle of mathematical induction. He also published in his *De sinibus, chordis et arcubus* (1342, On Sines, Chords, and Arcs) one of the earliest works on trigonometry.

Levi-Civita, Tullio (1873–1941) Italian mathematician who was the first, in 1896, to apply to dynamics the work of Ricci-Curbastro on the absolute differential calculus, better known as the tensor calculus. Further work with Ricci-Curbastro in 1900 led to their algorithm for the expression of physical laws in both Euclidean and Riemannian curved space, a result which later proved to be of value to Einstein.

L'Hôpital (or L'Hospital), Guillaume François Antoine, Marquis de (1661–1704) French mathematician noted for his *Analyse des infiniment petits* (1696, Analysis with Infinitely Small Quantities), the first textbook on differential calculus. It contains the first formulation of *L'Hôpital's rule for the limiting value of fractions whose numerators and denominators tend to zero. The rule was, in fact, devised by Jean Bernoulli (around 1694), who taught calculus to L'Hôpital, and later accused him of plagiarism. L'Hôpital also wrote a textbook on analytic geometry, *Traité analytique des sections coniques* (1707, Analytical Treatise on Conic Sections).

L'Hôpital's rule (L'Hospital's rule, de L'Hôpital's rule) A rule for finding the *limit of a ratio of two *functions each of which separately tends to zero. It states that for two functions $f(x)$ and $g(x)$ the limit of the ratio $f(x)/g(x)$ as $x \to a$ is equal to the limit of the ratio of the derivatives $f'(x)/g'(x)$ as $x \to a$. For example, the functions $x^2 - 4$ and $2x - 4$ have a ratio $(x^2 - 4)/(2x - 4)$. As $x \to 2$, this ratio takes the indeterminate form $0/0$, i.e. the limit of the ratio cannot be found directly. L'Hôpital's rule states that the limit of the ratio is equal to the limit of the ratio of the first derivatives, i.e. the limit of $2x/2$ as $x \to 2$, which is 2. If the ratio of the first derivatives is also indeterminate, higher-order derivatives can be used.

Li(x), li(x) *See* logarithmic integral.

liar paradox The *paradox that if someone says 'I am lying', then if what is said is true then it is false, and if what is said is false then it is true. Traditionally it is thought to have been put forward in the 6th century BC by the Cretan philosopher Epimenides. The liar paradox is an example of a sentence that may be grammatically correct yet is logically self-contradictory. *See* paradox.

Li Chih *See* Li Ye.

Lie, Marius Sophus (1842–99) Norwegian mathematician noted for his work on transformation groups, which he described in his major treatise, *Die Transformations-gruppen* (1888–93). He was also the first to make a methodical study of continuous groups, an important class of which have since become known as *Lie groups*.

life tables (J. Graunt, 1662) Tables that show, for a specific population of class of individuals (e.g. English males, Canadian females) and for a given number (e.g. 1000) alive at a specified age (e.g. 40), the numbers who live to or are expected to live to successive higher ages. Life tables may be based on retrospective studies of particular populations or groups of people, and actuaries use such information to produce tables to predict such outcomes for similar populations in the future. The latter are sometimes called *life expectancy tables*, and may be updated to take account of new factors that are expected to increase or decrease expectancy (e.g. improved health care, or greater exposure to accident risks) for a particular group. Life tables are also referred to as *mortality tables*. These tables have a useful role in comparing age-specific mortality rates for different illnesses, or for the same illness in different age groups.

lift An upward *force that is experienced by a body moving through a fluid, such as air or water, and that acts perpendicularly to the direction of motion. Lift thus acts at right angles to *drag and causes the body to rise. The amount of lift is given by $c\rho Av^2$, where ρ is the fluid density, A is a representative area of the body (such as the area of a wing), and v is the magnitude of the velocity of the body relative to the fluid. The coefficient c depends on the circulation around the body and is a function of the Reynolds number vl/v, where l is a repre-

sentative length of the body and v is the coefficient of kinematic viscosity. *Compare* drag.

light year A unit of distance used in astronomy equal to the distance travelled by light (electromagnetic radiation) in a vacuum in one year. 1 light year $= 9.4605 \times 10^{15}$ metres or approximately 5.88×10^{12} miles.

likelihood *See* likelihood function.

likelihood function The *frequency function of a continuous *random variable X belonging to a family of distributions dependent on a parameter θ may be written as $f(x, \theta)$ where x is variable and θ is fixed. However, if x_1 is an observed or sample value of X and we regard θ as a parameter that can be varied to specify different members of the family, then $L(\theta) = f(x_1, \theta)$, regarded as a function of θ for any given x_1, is called the *likelihood function*. The value of $L(\theta)$ for any particular value of θ is called the *likelihood*.

For a sample of n independent observations from the same distribution, the likelihood function is $L(\theta) = f(x_1, x_2, \ldots, x_n)$, and independence implies that

$$L(\theta) = f(x_1, \theta)f(x_2, \theta) \cdots f(x_n, \theta)$$

Then if, for two values θ_1 and θ_2 of θ, one finds that $L(\theta_2) < L(\theta_1)$, this implies that the sample has a smaller value of the joint frequency function if the unknown parameter is θ_2 rather than θ_1. This in turn implies that the sample is less likely to have come from a population where $\theta = \theta_2$ than from one where $\theta = \theta_1$. This reasoning leads to the concept of *maximum likelihood estimation of a parameter as determining the value of the parameter that maximizes the likelihood function.

The concept of a likelihood function can be extended to discrete random variables and to samples from distributions having more than one parameter. *See also* likelihood ratio.

likelihood ratio (J. Neyman and E.S. Pearson, 1928) If the *likelihood is L_1 when $\theta = \theta_1$, and L_2 when $\theta = \theta_2$, then the ratio L_2/L_1 may be used as a basis of a test of the null hypothesis $\theta = \theta_1$ against the alternative $\theta = \theta_2$. The concept may be extended to test the hypothesis that θ takes a value in a specified subset of all possible values by taking L_2 as the maximum for that subset. *See also* hypothesis testing.

Lim *See* limit.

limaçon of Pascal A type of plane curve. It is generated by first taking a fixed point O on a circle and drawing a variable line through this point. The limaçon is the locus of a point P that lies on the line and is a fixed distance a from Q, the other point of intersection of the line with the circle. If the fixed point O is taken to be the pole of a polar coordinate system, the equation of the limaçon is

$$r = d \cos \theta + a$$

where d is the circle's diameter. If $d = a$ the curve is a *cardioid. [French: snail; so named by Étienne Pascal (1588–1640)]

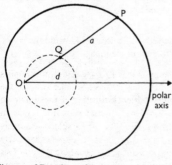

limaçon of Pascal: $a > d$.

limit 1. (of a function) A value that can be approached arbitrarily closely by the dependent variable when some restriction is placed on the independent variable of a *function. For example, as x increases, $f(x) = 1/x$ decreases, getting closer to zero. $f(x) = 1/x$ is said to *approach* or

tend to zero as x tends to infinity, written as $(1/x) \to 0$ as $x \to \infty$. Alternatively, this can be expressed as 'the limit of $1/x$ as x tends to infinity is zero', written as

$$\lim_{x \to \infty} \left(\frac{1}{x} \right) = 0$$

The function $\sin x/x$ also approaches zero as x tends to infinity but it alternates between positive and negative values.

In general, $f(x) \to l$ as $x \to \infty$ if, for every positive real number ε, there exists a positive real number N dependent on ε such that whenever $x > N$, then

$$|f(x) - l| < \varepsilon$$

In other words, by choosing a large enough value of x, $f(x)$ can be made as near to l as is required. Also $f(x) \to \infty$ as $x \to \infty$ if, for every positive real number M, there exists a real number N dependent on M such that whenever $x > N$, $f(x) > M$; i.e. by choosing a large enough value of x, $f(x)$ can be made arbitrarily large.

If $f(x)$ approaches a value l as x approaches a from the right (i.e. from ∞ to a) then the value l is said to be the *right-hand limit* of $f(x)$ at $x = a$; if $f(x)$ approaches a value k as x approaches a from the left (i.e. from $-\infty$ to a) then k is said to be the *left-hand limit* of $f(x)$ at $x = a$.

A function may become arbitrarily large when x is sufficiently close to a, written as $f(x) \to \infty$ as $x \to a$. Formally, for every real positive number M there exists a number δ dependent on M such that whenever

$$|x - a| < \delta, \quad \text{then } f(x) > M$$

Functions that tend to $-\infty$ as $x \to a$ or $x \to \infty$ are defined similarly to those that tend to $+\infty$.

See also continuous function.

2. (of a sequence) A number, A say, that an infinite *sequence

$$a_1, a_2, a_3, \ldots, a_n, \ldots$$

may approach (or *tend to*) as the number of terms n becomes very great, i.e. tends to infinity. This is written as

$$a_n \to A \text{ as } n \to \infty \quad \text{or} \quad \underset{n \to \infty}{\text{Lim}}\, a_n = A$$

A finite limit exists only if, given any positive number ε, however small, it is possible to find a term a_N such that all subsequent terms differ from A by less than ε, i.e.

$$|a_r - A| < \varepsilon \quad \text{for all } r > N$$

If an infinite sequence has a finite limit it is said to be *convergent*, otherwise it is *divergent*. *See also* convergent series.

3. One of the values of the variable between which a definite integral is evaluated. *See* integration.

limit inferior (of a sequence) *See* limit point.

limit of convergence *See* power series.

limit point (accumulation point, cluster point) 1. (of a sequence) A point associated with an infinite *sequence in whose neighbourhood lie an infinite number of terms of the sequence. In a sequence of real numbers, if there are an infinite number of terms greater (or less) than any number k, then $+\infty$ (or $-\infty$) is a limit point of the sequence. There may be more than one limit point. For a sequence of real numbers the largest limit point is known as the *limit superior*, and the smallest one as the *limit inferior*.

2. (of a set) A point P is a limit point of a *set A if every *neighbourhood of P contains a point that is distinct from P and is a member of A.

limit superior (of a sequence) *See* limit point.

Lindemann, Carl Louis Ferdinand von (1852–1939) German mathematician noted for his proof in 1882 that π is transcendental, thus finally demonstrating that it is impossible to square the circle using purely Euclidean constructions. He also

published several 'proofs' of Fermat's last theorem (since shown to be erroneous) and also propagated the views of Weierstrass on the arithmetization of calculus.

line 1. A *curve.
2. A *straight line*; i.e. a curve that, geometrically, is completely determined by two of its points. In plane *coordinate geometry a line is a set of points satisfying a *linear equation of the type

$$ax + by + c = 0$$

where a and b are not both zero. In simple rectangular Cartesian coordinates the equation of a straight line has various standard forms as follows:

Slope–intercept form. A line with the equation

$$y = mx + c$$

has a gradient m and an intercept of c on the y-axis. For instance, the line $y = 2x + 4$ has a gradient of 2 (the angle between the line and the x-axis is $\tan^{-1} 2$) and it cuts the y-axis at the point $(0, 4)$.

Intercept form. A line with an equation of the form

$$x/a + y/b = 1$$

intersects the x-axis at $(a, 0)$ and the y-axis at $(0, b)$. For example, the line

$$4y = 2x - 8$$

can be put in the form

$$x/4 - y/2 = 1$$

The intercept on the x-axis is 4 and the intercept on the y-axis is -2.

Point–slope form. A line with a slope m passing through a known point (x_1, y_1) has the equation

$$y - y_1 = m(x - x_1)$$

An example is the line with a gradient of 2 passing through the point $(5, 4)$. Its equation is

$$y - 4 = 2(x - 5)$$

which rearranges to give

$$y = 2x - 6$$

A negative value of m indicates a slope downwards from left to right.

Two-point form. A line passing through two known points (x_1, y_1) and (x_2, y_2) has an equation of the form

$$\frac{x - x_1}{x_2 - x_1} = \frac{y - y_1}{y_2 - y_1}$$

For example, the line passing through the points $(2, 1)$ and $(-6, 7)$ has the equation

$$\frac{x - 2}{-6 - 2} = \frac{y - 1}{7 - 1}$$

which rearranges to give

$$4y = -3x + 10$$

The forms above are the ones used in Cartesian coordinates in two dimensions. In three-dimensional Cartesian coordinates, the equation of a line in space may also have various forms:

Symmetric form (or *standard form*). The equation is written in terms of direction numbers l, m and n (*see* direction angles) together with one point on the line (x_1, y_1, z_1):

$$\frac{x - x_1}{l} = \frac{y - y_1}{m} = \frac{z - z_1}{n}$$

Two-point form. The equation is written in terms of two points on the line with coordinates (x_1, y_1, z_1) and (x_2, y_2, z_2). It has the form

$$\frac{x - x_1}{x_2 - x_1} = \frac{y - y_1}{y_2 - y_1} = \frac{z - z_1}{z_2 - z_1}$$

Parametric form. The line is described in terms of its direction cosines l, m, and n (*see* direction angles), a point on the line (x_1, y_1, z_1), and a variable parameter d. The parametric equations are

$$x = x_1 + ld$$

$$y = y_1 + md$$

$$z = z_1 + nd$$

Here, d is the distance of the variable point (x, y, z) from (x_1, y_1, z_1).

Vector form. The line through points with position vectors $\mathbf{a}$ and $\mathbf{b}$ has the parametric equation

$$\mathbf{r} = \mathbf{a} + t(\mathbf{b} - \mathbf{a})$$

or the equation

$$(\mathbf{r} - \mathbf{a}) \times (\mathbf{b} - \mathbf{a}) = 0$$

linear Describing an equation, expression, etc. that is of the first *degree. A *linear equation* is one in which all non-constant terms have degree 1. For example,

$$x + 3y + 2z = 7$$

is a linear equation in three variables.

A *linear combination* of variables x_1, x_2, x_3, ... is the sum

$$a_1 x_1 + a_2 x_2 + a_3 x_3 + \cdots$$

where a_1, a_2, a_3, ...are constants.

It is also possible to apply the term 'linear' to particular variables in an expression. Thus $3xyz^2$ is linear with respect to x and y.

See also vector space.

linear algebra 1. The branch of algebra that studies *linear equations, *matrices, *vector spaces, and *linear transformations.
2. A *vector space V over a *field F which is also a *ring and for which the law

$$(n\mathbf{u})\mathbf{v} = n(\mathbf{u}\mathbf{v}) = \mathbf{u}(n\mathbf{v})$$

holds for all $n \in F$ and $\mathbf{u}, \mathbf{v} \in V$ is called a *linear algebra* (or *associative algebra*) over F. For example, the set of all 2×2 matrices (with real or complex elements) is a linear algebra over $\mathbb{R}$, the field of real numbers. *See* algebra.

linear code *See* coding.

linear combination A sum of *scalar multiples of elements of a set. For example, $3\mathbf{u} + 4\mathbf{v}$ is a linear combination of the *vectors $\mathbf{u}$ and $\mathbf{v}$. *See also* vector space.

linear congruence A *congruence of the type $ax \equiv b \pmod{n}$ where n is a given natural number, a and b are given integers, and x is an unknown integer. Such a congruence can be solved for x if and only if b is divisible by the *highest common factor of a and n. If so, then $\mathrm{HCF}(a, n)$ gives the maximum number of solutions that are mutually incongruent modulo n. For example:

$2x \equiv 7 \pmod{18}$ is not solvable since $\mathrm{HCF}(2, 18) = 2$ does not divide 7; but $15x \equiv 6 \pmod{18}$ is solvable since $\mathrm{HCF}(15, 18) = 3$ does divide 6, and it has three incongruent solutions modulo 18, namely $x = 4$, 10, and 16.

$7x \equiv 8 \pmod{30}$ is solvable since $\mathrm{HCF}(7, 30) = 1$ divides 8, and it has a unique solution modulo 30, namely $x = 14$ (i.e. every solution will be congruent to 14 modulo 30).

linear convergence *See* order (**12**).

linear differential equation A *differential equation of the form

$$\mathrm{P}_0(x)y + \mathrm{P}_1(x)\frac{\mathrm{d}y}{\mathrm{d}x} + \cdots + \mathrm{P}_n(x)\frac{\mathrm{d}^n y}{\mathrm{d}x^n} = \mathrm{Q}(x)$$

which is *linear in y and its derivatives, and in which the coefficients of y and its derivatives are functions of x only. An example is

$$x\frac{\mathrm{d}y}{\mathrm{d}x} + y = \sin x$$

linear equation *See* linear.

linear form *See* form.

linear function A *polynomial function of *degree one. A linear function of one variable has the form

$$\mathrm{f}(x) = a_0 + a_1 x$$

where a_0 and a_1 are constants. The graph of the function is a straight line with gradient a_1 and intercept a_0 on the y-axis. A linear function of two variables has the form

$$\mathrm{f}(x, y) = a_0 + a_1 x + a_2 y + a_3 xy$$

where a_0, a_1, and a_2 are constants. Here, $\mathrm{f}(x, y)$ is linear in x and linear in y. A linear function of several variables is similarly defined.

linear hypothesis In general, a hypothesis concerning linear *functions of parameters; more specifically the term is applied to tests on linear functions of parameters in *regression analysis and *analysis of variance, e.g. a hypothesis that the difference between two treatment means τ_1 and τ_2 is zero, or takes a specific value, is a linear hypothesis about the function $\tau_1 - \tau_2$.

linear interpolation *See* interpolation; false position (rule of).

linearly dependent, independent *See* vector space.

linear mapping *See* linear transformation.

linear model In *statistics, a model in which the *expected value of a *random variable is a linear function of the *parameters in the model. *See* regression, generalized linear models.

linear momentum *See* momentum.

linear programming A method for determining optimum values of a *linear function subject to constraints expressed as linear equations or inequalities. In practice, functions to be maximized often represent profits or volume of goods that can be produced, while functions to be minimized may be production costs or production times. A practical problem may involve 100 or more variables, in which case it is usually solved by using the *simplex method and a computer.

Simple problems with only two variables may be solved graphically. For example, to minimize

$$U = 4x + 3y$$

subject to the constraints

$$x + y \leqslant 20, \quad 3x + y \leqslant 30,$$

$$x \geqslant 0, \quad y \geqslant 0$$

it is easily seen that the constraints require any permissible solution (usually called a *feasible solution*) to lie in or on the boundaries of the stippled area (the *feasible region*) in the diagram. Here the line AB represents the equation $x + y = 20$, and the line CD the equation $3x + y = 30$. These lines and the axes determine the boundaries of the region of feasible solutions. The dashed parallel lines represent the equations $4x + 3y = U$ for several values of U, these lines shifting to the right as U increases. Thus the optimum (maximum feasible) value of U occurs when U is chosen so that the line passes through the point E, where the lines $x + y = 20$ and $3x + y = 30$ intersect. Solving these equations gives $x = 5$ and $y = 15$, and so the maximum feasible value of U is $U = 4 \times 5 + 3 \times 15 = 65$, the required solution. *See* Karmarkar's algorithm.

linear programming

linear regression *See* regression.

linear scale *See* scales of measurement.

linear space *See* vector space.

linear transformation (linear mapping) 1. A *transformation of n variables expressed by n equations:

$$y_1 = a_{11}x_1 + a_{12}x_2 + \cdots + a_{1n}x_n$$
$$y_2 = a_{21}x_1 + a_{22}x_2 + \cdots + a_{2n}x_n$$
$$\vdots$$
$$y_n = a_{n1}x_1 + a_{n2}x_2 + \cdots + a_{nn}x_n$$

The *matrix of such a transformation is the $n \times n$ matrix **A** with elements a_{ij}. If **A** is a *nonsingular matrix, then $x_1, x_2, \ldots$ can be expressed as linear combinations of $y_1, y_2, \ldots$ with matrix $\mathbf{A}^{-1}$ (i.e. the inverse). If the x-variables are expressed in terms of a third variable z by linear equations having a matrix **B**, then the y-variables are linear combinations of the z-variables with a matrix **AB**.

In general, a linear transformation is a mapping from one *vector space into another, $L: V \rightarrow V'$, with the following properties:

(1) For any two vectors **u** and **v**

$$L(\mathbf{u} + \mathbf{v}) = L(\mathbf{u}) + L(\mathbf{v})$$

(2) If n is a number

$$L(n\mathbf{u}) = nL(\mathbf{u})$$

For a given transformation, there is an associated matrix **A** such that for any vector **u** in the space, $L(\mathbf{u}) = \mathbf{Au}$ (where **Au** denotes matrix multiplication of **A** and the column vector **u**). *See* transformation; affine transformation.

2. (homographic transformation, Möbius transformation) A *transformation of a complex variable z having the form

$$w = \frac{az + b}{cz + d}$$

and where $ad - bc \neq 0$.

line of apsides *See* apsis.

line of best fit Given a set of n points $(x_1, y_1), (x_2, y_2), \ldots, (x_n, y_n)$, the line of best fit $y = a + bx$ is often taken to mean the *least squares regression of y on x, obtained by choosing values a and b of α and β that minimize

$$L_2 = \sum_i (y_i - \alpha - \beta x_i)^2$$

leading to $b = s_{xy}/s_{xx}$, where $s_{xy} = \Sigma(x_i - \bar{x})(y_i - \bar{y})$, $s_{xx} = \Sigma(x_i - \bar{x})^2$, $a = \bar{y} - b\bar{x}$, $\bar{x} = \Sigma x_i/n$, and $\bar{y} = \Sigma y_i/n$. This choice of α and β minimizes the sum of squares of deviations from the line $y = \alpha + \beta x$ measured in the direction of the y-axis. If we wished instead to minimize the sum of the absolute deviations, the function to be minimized would be

$$L_1 = \sum_i |(y_i - \alpha - \beta x_i)|$$

Furthermore, the least-squares regression of x on y is not in general the same line as the least-squares regression of y on x, so what is meant by the line of best fit to a data set depends on what assumptions are made about the nature of any deviations from a fitted line.

line segment A portion of a straight line between two points. Note that strictly a *line* extends indefinitely in both directions; a *line segment* has a finite length. *See also* half-line.

Liouville, Joseph (1809–82) French mathematician noted as the editor of the *Journal de Mathématiques pures et appliquées*, launched in 1836 and more commonly known as *Liouville's Journal*. As a mathematician Liouville worked in the field of transcendental numbers. In 1844 he proved their existence and went on to construct an extensive class of *Liouville numbers. He also edited and published (1846) some manuscripts left by Galois on polynomial equations.

Liouville number An *irrational number θ with the property that for each natural number n there is at least one rational number $p/q \neq \theta$ with $|\theta - (p/q)| < 1/q^n$. All Liouville numbers are *transcendental. *See* Roth's theorem.

Liouville's theorem *See* entire function.

Lissajous figures Curves that are the *locus of a point in two dimensions with components that are simple *harmonic motions. The shape depends on the relative frequencies and phases of the two motions. The curves are named after the French mathematician Jules Antoine Lissajous (1822–80).

litre Symbol: l (alternatively, L). A *metric unit of capacity or volume, not an SI unit but used for some purposes as a special name for the cubic decimetre (dm^3). It is not recommended for use in high-precision measurements. The symbol *ml for millilitre is sometimes used as an alternative to cc. In *SI units, the symbol cm^3 is recommended for this quantity. The litre was formerly defined as the volume of 1 kilogram of pure water at $4\,°C$ and a pressure of 760 millimetres of mercury; by this definition the litre is equivalent to $1000.028\,cm^3$.

Littlewood, John Edensor (1885–1977) English mathematician best known for his long collaboration with G.H. Hardy during which they published nearly 100 papers. Littlewood worked on Fourier series, the Riemann zeta function, the partition of numbers, inequalities, the theory of functions, and the distribution of primes.

lituus *See* spiral.

Liu Hui (*c*. AD 263) Chinese mathematician whose inventive *Commentary on the Nine Chapters on the Mathematical Art* contained an interpolated value of 3.1416 for π based on a succession of regular polygons inscribed in a circle, and a proof of the formulae for the volume of a square pyramid and a tetrahedron which uses a form of *exhaustion. His *Haidao suanjing* (Sea Island Mathematical Manual) solved problems of surveying, and thus mapping, inaccessible objects using his 'method of double differences', which involves pairs of similar triangles.

Li Ye, Li Zhi (Li Chih) (1192–1279) Chinese mathematician whose *Ceyuan haijing* (1248, Sea Mirror of Circle Measurements) introduced the 'method of the celestial element' – a system of notation for polynomials in one variable (the 'celestial element'), and techniques for manipulating them and solving problems.

ln *See* logarithmic function.

load An *external force exerted on a body, such as a weight supported by a structure, or applied to a *machine.

Lobachevsky, Nikolai Ivanovich (1793–1856) Russian mathematician noted for his discovery in 1826, independently of Bolyai, of hyperbolic geometry, the first *non-Euclidean geometry to be described. Lobachevsky also worked on infinite series, probability, and algebraic equations.

local coordinates *See* manifold.

locally connected A set of points or a space X is *locally connected at a point* $a \in X$ if, within every *neighbourhood N of a, there is a subneighbourhood M of a contained in N such that any two points of M lie in a *connected subset of N. A set or space is *locally connected* if it is locally connected at each of its points.

located vector A *vector with a specified starting position.

location The notion of centrality in a sample or distribution measured by *mean, *median, or *mode.

locus (*plural* **loci**) A set of points satisfying given conditions. For instance, the locus of points in a plane that are all a distance r from a given point in the plane is a circle. The equation of the locus, in Cartesian coordinates, is

$$x^2 + y^2 = r^2$$

loess *See* lowess.

logarithm (log) For a positive number n, the logarithm of n (written as $\log n$) is the *power to which some number b must be raised to give n. Here b is the *base* of the logarithm, i.e.

$$\log_b n = x, \quad \text{if } b^x = n$$

An *antilogarithm is a number whose logarithm is a given number.

Logarithms obey certain laws:

$$\log(nm) = \log n + \log m$$

$$\log(n/m) = \log n - \log m$$

$$\log(n^m) = m \log n$$

Formerly, they were used extensively in computation, in the form of tables of logarithms to the base 10. Such logarithms are called *common logarithms* (or *Briggsian logarithms*). Logarithms to the base e (2.718 . . .) are *natural logarithms* (also called *Napierian* or *hyperbolic logarithms* – *see* Napier). By convention $\log_e n$ is often written as $\ln n$, and $\log_{10} n$ is often written as $\log n$ or $\lg n$.

Common logarithms for computation are used in the form of an integer (the *characteristic*) plus a positive decimal fraction (the *mantissa*). For example, to find the logarithm of 657.3, the number is written in standard form as 6.573×10^2. The logarithm of this is $\log 6.573 + 2 \log 10$, which is $2 + \log 6.573$, or 2.8178. Here 2 is the characteristic and 0.8178 the mantissa. For a number such as 0.06573, say, the standard form is 6.573×10^{-2}. The logarithm is then $-2 + \log 6.573$, which is written as $\bar{2}.8178$ (where $\bar{2}$ is read as 'bar two'). In tables of common logarithms, only the mantissae are tabulated. *See also* modulus (of logarithms).

logarithmic coordinate system A Cartesian coordinate system in which the axes are marked with logarithmic scales. *See also* graph.

logarithmic differentiation A method of finding *derivatives in which logarithms are taken before differentiating. For example, if $y = 2^x$, then, taking logarithms of both sides, $\ln y = x \ln 2$, and differentiation with respect to x gives

$$\frac{1}{y}\frac{dy}{dx} = \ln 2$$

Thus

$$\frac{dy}{dx} = y \ln 2 = 2^x \ln 2$$

The method is useful when differentiating a *continued product, e.g.

$$x(1 + 2x)(1 + 3x)$$

logarithmic distribution (R.A. Fisher, 1941) A *discrete distribution of a *random variable X for which

$$\Pr(X = r) = -\frac{\theta^r}{r \ln(1 - \theta)}, \quad r = 1, 2, \ldots$$

where θ is a parameter taking some value in the open interval $(0, 1)$. The distribution is sometimes known as the *log-series distribution* and is widely used in studies of species diversity.

logarithmic function The function $\ln x$ or $\log_e x$, defined, for $x > 0$, as the inverse function of the *exponential function so that $\ln x = y$, where $x = e^y$ and $\exp(\ln x) = x$. It is also defined by

$$\ln x = \int_1^x \frac{dt}{t}$$

The term is also used for functions of the type $\log_a x$, where $a > 0$, which satisfy $\log_a(a^x) = x$.

logarithmic graph *See* graph.

logarithmic integral The logarithmic integral of x is the *function Li(x) defined by

$$\text{Li}(x) = \int_2^x \frac{1}{\ln t}\, dt$$

where $\ln t$ is the *natural logarithm of t. Some authors use the function li(x) defined by

$$\text{li}(x) = \int_0^x \frac{1}{\ln t}\, dt$$

For $x > 1$ this is an *improper integral interpreted as the limit as $\varepsilon \to 0$ of the sum of integrals over the intervals $[0, 1-\varepsilon]$ and $[1+\varepsilon, x]$, where $\varepsilon > 0$. When $x > 2$ the two functions differ by just a constant. Both Li(x) and li(x) are *asymptotic to $\pi(x)$, the number of *primes less than or equal to x. For example, when $x = 10^6$ the values of Li(x) and $\pi(x)$ are 78 628 and 78 498, respectively.

logarithmic scale *See* scales of measurements.

logarithmic series The *power series

$$x - \frac{x^2}{2} + \frac{x^3}{3} - \frac{x^4}{4} + \cdots$$

The nth term is $(-1)^{n+1} x^n / n$.

If $-1 < x \leqslant 1$ the series converges and has the sum $\ln(1 + x)$, hence the name.

logarithmic spiral *See* spiral.

logarithmic transformation A transformation of a positive-valued *random variable X to $Y = \ln X$. In many situations Y has (or is well approximated by) a normal distribution. If Y has a normal distribution, X is said to have a *lognormal distribution*.

logic The study of deductive *argument. The central concept of logic is that of a valid argument where, if the premises are true, then the conclusion must also be true. In such cases the conclusion is said to be a *logical consequence* of the premises. Logicians are not, in general, interested in the particular content of an argument, but rather with those features that make an argument valid or invalid. So for the simple argument 'If Jones is a man then Jones is mortal; Jones is a man; therefore Jones is mortal' there is a structure 'If *A* then *B*; *A*; therefore *B*'. This argument form (called *modus ponens*) is valid no matter what sentences are substituted for *A* and *B*. This focus on structure leads to the logician's concern with the logical form of sentences irrespective of their content.

This distinction between form and content mirrors closely the distinction between a formal language and its *interpretation. A formal language is built from

(1) a set of symbols organized by syntactic rules that delineate a class of *wffs; and
(2) a set of rules of inference that permit us to pass from a set of wffs (intuitively, the premises) to another wff (intuitively, the conclusion).

The specific way in which (1) and (2) are met determines the type of arguments that we can analyse in a formal language. The *propositional calculus was devised to analyse arguments whose only logical constants are truth-functional connectives, such as '&' (*see* and) and '⊃' (*see* implication). But such a language is not sufficiently refined to capture all those arguments that we intuitively recognize as valid.

Consider 'All men are mortal; John is a man; therefore John is mortal.' Although valid, this argument cannot be represented by means of truth-functional connectives alone: we also need *quantifiers. The above argument would then be formalized as

$$(\forall x)(\text{Man}(x) \supset \text{Mortal}(x))$$

$$\text{Man}(\text{John})$$

$$\therefore \text{Mortal}(\text{John})$$

The rules of inference that permit the passage from premises to conclusion in

this argument are *universal instantiation* (from '$(\forall x)\text{F}(x)$' we can infer '$\text{F}(a)$') and *modus ponens*.

The *predicate calculus is a language that can be used to analyse sentences containing quantifiers. For more complex types of argument we need to construct other languages, for example *modal logic.

The branch of logic concerned with the study of formal languages independently of any content the symbols may have is called *proof theory*. From a proof-theoretic standpoint there is no way of telling whether a rule of inference will allow us to pass from true premises to a false conclusion. In order to judge the adequacy of a formal language as a tool for reasoning we need to turn to the branch of logic called *model theory*, which is concerned with the interpretations of formal languages. For example, the propositional calculus is interpreted by assigning truth values to wffs. More complex languages require more complex types of interpretation. A valid argument can be defined in model-theoretic terms as one where the conclusion is true in all those interpretations under which the premises are true. Those formal languages in which the rules of inference preserve truth in that we cannot pass from true premises to false conclusions are called *sound*. A formal language is *complete* if there are no valid arguments expressible in the language that cannot be proved by use of the rules of inference. By linking proof theory with model theory, completeness and soundness proofs are two of the most important ways of showing that a formal language is satisfactory.

logical consequence *See* consequence.

logical constant *See* constant.

logical equivalence *See* equivalence.

logical form The logical structure that an *argument or sentence possesses independently of its content. For example, consider:

(1) All men are mortal; Alfred is a man; therefore Alfred is mortal.

(2) All dogs are four-legged; Rover is a dog; therefore Rover is four-legged.

Both (1) and (2) have the same logical form, and are instances of the (valid) *argument form*:

(3) $(\forall x)(M(x) \supset F(x))$; $M(a)$; therefore $F(a)$.

The validity or invalidity of an argument is thus seen to follow from its logical form (in the above cases, the logical form as given by (3)), and in particular the distribution of the logical *constants, rather than from any specific content. *See* logic; quantifier.

logical syntax *See* proof theory.

logical truth An instance of a *valid *wff. For example, from the valid wff '$A \vee \sim A$' we can obtain as a logical truth 'snow is white $\vee \sim$ snow is white'. Logical truths are thus true by virtue of their *logical form rather than their content.

logicism The thesis, first propounded by Frege, that mathematics is reducible to *logic in the sense that (1) mathematical concepts can be explicitly defined in terms of logical concepts, and (2) the theorems of mathematics can be derived through logical deduction. The truth of logicism would show that mathematical truths are analytic (that is, true by virtue of meaning) and thus known *a priori*. *See* formalism; intuitionism.

logistic curve *See* sigmoid curve.

logistic map An *iterated map of the interval [0, 1] of the form $x \mapsto ax(1 - x)$, with parameter a where $0 < a \leqslant 4$. For $0 < a < 1$, the point 0 is an attracting fixed point (for all starting points in the interval). For $1 < a < 3$, the point 0 is a repelling fixed point and $1 - 1/a$ is an attracting fixed point. For $3 < a < 1 + \sqrt{6}$, the attracting fixed point is replaced by two repelling *periodic points of period 2. For increasing values of a the periodic orbits bifur-

cate (*see* bifurcation) into orbits of period 2, 4, 8, …. *See* Feigenbaum number.

logistic method The study of formal logic through the construction of *logistic systems.

logistic regression Many experiments are effectively sets of independent *Bernoulli trials, the ith trial giving rise to a binary variable Y_i which may take only the value 0 or 1. If $\Pr(Y_i = 1) = p_i$, this probability often depends on one or more explanatory variables, for example the treatment level or dose level x_i of an insecticide when the response of interest Y_i is death ($Y_i = 1$) or survival ($Y_i = 0$). The *odds of death for the ith individual are then $\theta_i = p_i/(1 - p_i)$, and it is often found that the empirical relationship between x_i and θ_i is well described by the logistic regression equation

$$\ln(\theta_i) = \alpha + \beta x_i$$

Since the expected value of Y_i is p_i, this is a special case of a *generalized linear model. *See also* logit.

logistic spiral *See* spiral.

logistic system A *formal system that contains only logical axioms. The *predicate calculus, for example, is a logistic system. *See* logic.

logit (J. Berkson, 1944) The quantity $Y = \ln[p/(1 - p)]$, i.e. the logarithm of the *odds, is called the logit of p. If p satisfies a logistic relationship with an *explanatory variable x of the form $p - [1 + \exp\{-(\alpha + \beta x)\}]^{-1}$, it follows that $Y = \alpha + \beta x$. *See also* logistic regression; probit analysis.

loglinear model A model widely used in the analysis of association between categories in a *contingency table. In an $r \times c$ table with independence between row and column categories, the expected frequency in cell (i, j) is $m_{ij} = n_{i+}n_{+j}/N$, where N is the total for the table, and n_{i+} and n_{+j} denote the totals for the ith row and jth column. Taking logarithms gives

$$\ln m_{ij} = \ln n_{i+} + \ln n_{+j} - \ln N$$

i.e. under independence, the logarithm of the expected number in any cell is a linear function of the logarithms of the row, column, and grand totals. Further additive terms may be used to represent various kinds of association (the analogue of interactions in *factorial experiments). For 2×2 tables, it is easily verified that under independence

$$\ln \left(\frac{m_{11} m_{22}}{m_{12} m_{21}} \right) = 0$$

in accordance with the condition for independence that the *odds ratio $\theta = (m_{11} m_{22} / m_{12} m_{21}) = 1$.

lognormal distribution *See* logarithmic transformation.

log-series distribution *See* logarithmic distribution.

long arc *See* arc.

longitude 1. The angle by which a point is east or west of the prime *meridian (the meridian through Greenwich) taken as the angle measured along the equator between the prime meridian and the meridian through the point. Longitude is measured from Greenwich, from $0°$ to $180°$ east and from $0°$ to $180°$ west.
2. *See* celestial longitude.
3. *See* galactic longitude.

longitudinal wave A form of *wave motion in which energy is propagated by the displacement of the transmitting medium along the direction of propagation. The wave velocity depends on the elastic properties of the medium and on its density. There is no propagation in a vacuum. Sound waves are longitudinal. *Compare* transverse wave.

long radius *See* polygon.

loop 1. A part of a plane *curve that intersects itself, so that it encloses a bounded set of points.
2. *See* graph.

Lorentz–Fitzgerald contraction The apparent contraction of a moving object in the direction of motion that is observed by someone in a different inertial *frame of reference. If v is the magnitude of the relative velocity of the two frames and c is the speed of light, the contraction amounts to a factor of $\sqrt{(1 - v^2/c^2)}$, i.e. the contraction is negligible at speeds considerably less than c. It was predicted independently by G.F. Fitzgerald (1889) and H.A. Lorentz (1895), and was later explained by the special theory of *relativity.

Lorentz transformation *See* relativity.

Lorenz attractor *See* chaos.

Löwenheim, Leopold (1878–c.1940) German mathematician noted for his proof in 1915 of the *Löwenheim–Skolem theorem*, which showed that any formula valid in a denumerably infinite domain is universally valid.

lower bound *See* bound.

lower limit (of integration) *See* integration.

lower triangular matrix *See* triangular matrix.

lowess (W.S. Cleveland, 1979) A method for fitting smooth curves to large data sets that is resistant to *outliers. It extends the concept of a weighted *moving average widely used in *time series analysis. A low-degree *polynomial in x is fitted at each data point x_i using a generalization of *least squares called *weighted least squares*, where weights are allocated in a way that reduces the influence of points as x moves away from x_i, these weights being zero outside a distance determined by a *bandwidth*. Further iterations with a different choice of weights then markedly reduce the influence of outliers or a few extreme observations. The method is computer intensive.

The name *lowess* is an acronym derived from 'locally weighted smoothing scatterplots'. The word *loess* is sometimes used as an alternative, though some confusion can arise as *loess* is also used for a modification that does not invoke the separate process for reducing the influence of outliers.

loxodrome A curve on the surface of a sphere that cuts *meridians at a constant angle. It is also called a *rhumb line*.

l.u.b. *Abbreviation for* *least upper bound.

Lucas sequence The *sequence 1, 3, 4, 7, 11, 18, 29, ..., in which each term after the first two is the sum of the preceding pair of terms. It is named after the French mathematician François Edouard Anatole Lucas (1842–91). *See also* Fibonacci sequence.

LU factorization For a square matrix A, a factorization $A = LU$ into the product of a *lower triangular matrix L and an *upper triangular matrix U. Usually, either L or U is taken to have a unit diagonal. An example of an LU factorization is

$$\begin{pmatrix} 2 & 1 \\ -2 & 2 \end{pmatrix} = \begin{pmatrix} 1 & 0 \\ -1 & 1 \end{pmatrix} \begin{pmatrix} 2 & 1 \\ 0 & 3 \end{pmatrix}$$

The method of *Gaussian elimination effectively computes an LU factorization: U is the reduced upper triangular matrix, and the elements of L are the multipliers.

Łukasiewicz, Jan (1878–1956) Polish logician, and one of the founders of the important school of mathematics and logic that flourished in Poland during the interwar years. Łukasiewicz left Poland in 1944 and finally settled in Ireland, where he spent the remainder of his life at the Royal Irish Academy, Dublin. Among his important contributions to logic are the development of *three-valued logic, the construction of a novel system of modal logic, the creation of a new logical notation, and an important study: *Aristotle's Syllogistic* (1951).

lumen Symbol: lm. The *SI unit of luminous flux, equal to the amount of light emitted in 1 second into a solid angle of 1 steradian by a uniform point source of 1 candela intensity.

lune 1. One of the parts of the surface of a sphere bounded by two intersecting *great circles. The area of a lune is $4\pi r^2 \theta / 360$, where θ is the spherical angle (in degrees) between the great circles and r is the radius of the sphere.
2. The area enclosed between the arcs of two intersecting circles.

lux Symbol: lx. The *SI unit of illuminance, equal to the illumination of 1 lumen uniformly spread over an area of 1 square metre.

M

machine Any system that replaces or augments human or animal effort in order to accomplish a physical task. Machines vary widely in function and complexity, but in general the performance of useful work is achieved by means of the motions of interconnected components – gears, cranks, levers, pulleys, screws, etc. A force known as the *effort* is applied to one component and produces an effective force of different magnitude at some other part of the system. This effective force is applied to a *load*. The ratio load/effort is called the *mechanical advantage; the ratio of the distance moved by the effort to the distance moved by the load is called the *velocity ratio*. The machine's performance can be measured in terms of *efficiency.

Mach number Symbol: M or Ma. The ratio of the speed of a body in a fluid to the speed of sound in that fluid. The speed of sound in air at ground level is about $330 \, \text{m s}^{-1}$. A Mach number in excess of unity thus indicates a super-sonic speed. A high Mach number will affect the motion of a body through a fluid. [After E. Mach (1836–1916)]

Maclaurin, Colin (1698–1746) Scottish mathematician who, in his *Geometrica organica* (1720, Organic Geometry) and *Treatise of Fluxions* (1742), made a number of contributions to the newly developed calculus of Newton. His best-known result is the expansion since referred to as the Maclaurin series.

Maclaurin series *See* Taylor's theorem.

McNemar's test (Q. McNemar, 1947) A nonparametric test for differences in proportions in related samples. It is often used to test whether a stimulus has produced a response in a particular direction. For example, the political allegiance of a sample of voters to party A or B may be determined prior to a party political broadcast; after the broadcast any changes in allegiance are noted and the test is used to indicate whether the proportion changing from A to B differs significantly from that changing from B to A. *See* nonparametric methods.

Madhava of Sangamagramma (*c*. AD 1400) Indian astronomer-mathematician. All his works that have been discovered so far are astronomical treatises. His mathematical contributions – which include infinite series expansions of trigonometric and inverse trigonometric functions and finite series approximations which foreshadowed results usually attributed to Leibniz, Newton and Gregory, and Taylor – are known only from reports by his contemporaries and successors.

magic constant *See* magic square.

magic square A square *array of numbers in which the numbers in any row, column, or full diagonal have the same sum. This sum is called the *magic constant* of the square. The earliest known example is the Luo-shu (Lo-shu) square

4	9	2
3	5	7
8	1	6

found in ancient Chinese writings. Another well-known magic square is

16	3	2	13
5	10	11	8
9	6	7	12
4	15	14	1

which is included in an engraving, *Melancholia*, by Albrecht Dürer (1514).

A square is *semi-magic* if the numbers in just any row or column have the same sum. A magic square is *pandiagonal* or *diabolic* if, in addition to the usual magic square properties, every diagonal (including the broken ones) adds up to the same magic constant. For example, in the pandiagonal square

1	8	11	14
12	13	2	7
6	3	16	9
15	10	5	4

all the rows, columns, full diagonals, and broken diagonals, such as 11, 7, 6, 10, add up to the same total, 34.

Mahavira (*fl.* AD 850) Indian mathematician. In his *Ganita Sara Samgraha* (The Compendium of Arithmetic) there is a detailed examination of operations with fractions, permutations and combinations, and mathematical series, as well as – something unusual in Indian mathematics – an (unsuccessful) attempt to derive formulae for the area and perimeter of an ellipse.

main diagonal, main antidiagonal *See* diagonal.

major arc *See* arc.

major axis The longest diameter of an *ellipse or *ellipsoid.

major segment *See* segment.

Malthus, Thomas Robert (1766–1834) English sociologist, classicist, and mathematician famous for his theory, expressed in *An Essay on the Principle of Population* (1798), that population growth will always tend to outgrow food resources unless strict limitations are placed on human reproduction. His theory had a profound influence on social policy, for it had previously been regarded as almost axiomatic that high birth rates added to natural wealth.

Mandelbrot set (B.B. Mandelbrot, 1980) A subset of the *complex plane associated with complex numbers c for the family of maps on the complex plane given by $T_c: z \mapsto z^2 + c$ *see* diagram (a)). In particular, c is in the Mandelbrot set if the *orbit $0, T_c(0), T_c^2(0), \ldots$ of 0 is bounded. The boundary is *fractal and self-similar.

(a)

(b)

Mandelbrot set (a) The Mandelbrot set and (b) the analogous set for the map $z \mapsto z^4 + c$.

The Mandelbrot set is closed, *connected, and lies in the disc $|z| \leq 2$. On the real line it contains the interval $[-2, \frac{1}{4}]$. If c lies in the central, cardioid-like region of the set, then T_c has an attracting fixed point. The other regions correspond to where T_c has different attracting *periodic points. For example, if c lies in the circular region to the left of the central region, then T_c has a pair of attracting points of period 2, whereas if c lies in the regions directly above or below the central region, then T_c has three points of period 3. It is not certain whether the Mandelbrot set is *locally connected. An analogous set can be defined for other families of maps such as $z \mapsto z^4 + c$ (*see* diagram (b)). *See also* Julia set.

manifold A *topological space M is called an *n-manifold* (or *manifold of dimension n*) if it 'looks locally like' n-dimensional Euclidean space $\mathbb{R}^n$. More precisely, M is an *n*-manifold if for each point $x \in M$ there is an open *neighbourhood U_x of x in M and a *homeomorphism f_x from U_x to an open set in $\mathbb{R}^n$. The coordinates defined by f_x are called *local coordinates*.

Thus $\mathbb{R}^n$ itself is an *n*-manifold, as also are the *n*-sphere S^n and any open subset of an *n*-manifold. The torus, projective plane, and Klein bottle are all examples of 2-manifolds (sometimes called *surfaces*).

More specialized classes of manifold may be defined by restricting the homeomorphisms f_x in various ways. For example, whenever two of the open neighbourhoods U_x and U_y meet, $f_y f_x^{-1}$ is a homeomorphism between open sets in $\mathbb{R}^n$; if these homeomorphisms are all required to be infinitely continuously differentiable, the manifold M is said to be a *differential* (or *smooth*) *n-manifold*.

An important topic in topology is the classification of manifolds to within homeomorphism. The problem has been solved (by M. Dehn and P. Heegaard, 1907) for (compact) 2-manifolds: each of them is homeomorphic to one of either of two sets of 'standard' 2-manifolds, M_g ($g \geq 0$) and N_h ($h \geq 1$). Here, M_g (the 'orientable 2-manifold of *genus g*') is obtained from S^2 by attaching g handles, where a handle is a homeomorphic copy of the cylinder

$$\{(x_1, x_2, x_3) \in \mathbb{R}^3; x_1^2 + x_2^2 = 1, |x_3| \leq 1\}$$

the two circles $x_3 = \pm 1$ being identified with the boundary circles of two discs removed from S^2 (*see* diagram). Similarly, N_h is obtained from S^2 by attaching h *cross-caps*, where a cross-cap is a homeomorphic copy of the *Möbius strip, whose boundary circle is identified with the boundary circle of a single disc removed from S^2. Thus S^2 itself is M_0, the torus is M_1, the projective plane is N_1, and the *Klein bottle is N_2.

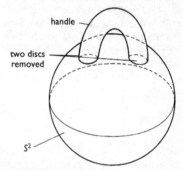

manifold The 2-manifold M_1.

Mann–Whitney test *See* Wilcoxon rank sum test.

mantissa (*plural* **mantissae**) The decimal part of a logarithm, or the fractional part of the *floating-point representation of a number.

many–one correspondence A *correspondence between two sets X and Y in which some element y of Y is paired with more than one element x of X. For example, the correspondence defined by $x^2 = y$ between the set of real numbers and the set of non-negative numbers is many–one. A *one–many correspondence* between sets X and Y

has the property that some element x of X is paired with more than one element y of Y. For example, the correspondence defined by $x = \sin y$, where $x \in [-1, 1]$ and $y \in \mathbb{R}$, is one–many. *See also* one-to-one correspondence.

many-valued function *See* multiple-valued function.

many-valued logic A *formal system in which more than two *truth values are permitted.

Classical logic is essentially two-valued logic, and is committed to the view that all propositions must be either true or false; or, more formally, that they take the value 1 or 0 (*see* truth table). In 1921 Emil Post introduced an alternative approach which allowed variables to take any one of $n > 2$ different values. Since then, numerous many-valued systems have been constructed, and, as formal systems, they contain nothing exceptional. Thus, one approach would be to assign values on the basis of the following rules:

$$|\sim A| = 1 - |A|$$

$$|A \vee B| = \max\{|A|, |B|\}$$

$$|A \,\&\, B| = \min\{|A|, |B|\}$$

$$|A \rightarrow B| = 1 \quad \text{iff } |A| \leqslant |B|$$

$$|A \rightarrow B| = 1 - |A| + |B| \text{ iff } |A| > |B|$$

where $|A|$ denotes the value of A.

The main problem with many-valued logics has been to find an acceptable and interesting interpretation. *See also* three-valued logic.

mapping (map) *See* function.

marginal distribution *See* bivariate distribution; multivariate distribution.

Markov, Andrei Andreevich (1856–1922) Russian mathematician noted for his work in probability theory and his introduction in 1906 of what has since become known as a *Markov chain.

Markov chain A *stochastic process in which a discrete *random variable $X(t)$ may change state (i.e. value) at times t_1, $t_2, t_3, \ldots$ (usually equally spaced) is called a Markov chain if the *conditional distribution of $X(t_{i+1})$ at t_{i+1} depends only on $X(t_i)$ and not on the value of X at any earlier time. This is often expressed by saying that the state of the system in the future is unaffected by its history. The simplest case is that in which X takes values 0 and 1 only (corresponding, for example, to a circuit in which current may be either off (0) or on (1)). At any transition time t_i there is a probability p_{rs} that the system, if in state r, will change to state s, where $r, s = 0, 1$. Thus p_{01} is the probability of a change from state 0 to state 1, and p_{11} is the probability of the system remaining in state 1 if it is already there (*see* diagram). The matrix

$$\begin{pmatrix} p_{00} & p_{01} \\ p_{10} & p_{11} \end{pmatrix}$$

where $p_{00} + p_{01} = p_{10} + p_{11} = 1$ is called the *transition matrix*. If $p_{10} = p_{01} = 1$ this implies that $p_{00} = p_{11} = 0$, and the system oscillates repeatedly from state 0 to state 1. If $p_{00} = p_{11} = 1$ the system never changes state, remaining in one of them at all times, and that state is called an *absorbing state*. If all the elements in the transition matrix are nonzero then, for equally spaced t_i, the system approaches a condition in which at any time it has a probability $p_{10}/(p_{01} + p_{10})$ of being in state 0 and a probability $p_{01}/(p_{01} + p_{10})$ of being in state 1. This property is called *ergodicity. If changes may take place continuously in time, but future behaviour depends only on the present state, the system is called a *Markov process*. *See also* random walk.

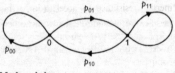

Markov chain

martingale A *stochastic process in which, if observations x_i are taken at times t_i, where $t_1 < t_2 < \cdots < t_n < t_{n+1}$, and

$$E(x_{n+1}|x_1, x_2, \ldots, x_n) = x_n$$

for all $n \geqslant 1$, the mean is finite for all t. It is a process in which the expected value of the observation at any stage, conditional on all earlier observations, is equal to the last of these earlier observations. For a simple *random walk with equal probabilities $p = 0.5$ of a unit step left or right, the sequence of positions x_i at times t_i form a martingale because, if the walk is at position x_n at time t_n, then at time t_{n+1} we have $x_{n+1} = x_n \pm 1$, each with probability 0.5, whence

$$E(x_{n+1}) = \tfrac{1}{2}[(x_n + 1) + (x_n - 1)] = x_n$$

Mascheroni, Lorenzo (1750–1800) Italian mathematician who in his *Geometria del compasso* (1797, Geometry with the Compass) demonstrated that all Euclidean constructions can be made with the compass alone. Such compass-only constructions are sometimes called *Mascheroni constructions*. A Danish mathematician, Georg Mohr, had in 1672 covered the same ground in an obscure book.

mass Symbol: m. A fundamental characteristic of a body related to the quantity of matter in the body. In classical mechanics it is considered constant, unlike volume or weight. The SI unit of mass is the kilogram; mass is also measured in pounds. The mass of a body characterizes its interactions with other bodies. A body's *momentum is partly determined by its mass. Mass is also the constant of proportionality between the force $\mathbf{F}$ on a body and the resulting acceleration $\mathbf{a}$, i.e. $\mathbf{F} = m\mathbf{a}$. Mass can thus be considered as a measure of a body's *inertia (resistance to acceleration); this is known as *inertial mass*. Mass can also be considered in terms of the gravitational field produced by the body; this is known as *gravitational mass*. The inertial mass of an object is equal to its gravitational mass.

Einstein's special theory of *relativity predicts that the mass of a body is not constant, but increases with speed v:

$$m = m_0/\sqrt{(1 - v^2/c^2)}$$

m_0 being the *rest mass and c the speed of light. This has been verified experimentally but is significant only at very high velocities. *See also* mass–energy equation; relativistic mass; conservation of mass; weight.

mass centre *See* centre of mass.

mass–energy equation (Einstein's equation) The equation stating the relationship between *mass m and *energy E:

$$E = mc^2$$

where c is the speed of light in vacuum. It was proposed by Albert Einstein as part of the special theory of *relativity and has since been verified experimentally. It indicates the equivalence of mass and energy. Mass can be considered as a form of energy, there being conservation of mass–energy in an isolated system (*see* conservation of energy), and can be converted to energy and vice versa. For example, the *rest mass of an atomic nucleus is somewhat less than the masses of the constituent neutrons and protons, where the mass difference is equivalent to the energy required to bind neutrons and protons together. Again, under the right conditions, an electron and its antiparticle, the positron, can form simultaneously from a high-energy photon, the photon having no rest mass. *See also* relativistic mass.

matched pairs The pairing of units in an experiment so that each member of a pair is as close as possible to the other in characteristics that might influence response to a treatment. If two treatments are being compared, one is allocated at random to each member of a pair. For example, in a test to determine whether one method of teaching reading is superior to another, pupils might be matched in pairs on the basis of age, or of IQ or some similar measure of aptitude. The procedure can be extended to form matched groups for comparing more than two treatments. *See* randomized blocks.

material Consisting of or relating to matter; having mass.

material equivalence *See* equivalence.

material implication *See* implication.

mathematics The study of numbers, shapes, and other entities by logical means. It is divided into *pure mathematics* and *applied mathematics*, although the division is not a sharp one and the two branches are interdependent. Applied mathematics is the use of mathematics in studying natural phenomena. It includes such topics as *statistics, *probability, *mechanics, *relativity, and *quantum mechanics. Pure mathematics is the study of relationships between abstract entities according to certain rules. It has various branches, including *arithmetic, *algebra, *geometry, *trigonometry, *calculus, and *topology.

Mathieu's equation A second-order *differential equation of the form

$$\frac{\mathrm{d}^2 y}{\mathrm{d}x^2} + (a + b \cos 2x)y = 0$$

The general solution is

$$A\mathrm{e}^{rx}\phi(x) + B\mathrm{e}^{-rx}\phi(-x)$$

where r is a constant and ϕ a periodic function (period 2π). The equation was studied by the French mathematician and physicist Émile Léonard Mathieu (1835–90).

matrix (*plural* **matrices**) A set of quantities (called *elements* or *entries*) arranged in a rectangular *array, with certain rules governing their combination. Conventionally, the array is enclosed in round brackets or, less commonly, in square brackets. Unlike a determinant, a matrix does not have a

numerical value, but matrices can be used to treat problems involving relationships between the elements.

The horizontal lines of elements are *rows* and the vertical lines are *columns*. A *square matrix* has the same number of rows as columns. The element in the ith row and jth column of a matrix A is usually denoted by a_{ij}, and may be referred to as the (i, j) element or entry. A diagonal line of elements in a square matrix is a *diagonal*. The elements in the positions from top left to bottom right form the *main* or *principal diagonal*. The diagonals lying above the main diagonal are the *superdiagonals*, and those lying below the main diagonal are the *subdiagonals*. Diagonal lines of elements perpendicular to the main diagonals are called *antidiagonals*. Elements not lying on the main diagonal are called *off-diagonal*.

The *dimension* or *order* of a matrix is expressed as $m \times n$, where m is the number of rows and n the number of columns. A square matrix of dimension $n \times n$ is sometimes said to be 'of dimension n'. A matrix consisting of a single row is a *row matrix* or *row vector*; one consisting of a single column is a *column matrix* or *column vector*.

The rules of combination for matrices are as follows:

(1) *Multiplication of a matrix by a number* k. Each element a_{ij} of the matrix is multiplied by the number:

$$k\begin{pmatrix} a & b \\ c & d \end{pmatrix} = \begin{pmatrix} ka & kb \\ kc & kd \end{pmatrix}$$

For two matrices A and B, $k(A + B) = kA + kB$.

(2) *Addition of two matrices*. The sum of the matrices is a matrix in which the elements are obtained by adding corresponding elements. Thus, if the elements of A are a_{ij} and those of B are b_{ij}, then the elements of $C (= A + B)$ are $a_{ij} + b_{ij}$, where i is the row number and j the column number:

$$\begin{pmatrix} a & b \\ c & d \end{pmatrix} + \begin{pmatrix} e & f \\ g & h \end{pmatrix} = \begin{pmatrix} a+e & b+f \\ c+g & d+h \end{pmatrix}$$

Two matrices can be added only if they have the same number of rows and columns, i.e. they must be of the same *type*.

(3) *Multiplication of two matrices.* A has elements a_{ij} with $i = 1, 2, \ldots$ and $j = 1, 2, \ldots$; similarly, B has elements b_{ij}. The elements of $C (= AB)$ are given by

$$c_{ij} = a_{i1}b_{1j} + a_{i2}b_{2j} + \cdots + a_{in}b_{nj}$$

provided that n, the number of columns of A, equals the number of rows of B (i.e. the matrices are *conformable*). If the dimensions of A and B are $m \times n$ and $n \times p$, respectively, then the dimension of C will be $m \times p$ (*see* figure below).

A square matrix can be converted into another *equivalent matrix* by a combination of any of the following operations:

(1) Interchange of two rows or two columns.

(2) Multiplication of a row or column by a nonzero scalar.

(3) Addition to the elements of one row (or column) multiples of the corresponding elements of another row (or column).

It is often convenient to simplify a matrix by putting it into an equivalent form, especially one in which the only nonzero elements appear along the leading diagonal. It can be shown that any square matrix is equivalent to some *diagonal matrix. A change from one matrix B to an equivalent matrix A is an *equivalence transformation*. Such transformations can be effected by multiplying B by other nonsingular matrices X and Y, such that $A = XBY$. There are certain special transformations

depending on the connection between X and Y, as follows:

(1) *Collinearity* (or *similarity*) *transformation* (*collineation*) in which X is the inverse of Y, i.e. a transformation of the type $A = Y^{-1}BY$. In this case, A and B are said to be *similar matrices*.

(2) *Congruence transformation* in which X is the *transpose of Y, i.e. a transformation of the type $A = Y^{T}BY$. A and B are *congruent matrices*.

(3) *Conjunctive transformation* in which X is the *Hermitian conjugate of Y, i.e. a transformation of the type $A = Y^{*}BY$.

(4) *Orthogonal transformation* in which X is the inverse of Y, and Y is an *orthogonal matrix.

(5) *Unitary transformation* in which X is the inverse of Y, and Y is a *unitary matrix.

The *determinant of a square matrix is the determinant of the elements of the matrix.

See also adjacency matrix; augmented matrix; block matrix; complex conjugate; correlation matrix; covariance matrix; diagonal matrix; elementary matrix; Hankel matrix; Hessian; Hilbert matrix; identity matrix; inverse; Jacobian; Jordan matrix; permutation matrix; symmetric matrix; triangular matrix; Vandermonde matrix.

matrix of coefficients *See* augmented matrix.

Maupertuis, Pierre Louis Moreau de (1698–1759) French mathematician and astronomer who formulated the principle of least action. He also led an expedition to measure the length of a degree along a meridian; the result verified that the earth is an oblate spheroid.

max *Abbreviation for* *maximum.

$$\begin{pmatrix} a_{11} & a_{12} & a_{13} \\ a_{21} & a_{22} & a_{23} \end{pmatrix} \begin{pmatrix} b_{11} & b_{12} \\ b_{21} & b_{22} \\ b_{31} & b_{32} \end{pmatrix} = \begin{pmatrix} (a_{11}b_{11} + a_{12}b_{21} + a_{13}b_{31}) & (a_{11}b_{12} + a_{12}b_{22} + a_{13}b_{32}) \\ (a_{21}b_{11} + a_{22}b_{21} + a_{23}b_{31}) & (a_{21}b_{12} + a_{22}b_{22} + a_{23}b_{32}) \end{pmatrix}$$

matrix multiplication of two matrices.

maximal An element x is said to be a maximal element of a partially ordered set A if there is no element $y \in A$ such that $y > x$.

There may be more than one maximal element in a partially ordered set. For example, consider the diagram, in which $x > y$ if there is a sequence of lines successively sloping downwards from x to y, and x cannot be compared with y if there is no line joining them. So $a > e$ and $b > c$, but neither $d > c$ nor $c > d$ is true. There are two maximal elements, namely a and b.

For another example, take the set of five numbers $\{1, 3, 5, 7, 9\}$ and consider those subsets that do not contain both 3 and 5, where two such subsets A and B satisfy $A > B$ if A contains B. In this collection of partially ordered subsets both the subsets $\{1, 3, 7, 9\}$ and $\{1, 5, 7, 9\}$ are maximal.

Similarly, an element x in a partially ordered set A is called *minimal* if there is no element $y \in A$ with $y < x$. In the diagram there are two minimal elements. c and d. *See* partial order.

maximin criterion *See* game theory.

maximum (*plural* **maxima**) Greatest possible. The *maximum value* of a function is the greatest value that it attains. A point x is a *local maximum* of a function f if $f(x) \geqslant f(y)$ for all points y in a neighbourhood of x (sometimes a *strict local maximum* is considered, which means that $f(x) > f(y)$). Often a local maximum is found by the study of *stationary points. *See* turning point.

maximum likelihood estimation The procedure whereby the value of an *estimator of a parameter is chosen to maximize the likelihood. Maximum likelihood estimators are usually *consistent and efficient (*see* efficiency), though not always *unbiased. The estimates are usually obtained by differentiating the logarithm of the *likelihood function with respect to θ, equating the derivative to zero, and solving the resulting equation to determine any extremum and selecting the maximum.

An iterative solution may be needed. The maximum likelihood estimator of the mean μ of a normal distribution is unbiased and is the sample mean $\bar{x}$, but the maximum likelihood estimator of σ^2 is biased and must be multiplied by $n/(n-1)$ to produce an unbiased estimator.

maximal

maximum modulus theorem If the *holomorphic function $f(z)$ is not constant on an *open *connected set, then its *modulus $|f(z)|$ does not attain a *maximum value on the set.

Maxwell, James Clerk (1831–79) Scottish mathematical physicist who, in his *A Dynamical Theory of the Electromagnetic Field* (1865), first presented his famous field equations (*see* Maxwell's equations), to appear later in the form described below in his *Treatise on Electricity and Magnetism* (1873). Maxwell was also one of the founders of statistical mechanics and in 1860 published his distribution law. Such work suggested a statistical interpretation of thermodynamics.

Maxwell's equations *Differential equations relating the magnetic field strength (**H**), the electric displacement (**D**), the magnetic flux density (**B**), the electric field strength (**E**), and the current density (**j**) at any point in a region containing a varying electromagnetic field:

$$\text{curl } \mathbf{H} = \mathbf{j} + \frac{\partial \mathbf{D}}{\partial t}, \quad \text{div } \mathbf{B} = 0$$

$$\text{curl } \mathbf{E} = -\frac{\partial \mathbf{B}}{\partial t}, \quad \text{div } \mathbf{D} = q$$

where t is the time and q the volume charge density.

mean 1. The *arithmetic mean* or common *average* of a set of observations is their sum divided by the total number of observations. A *weighted mean* is one in which each observation x_i is given a weight w_i and is defined as

$$\bar{x}_w = \sum w_i x_i / \sum w_i$$

In a *frequency table, if x_i occurs f_i times the ordinary mean is obtainable by putting $w_i = f_i$, whence

$$\bar{x} = \sum f_i x_i / \sum f_i$$

It is a measure of *location.
2. The *geometric mean* of n observations is the nth root of their product. For two observations the geometric mean is the square root of their product and is sometimes called their *mean proportional*.
3. The *harmonic mean* is the reciprocal of the arithmetic mean of the reciprocals of the observations. It is not widely used in statistics.
4. The *arithmetic–geometric mean* of two positive numbers a and b is the common limit of the sequences $a_1, a_2, \ldots$ and $b_1, b_2, \ldots$ formed as follows:

$$a_1 = \frac{1}{2}(a + b), \quad b_1 = \sqrt{(ab)}$$

$$a_2 = \frac{1}{2}(a_1 + b_1), \quad b_2 = \sqrt{(a_1 b_1)}$$

etc.
5. The mean of a *random variable is its *expectation, i.e. its first *moment about the origin. *See also* arithmetic–geometric mean inequality; centrality.

mean absolute deviation The *mean of the modulus or magnitude of the deviation of observations from some measure of centrality, usually the mean but sometimes the median. It is a measure of *dispersion. If $x_1, x_2, \ldots, x_n$ have mean $\bar{x}$ then the mean absolute deviation about the mean is

$$\sum_1^n \frac{1}{n} (|x_i - \bar{x}|)$$

If X is a random variable, the mean absolute deviation is the first absolute moment about the chosen measure of centrality. Thus if X is continuous with *frequency function $f(x)$ the mean absolute deviation about the median m is

$$\int_{-\infty}^{+\infty} |x - m| \, f(x) \, dx$$

mean axis *See* ellipsoid.

mean deviation For a distribution or sample the mean deviation about the mean (first *moment about the mean) is identically zero, but unless the mean and median coincide the mean deviation about the median is not zero. *See* mean absolute deviation.

mean proportional *See* mean.

mean squared error The expected value (*see* estimation) of the square of the difference between an *estimator T and the true parameter value θ. For an *unbiased estimator it is equal to the variance of T. For a biased estimator the mean squared error is the sum of the variance and the square of the bias. *See* estimation.

mean square deviation The second *moment about a point a is the mean square deviation about that point. In statistics, the mean square deviation about the mean is of particular interest. For a set of n observations x_i, the mean square deviation about the mean $\bar{x}$ (i.e. the *variance) is less than the mean square deviation about any other point.

mean value (of a function) For a *function $f(x)$, the value

$$\frac{1}{b-a}\int_a^b f(x)\,dx$$

where $f(x)$ is defined on the real *interval $[a, b]$. In general, if f is a function with domain D and m is a *measure, the mean value of f is

$$\frac{1}{m(D)}\int_D f\,dm$$

mean-value theorem The theorem that if a *function $f(x)$ is *continuous for $a \leqslant x \leqslant b$ and $f'(x)$ exists for $a < x < b$, then there exists some value of x between a and b for which

$$f'(x) = \frac{f(b) - f(a)}{b-a}$$

The *second* (or *extended*) *mean-value theorem* states that if $f(x)$ and $f'(x)$ are continuous for $a \leqslant x \leqslant b$, and $f''(x)$ exists for $a < x < b$, then

$$f(b) = f(a) + (b-a)f'(a)$$
$$+ \frac{1}{2!}(b-a)^2 f''(x_2)$$

where $a < x_2 < b$. The mean-value theorem for integrals states that there exists some value of x (ε, say) between a and b for which

$$\int_a^b f(x)dx = (b-a)f(\varepsilon)$$

See continuous function.

measurable function A *function f with *domain D that is a measurable set (*see* measure) contained in a space in which an outer measure is defined, and *range

R contained in a *topological space, such that for every *open set A in R, its *pre-image* $f^{-1}(A)$ is measurable. In particular, if f is a finite real-valued function it is measurable if the set $\{x: a < f(x) < b\}$ is measurable for arbitrary $a < b$.

measure A property, akin to area or volume, associated with *sets. For a collection of *subsets $A_1, A_2, \ldots$, a measure μ is a set function associating a non-negative real number (or $+\infty$) with each subset, such that

(1) $\mu(\varnothing) = 0$
(2) If $A_1 \cap A_2 = \varnothing$, then

$$\mu(A_1 \cup A_2) = \mu(A_1) + \mu(A_2)$$

(3) $\mu(A_1 \cup A_2 \cup A_3 \cup \cdots)$

$$= \mu(A_1) + \mu(A_2) + \mu(A_3) + \cdots$$

where the A_i are disjoint. A set which has measure is called *measurable*. Various types of measure may be defined, the most important being *Lebesgue measure* defined in Euclidean space. Measure theory is important in the theory of integration and probability.

The theory of measure is complicated by the existence of *non-measurable sets*. They can be constructed only indirectly using the *axiom of choice. Examples are subsets A_n, $n \geqslant 1$, of the interval $[0, 1]$ which are disjoint and whose union is the whole of $[0, 1]$ but, by their construction, if they were measurable would have to have the same measure. It is then straightforward to obtain the contradiction that $0 = \infty$, and so the sets A_n cannot be measurable. The subsets of three-dimensional space involved in the *Banach–Tarski paradox are necessarily non-measurable.

See Lebesgue integral.

measurement The assignment of a number to an object or observation according to some *scale of measurement.

measures of dispersion *See* dispersion.

measures of location *See* location.

mechanical advantage Of a *machine, the ratio of the force exerted by the machine to the force exerted on the machine, i.e. the ratio of load to effort. It expresses the ability of an available force to overcome a resisting force: if an effort E balances a load W then the mechanical advantage is W/E. For a simple machine, such as a lever or pulley system, mechanical advantage is used as an indicator of effectiveness.

mechanics The study of the behaviour of systems under the action of *forces, i.e. the study of *motion and *equilibrium. *Classical or Newtonian mechanics is concerned with systems that can be adequately described by *Newton's laws of motion. When speeds approach the speed of light then the principles of *relativity must be taken into account. Such systems are the subject of *relativistic mechanics: the equations reduce to those of classical mechanics for speeds which are very much less than that of light. The behaviour of systems of extremely small particles – atoms, molecules, nuclei, etc. – cannot be described by Newton's laws alone but requires the principles of *quantum mechanics, primarily that certain quantities such as energy can change only in discrete steps, and not continuously. These systems can be relativistic in nature. When there are a large number of particles in a system, the equations of motion are treated on a statistical basis rather than by considering individual particles. These systems are the subject of *statistical mechanics*.

median (midline) **1.** A line joining the vertex of a triangle to the mid-point of the opposite side. A triangle has three medians, which intersect at a single point (called the *centroid*).
2. The line joining the mid-points of the two nonparallel sides of a trapezium.
3. A measure of *centrality or location. For a *random variable with *distribution function $F(x)$ the median is the value m

such that $\Pr(X \leqslant m) = F(m) = 0.5$. It equals the 50th percentile. Special conventions are needed for uniqueness in discrete distributions. For a sample of n observations arranged in ascending order the median is the $\frac{1}{2}(n + 1)$th observation if n is odd, and the mean of the $\frac{1}{2}n$th and $(\frac{1}{2}n + 1)$th observations if n is even. *See* quantiles.

median formula A formula for finding the length of a *median of a triangle. If, in a triangle ABC, the side BC has mid-point D, then the length of the median AD can be calculated from the lengths of the sides by means of the formula $AD^2 = \frac{1}{2}AB^2 + \frac{1}{2}AC^2 - \frac{1}{4}BC^2$.

median test A distribution-free test for whether two *populations have the same *median. A sample is taken from each population, the median of all values in both samples (the combined sample) is calculated, and a 2×2 table is formed with rows corresponding to each sample and columns corresponding to numbers of sample values above and below the median of the combined sample. A *Fisher's exact test is performed to test for any difference in proportions above and below the combined median; differing proportions indicate that the populations have different medians. The test extends readily to more than two samples. *See* distribution-free methods.

mediator The perpendicular *bisector of a line segment.

meet *See* intersection.

mega- *See* SI units.

member (element) Any of the individual entities belonging to a *set. The membership relation is denoted by the symbol $\in$. Thus the expression $x \in A$ is read as 'x is a member of A' (or 'x is an element of A', or 'x belongs to A'), while the expression $x \notin A$ is read as 'x is not a member of A' (or 'x is not an element of A', or 'x does not belong to A').

Menaechmus (*fl.* 350 BC) Greek mathematician who is traditionally supposed to have been the first to describe the conic sections.

Menelaus of Alexandria (*fl.* AD 100) Greek mathematician noted for his *Sphaerica* (Spheres), which contains the earliest known theorems of spherical trigonometry, and also the theorem since known as *Menelaus' theorem (rediscovered by Giovanni Ceva in 1678). Menelaus is also reported to have written *Chords in a Circle* and *Elements of Geometry*, neither of which has survived.

Menelaus' theorem In a triangle ABC, L, M, and N are points on the sides AB, BC, and CA, respectively. The theorem states that the *necessary and sufficient condition for L, M, and N to be collinear is:

$$(AL/LB) \cdot (BM/MC) \cdot (CN/NA) = -1$$

Compare Ceva's theorem.

Mengoli, Pietro (1626–82) Italian mathematician who worked on infinite series. In 1650 he established that the harmonic series is divergent, and that the series formed by the reciprocals of triangular numbers is convergent.

mensuration The measurement of angles, lengths, areas, or volumes of geometric figures.

Mercator's projection A *projection from a sphere onto a plane, often used for maps of the earth's surface. It is obtained by placing a cylinder around the sphere (for the earth, the axis of the cylinder lies along the earth's axis). The projection of a point on the sphere is obtained by a line drawn through the point from the centre of the sphere to cut the cylinder. In Mercator's projection, lines of longitude are the same distance apart, but lines of latitude get farther apart farther from the equator. It is named after the Flemish geographer Gerhardus Mercator (1512–94).

Mercator's series The series *expansion for $\ln(1 + x)$. It is named after the Danish mathematician Nicolaus Mercator (*c*.1619–87), who published it in 1668. *See* logarithmic series.

meridian 1. A *great circle on the earth passing through the geographical poles. The *principal meridian* is the one through Greenwich from which longitude is measured.
2. *See* celestial meridian.

meridian section A *section of a *surface of revolution made by a plane that contains the axis of revolution. For example, a meridian section of a paraboloid of revolution is a parabola.

meromorphic function A *function whose only singularities are *poles. *See* singular point.

Mersenne, Marin (1588–1648) French mathematician and philosopher noted for his introduction into number theory of *Mersenne numbers in his *Cogitata physico-mathematica* (1644, Physico-Mathematical Thoughts).

Mersenne numbers Numbers M_n of the form $2^n - 1$ where n is a natural number. Much effort has gone into finding *Mersenne primes* – those Mersenne numbers that are prime; Mersenne's own guess as to which M_n are prime with $n \leqslant 257$ was incorrect. It is known that for $M_n = 2^n - 1$ to be prime the number n must itself be prime, but not every prime p leads to a Mersenne prime M_p (e.g. $M_{11} = 2^{11} - 1 = 23 \times 89$). After the first few values, the primes p leading to Mersenne primes M_p start to occur very infrequently and show no discernible pattern. At present (2008) there are 44 known Mersenne primes with values of p ranging from 2 to 32 582 657. Every Mersenne prime is associated with an even perfect number, and vice versa. *See* perfect number.

mesokurtic *See* kurtosis.

***m*-estimator** An *estimator that behaves like an optimum estimator when the assumptions for that estimator hold, and gives estimates which are almost as efficient when there is some breakdown in assumptions, is said to be robust. Robust estimators that give *maximum likelihood estimations when they are optimum, and are little influenced by departures from requirements for optimality, are called *m-estimators*. Typically, they reduce or eliminate the influence of *outliers. *See also* robustness.

metalanguage When we *use* a language ML to *discuss* a language OL, then ML is called the *metalanguage* and OL the *object language*. An OL is to be thought of as a *formal language, and quotation marks are used to indicate that the expressions of a language are under consideration independently of anything that the expressions may stand for. An ML is used to talk about the world, including expressions. An OL may be, but need not be, different from the ML; we can use English as a metalanguage to talk about either German or English as an object language. For example, to say that 'Arthur' is a word is to say something about an English word using English as the metalanguage, and not anything about the person Arthur.

metamathematics (metalogic) *See* proof theory.

metatheorem A *theorem in the *metalanguage *about* a *formal system, rather than a theorem *of* a formal system. For example, '$\sim\sim p \supset p$' is a theorem of the *propositional calculus, while the completeness theorem for the propositional calculus is a metatheorem proved in the metalanguage.

method of false position *See* false position (rule of).

method of least squares *See* least squares.

method of moments *See* moments, method of.

metre Symbol: m. The *SI unit of length, equal to the length of the path travelled by light in vacuum during a time interval of 1/299 792 458 of a second. The original unit was defined by the Paris Academy of Sciences in 1791 as one ten-millionth of the length of the quadrant of the earth's meridian that passes through Dunkirk. This definition was replaced in 1927 by one based on the length of a 'standard' platinum–iridium bar, and in 1963 it was redefined in terms of the wavelength of an electronic transition in the krypton-86 atom. The present definition dates from 1983.

metric (distance function) A measure of distance between points that can be used to form a *metric space.

metric space A set of points is a metric space if there is a *metric d which gives to any pair of points x and y a non-negative number $\mathrm{d}(x, y)$, their distance (or separation), and is such that

(1) $\mathrm{d}(x,y) = 0$ if and only if $x = y$;
(2) $\mathrm{d}(x,y) = \mathrm{d}(y,x)$; and
(3) $\mathrm{d}(x,y) + \mathrm{d}(y,z) \geqslant \mathrm{d}(x,z)$ for any points x, y, and z of the set.

This last condition is known as the *triangle inequality*.

A *Cauchy sequence* or *regular sequence* is a set of points $x_1, x_2, \ldots$ of a metric space such that for any $\varepsilon > 0$ there is an integer N such that $\mathrm{d}(x_i, x_j) < \varepsilon$ for all $i, j \geqslant N$. A metric space is *complete* if every Cauchy sequence converges to a point of the space. For example, with the metric $\mathrm{d}(x,y) \equiv |x - y|$, the space of all real numbers is complete, but the space of all rational numbers is not. Examples of metric spaces are spaces of functions such as $L^2(\mathbb{R})$, which consists of *square-integrable* real-valued functions f (i.e. such that $\int_{-\infty}^{\infty} \mathrm{f}(x)^2 \, \mathrm{d}x$ is finite); the metric d is defined by

$$d(f, g) = \sqrt{\left[\int\limits_{-\infty}^{\infty} (f(x) - g(x))^2 \, dx \right]}$$

See Euclidean space; Riemannian geometry; topological space.

metric system A system of units based on the decimal *number system. First suggested in 1585 by Simon Stevin, it later found a champion in Lagrange, and was formally adopted in 1795 when French laws gave basic definitions for various metric units, including the metre, litre, and gram. During the first quarter of the 19th century the metric system was adopted in most European countries. The UK, however, persisted with its own *imperial units until 1963, when the yard was formally defined in terms of the metre. Since then there has been in the UK a gradual change to the metric system. *See also* m.k.s. units; SI units.

metric ton *See* tonne.

micro- *See* SI units.

micron A former name for the micrometre (10^{-6} metre).

midline *See* median.

mid-point *See* bisect.

mid-point theorem If a line joins the midpoints of two sides of a triangle, then it is parallel to the third side, and is half its length. The *converse is also true. *See also* intercept theorem.

mil An *imperial unit of length, equal to one-thousandth of an inch. This unit, which was used in engineering, is also called a *thou*. 1 mil = 2.54×10^{-5} metre.

mile A *British unit of length equal to 1760 yards. This unit is also called the *statute mile*. 1 mile = 1.609 344 kilometres. *See also* nautical mile.

Millennium Prize problems A set of seven research problems chosen by the Clay Mathematics Institute and announced in 2000. A significant prize would be awarded for the solution of any of them. The announcement in 2000 was influenced by the fact that David *Hilbert had posed his famous set of 23 problems in 1900. The titles given to the seven problems are:

(1) P versus NP;
(2) the Hodge conjecture;
(3) the Poincaré conjecture;
(4) the Riemann hypothesis;
(5) Yang–Mills existence and mass gap;
(6) Navier–Stokes existence and smoothness;
(7) the Birch and Swinnerton-Dyer conjecture.

The third problem was solved by Grigori Perelman in 2004. *See* NP problem; Poincaré conjecture; Riemann hypothesis; Navier–Stokes equations.

milli- *See* SI units.

millimetre of mercury Symbol: mmHg. A *metric unit of pressure, equal to the pressure that will support a column of mercury (density 13 595.1 kg m^{-3}) 1 millimetre high under the standard acceleration of free fall. 1 millimetre of mercury = 133.322 pascals.

million One thousand thousand (10^6).

min *Abbreviation for* *minimum.

minimal *See* maximal.

minimal polynomial For an element a of a *ring R, its minimal polynomial over a *field F contained in R is the *monic polynomial $m(x)$ of least degree such that $m(a) = 0$. For example, the minimal polynomial of $\sqrt{2}$ over the *rational numbers is $x^2 - 2$. The minimal polynomial of a square *matrix A is the polynomial of least degree such that $m(A) = O$, where O is the *zero matrix. For example, if A is a 2×2 matrix and A is neither a multiple of the *identity matrix I nor the zero matrix, then its minimal polynomial is $x^2 - ax + b$, where $a = \text{trace}(A)$ and $b = \det(A)$. (One can easily check that $A^2 - aA + bI = O$.) If A is a multiple of the identity, say $A = kI$, then its minimal polynomial is

$x - k$. The minimal polynomial of a matrix divides the *characteristic polynomial.

minimal surface A surface on which the mean *curvature vanishes, i.e. the principal curvatures are equal but of opposite sign. Minimal surfaces are well illustrated by soap bubbles, and are configurations of least energy (*see* calculus of variations). Computer graphics have enabled the discovery of many new examples. *See* catenoid.

minimax principle (A. Wald, 1939) In *decision theory, the rule that one minimizes the maximum risk in making a wrong decision. It is generally regarded as undue pessimism.

minimax theorem *See* game theory.

minimum (*plural* **minima**) Least possible. The *minimum value* of a function is the least value that it attains. A point x is a *local minimum* of a function f if $f(x) \leqslant f(y)$ for all points y in a neighbourhood of x (sometimes a *strict local minimum* is considered; then $f(x) < f(y)$). Often a local minimum is found by the study of *stationary points. *See* turning point.

Minkowski, Hermann (1864–1909) Russian-German mathematician best remembered for his *Raum und Zeit* (1907, Space and Time) in which he argued for the need to think in terms of a four-dimensional spacetime continuum. He also made important contributions to number theory.

Minkowski universe *See* spacetime.

minor *See* cofactor.

minor arc *See* arc.

minor axis The shortest diameter of an *ellipse or *ellipsoid.

minor segment *See* segment.

minuend The quantity from which another quantity is subtracted in finding a difference. *See* subtraction.

minus sign 1. The sign − used to denote a *negative number, as in −7. It first occurs in a book by Johannes Widmann in 1489. More generally it is the sign used to denote the *additive inverse of an element a, as $-a$.
2. The sign − denoting *subtraction, as in $5 - 3$. It was first used in this sense by Henricus Grammateus in 1518. *See also* plus sign; plus/minus sign.

minute 1. Symbol: ′. A unit of angle equal to 1/60 of a degree. *See* angular measure.
2. A unit of time equal to 60 seconds.

minute of arc *See* degree of arc.

missing-plot techniques Techniques for simplifying the *analysis of variance of designed experiments when some planned observations are lost, e.g. by accident or failure of equipment.

mixed decimal *See* decimal.

mixed fraction A fraction consisting of an integer together with a proper fraction, for example $1\frac{1}{2}$.

mixed strategy *See* game theory.

mixed surd *See* surd.

mixed tensor *See* tensor.

m.k.s. units A system of units based on the metre, kilogram, and second. In its extended form, the m.k.s.A. or *Giorgi system* (after Giovanni Giorgi (1871–1950)), the ampere was introduced; this eventually became the SI system (*see* SI units) now widely used for scientific purposes.

ml A *metric unit of capacity or volume equal to 1 millilitre (of which it is a contracted form). This unit is used for some pharmaceutical purposes, but for scientific work the cubic centimetre is preferred. *See* litre.

Möbius, August Ferdinand (1790–1868) German mathematician noted for his work in geometry and topology, in which latter discipline he first described the one-sided surface since known as the *Möbius strip.

Möbius function The *function $\mu(n)$ defined for each positive integer as follows:

$$\mu(n) = \begin{cases} 1 & \text{if } n=1 \\ (-1)^k & \text{if } n \text{ is a product of } k \text{ distinct primes} \\ 0 & \text{otherwise.} \end{cases}$$

Thus $\mu(1) = 1, \mu(2) = -1, \mu(4) = 0, \mu(6) = 1$. One of the most important properties of this function is that it is *multiplicative.

Möbius inversion formula If $f(n)$ is a function defined on the positive integers and $g(n)$ is defined by

$$g(n) = \sum_{d|n} f(d)$$

where the sum is over all the positive *divisors d of n, then

$$f(n) = \sum_{d|n} \mu(d)g(n/d)$$

where $\mu(n)$ is the *Möbius function.

Möbius strip (Möbius band) A one-sided surface that can be formed by taking a strip of paper, giving it a half-twist, and sticking the ends together.

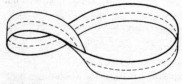

Möbius strip

Möbius transformation *See* linear transformation.

mod *See* modulus.

modal class The class (not always unique) that has the greatest frequency in classified data. *See* class intervals; mode.

modal logic The *logic of necessity and possibility. Systems of modal logic are constructed by taking the notion of strict implication as primitive, or by using the modal operators '□' (or 'L') and '◇' (or 'M'). □A and ◇A are to be read as 'It is necessarily the case that A' and 'It is possibly the case that A', respectively. Only '□' need be taken as primitive, '◇' being definable in terms of '□' through the definition '◇A' is equivalent to and replaceable by '$\sim \square \sim A$'. Also, the notions of strict implication and necessity are interdefinable: $A \Rightarrow B$ is equivalent to and replaceable by $\square(A \supset B)$. Modal logics are *intensional* in that the truth value of a wff A does not determine the truth value of $\square A$.

As an example of a modal system, consider S5, which is the system that contains the axioms and rules of inference of the *propositional calculus together with the axioms:

(1) $\square A \supset A$;
(2) $\square(A \supset B) \supset (\square A \supset \square B)$; and
(3) $\Diamond A \supset \square \Diamond A$.

In addition, S5 has a rule of inference: if $\vdash A$ then $\vdash \square A$ (*see* theorem). In order to interpret S5 we need to state the conditions under which a wff of the form $\square A$ is to be assigned the truth value 'True'. We can do this through the clause '$\square A$ is true if and only if A is true in all possible worlds'. This clause, together with the definition of $\Diamond$, leads to: '$\Diamond A$ is true if and only if A is true in some possible world'.

mode A measure of centrality or *location. For a *random variable X, the modes are the values of X corresponding to any maxima of the *frequency function; thus a distribution may have more than one mode. For a sample, the mode is the observation with the greatest *frequency, or for *grouped data the class with the greatest frequency. Again, there may be more than one mode. The term *bimodal* is used for a distribution or sample with two modes. *See* bimodal distribution.

model 1. (of a set of wffs) An *interpretation *I* of a *set of *wffs such that each member of the set is true in *I*.
2. (of a formal system) An *interpretation.
3. (mathematical) Any system of definitions, assumptions, and equations set up to discuss particular natural phenomena. Thus, Newtonian mechanics is a mathematical model of the motion and equilibrium of physical bodies.

A model which incorporates random elements or processes is called a *stochastic model*; otherwise it is said to be a *deterministic model*.

model theory The study of the *interpretations (models) of *formal systems. Of particular importance in model theory are the notions of logical consequence, validity, completeness, and soundness. *See* logic.

modular arithmetic *See* congruence modulo *n*.

module An *Abelian group, with operation written as addition, whose elements can be 'multiplied' by the elements of a *ring *R*. There are two closely related kinds of module.

A *left R-module* is a set *M* that forms an Abelian group with respect to an operation + such that each element *x* in *M* can be combined with any element *a* in *R* to form another element *ax* in *M*. This 'left multiplication' by ring elements has to satisfy each of the following conditions:

(1) $a(x + y) = ax + ay$, for *a* in *R* and *x* and *y* in *M*;
(2) $(a + b)x = ax + bx$, for *a* and *b* in *R* and *x* in *M*;
(3) $(ab)x = a(bx)$.

If the ring *R* has a (multiplicative) identity 1 and if it satisfies

(4) $1x = x$ for each *x* in *M*

then *M* is called a *unitary* left *R*-module.

A *right R-module* is similarly an Abelian group *M'* with respect to an operation written as +, together with a way of combining any element *x* in *M'* with any *a* in *R* to give another element, denoted by *xa*, in *M'*. For any *a* and *b* in *R* and *x* and *y* in *M'* the right multiplication must satisfy

(1) $(x + y)a = xa + ya$;
(2) $x(a + b) = xa + xb$;
(3) $x(ab) = (xa)b$.

Again, if *R* has an identity 1 and if it satisfies

(4) $x1 = x$ for each *x* in *M'*

then *M'* is a unitary right *R*-module.

A *vector space over a field *F* is an *F*-module (both left and right) since the axioms for a vector space are the same as those for a unitary module, except that in the former case the multiplying numbers come from a field and not just a ring. Also, any Abelian group *A*, written additively, can be regarded as a module over the integers where, for *x* in *A* and *n* a natural number, *nx* means $x + x + \cdots + x$ (*n* summands), $(-n)x = -(nx)$, and $0x$ is the zero element of *A*.

moduli space *See* gauge theory.

modulo *See* congruence.

modulus (*plural* **moduli**) 1. (**absolute value**) The magnitude of the length of a *vector representing a given *complex number. For example, the modulus of $a + ib$ is $\sqrt{(a^2 + b^2)}$. If the complex number is put in the form $r(\cos\theta + i\sin\theta)$, then the modulus is *r*. The modulus of a complex number $a + ib$ is written using the notation $|a + ib|$; for example, $|7 + i24|$ is $\sqrt{(7^2 + 24^2)} = 25$.
2. (of logarithms) The number by which *logarithms to one base are multiplied to give logarithms to a different base. The value of the modulus can be obtained from the formula for change of base. Thus, in converting logarithms to base *a* into those to base *b*,

$$\log_b n = \log_a n \cdot \log_b a$$

the multiplying factor, $\log_b a$, is called the modulus of base b (the resulting logarithm) with respect to base a (the original one).

Most frequently, interconversion is between common logarithms (base 10) and natural logarithms (base e). Thus, natural logarithms can be converted into common logarithms by multiplying by $\log_{10} e$ (0.434 294 . . .); this is the modulus of common logarithms with respect to natural logarithms. Conversely, common logarithms can be converted into natural logarithms by multiplying by $\log_e 10$ (2.302 585 . . .); this is the modulus of natural logarithms with respect to common logarithms.

3. A number by which another number is divided in a *congruence. Division by a number n is expressed as 'modulo n'.

4. *See* elliptic integral.

5. (elastic modulus) The ratio of stress to strain for a body or material obeying *Hooke's law: this is the slope of the linear region of the stress–strain diagram. Different moduli apply to different types of strain. These include *Young's modulus (longitudinal strain), *bulk modulus (volume strain), and *rigidity modulus (shear).

modulus sign The symbol $|\ \ |$ used to denote the *absolute value of a number or *vector.

modus ponens Either the rule of *inference that permits us to infer from $A \supset B$ and A that B, or an argument that takes this form. *See* logic. [Latin: method of affirming]

modus tollendo ponens Either the rule of *inference that permits us to infer from $A \lor B$ and $\sim A$ that B, or an argument that takes this form. [Latin: method of denying and affirming]

modus tollens Either the rule of *inference that permits us to infer from $A \supset B$ and $\sim B$ that $\sim A$, or an argument that takes this form. [Latin: method of denying]

mole Symbol: mol. The *SI unit of amount of substance, equal to the amount of substance that contains as many elementary units as there are atoms in 0.012 kilogram of carbon-12. The elementary unit must be specified and may be an atom, molecule, ion, radical, electron, photon, etc., or a specified group of such entities.

molecular sentence *See* compound sentence.

moment For a *random variable X the rth moment about the origin is the *expectation of $g(X) = X^r$, written as $E(X^r)$. The rth moment about a point a is $E(X - a)^r$. The moments most frequently encountered in statistics are moments about the origin or moments about the mean. The rth moment about the mean is often denoted by μ_r. The first moment about the origin is the mean and the second moment about the mean, μ_2, is the variance. For bivariate distributions product moments may also be defined. If X and Y are random variables with means μ_x and μ_y, respectively, then

$$E(X - \mu_x)(Y - \mu_y)$$

is the *covariance of X and Y, written as $\text{Cov}(X, Y)$. For a random sample of size n the kth sample moment about the origin is

$$\frac{1}{n} \sum_i x_i^k$$

and the corresponding moment about the sample mean $\bar{x}$ is

$$\frac{1}{n} \sum_i (x_i - \bar{x})^k$$

The kth moment about the sample mean is often denoted by m_k.

momenta *Plural of* momentum.

moment generating function For a *random variable X, the *expectation of $\exp(tX)$, where t is a constant. It is denoted by $M(t) = E[\exp(tX)]$. If the associated sum or integral is convergent for some $t > 0$, the coefficient of $t^r/r!$ is the rth moment about the origin, $E(X^r)$, of X. This is also the value of the rth derivative of $M(t)$ at $t = 0$. In particular, $E(X) = M'(0)$ and $Var(X) = M''(0) - [M'(0)]^2$. When it exists, the moment generating function characterizes a distribution uniquely.

For the exponential distribution with parameter λ, if $t < \lambda$ then

$$M(t) = \int_0^\infty e^{tx} \lambda e^{-\lambda x} dx = \int_0^\infty \lambda e^{-x(\lambda-t)} dx$$

$$= \frac{1}{(1 - t/\lambda)}$$

giving the series expansion

$$M(t) = 1 + \frac{t}{\lambda} + \left(\frac{t}{\lambda}\right)^2 + \cdots$$

$$+ \left(\frac{t}{\lambda}\right)^r + \cdots$$

where the coefficient of $t^r/r!$ is $(r!)/\lambda^r$, so that $E(X^r) = (r!)/\lambda^r$. In particular,

$$E(X) = \frac{1}{\lambda}$$

and

$$Var(X) = E(X^2) - [E(X)]^2$$

$$= \frac{2}{\lambda^2} - \frac{1}{\lambda^2} = \frac{1}{\lambda^2}$$

Moment generating functions are also defined for *multivariate distributions. See also characteristic function; probability generating function.

moment of a couple See couple.

moment of a force (torque) A measure of the turning power of a *force. For a force $\mathbf{F}$ acting at a point P on a body and causing it to turn about a point O, the moment of

the force about O is the *vector product of the vector $\overrightarrow{OP}(= \mathbf{r})$ and the force $\mathbf{F}$, i.e. $\mathbf{r} \times \mathbf{F}$ (see diagram). Its magnitude is $|\mathbf{F}|\,|\mathbf{r}|\sin\theta$ or $|\mathbf{F}|\,d$, where d is the perpendicular distance from the turning point O to the line of action of the force. Its direction is perpendicular to the plane containing O and the line of action of $\mathbf{F}$.

moment of a force

moment of inertia Symbol: I. A rotating body consisting of a collection of n particles of mass $m_i (i = 1, 2, \ldots, n)$, whose perpendicular distance from the *axis of rotation is r_i, has moment of inertia I about that axis given by

$$I = \sum_{i=1}^n m_i r_i^2$$

The *angular momentum and *kinetic energy of the body are equal to $I\omega$ and $\frac{1}{2}I\omega^2$, where $\mathbf{\omega}$ is the angular velocity about the axis and ω is its magnitude. In many ways the role of moment of inertia in rotational motion is similar to that of mass in translational motion; the distribution of mass does, however, play a major part in rotation.

Moments of inertia can be calculated about coordinate axes Ox, Oy, and Oz passing through a point O on the rotational axis, and are denoted by I_{xx}, I_{yy}, and I_{zz}. If the particles comprising the body have coordinates (x_i, y_i, z_i) these

moments of inertia are given by

$$I_{xx} = \sum m_i(y_i^2 + z_i^2)$$
$$I_{yy} = \sum m_i(z_i^2 + x_i^2)$$
$$I_{zz} = \sum m_i(x_i^2 + y_i^2)$$

For a continuous mass distribution, these sums will be replaced by integrals.

There are also additional quantities, known as *products of inertia*, given by

$$I_{yz} = \sum m_i y_i z_i$$
$$I_{zx} = \sum m_i z_i x_i$$
$$I_{xy} = \sum m_i x_i y_i$$

The moment of inertia I about the axis of rotation is then

$$I_{xx}l^2 + I_{yy}m^2 + I_{zz}n^2 - 2I_{yz}mn - 2I_{zx}nl$$
$$-2I_{xy}lm$$

where l, m, and n are the direction cosines of the rotational axis with respect to the coordinate axes (*see* direction angles). There always exists a set of axes for which the products of inertia are zero. These are called the *principal axes*, and the associated moments I_{xx}, I_{yy}, and I_{zz} are called the *principal moments of inertia*.

A table listing the moments of inertia of certain bodies is given in the Appendix. *See*

moment of inertia

parallel axes theorem; perpendicular axes theorem.

moment of mass The moment of mass of a particle about a point, line, or plane is the product of the mass of the particle and its perpendicular distance from the point, line, or plane.

moment of momentum *See* angular momentum.

moments, method of The estimation of k parameters of a statistical *distribution by equating the first k sample *moments to their population equivalents and solving the resulting equations. Although more efficient methods of estimation are often available, the method has intuitive appeal and in some situations provides a relatively simple means of obtaining reasonably simple estimators. *See also* plug-in estimator.

momentum (linear momentum) (*plural* **momenta**) Symbol: **p**. The product of the mass m of a particle and its velocity **v**. It is a *vector quantity that acts in the direction of motion.

The momentum of a system of particles, i.e. a body, is the vector sum of the momenta of the component particles. If a particle is subject to a force **F**, there is a change in its momentum, known as the *impulse of the force. By Newton's second law of motion the rate of change of momentum is equal to the force experienced by the particle:

$$\mathbf{F} = \frac{d\mathbf{p}}{dt} = m\frac{d\mathbf{v}}{dt} = m\mathbf{a}$$

where **a** is the acceleration of the particle, and **v** its velocity, at time t. *See also* angular momentum; conservation of momentum.

Monge, Gaspard (1746–1818) French mathematician noted for his *Géométrie descriptive* (1799) in which he demonstrated the value of geometry in showing how three-dimensional objects could be represented accurately on the two-dimensional plane. Instead of using numerous *ad* *hoc* constructions, Monge worked exclusively from general principles in a rigorous manner. He also contributed to the development of analytical geometry.

monic polynomial   A *polynomial of the form $x^n + a_1 x^{n-1} + \cdots + a_{n-1} x + a_n$, where the coefficient of the highest-degree term is +1.

mono-alphabetic substitution cipher   A *substitution cipher in which each individual character is replaced by a character (or group of characters) chosen from a single *alphabet. *Compare* polyalphabetic substitution cipher.

monoid   A *set M together with a *binary operation $\circ$ on it that satisfies the following two conditions:

(1) the operation is associative: given any three elements a, b, and c in M,

$$a \circ (b \circ c) = (a \circ b) \circ c$$

(2) there is an element I in M (the identity element) such that for any element a in M,

$$a \circ I = I \circ a = a$$

So a monoid is a *semigroup which possesses an identity element.

The set $\mathbb{N}$ of natural numbers, with the operation multiplication, forms a monoid. A more complicated, but still typical, example is the monoid whose elements are the real *continuous functions f with domain the interval [0, 1] and range [0, 1] (i.e. $0 \leqslant x \leqslant 1$ and $0 \leqslant f(x) \leqslant 1$). The operation $\circ$ here is function composition (i.e. $f \circ g(x) = f(g(x))$) and the identity is the function I, where $I(x) = x$ for each x in [0, 1].

monomial   An algebraic expression with a single term.

monotonic (monotone)   Changing always in the same direction. *See* monotonic decreasing function; monotonic increasing function.

monotonic decreasing function   A *function f with *domain and *codomain that are sets of real numbers such that the dependent variable decreases or stays the same as the independent variable increases. Formally, if for every x_1 and x_2 such that $a \leqslant x_1 < x_2 \leqslant b$ we have $f(x_1) \geqslant f(x_2)$, then f is said to be *monotonic decreasing* on [a, b]. If $f(x_1) > f(x_2)$, then f is said to be *strictly monotonic decreasing* on [a, b]. If $f(x)$ is differentiable and $f'(x) \leqslant 0$ in [a, b], then $f(x)$ is monotonic decreasing on the interval. If $f'(x) < 0$ then $f(x)$ is strictly monotonic decreasing on the interval. *Compare* monotonic increasing function.

monotonic increasing function   A *function f with *domain and *codomain that are sets of real numbers such that the dependent variable increases or stays the same as the independent variable increases. Formally, if for every x_1 and x_2 such that $a \leqslant x_1 < x_2 \leqslant b$ we have $f(x_1) \leqslant f(x_2)$, then f is said to be *monotonic increasing* on [a, b]. If $f(x_1) < f(x_2)$, then f is said to be *strictly monotonic increasing* on [a, b]. If $f(x)$ is differentiable and $f'(x) \geqslant 0$ in [a, b], then $f(x)$ is monotonic increasing on the interval. If $f'(x) > 0$ then $f(x)$ is strictly monotonic increasing on the interval. *Compare* monotonic decreasing function.

monotonic sequence   *See* decreasing sequence; increasing sequence.

monster group   The largest of the *sporadic simple groups*; it has

$$2^{46} \times 3^{20} \times 5^9 \times 7^6 \times 11^2 \times 13^3 \times 17 \times 19$$

$$\times 23 \times 29 \times 31 \times 41 \times 47 \times 59 \times 71$$

elements, but it has no nontrivial *normal subgroups. It is closely related to the group of symmetries of the *Leech *lattice*, a subset in 24-dimensional Euclidean space.

Monte Carlo methods 1. The solution of a problem by sampling experiments. For example, to estimate the area of a bounded region A it might be enclosed by a square of side length a, area a^2; n points are selected at random from inside the square; if r of these fall in the region A then an estimate of the area of A is ra^2/n. The technique is useful in numerical problems such as evaluation of multiple integrals.

2. A method used for making inferences or exploring distribution properties by repeated sampling; it is especially useful when an analytic solution is difficult to obtain. For example, to estimate the probability of the event *at least one of four bridge players will hold more than 6 cards of any one suit when each is dealt a hand of 13 cards from a well-shuffled pack*, we could obtain a Monte Carlo estimate of this probability by dealing, say, 1000 such sets of hands and recording the number of occasions, r, at which the event occurred. The probability of the event is estimated as $p = r/1000$ or, more generally, if N samples are used, by $p = r/N$. It is often possible to obtain a *confidence interval that indicates the precision of a Monte Carlo estimate; this interval may be shortened by increasing N, the number of samples. Most practical applications are carried out on computers using random number generators.

mood *See* syllogism.

Moore–Penrose conditions, Moore–Penrose pseudoinverse *See* pseudoinverse.

Mordell's conjecture The conjecture, made by L.J. Mordell in 1922, that algebraic equations with *genus greater than 1 have only a finite number of rational solutions. It was eventually proved by G. Faltings in 1983.

Morley's theorem (F. Morley, 1899) If in a triangle ABC, the adjacent *trisectors of angles B and C meet at P, of angles C and A meet at Q, and of angles A and B meet at R, then P, Q, and R are the vertices of an equilateral triangle.

Morley's theorem The triangle PQR is equilateral.

morphism A *mapping between mathematical objects that preserves some structure, for example a *homomorphism, a *homeomorphism, or an *isomorphism. More generally, a morphism is part of the definition of a *category.

Morse theory A theory, first introduced by the American mathematician Marston Morse (1892–1977) in the 1920s. It gives a means by which the topological properties of a manifold M can be described in terms of the singularities of (almost) any function defined on M. For example, there is an algorithm for calculating the *homology groups of M in this way. *See* singularity theory.

mortality rate *See* death rate.

mortality tables *See* life tables.

motion A change in the position of a particle or a system of particles (i.e. a body), as seen by a particular observer. The motion may be along a straight line or along a curve, and may be periodic in nature. *See also* Newton's laws of motion; equation of motion.

moving average A method of smoothing *time series by replacing each observation by a (usually weighted) *mean of that observation and its near neighbours. For example, given a series $y_1, y_2, \ldots, y_t, \ldots$, with observations equally spaced in time, a possible *three-point moving average* would replace y_s by $y_s' = \frac{1}{4}(y_{s-1} + 2y_s + y_{s+1})$ for all $s \geqslant 2$. More elaborate weighting systems are often used in practice. *See* lowess.

Moxon, Joseph (1627–1700) English mathematical lexicographer. He wrote a number of elementary textbooks on such subjects as astronomy, geography, and mechanics. Moxon's best-known work, however, remains his *Mathematicks made Easie: or, a Mathematical Dictionary Explaining the Terms of Art, and Difficult Phrases used in Arithmetick, Geometry, Astronomy, Astrology, and other Mathematical Sciences* (1679), the first mathematical dictionary to be published in English.

Müller, Johann See Regiomontanus.

multinomial An algebraic expression that is a sum of two or more terms. See also polynomial.

multinomial distribution A generalization of the *binomial distribution to r (>2) possible outcomes with *probabilities $p_1, p_2, \ldots, p_r$ at each of n trials, where $\Sigma p_i = 1$. The probabilities of the various outcomes are given by the terms of the multinomial expansion of $(p_1 + p_2 + \cdots + p_r)^n$.

multinomial theorem A generalization of the *binomial theorem for positive integral n which states that $(x_1 + x_2 + \cdots + x_r)^n$ may be expressed as

$$\sum \frac{n!}{a!b!c! \ldots k!} x_1^a x_2^b \ldots x_r^k$$

where the summation is taken over terms with all possible integral values of a, b, c, $\ldots$, k between 0 and n, subject to the constraint that $a + b + c + \ldots + k = n$.

multiple A number that is a product of a given number and an integer. For example, 6 is a multiple of 2, and 5.6 is a multiple of 1.4. See also common multiple.

multiple-angle formulae Formulae in plane trigonometry that give trigonometric functions of multiple angles in terms of functions of the angles, for example

$$\sin 3x = 3 \sin x - 4 \sin^3 x$$

$$\cos 3x = 4 \cos^3 x - 3 \cos x$$

See also double-angle formulae.

multiple comparisons When a number of non-independent comparisons are made between treatment means (i.e. means for all units receiving the same treatment) in a designed experiment (*see* experimental design), significance tests and interval estimates based on the *t- or *F-distributions are no longer valid. Similar difficulties arise if a large number of comparisons are possible and attention is confined to those that look interesting because they appear to indicate large differences. Multiple-comparison tests overcome these difficulties.

multiple correlation coefficient In *multiple regression, if

$$\mathrm{E}(y) = \beta_0 + \beta_1 x_1 + \beta_2 x_2 + \ldots$$

and the *least-squares regression estimators of y corresponding to the observed y_i are $\hat{y}_i$, then the multiple correlation coefficient is the product moment *correlation coefficient between the y_i and $\hat{y}_i$ and is denoted by R. No other linear function of the x_i has a higher correlation with the y_i. R^2 is sometimes referred to as the *coefficient of multiple determination*.

multiple integral (iterated integral) An integral involving two or more successive *integrations, in which one variable is integrated at a time, the others being kept constant. Multiple integration is the inverse process of successive *partial differentiation. A multiple integral involving two integrations (called a *double integral*) is written as

$$\iint f(x, y) \, dx \, dy$$

which is the same as

$$\int \left[\int f(x, y) \, dx \right] dy$$

An iterated integral having three integrations is a *triple integral*. *See also* area; volume.

multiple point (*k*-tuple point) A *singular point on a curve at which two or more (*k*) *arcs (or branches) of the curve intersect. The simplest type involves two arcs (*see* double point).

multiple regression *See* regression.

multiple root A repeated *root of an equation. For example, the cubic equation

$$x^3 - 3x^2 + 4 = 0$$

has factorized form

$$(x - 2)(x - 2)(x + 1) = 0$$

The roots are -1 and 2, and the value 2 appears twice, i.e. it is a *double root*. In general, if $(x - r)^n$ is a factor of a polynomial equation, then r is an *n-tuple root* or *root of multiplicity n* of the equation.

For any equation $f(x) = 0$ a multiple root is also a root of the first derived equation $f'(x) = 0$. A double root is a root of the equation and of the first derived equation, but not of the second derived equation. A triple root is a root of the equation and of the first and second derived equations, but not of the third derived equation. In general, an *n*-tuple root is a common root of the equation itself and of all the derived equations up to the $(n - 1)$th, but not of the *n*th derived equation. *Compare* simple root.

multiple-valued function (many-valued function) A one-to-many mapping from one *set to another set. Each element x of the first set can be mapped to more than one element $y_1, \ldots, y_r$ of the second set. A multiple-valued function is not a true function but consists of single-valued branches that are separate functions. If the graph of a multiple-valued function is drawn, some parallels to the y-axis cut the resultant curve at more than one point. The circle $x^2 + y^2 = 1$ may be regarded as the graph of a multiple-valued function consisting of two branches:

$$y = +\sqrt{(1 - x^2)} \quad \text{and} \quad y = -\sqrt{(1 - x^2)}$$

See function.

multiplicand The number or term that is multiplied by another (the *multiplier*) in a multiplication.

multiplication A mathematical operation in which two numbers are combined to give a third number (the *product*). It is denoted by $a \times b$ or $a \times b$, or (for symbols) by ab. Multiplication of integers can be regarded as repeated addition: for example, $2 \times 3 = 6$ is the integer obtained by adding three 2's $(2 + 2 + 2)$. This is the same as adding two 3's $(3 + 3)$, a demonstration of the *commutative nature of multiplication of numbers. Fractions are multiplied by multiplying the numerators and denominators separately:

$$\frac{a}{b} \times \frac{c}{d} = \frac{ac}{bd}$$

For irrational numbers a more formal, set-theoretic definition must be used (*see* Dedekind cut). Multiplication can be regarded as the process of multiplying one number (the *multiplicand*) by another (the *multiplier*), although the result is the same whichever number is chosen for the multiplicand.

Polynomials are multiplied by using the *distributive law (*see also* expansion). *Complex numbers can be multiplied similarly:

$$(a + ib)(c + id) = ac + iad + ibc + i^2bd$$

$$= (ac - bd) + i(ad + bc)$$

The concept of multiplication has been extended to other entities, such as *vectors, *matrices, and sets (*see* Cartesian product).

multiplication sign The sign ×, or . denoting *multiplication. The cross × appeared in an anonymous article of 1618, was advocated by William Oughtred in 1631, and eventually became common in arithmetic. Christopher Clavius used a dot for multiplication in 1583, as did Thomas Harriot in a posthumous work of 1631. The dot came to be used more often in algebra since it did not resemble the letter x, but now the usual practice is to indicate multiplication by writing unknowns side by side, as in 'xy'.

multiplication table A rectangular table giving the results of multiplying together pairs of numbers, or combining pairs of elements in a *group or other *algebraic structure. A multiplication table for a group is sometimes called a *Cayley table* after A. Cayley, who proposed the idea in 1854.

multiplicative function 1. A *function f is multiplicative if $f(xy) = f(x) . f(y)$ for all x and y in its domain. *Compare* additive function.

2. An *arithmetic function f is multiplicative if $f(mn) = f(m) . f(n)$ whenever m and n are *coprime. It is *completely multiplicative* if $f(mn) = f(m) . f(n)$ for every m and n.

multiplicative group A *group where the result of combining a and b is written as ab and the group identity is denoted by 1. In a *field, the phrase is often used to refer to the group obtained by just considering the field's nonzero elements with respect to its operation of multiplication.

multiplicative inverse *See* inverse.

multiplicity *See* multiple root.

multiplier The number or term by which another (the *multiplicand*) is multiplied in a *multiplication.

multivariate data *See* data.

multivariate distribution An extension of *bivariate distribution concepts of joint, marginal, and conditional distributions to more than two *random variables.

mutually exclusive events *See* probability.

mutual variation *See* variation.

N

ℕ A symbol for the set of all natural numbers.

nabla *See* del.

nadir A point on the *celestial sphere directly below the observer. The nadir is one of the poles of the horizon. *Compare* zenith.

nano- *See* SI units.

Napier, John (1550–1617) Scottish mathematician who worked on trigonometry and methods of computation. In 1614 he published his *Mirifici logarithmorum canonis descriptio* (Description of the Marvellous Rule of Logarithms) – the first tables of logarithms for aiding calculation. Napier started work on this around 1594. His method was based on geometric principles and his logarithms could be obtained from the formula

$$N = 10^7(1 - 1/10^7)^L$$

where L is the logarithm of N (the 10^7 was used to avoid decimals). Natural logarithms (to base e) are sometimes called *Napierian logarithms* in his honour, although the logarithms invented by Napier actually had a base close to 1/e. The device known as *Napier's bones* is an early mechanical calculator. *Napier's analogies and *Napier's rules of circular parts are formulae in spherical trigonometry.

Napier's analogies Relations between the sides and angles of a *spherical triangle:

$$\frac{\sin\frac{1}{2}(A - B)}{\sin\frac{1}{2}(A + B)} = \frac{\tan\frac{1}{2}(a - b)}{\tan\frac{1}{2}c}$$

$$\frac{\cos\frac{1}{2}(A - B)}{\cos\frac{1}{2}(A + B)} = \frac{\tan\frac{1}{2}(a + b)}{\tan\frac{1}{2}c}$$

$$\frac{\sin\frac{1}{2}(a - b)}{\sin\frac{1}{2}(a + b)} = \frac{\tan\frac{1}{2}(A - B)}{\cot\frac{1}{2}C}$$

$$\frac{\cos\frac{1}{2}(a - b)}{\cos\frac{1}{2}(a + b)} = \frac{\tan\frac{1}{2}(A + B)}{\cot\frac{1}{2}C}$$

where A, B, and C are the angles and a the side opposite A, b the side opposite B, and c the side opposite C. Napier's analogies are used in solving oblique spherical triangles.

Napier's bones *See* Napier.

Napier's rules of circular parts A pair of rules used for remembering the formulae for solving right *spherical triangles. Suppose the triangle has angles A, B, and C, with C as the right angle, and sides a, b, and c (a is opposite angle A, etc.). The method is to omit the right angle and to take the two sides a and b together with the complements of angles A and B and side c. These are then arranged on a circle in the order in which they occur in the triangle (a, $90° - B$, $90° - c$, $90° - A$, b). Each circular part has two adjacent parts and two opposite parts on the circle. The rules are:

(1) The sine of a part is equal to the products of the tangents of the two adjacent parts.

(2) The sine of a part is equal to the products of the cosines of the two opposite parts.

Applying the two rules to each of the five parts generates the ten formulae required.

Napoleon's theorem If on the sides of a triangle ABC, equilateral triangles with sides equal to BC, CA, and AB are drawn outwardly (*see* diagram), then O_1, O_2, and O_3, the centres of the triangles opposite A, B, and C, are the vertices of an equilateral triangle.

Napoleon's theorem The triangle $O_1O_2O_3$ is equilateral.

If the triangles are drawn inwardly, their centres also form an equilateral triangle. The origin of the attribution of this theorem to Napoleon Bonaparte (1769–1821) is uncertain. The earliest known statement of the result is by W. Rutherford in 1825.

nappe Either of the two parts into which a *conical surface is divided by the vertex.

***n*-ary relation** *See* relation.

Nash equilibrium (J.F. Nash, 1950) In *game theory if a set of strategies has the property that no player can benefit by changing his or her strategy while the other players keep their strategies unchanged, then that set of strategies and the corresponding payoffs are in Nash equilibrium. Nash equilibria exist for all finite games with any number of players.

natural deduction A *formal system that uses a large set of rules of *inference, and permits the deduction of conclusions from premises rather than from a set of *axioms. As rules of inference and axioms are closely related, natural deduction systems and logistic systems share many attributes. The first system of rules for natural deduction was proposed by Gerhard Gentzen in 1934. *See also* logic.

natural logarithm *See* logarithm.

natural number *See* integer.

nautical mile A unit of length used in navigation, originally defined in the UK as the mean length of one *minute of longitude. The value 6082 feet was later adopted. The international nautical mile was defined in 1929 as 1852 metres. 1 international nautical mile = 0.999 363 UK nautical mile.

Navier–Stokes equation Partial differential equations that model the flow of a fluid such as a liquid or a gas in three-dimensional space. If $\mathbf{v}(\mathbf{x}, t)$ is the velocity of the fluid at point $\mathbf{x}$ and time t and $p(\mathbf{x}, t)$ is the pressure, then, assuming that the fluid is incompressible, the Navier–Stokes equations are the three components of the vector equations

$$\rho\left(\frac{\partial \mathbf{v}}{\partial t} + (\mathbf{v} \cdot \nabla)\mathbf{v}\right) = -\nabla p + \nu\nabla^2\mathbf{v} + \mathbf{f}$$

where ρ is the density of the fluid, ν is the viscosity, and $\mathbf{f}$ represents external force; ∇ is the operator *del. The additional equation

$$\nabla \cdot \mathbf{v} = 0$$

corresponds to the incompressibility of the fluid. One of the *Millennium Prize problems is to show that, from a certain wide class of initial conditions, there are smooth solutions $\mathbf{v}$ and p that are valid throughout space and for all time. The equations are named after Claude-Louis Marie Henri Navier (1785–1836) and G.G. Stokes.

nearest–neighbour decoding A *decoding method by which an inadmissable codeword (i.e. one that has probably been corrupted during transmission) is replaced by the codeword that is closest to it in *Hamming distance.

necessary condition Statement A is a *necessary* condition for statement B if A is true whenever B is true. A is a *sufficient* condition for B if B is true whenever A is true. A is a *necessary and sufficient* condition for B if A and B are both true (or both false) together. This is often written as 'A if and only if B' or 'A iff B'.

Thus, for an integer to be divisible by 6 a necessary (but not sufficient) condition is that the integer be even; a sufficient (but not necessary) condition is that the integer be divisible by 12; and a necessary and sufficient condition is that it be even and divisible by 3.

needle problem *See* Buffon's needle problem.

negation A sentence of the form 'It is not the case that A', often symbolized in a formal language as '$\sim A$'. In practice, it is not necessary for the term 'not' to occur. Thus, for example, the negation of $x > y$ is $x \leqslant y$. *See* not.

negative angle A rotation angle measured from an initial axis in a clockwise sense.

negative binomial distribution The *distribution of the number of failures, X, prior to the kth success in a sequence of *Bernoulli trials. If p is the probability of success and $q(= 1 - p)$ the probability of failure, then X has frequency function

$$\Pr(X = r) = \binom{r + k - 1}{r} p^k q^r, \quad r \geqslant 0$$

It has mean kq/p and variance kq/p^2. The case $k = 1$ is a *geometric distribution. *See also* binomial distribution.

negative number A real number that is less than zero.

negative series A *series whose terms are all negative real numbers.

neighbourhood 1. The neighbourhood of a point P is the set of all points whose distance (*see* metric space) from P is less than some arbitrarily chosen distance. For example, the ε-neighbourhood of a point P is the set of all points whose distance from P is less than ε.
2. In a *topological space X, a neighbourhood of a point $x \in X$ is a subset A of X which contains an open set U such that $x \in U$. *See also* Hausdorff metric.

nephroid A plane curve that is the locus of a point on the circumference of a circle that rolls on the outside of a fixed circle of twice its radius. It has two *cusps and is an example of an *epicycloid.

nested multiplication *See* Horner's method.

nested sets A family of *sets A is nested if and only if for any two sets B and C in A, either B is included (*see* inclusion) in C or C is included in B. For example, the family of sets $A = \{\{1\}, \{1, 2\}, \{1, 2, 3\}\}$ constitutes a nest. Such a family of sets is also known as a *chain* or *tower*.

net 1. Remaining after all deductions. *Net profit*, for instance, is the profit after taking away all operating costs. *Compare* gross.

net of a pyramid.

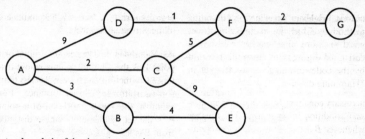

network analysis of the cheapest routes between towns.

2. The *net weight* of an object is the weight remaining after subtracting an allowance (the *tare*) for the weight of any wrapper, vessel, vehicle, etc. in which the object is when its weight is measured. *Compare* gross.

3. A plane figure composed of polygons from which, by folding along certain edges and joining others, a *polyhedron can be constructed (*see* diagram).

network *See* graph.

network analysis A class of procedures for solving optimization problems used in *operational research; it is especially relevant to scheduling problems and to routing and capacity problems in communication. One of the best-known applications is *critical path analysis. More generally, specific problems in network analysis may be solved by *dynamic programming or by algorithms developed for particular applications.

In the network shown in the diagram, the nodes represent towns and the arcs are permissible routes between each. The problem of interest is to determine the cheapest routes from town A to each of the towns B, C, D, E, F, and G. The figures beside each arc indicate the cost for a journey over that route. Clearly, it is cheaper to go from A to D via C and F than to go directly.

In this simple example the optimum routes can be determined by inspection, but for more complex situations with many towns and routes, algorithms are needed for the solution. One appropriate

algorithm in this case is *Dijkstra's algorithm* (E.W. Dijkstra, 1959). If optimum routes from one town to any number of other towns are all unique (i.e. if there are no alternatives with the same cost), the optimum routes may be displayed on a *tree diagram.

The cost associated with routes may be distances or times taken to traverse routes rather than monetary sums, and are sometimes referred to as *penalties*. Another problem amenable to analysis as a network problem is that where the penalties are maximum altitudes on each segment of a number of routes over a mountain, and it is wished to select a route such that the maximum altitude over all segments is a minimum. This may be relevant to a transport company in winter, if the probability of passes being blocked by snow increases with altitude. A similar class of problem is one where there are load restrictions on each segment, and a trucking company may wish to identify the route for which the minimum of all restrictions is as large as possible, so that each truck may carry as large a load as possible between any two nodes.

In other network analyses there may be a requirement that every node in a network be visited, and that this is to be done with the least possible travel (all possible route distances between nodes being known). This problem is commonly called the *travelling salesman problem. The complementary problem, where every arc between nodes must be traversed with the total distance covered a minimum, is sometimes called the *Chinese postman problem. Many problems in the economic design of pipelines or other distribution systems are amenable to network analysis.

Neumann function A *Bessel function of the second kind. Named after the German mathematician Karl Gottfried Neumann (1832–1925).

Newton, Sir Isaac (1642–1727) English mathematician and physicist who, in work beginning in the late 1660s, developed for the first time the principles and methods of both the differential and integral calculus. Although some of his results were shown to friends and reported in letters, nothing of any substance was published by Newton before his *De quadratura curvarum* (On the Quadrature of Curves) appeared as an appendix to his *Opticks* (1704). Fuller details were published in his *Analysis per quantitatum series* … (1711, Analysis by Means of Various Series) and the posthumously published *The Method of Fluxions and Infinite Series* (1736). Other important mathematical work by Newton includes his discovery of the binomial theorem, announced in letters written in 1676, his discovery of 72 of the possible 78 cubic curves, published in his *Enumeratio linearum tertii ordinis* (1704, Enumeration of Lines of the Third Order), and his work in algebra collected in his *Arithmetica universalis* (1707). In his major work, *Philosophiae naturalis principia mathematica* (1687, The Mathematical Principles of Natural Philosophy, known as *Principia*), Newton formulated his laws of motion, derived his law of universal gravitation, and presented a system of mechanics capable of precise and accurate descriptions of the motions of all bodies, whether celestial or terrestrial.

newton Symbol: N. The *SI unit of force, equal to the force required to impart to a mass of 1 kilogram an acceleration of 1 metre per second per second.

1 newton = 10^5 dynes = 7.233 poundals. [After Sir Isaac Newton]

Newton–Cotes rule A rule for *numerical integration which approximates $\int_a^b f(x)\,dx$ by a formula $\Sigma_{i=1}^n w_i f(x_i)$ obtained by integrating a polynomial that takes the same values as f at equally spaced points $x_1 < x_2 < \cdots < x_n$ in the interval $[a, b]$, where $x_{i+1} - x_i = h$. The rule is *closed* if $x_1 = a$ and $x_n = b$, and *open* if $x_1 = a + h$ and $x_n = b - h$. Examples of Newton–Cotes rules are *Newton's rule, *Simpson's rule, and the *trapezoidal rule.

Newton–Gregory interpolation *See* Gregory–Newton interpolation.

Newtonian frame of reference *See* frame of reference.

Newtonian mechanics *See* classical mechanics.

Newton–Raphson method *See* Newton's method.

Newton's identities (I. Newton, 1707) A set of *identities relating sums of powers of the roots of a *polynomial equation to the coefficients of the polynomial. Suppose that the equation is

$$a_n x^n + a_{n-1} x^{n-1} + \cdots + a_1 x + a_0 = 0$$

If s_1 is the sum of the roots of the equation, s_2 is the sum of the squares of the roots, and in general s_k is the sum of the kth powers of the roots, then Newton's identities are

$$a_n s_1 + a_{n-1} = 0$$
$$a_n s_2 + a_{n-1} s_1 + 2a_{n-2} = 0$$
$$\vdots$$
$$a_n s_k + a_{n-1} s_{k-1} + \cdots$$
$$+ a_{n-k+1} s_1 + k a_{n-k} = 0$$

where a_i is taken to be 0 if $i < 0$.

Newton's law of gravitation *See* gravitation.

Newton's law of restitution The relative velocities before and after the impact of two bodies, resolved along the common normal at the point of contact, are in the ratio $1:-e$, where e is the *coefficient of restitution* and $0 \leqslant e \leqslant 1$. For a perfectly elastic collision, $e = 1$. When $e = 0$, the collision is perfectly inelastic and the bodies coalesce.

If the resolved components of velocity along the common normal of the two bodies are u_1, v_1 and u_2, v_2 before and after impact, the law states that

$$v_2 - v_1 = -e(u_2 - u_1)$$

This empirical law was stated at the same time as Newton's laws of motion.

Newton's laws of motion Three fundamental laws that are the basis of *classical mechanics as expounded in Newton's *Principia* (1687):

(1) Every particle remains at rest or moves with uniform motion (i.e. at constant speed) in a straight line unless or until acted upon by an external force.
(2) The rate of change of momentum is proportional to the applied force, and takes place in the direction in which the force is applied.
(3) For every force (the *action*) acting on a particle there is a corresponding force (the *reaction*) of the same magnitude exerted by the particle in the opposite direction.

The first law is concerned with *inertia, and the second and third are concerned with *force. Since the momentum **p** of a particle is the product of its mass m and velocity **v**, the second law can be restated thus – the acceleration **a** of a particle is directly proportional to the applied force **F**:

$$\mathbf{F} = \frac{d\mathbf{p}}{dt} = m\frac{d\mathbf{v}}{dt} = m\mathbf{a}$$

The third law is known as the principle of action and reaction. The laws can be extended to systems of particles, and to continuous bodies by the assumption that such bodies are collections of particles.

Newton's laws have proved valid in most circumstances but are limited to cases in which speeds are small compared with the speed of light (3×10^8 m s^{-1}) and to systems that do not involve atomic or nuclear particles. *See also* mechanics.

Newton's method (Newton–Raphson method) (J. Raphson, 1690) A method for solving an equation in one variable, $f(x) = 0$, by carrying out the iteration

$$x_{n+1} = x_n - \frac{f(x_n)}{f'(x_n)}, \quad n = 0, 1, 2, \ldots$$

where x_0 is a first approximation to the root (*see* diagram). Convergence can be slowed or prevented if x_0 is not chosen appropriately, or if the desired root α is a multiple root (i.e. $f'(\alpha) = 0$).

For a system of n equations with n variables, $f_i(x_1, x_2, \ldots, x_n) = 0$, the Newton–Raphson iterative formula in matrix notation is

$$\mathbf{x}_{m+1} = \mathbf{x}_m - \mathbf{J}^{-1}(\mathbf{x}_m)\mathbf{f}(\mathbf{x}_m)$$

where $m = 0, 1, 2, \ldots$. Here $\mathbf{x}_m$ and $\mathbf{f}(\mathbf{x}_m)$ are $n \times 1$ column vectors, $\mathbf{J}^{-1}(\mathbf{x}_m)$ is the inverse of the *Jacobian matrix evaluated at $\mathbf{x}_m$, and $\mathbf{x}_0$ is the column vector of the initial values of $x_1, x_2, \ldots, x_n$.

Newton's method Newton–Raphson iteration to solve $f(x) = 0$.

Newton's rule A rule for *numerical integration. The integration of a real *function $y = f(x)$ from a to b is approximated by first dividing the interval $[a, b]$ into $3n$ equal parts at points $x_1, x_2, \ldots, x_{3n-1}$ lying between a and b. The ordinates at these points are $y_1, y_2, \ldots, y_{3n-1}$. The width of each strip so formed is $h = (b - a)/3n$. An approximate value of the area under the curve of the function between a and b is then given by

$$A = \frac{3}{8}h(y_a + 3y_1 + 3y_2 + 2y_3 + 3y_4$$
$$+ 3y_5 + 2y_6 + \cdots + y_b)$$

The rule is sometimes known as *Newton's three-eighths rule. See also* Simpson's rule; trapezoidal rule.

Neyman–Pearson lemma (J. Neyman and E.S. Pearson, 1937) A theorem giving the best critical region of size α for testing a simple null hypothesis H_0 against a simple alternative H_1, based on the *likelihood ratio. See* hypothesis testing.

n-gon A *polygon with n sides.

nilpotent Describing a *matrix, A, that vanishes when raised to some power, i.e. $A^n = 0$ for some value of n.

Nine Chapters on the Mathematical Art *See: Jiuzhang suanshu.*

nine-point circle (C.J. Brianchon and J.V. Poncelet, 1820; K.W. Feuerbach 1822) A circle associated with a triangle and passing through nine points:

(1) the mid-points of the three sides;
(2) the feet of the three altitudes;
(3) the mid-points of the three line segments between the vertices and the orthocentre.

Feuerbach also proved that it touches the *incircle and the three *excircles of the triangle.

nine-point circle

node 1. (crunode) A *singular point at which a curve intersects itself such that there are two different *tangents at the point. A node is a special case of a *double point in which the tangents are not coincident.
2. *See* graph.
3. *See* approximation theory.
4. *See* decision tree.

Noether, Amalie (Emmy) (1883–1935) German mathematician who developed the study of *ideals in abstract *rings, and was largely responsible for directing algebra away from detailed arithmetical calculations to the study of structure in *axiom systems. As a woman she found few willing to accept her as either pupil or colleague. It took all of *Hilbert's influence to get her appointed to a position (honorary until 1922) at Göttingen.

noise Alternative name for random disturbance or error, commonly used in communications engineering.

nominal data Data that cannot be ordered, for example the eye colours of ten children, or the marital status of groups of individuals as single, married, widowed, or divorced. *Compare* ordinal data.

nominal rate (of interest) *See* interest.

nomogram (alignment chart) A chart usually consisting of three or more parallel lines, each graduated with a scale. The scales are chosen so that relationships between three (or more) variables can be read by placing a straightedge across the chart.

nonagon (enneagon) A *polygon that has nine interior angles (and nine sides).

nonconstructive Describing a proof or definition that fails to be *constructive. Normally at some stage in a nonconstructive proof, reference will be made to a set or number with certain properties without giving an *effective procedure for constructing the set or number. For example, the *axiom of choice permits the formation of a set consisting of single elements taken from an infinite number of sets without indicating how such a choice could be made. Thus proofs relying on the axiom of choice are often held to be nonconstructive.

nondenumerable Describing an infinite set that cannot be put into a *one-to-one correspondence with the set of positive integers. An example of such a set is the set of *real numbers. *See* countable; diagonal argument.

nondeterministic polynomial time *See* NP problem.

non-Euclidean geometry Any of various forms of *geometry based on a set of *axioms other than those of *Euclidean geometry. In particular, non-Euclidean geometry does not depend on the fifth (parallel) postulate of Euclid. This postulate is often stated in the form: for a given point outside a given line, only one line can be drawn through the point parallel to the given line. To many mathematicians, this seemed less fundamental than the other axioms, and numerous attempts were made to derive it from the others (*see* Saccheri; Lambert). In the 19th century, three mathematicians independently came to the conclusion that the postulate could not be proved, and that quite self-consistent geometries could be constructed using alternative axioms.

Lobachevsky, between 1826 and 1829, developed a version of geometry based on the axiom that more than one line can be drawn through the point not meeting the given line. Bolyai, around 1829, also developed similar ideas, based on the postulate that an infinite number of lines can be drawn through the point. Gauss had come to similar conclusions earlier, although he did not publish his results. An alternative form of non-Euclidean geometry was put forward by Riemann in the 1850s (*see* Riemannian geometry). Riemann's geometry involves the postulate that no line can be drawn through the point parallel to the given line.

The geometry of Riemann (sometimes known as *elliptic geometry*) is one in which the 'plane' can be thought of as the surface of a sphere, with lines as great circles on the sphere. The angle sum of a 'plane' triangle (i.e. a spherical triangle) is greater than $180°$. In the geometry of Lobachevsky and Bolyai (sometimes called *hyperbolic geometry*), the opposite is the case – the angle sum of a triangle is less than $180°$. A model for this type of geometry is the pseudosphere (*see* tractrix). Euclidean geometry, in which the angle sum of a triangle is $180°$, can be regarded as intermediate between the two. *See also* relativity.

nonlinear Describing an equation, expression etc. that is not of the first degree. For example, the equation $y = x^2$ is a nonlinear equation in the variables x and y.

nonlinear dynamics The study of dynamical systems for which the underlying equations are nonlinear. *See* chaos.

non-negative number A real number which is greater than or equal to zero.

nonparametric methods Inference procedures in which no assumptions are made about any population parameter. The term is often taken to be synonymous with *distribution-free methods, and is widely used in this way, but it is better restricted to the above definition. With this usage, if we have random samples from two populations with unspecified continuous cumulative distribution functions $F(x)$ and $G(y)$, then a test of the hypothesis $G(x) = F(x)$ for all x against the alternative $G(x) < F(x)$ for all x would be both distribution-free and nonparametric; but a test of the hypothesis $G(x) = F(x)$ against the alternative $G(x) = F(x - \theta)$, where θ is some specified nonzero parameter, would be distribution-free but it would not be nonparametric because it involves a parameter θ.

The distinction is more one of logic than one of practical importance, so the term 'nonparametric' is commonly used somewhat loosely when 'distribution-free' would be logically more appropriate.

See also coefficient of concordance; correlation coefficient; Friedman's test; Jonckheere–Terpstra test; Kolmogorov–Smirnov tests; Kruskal–Wallis test; median test; Page test; sign test; Wilcoxon rank sum test; Wilcoxon signed rank test.

nonperiodic decimal *See* decimal.

nonperiodic tiling *See* periodic tiling.

nonrepeating decimal *See* decimal.

nonsingular matrix A square *matrix whose *determinant is not equal to zero; a matrix that has an inverse.

nonstandard analysis A theory of the foundations of analysis, invented by Abraham Robinson and based on the idea of an *infinitesimal. The proof of the existence of infinitesimals is technical, but it leads to an intuitive theory based on the properties of an extended set $\mathbb{R}^*$ of real numbers.

nonterminating decimal *See* decimal.

nonterminating fraction An infinite *continued fraction.

norm A generalization to *vector spaces of the *modulus (or *absolute value) of a complex number. The purpose of a norm is to give a single number that indicates the size of an element in a vector space.
1. (of a vector space) A mapping that assigns a real number to every element in a *vector space. The norm of a vector $\mathbf{v}$ is

denoted by $\|\mathbf{v}\|$, and is required to have the following properties:
(1) $\|\mathbf{v}\| \geqslant 0$ for all $\mathbf{v}$ and $\|\mathbf{v}\| = 0$ only for $\mathbf{v} = 0$.
(2) If n is a number, $\|n\mathbf{v}\| = |n|\,\|\mathbf{v}\|$, where $|n|$ denotes the absolute value of n. This applies to all $\mathbf{v}$ in the vector space and all n in the field.
(3) $\|\mathbf{u} + \mathbf{v}\| \leqslant \|\mathbf{u}\| + \|\mathbf{v}\|$ for all $\mathbf{u}$ and $\mathbf{v}$ in the space. This is known as the *triangle inequality*.

These axioms give a general definition for a norm of a vector space. A vector space with a norm is a *normed space*. A norm is used to define a *metric, i.e.

$$d(\mathbf{x}, \mathbf{y}) = \|\mathbf{x} - \mathbf{y}\|$$

The *Euclidean norm*, in particular, is defined by $\|\mathbf{v}\| = \sqrt{(\mathbf{v} \cdot \mathbf{v})}$, where $\mathbf{v} \cdot \mathbf{v}$ is a scalar product and the positive value of the square root is taken. This gives a length in n-dimensional Euclidean space. Any *inner product space can be given a norm in this way.
2. (of a matrix) A norm on the *vector space of matrices. One commonly used norm is the *Frobenius* (or *Euclidean*) *norm*, defined as the positive square root of the *trace of $\mathbf{A}^*\mathbf{A}$, where $\mathbf{A}$ is the given matrix and $\mathbf{A}^*$ is its *Hermitian conjugate (the Frobenius norm is just the square root of the sum of squares of moduli of elements of the matrix). The general class of subordinate (or operator) matrix norms is defined by

$$\|\mathbf{A}\| = \max_{\mathbf{x} \neq \mathbf{0}} \frac{\|\mathbf{A}\mathbf{x}\|}{\|\mathbf{x}\|}$$

where on the right-hand side of the equation $\|\cdot\|$ denotes a given vector norm.

normal 1. In general, perpendicular; at right angles.
2. (normal line) A line through a given point on a curve (or surface) perpendicular to the *tangent line (or tangent plane) at that point.

normal approximation The use of a *normal distribution as an approximation to a given *distribution. Many distributions, including some discrete ones, approach a normal distribution for certain parameter values or combinations of parameter values. For example, if both np and nq are large, the *binomial distribution may be approximated by a normal distribution with mean np and variance npq. For large λ, the *Poisson distribution may be approximated by a normal distribution with mean and variance both equal to λ. For discrete distributions, the approximations may usually be improved by using a *continuity correction.

normal component *See* acceleration.

normal deviate *See* normal distribution.

normal distribution An important *distribution in *statistics, also sometimes called the *Gaussian distribution*. It is a two-parameter distribution of a *random variable X with frequency function

$$f(x) = \frac{1}{\sigma\sqrt{(2\pi)}} \exp\left(\frac{-(x-\mu)^2}{2\sigma^2}\right)$$

The mean $E(X) = \mu$, and the variance $\mathrm{Var}(X) = \sigma^2$, and X is described as $N(\mu, \sigma^2)$. The frequency function $f(x)$ is symmetric about the ordinate $x = \mu$,

normal distribution Standard normal distribution curve.

and is often described as *bell-shaped* (*see* diagram). If $Z = (X - \mu)/\sigma$, then Z is $N(0, 1)$ and is called a *standard* (or *standardized) normal variable*. Tables of the distribution function of Z are widely available, and probabilities associated with X may be derived from these tables using the above transformation.

The importance of the distribution lies not only in the fact that much experimental data exhibit properties of a random sample from a normal distribution (sometimes after appropriate transformation), but also in its key role in the *central limit theorem. As a consequence of this theorem, we may often make inferences about population means on the basis of sample means, even for non-normal populations. However, for small samples asymptotic theory based on this theorem may give misleading results, and *nonparametric methods are often preferred. Normal distribution theory inference is generally inappropriate for survival-time studies where data tend to be highly skewed (*see* skewness), or for studies involving counts where the *Poisson distribution plays a key role.

normal form (of a matrix) *See* canonical form.

normal functions (normalized functions) *See* orthogonal function.

normalized number *See* floating-point representation.

normalizing transformation A *transformation $Y = f(X)$ of a *random variable X so that Y is normally distributed. *See* logarithmic transformation.

normal modes In general, once disturbed from equilibrium, an oscillating system will have a complex motion that may be regarded as a combination of a number of independent normal modes of vibration. The contribution of each mode to the motion is determined by the initial disturbance. This disturbance can be chosen so as to make the system vibrate exclusively in any one of these modes, with all the elements of the system performing simple *harmonic motion and passing simultaneously through their equilibrium positions. The period of each mode depends solely on the constitution of the system, and not on the initial disturbance.

The number of modes is equal to the number of *degrees of freedom* of the system, i.e. the number of independent variables needed to specify completely the configuration of the system at any particular time.

normal number A real number whose *expansion to a given base b is such that all possible blocks of digits of equal length occur and are equally likely is said to be *normal* to base b. In 1909 Borel showed that almost all real numbers are normal. If a number is normal to every base b it is *absolutely normal*. Almost all real numbers are absolutely normal. *See also* Champernowne's number; Copeland–Erdős number.

normal section A *section of a figure made by a plane perpendicular to its surface.

normal series A sequence $H_0, H_1, \ldots, H_n$ of *normal subgroups of a *group G with *identity element e such that $\{e\} = H_0 \subset H_1 \subset \cdots \subset H_n = G$. *See also* composition series.

normal space A *topological space in which every pair of *disjoint *closed sets A and B is contained in a pair of disjoint open sets U and V, i.e. $A \subset U, B \subset V$, and $U \cap V = \emptyset$.

normal subgroup If the operation in the *group G is indicated by juxtaposition, the *subgroup H is normal in G if and only if $g^{-1}Hg = H$ for each element g of G. This means that for every h in H the element $g^{-1}hg$ must also be in H. Alternatively the condition can be expressed as $gH = Hg$ for each g, and this says that each left *coset of H is also a right coset. The importance of the concept is that when

H is a normal subgroup of G, then – and only then – do the cosets of H themselves form a group with the operation of combining the cosets g_1H and g_2H:

$$(g_1H)(g_2H) = g_1g_2H$$

This group of cosets is called the *factor group* or *quotient group* 'G over H', written as G/H. The subgroup H is itself a coset and is the identity element of G/H.

A group that has no normal subgroups other than itself and the subgroup consisting of the identity element is said to be *simple*. Every group can be built of simple groups. A complete list of finite simple groups is known. They consist of 16 infinite families and 26 '*sporadic simple groups*'.

normed space *See* norm.

north polar distance (NPD) *See* declination.

northwest-corner rule *See* transportation problem.

not A truth-functional connective (*see* truth function), often symbolized in a *formal system as '$\sim$', '$-$', or '$\neg$', and whose meaning is given by the following *truth table:

A	$\sim A$
T	F
F	T

See negation.

NP complete Describing a problem that belongs to the class of *NP problems.

NP problem The abbreviation NP stands for '*nondeterministic polynomial time*', and is used in connection with *decision problems:* those for which a yes/no answer is required. Such a problem is then said to be of type P if it can be solved by an algorithm running in *polynomial time.

For some problems no such simple algorithm is known. The *travelling salesman problem, to find the shortest route visiting a number of cities exactly once, is one such problem and can be put in the form 'Is there a route shorter in length than some number C?' An algorithm which systematically generated and checked all routes while looking for a route shorter than C would run in exponential time. However, whether a particular randomly chosen or 'nondeterministic' route is shorter than C can be decided in polynomial time. Thus, while particular examples of the travelling salesman problem – produced at random or by guessing – can be decided nondeterministically in polynomial time, no solution is available for the general problem itself. It is thus said to be an NP problem.

Many of the problems for which no general polynomial time algorithm is known are NP problems. Moreover, as demonstrated by Stephen Cook in 1971, many of them – the *NP-complete* problems – are 'equally hard' in the sense that, if any one of them can be shown to be solvable by a polynomial time algorithm, then so, in theory, must all the others. However, whether or not any NP-complete problems possess a general polynomial time algorithm (the P = NP problem) remains an open question.

n-tuple A *set of n items listed in a particular order. The n-tuple consisting of the numbers $x_1, x_2, \ldots, x_n$ in that order is written as $(x_1, x_2, \ldots, x_n)$. *See* ordered pair.

nuisance parameter A *parameter that, despite it being needed to specify a *population distribution, is a nuisance in formulating statements about other parameters. In the formation of a confidence interval for the mean of a normal population when the *variance σ^2 is unknown, the latter is a nuisance parameter. The difficulty this causes is overcome by basing the interval on the *t-distribution, which does not involve σ^2.

null angle (zero angle) An angle of $0°$.

null element An element n of a *partially ordered set S (*see* partial order) which is such that $n \leqslant a$ for all $a \in S$. If a null element exists, it is unique, and is usually denoted by O (or 0).

null hypothesis *See* hypothesis testing.

nullity *See* null space.

null matrix A *matrix in which all the elements are zero.

null sequence An infinite *sequence whose *limit is zero.

null set (empty set) The null *set, denoted by $\emptyset$, is the set lacking all members:

$$\emptyset = \{x : x \neq x\}$$

It follows from the definition of a null set that it is included in every set, and from the *axiom of extensionality that it is unique. *Compare* universal set.

null space The null space (or *kernel) of a *linear transformation is the set of all vectors that are mapped into zero. Thus the null space of a matrix **A** consists of all vectors **x** such that $\mathbf{Ax} = \mathbf{0}$. The dimension of the null space is called the *nullity*.

null vector *See* zero vector.

number 1. (natural number) A positive *integer.
2. A member of the *set of all *complex numbers. The real numbers are numbers that do not involve $\sqrt{-1}$. These are classified into *rational numbers and *irrational numbers.
3. *See* cardinal number.
4. *See* ordinal number.

number field A *field whose elements are numbers.

number line *See* real number.

number sieve A numerical procedure for finding *factors of large numbers. *See* sieve of Eratosthenes.

Egyptian

1	10	10^2	10^3	10^4	10^5	10^6

Greek

1	2	3	4	5	6	7	8	9	10	50	100	500	1000
α	β	γ	δ	ε	ς	ζ	η	θ	ι	ν	ρ	ϕ	$'\alpha$

Roman

1	2	3	4	5	6	7	8	9	10	50	100	500	1000
I	II	III	IV	V	VI	VII	VIII	IX	X	L	C	D	M

Indian

1	2	3	4	5	6	7	8	9

number system Some early numeral systems.

number system A method of writing numbers. The earliest systems probably simply used the requisite number of marks: I, II, III, etc. Very early in the development of mathematics, groupings of numbers were given special symbols. Around 5000 BC the Egyptians had a number system based on 10. Different symbols were used for 10, 100, 1000, etc., and numbers were written by drawing the symbol the required number of times. An ancient Egyptian would write 764 by drawing seven snares, six heel bones, and four vertical strokes.

The fundamental grouping unit is called the *base* of the number system. In common with the Egyptians, most peoples have used a base of 10 in their counting, simply because they have eight fingers and two thumbs. The *quintal system*, based on 5, was also quite common. The V for 5 in Roman numerals probably represented a hand with the fingers together and thumb outstretched. Twenty (*vigesimal system*) was also used as a base, and a remnant of this can still be seen in some present-day names. For example, in Welsh 20 is *ugain*, 30 is *deg ar ugain* (ten and twenty), 40 is *deugain* (two twenties), etc.

The next development in notation occurred in Mesopotamia around 3000 BC. Babylonian numbers were written using wedge-shaped (cuneiform) marks impressed in clay. The Babylonians developed a notation, using the two symbols Γ and ⟨ for 1 and 10, in which sets of symbols were used in different positions to represent different numbers. For example, ΓΓ⟨Γ indicated two 60's plus eleven units (131). The base of 60 (*sexagesimal system*) survives in our units of time and angle. The Babylonian notation had the drawback that there was no way of representing an empty position. Around 300 BC a special symbol came into use to indicate an empty place between groupings (i.e. to indicate a *zero*). A system of this type, in which a small set of symbols is used and the grouping is shown by relative position, is called a *positional notation*.

Our present number system is a positional notation with the base ten. It was first used in India – the earliest recorded occurrence is in AD 595, and the earliest record of the system with a zero is from AD 876. The system was taken up by the Arabs and introduced into Europe later, largely through 12th-century translations of the book *Algebra* written by the Arab mathematician al-Khwarizmi. It is known as the *Hindu–Arabic system*. The use of positional notation to indicate fractions was introduced around 1579 by François Viète. The dot for a decimal point came a few years later, but did not become popular until its use by Napier.

In our present method of writing numbers, positions to the left of the point represent numbers of increasing powers of 10. Numbers to the right of the point represent successive numbers of tenths, hundredths, thousandths, etc. For example, 6735.249 is a shorthand way of writing (6×10^3) + (7×10^2) + (3×10^1) + (5×10^0) + (2×10^{-1}) + (4×10^{-2}) + (9×10^{-3}). The same method for denoting numbers can be used for other bases. The number of characters required is equal to the base: a *binary system (base 2) requires two characters (0 and 1); an *octal system (base 8) requires eight characters (0–7); a *duodecimal system (base 12) requires twelve characters (0–9 and two other characters); *see also* hexadecimal system.

A number written in a base other than 10 can be changed to its decimal equivalent by writing it in powers of the base in decimal. For example, the binary number 1101 is equivalent to (1×2^3) + (1×2^2) + (0×2^1) + (1×2^0) = 13. The number 215 in octal is, in decimal, (2×8^2) + (1×8^1) + (5×8^0) = 141.

The opposite process – conversion from decimal notation to some other base – is accomplished by successive divisions. For example, to convert 19 in decimal into binary:

$$19 = 18 + 1$$
$$= (2 \times 9) + 1$$
$$= [2 \times (8 + 1)] + 1$$
$$= [2 \times (2^3 + 1)] + 1$$
$$= 2^4 + 2 + 1$$

This can be written as

$$(1 \times 2^4) + (0 \times 2^3) + (0 \times 2^2)$$
$$+ (1 \times 2^1) + (1 \times 2^0)$$

and hence the binary equivalent of 19 is 10011.

number theory The study of the arithmetic properties of *integers and closely related *number systems. Ancient Babylonian, Chinese, and Greek mathematicians were among the first to investigate numbers as interesting objects in themselves. Nowadays number theory is a large and many-sided discipline using, and stimulating the development of, sophisticated methods in several other areas of mathematics such as algebra and analysis. *See* Diophantine equation.

numerator The dividend in a fraction, i.e. the number on the top. In $\frac{3}{4}$, 3 is the numerator (4 is the *denominator*).

numerical analysis The branch of mathematics concerned with finding numerical solutions to problems, especially those for which analytical solutions do not exist or are not readily obtainable. Many methods currently in use depend heavily on the concepts of *interpolation, *iteration, and *finite differences. Typical applications include:

(1) *Interpolation (*See* Gregory–Newton interpolation; Lagrange's interpolation formula; extrapolation).
(2) Approximations to functions whose values are known only at certain points or to complicated functions by methods such as *least squares (*see* approximation theory).
(3) *Numerical differentiation, usually based on interpolation formulae or function approximations.
(4) *Numerical integration using the *trapezoidal rule, *Simpson's rule, *Newton's rule, or more sophisticated methods.
(5) Solution of an equation by *iteration.

(6) Solution of *simultaneous linear equations by *Gaussian elimination or by the *Gauss–Seidel method.

(7) Solution of differential equations by, for example, the *Runge–Kutta method.

(8) Solution of *integral equations, often by using numerical integration formulae to convert the integral equations into systems of equations that may be solved numerically.

(9) Optimization, which often effectively involves solutions of nonlinear systems of equations by interative methods (*see* iteration), for example the Newton–Raphson method.

Pure-mathematics aspects of numerical analysis are concerned with approximation *errors when functions are approximated by simpler functions, and truncation errors when, for example, a Taylor series expansion is terminated after a few terms (*see* Taylor's theorem). Practical difficulties include round-off errors, and the nonconvergence of algorithms or convergence to inappropriate values (e.g. local rather than global optima). Speed of convergence may also be of importance, especially for procedures that require a considerable amount of computer time.

numerical differentiation The use of formulae, often expressed in terms of *finite differences, for estimating the *derivatives of a *function f(x), given the value of x.

Examples are the forward-difference estimate of the first derivative:

$$f'(x) \approx \frac{f(x+h) - f(x)}{h}$$

and the second difference estimate of the second derivative:

$$f''(x) \approx \frac{f(x+h) - 2f(x) + f(x-h)}{h^2}$$

where h is a small positive increment.

numerical equation An equation in which the coefficients and constant term are numbers (rather than symbols). Thus, $2x^2 = 5$ is a numerical equation; $ax^2 = k$ is not.

numerical integration The numerical approximation of a *definite integral. The integration formula is known as a *rule*. Simple well-known rules are the *trapezoidal rule, *Newton's rule, and *Simpson's rule. Most integration rules are derived by the integration of polynomials fitted to function values (*see* Newton–Cotes rule, of which the three rules above are examples) or by choosing the rule in an optimal way (*see* Gaussian integration rule). An analysis of the error is possible for most rules. A *repeated rule*, e.g. the repeated trapezoidal rule, breaks the range of integration into equally sized subintervals and applies the basic rule over each subinterval.

O

𝕆 Symbol for the set of all *octonions.

object *See* category.

objective function In *operational research, a function which is to be maximized or minimized subject to specified constraints on the variables. In *linear programming, both the objective function and the constraints are linear functions of the variables.

object language *See* metalanguage.

oblate *See* ellipsoid.

oblique Not at right angles; not containing a right angle.

oblique angle An angle that is not a multiple of 90°.

oblique cone A *cone with a vertex that is not directly above the centre of its base.

oblique coordinate system A *coordinate system in which the axes are not at right angles. *See* Cartesian coordinate system.

oblique prism A *prism with lateral edges that are not perpendicular to its bases.

oblique pyramid A *pyramid with a vertex that is not directly above the centre of its base.

oblique triangle A triangle that does not contain a right angle.

obliquity of the ecliptic *See* ecliptic.

oblong A *rectangle with adjacent sides unequal.

obtuse angle An angle between 90° and 180°.

obtuse triangle A triangle that has one interior angle greater than 90°.

octagon A *polygon that has eight interior angles (and eight sides).

octahedron (*plural* **octahedra**) A *polyhedron that has eight faces. A regular octahedron, in which all the faces are equilateral triangles, is one of the five regular polyhedra.

octal notation The method of positional notation used in the *octal system.

octal system A *number system using the base eight. The eight numerals 0–7 are used. Eight is written as 10, nine as 11, etc. For example, the number 273 in octal would, in the decimal system, be $(2 \times 8^2) + (7 \times 8^1) + (3 \times 8^0) = 187$. Octal numbers are commonly used in computer systems to represent bytes of information, since one byte equals eight bits.

octant One of the eight regions into which three-dimensional space is divided by the three planes containing the axes in a *Cartesian coordinate system.

octonion *See* Cayley algebra.

odd function A *function f such that for every x in the *domain, $f(-x) = -f(x)$. For example, $f(x) = x^3$ is an odd function. The *graph of an odd function has the origin as its centre of symmetry. *Compare* even function.

odd number

odd number An integer that is not divisible by 2.

odd permutation A *permutation equivalent to an odd number of *transpositions. For example, 321 is an odd permutation of 123 since it is equivalent to the single transposition (13). *Compare* even permutation.

odds If an event A has associated probability p, the probability of the event not occurring is $1 - p$. The quotient $p/(1 - p)$ specifies the *odds on* A. Thus, if $p = \frac{1}{3}$, the odds on A are $\frac{1}{2}$, or 'one to two on', sometimes inverted to 'two to one against'. The concept is in everyday use in gambling parlance, but statistical statements are better framed in terms of probability rather than odds, except in the context of *odds ratios.

odds ratio If the *odds on an event A are $\theta_A = p_A/(1 - p_A)$ and the odds on an event B are $\theta_B = p_B/(1 - p_B)$, the ratio $\theta = \theta_A/\theta_B$ represents the odds ratio of A to B. An alternative name sometimes used is *relative risk*, but this term is more commonly used for the ratio p_A/p_B. The *empirical odds ratio* θ^* is the ratio of the observed proportions A : not A and B : not B. For example, if among 150 smokers 12 have a certain disease and 138 do not, whilst among 100 nonsmokers 2 have the disease and 98 do not, the empirical odds ratio of having the disease relative to smoking or nonsmoking is $\theta^* = (12/138)/(2/98) = 4.26$. A common *hypothesis test is whether $p_A = p_B$, and this is equivalent to testing whether θ^* is significantly different from $\theta = 1$. In *contingency table format:

	Diseased	Not diseased
Smokers	12	138
Nonsmokers	2	98

and, more generally, for a 2×2 table:

a	b
c	d

the empirical odds ratio θ^* is then the ratio ad/bc. If the expected numbers in the four cells are m_{11}, m_{12}, m_{21}, and m_{22}, then $\theta = m_{11}m_{22}/m_{12}m_{21}$, and $\theta = 1$ implies no association, equivalent to $p_A = p_B$ in the disease and smoking example. For any 2×2 table, $0 \leqslant \theta^* \leqslant \infty$. It is often more convenient to work with $\ln \theta^*$, which is symmetrically distributed about zero (the condition for independence). *See also* loglinear model.

ODE *Abbreviation for* *ordinary differential equation.

off-diagonal *See* matrix.

ogive The graph of the *distribution function of a *random variable. Its literal meaning implies a *sigmoid type curve, and the term is unnecessary and best avoided.

ohm Symbol: Ω. The *SI unit of electrical resistance, equal to the resistance between two points on a conductor when a constant potential difference of 1 volt between these points produces a current in the conductor of 1 ampere. [After G. Ohm (1787–1854)]

Omar Khayyam (*c*.1048–*c*.1122) Persian mathematician, astronomer, and poet, best known for his poems freely translated and adapted in 1859 by Edward FitzGerald (*The Rubaiyat of Omar Khayyam*). His *Algebra* included rules for solving quadratic equations by both algebraic and geometric methods. More originally, he gave a discussion of the general solution of cubic equations by geometric methods (using conics), although he did not recognize the existence of negative roots and believed that these equations could not be solved algebraically.

one–many correspondence *See* many–one correspondence.

one-tail test *See* hypothesis testing.

one-time pad An encryption method that relies on a random key that is used for a single message only. Use of a one-time pad has been proved to be unbreakable and is the only known method with this property.

one-to-one correspondence (one–one *or* **1–1 correspondence)** A correspondence between two *sets in which each member of either set is paired with one and only one member of the other set. The two sets must have the same number of members: for example, the elements of the set $A = \{2, 4, 6, 8\}$ can be paired with the elements of the set $B = \{3, 5, 7, 9\}$, but not with the elements of $C = \{1, 2, 3\}$ or $D = \{1, 2, 3, 4, 5\}$. A can be put into a one-to-one correspondence with B, but not with C nor with D. *See also* bijection.

one-to-one function (one-to-one mapping, one-to-one map) A *function f whose domain is the *set X and whose range is the set Y is *one-to-one* if, for each element y of Y, there is only one element x in X such that $f(x) = y$. For example, the function $y = x^3$ with domain and range $\mathbb{R}$ is a one-to-one function. A function which is one-to-one has an *inverse function. A one-to-one function is sometimes called a *one–one function* or 1–1 *function*.

one-way classification *See* analysis of variance.

***O* notation, *o* notation** *See* order notation.

onto *See* surjection.

open curve (arc) A curve that has *end points; one that is a continuous transformation of an *interval $[a, b]$ in which the *images of a and b do not coincide. *Compare* closed curve.

open interval A *set of real numbers $\{x : a < x < b\}$ written as (a, b). The interval does not contain the end points a and b. In n-dimensional space, if $a = (a_1, \ldots, a_n)$ and $b = (b_1, \ldots, b_n)$ are two distinct points with $a_j \leqslant b_j (j = 1, 2, \ldots, n)$, then the open interval (a, b) is the set $\{(x_1, \ldots, x_n) : a_j < x_j < b_j\}, j = 1, 2, \ldots, n$. An interval is partly open and partly

closed if it contains just one of its end points, and is written as $(a, b]$ if it does not contain a and as $[a, b)$ if it does not contain b. *Compare* closed interval.

open mapping A *function f with *domain X and *codomain Y that are both spaces such that $f(A)$ is an *open set in Y whenever A is an open set in X.

open region *See* region.

open sentence *See* variable.

open set (of points) A *set of points A is open if every point that is a member of A has a *neighbourhood completely in the set A. For example, the points corresponding to the real numbers greater than 0 and less than 1 constitute an open set. An open set is the complement of a *closed set. *See also* topological space.

operand *See* operator.

operating characteristic curve *See* acceptance sampling.

operation For any *natural number n, an n-ary operation on a *set S is a *function f whose domain is $S \times S \times \cdots \times S$ (the set of all n-tuples of elements of S), and whose codomain is S. An example of an n-ary operation on the set of real numbers is the function which maps the n-tuple of real numbers $(x_1, x_2, \ldots, x_n)$ to their largest value. *See* binary operation; unary operation.

operational research (OR) (*US:* **operations research**) The application of mathematics and statistics to problems of management, business, organization, and production in services, commerce, and industry. Optimization problems arise in minimizing costs, maximizing profits, scheduling interrelated tasks, and determining stocking and replacement policy for perishable items, machinery, and so on. *Linear programming and *dynamic programming are widely used techniques. Problems about reliability and about stock control with variable demand have a high statistical content. *Critical path analysis is used to determine optimal allocation of resources and scheduling of tasks subject to constraints on order of performance. *Network analysis is widely used to study the optimal loading of pipelines, roads, and distribution systems, subject to capacity or cost constraints. *Game theory and *simulation studies also have applications in operational research.

operator A symbol indicating that a mathematical operation is to be performed on an associated symbol or expression (the *operand*). Examples are the differential operator d/dx, indicating differentiation with respect to x, and the symbol $\sqrt{\ }$, meaning 'take the square root of'.

Strictly, operators are the same as *functions in the sense that they define a mapping between one set and another. The concept of operators is used particularly when it is possible to treat them as entities obeying laws similar to the laws of ordinary algebra. *See* differential operator.

opposite **1.** In a triangle, a side and an angle are said to be opposite if the side is not one of the sides forming the angle.
2. In a *quadrilateral, two sides are said to be opposite if they join entirely different pairs of vertices, and two angles are opposite if they are formed by entirely different pairs of sides. Thus in a quadrilateral ABCD, the sides AB and CD are opposite and the angles ABC and CDA are opposite.
3. In a figure having a centre of *symmetry, two sides, angles, etc. are opposite if they are joined by a line through the centre. In the figure formed by two intersecting lines, opposite angles are commonly called *vertically opposite* and are equal.

oppositely congruent *See* congruent.

optical property The *focal property of a conic. *See* ellipse; hyperbola; parabola.

optimization theory The mathematics of determining maxima and minima of *functions (*see* turning point). *Constrained optimization* applies to problems with restrictions on values that may be taken by certain variables or combinations of variables, with consequent restrictions on permissible values of the function itself. *Linear programming problems are typical constrained optimization problems. If there are no constraints, we speak of *unconstrained optimization*.

The vast majority of optimization problems arising in modern science and engineering have no analytic solution, so *numerical analysis plays a key role in their solution. *See also* Lagrange multipliers; dynamic programming.

or A truth-functional connective (*see* truth function), often symbolized in a *formal language as '$\vee$', and whose meaning, in its inclusive sense (*see* disjunction), is given by the following *truth table:

A	B	$A \vee B$
T	T	T
T	F	T
F	T	T
F	F	F

The connective is both commutative and associative, and thus obeys the following laws:

$$A \vee B \leftrightarrow B \vee A$$

$$A \vee (B \vee C) \leftrightarrow (A \vee B) \vee C$$

The connective 'or' can also be defined in terms of conjunction ($\&$) and negation ($\sim$) alone in accordance with *De Morgan's laws by the following equivalence:

$$A \vee B \leftrightarrow \sim(\sim A \,\&\, \sim B)$$

The use of '$\vee$' between two statements indicates only that at least one of the statements is true. Thus, while the propositions 'Either $2 + 2 = 4$ or 6 is an odd number' and 'Either $2 + 2 = 4$ or 6 is not an odd number' are both true, the proposition 'Either $2 + 2 = 5$ or 6 is an odd number' is false.

See also disjunction.

OR *Abbreviation for* *operational research.

orbit **1.** A path followed by a particle or body under the influence of a *central force.
2. *See* dynamical system.

order **1.** (of a derivative) The number of times a *differentiation is performed. If $y = f(x)$, the first-order derivative (or first derivative) is $dy/dx = f'(x)$. The second-order derivative is

$$\frac{d}{dx}\left(\frac{dy}{dx}\right) = \frac{d^2y}{dx^2} \text{ or } f''(x)$$

The nth-order derivative is written as d^ny/dx^n or $f^{(n)}(x)$.
2. (of a differential equation) The order of the highest-order *derivative in a *differential equation.
3. (of a curve or surface) The *degree of the equation representing the curve or surface.
4. (of a determinant) The number of rows (or columns) in the *determinant.
5. (of infinitesimals) Two *variables x and y, both of which tend to the *limit zero, are *infinitesimals of the same order if the ratio x/y is finite. If $x/y \to 0$, x is an infinitesimal of higher order than y, and if $x/y \to \infty$, x is of lower order than y. If the limit of x/y^n is finite and nonzero, x is said to be an infinitesimal of the nth order if y is taken as being of the first order.
6. (of a group or element) If the number of *elements in a *group is finite, the group is *finite* and the number of elements is the *order* of the group. The order of an element a of a group is the least positive integer m such that $a^m = e$, where e is the identity element and the group operation is denoted by juxtaposition. If no such integer exists, then a has *infinite order*.
7. (of a modulo m) If a is *coprime to m, the least positive *integer k such that $a^k \equiv 1$ (mod m) is called the order of a modulo m. *See also* congruence modulo n.
8. (of a matrix) The dimension of a *matrix.
9. (of a polynomial) The degree of a *polynomial.
10. (of a graph) The number of vertices of a *graph.
11. *See* symmetry.
12. (of convergence of a sequence) If x_1, x_2, x_3, ... is a sequence of numbers (or vectors in a normed vector space) converging to a limit l, then the order of convergence of the sequence is the largest integer p such that $|x_{n+1} - l| \leqslant c|x_n - l|^p$ for all sufficiently large n and for some nonzero constant c. Important special cases are *first-order* or *linear convergence* ($p = 1$) and *second-order* or *quadratic convergence* ($p = 2$). The order of convergence is also called the *rate of convergence*.

ordered field *See* order properties.

ordered pair A *pair set in which x is designated the first element and y the second, denoted by (x, y) or $\langle x, y \rangle$. It was defined by N. Wiener in 1914 as

$$(x, y) = \{\{x\}, \{x, y\}\}$$

which leads to the result

$$(x, y) = (u, v) \Leftrightarrow (x = u) \,\&\, (y = v)$$

An *ordered triple* (x, y, z) has x, y, and z as its first, second, and third elements; an *ordered n-tuple*, or simply *n-tuple*, $(x_1, x_2, \ldots, x_n)$ has x_i as its ith element for $i = 1, 2, \ldots, n$.

ordered set A set with an order relation between its elements. *See* partial order.

ordered triple *See* ordered pair.

order notation A notation using the symbols O and o for comparing the values of *functions as the independent variable tends to infinity or to a limit.

Suppose that $f(x)$ and $g(x)$ are two functions. Usually $g(x)$ is a simple function such as a power of x.

(i) *O notation.* The statement $f = O(|g|)$ when $x \to \infty$ means that there is a constant K such that $|f(x)| < K|g(x)|$ for all sufficiently large values of x, and $f = O(|g|)$ when $x \to a$ means that $|f(x)| < K|g(x)|$ for all x differing from but sufficiently near to a. For example, $x^2 + x = O(x^2)$ when $x \to \infty$, and $x^2 + x = O(x)$ when $x \to 0$. The 'big O' notation is often used to describe the *complexity of algorithms. For example, if n is a measure of the size of an imput to an algorithm and the number of steps needed to carry out the algorithm is $3n^4$, then the complexity of the algorithm is said to be $O(n^4)$. *See also* polynomial time.

(ii) *o notation.* The statement $f = o(g)$ when $x \to \infty$ (or when $x \to a$) means that $f/g \to 0$ when $x \to \infty$ (or when $x \to a$). For example, $x^2 + x = o(x^3)$ when $x \to \infty$, and $x^2 + x = o(1)$ when $x \to 0$. *See also* asymptotic.

order properties (of real numbers) The properties satisfied by the relation $<$ ('less than') in the *field $\mathbb{R}$ of real numbers. The basic properties are:

(1) *Trichotomy law*: if r and s are real numbers, then one and only one of the statements $r < s$, $r = s$, and $s < r$ holds.
(2) *Transitive law*: if r, s, and t are real numbers with $r < s$ and $s < t$, then $r < t$.
(3) If $r < s$, then $r + u < s + u$ for any real number u.
(4) If $r < s$ and u is a real number, then $ru < su$ if $u > 0$.
(5) *Completeness property*: any nonempty set of numbers that is bounded above has a least upper bound.

The first four properties above are summarized by saying that $\mathbb{R}$ is an *ordered field*. There are other ordered fields. For instance, the rational numbers satisfy (1)

to (4) (reading 'rational' for 'real' each time), but $\mathbb{R}$ is the only ordered field which also has the completeness property (5), i.e. is a *complete field*.

In $\mathbb{R}$ a set S of real numbers is bounded above if there is a real number m that is greater than or equal to every member of S. Such an m is an *upper bound* for S and it is a *least upper bound* if there is no number less than m which is also an upper bound (*see also* partial order). Similarly, a number l is a *lower bound* for S if it is less than or equal to every member of S; and it is a *greatest lower bound* if no larger number is a lower bound. It follows from the above properties that every nonempty set of real numbers that is bounded below (has a lower bound) must have a greatest lower bound.

The notation '$s > r$' ('s is greater than r') means the same as '$r < s$'; '$r \leqslant s$' means that either $r = s$ or $r < s$. Properties (1) and (2) above imply that $\leqslant$ is a *partial order. All the other order properties of the field of real numbers follow from (1) to (5) above. For example, if $r < s$ and $u < 0$ then $ru > su$; the square of any nonzero number x is positive (i.e. $x^2 > 0$); and there is a rational number between any two distinct real numbers.

order statistics When a sample of n observations is arranged in ascending order, the ith value is called the ith *order statistic*. It is often written as $x_{(i)}$ to distinguish it from the observation labelled x_i before ordering. If there are $2n + 1$ observations, $x_{(n+1)}$ is the *median. Other examples of order statistics are the *quantiles, the least observation $x_{(1)}$, and the greatest observation $x_{(n)}$. *See* extreme value distribution; five-number summary; rank; sample distribution function.

ordinal data Data that can be ordered, for example the heights of seven children or the weekly wages of each of 15 workers. *See* rank; *compare* nominal data.

ordinal number 1. A number denoting position in a sequence e.g. 'first', 'second', 'third'.

2. A number that describes the order property of a set as well as its *cardinal number. Two ordered sets that can be put into *one-to-one correspondence in a way that preserves the ordering property have the same ordinal number. The ordinal number of the set of all positive integers is given the symbol ω.

ordinary differential equation (ODE) *See* differential equation.

ordinate The y-coordinate, measured parallel to the y-axis in a *Cartesian coordinate system. *Compare* abscissa.

Oresme, Nicole (*c*.1323–82) French mathematician and the author of a number of texts on the subject, including *De proportionibus proportionum* (*c*.1350), which gave rules similar to the present laws of exponents, and his *Algorismus proportionum* (*c*.1350), which contained the first known use of fractional exponents. He also suggested that it was possible to use irrational powers. Oresme's most influential discovery was that a uniformly varying quantity (e.g. a body with uniform acceleration) may be represented by a graph (velocity against time), and that distance is given by the area under the line.

orientation A sense of a handedness in a real *vector space. A plane has two possible orientations: clockwise and anticlockwise. Three-dimensional space has right-handed and left-handed orientations. An orientation of a vector space is determined by an ordered basis. Two ordered bases determine the same orientation if the matrix representing one of them in terms of the other has positive determinant.

An orientation of a differential *manifold is a continuous choice of orientations for each of its *tangent spaces. Some manifolds (e.g. *Möbius strip, real projective plane) are *non-orientable*: it is impossible to find a continuous orientation for them since they are 'one-sided'.

origin The point from which distances are measured in a *coordinate system.

orthocentre The point of intersection of three lines drawn from each of the vertices of a triangle perpendicular to the opposite sides, i.e. the point of intersection of the altitudes of the triangle.

orthogonal At right angles. For instance, two curves are said to be orthogonal at a point of intersection if their tangents at that point are perpendicular.

orthogonal basis A *basis of a *vector space in which the elements of the basis are orthogonal. If the lengths of the elements are all unity, the basis is also *orthonormal*.

orthogonal complement For a given vector in a *vector space, the orthogonal complement is the set of all vectors that are orthogonal to the given vector. *See* orthogonal vectors.

orthogonal functions A system of *functions $\{f_1, f_2, f_3, \ldots\}$, integrable on the interval $[a, b]$ such that the inner product, denoted by (f_m, f_n), is such that

$$(f_m, f_n) = \int_a^b f_m(x) f_n(x)\, \mathrm{d}x = 0$$

when $m \neq n$. If, for all m,

$$\int_a^b (f_m(x))^2 \mathrm{d}x = 1$$

the functions are said to be *normal* or *normalized*. Normal orthogonal functions are called *orthonormal*.

orthogonal group The *group $O(n)$ of all orthogonal $n \times n$ real matrices.

orthogonal matrix A square matrix that is equal to the inverse of its transpose. Two examples of 2×2 orthogonal matrices are

$$P = \begin{pmatrix} \cos\theta & -\sin\theta \\ \sin\theta & \cos\theta \end{pmatrix}$$

$$Q = \begin{pmatrix} \cos\theta & \sin\theta \\ \sin\theta & -\cos\theta \end{pmatrix}$$

The matrix P is called a *rotation matrix*, because premultiplying a vector $\begin{pmatrix} x \\ y \end{pmatrix}$ by P rotates the vector θ radians anticlockwise. The matrix Q is called a *reflection matrix*, because its effect on vectors $\begin{pmatrix} x \\ y \end{pmatrix}$ is to reflect them in the line $y = x \tan\frac{1}{2}\theta$.

orthogonal polynomials A set of *orthogonal functions, with respect to a continuous or a discrete (weighted) inner product, comprising polynomials of degree 0, 1, 2, ... In general, the polynomials $p_0(x)$, $p_1(x)$, $p_2(x)$, ..., where p_k has degree k, are said to be orthogonal with respect to an interval $[a, b]$ and a continuous non-negative weight function $w(x)$ on $[a, b]$ if

$$\int_a^b w(x) p_i(x) p_j(x)\, dx = 0, \quad \text{for } i \neq j$$

Orthogonal polynomials can be generated from a three-term *recurrence relation. The *Tchebyshev polynomials are an example of a set of orthogonal polynomials. Orthogonal polynomials are used in many applications, including quadrature (*see* Gaussian integration rule) and *least-squares data fitting (for example, in *regression).

orthogonal projection A *projection that involves perpendiculars. The orthogonal projection of a point P onto a line or plane is the point P′, where PP′ is the perpendicular from P to the line or plane. The orthogonal projection of a line or figure is formed by orthogonal projection of the points of the line or figure.

orthogonal transformation *See* matrix.

orthogonal vectors Two elements of a *vector space that have a *scalar product equal to zero. In the case of simple geometric vectors in Euclidean space, orthogonal vectors are perpendicular.

orthonormal Describing mathematical entities that are both orthogonal and normalized. *See* orthogonal functions.

orthonormal basis *See* orthogonal basis.

orthonormal functions *See* orthogonal functions.

oscillating product An *infinite product that alternates between two values and does not converge or diverge.

oscillating sequence A *sequence that tends neither to a finite limit nor to infinity as the number of terms in the sequence tends to infinity. It is a *divergent sequence that is not properly divergent. An example is

$$1, -1, 1, -1, 1, \ldots$$

oscillating series (oscillating divergent series A *divergent series that is not properly divergent. Examples are:

$$1 - 2 + 3 - 4 + \cdots$$
$$1 - 1 + 1 - 1 + \cdots$$

oscillation A regular fluctuation in the magnitude of the displacement about a mean or reference position, value, or state. Common examples are the oscillations that occur in mechanical and electrical systems. Mechanical oscillations include the swinging motion of a pendulum and the very much faster motion of a tuning fork. Oscillation is usually considered synonymous with *vibration*, although the latter is sometimes restricted to mechanical systems.

osculating circle The circle of *curvature of a curve at a given point.

osculating plane For a given twisted *curve at a point P, the osculating plane is the limiting position of a plane through P and two other points on the curve, P′ and P″, as P′ and P″ approach P.

osculation *See* cusp.

osculinflection *See* cusp.

Oughtred, William (1575–1660) English mathematician who, in his popular *Clavis mathematicae* (1631, The Key to Mathematics), introduced × as the familiar sign of multiplication and the abbreviations 'sin' and 'cos' into trigonometry. He also invented the slide rule in 1622, although it was not until 1632 that he announced his discovery.

ounce *See* avoirdupois; apothecaries' system; troy system.

outer product For two column vectors **x** and **y**, the matrix $\mathbf{xy}^T$, whose (i, j) element is $x_i y_j$. This matrix has *rank 1. *See also* scalar product.

outlier An observation that departs in some way from the general pattern of a data set. For example, in the set $\{7, 9, 3, 5, 4, 202\}$ the observation 202 is an outlier. Outliers may be correct observations reflecting some abnormality in the measured characteristic for a unit, or they may result from an error in measurement or recording; for example, in the above example 202 could be a mistyping of 2, 2. *See* robustness.

oval A closed curve like an elongated circle; an elliptical curve or an egg-shaped curve. *See also* Cassini's ovals.

P

Paasche index *See* index.

Pacioli, Luca (*c*.1445–1517) Italian mathematician who in his *Summa* (1494) published a compilation of the mathematics of his day, the first such work to appear in Europe since Fibonacci's *Liber abaci* of 1202.

Padé approximation (H.E. Padé, 1892) The approximation of a *function by a *rational function where the numerator and denominator polynomials have a specified *degree and are chosen so that the *Maclaurin series of the function and the approximation agree to as many terms as possible. The approximation is called a *Padé approximant* to the function. For example,

$$\frac{x^2 + 6x + 12}{x^2 - 6x + 12}$$

is a Padé approximant to e^x with numerator and denominator polynomials of degree 2.

Page test (E.B. Page, 1963) A *nonparametric test for monotonic trends in *means using ranks in a *randomized block design. The hypotheses involved are the same as those in the *Jonckheere–Terpstra test.

paired observations *See* matched pairs.

pair set Given any two elements x and y, it is possible to form the pair *set, denoted by $\{x, y\}$, consisting of just the two elements x and y:

$$\{x, y\} = \{z : (z = x) \vee (z = y)\}$$

pair-wise disjoint Describing a collection of sets, $A, B, C, D, \ldots$, in which each pair of sets is *disjoint.

pandiagonal *See* magic square.

Pappus of Alexandria (*c*. AD 320) Greek mathematician who produced valuable commentaries on Euclid and Ptolemy, parts of which are extant. His most important work, however, remains his *Synagoge* (Collections), of which Books III–VII of the original eight have survived, providing an indispensable guide to much of the lost mathematics and astronomy of late antiquity. His name has also survived as the discoverer of *Pappus' theorems. *See also* problem of Pappus.

Pappus' theorems Three theorems named after Pappus of Alexandria:

(1) If a plane *curve is revolved about a line in its plane (not cutting the curve), then the area of the surface of revolution is equal to $2\pi rs$, where s is the length of the curve and r the radius of the circle described by its *centroid.

(2) If a plane area is revolved about a line in its plane (the line not cutting the plane area), then the volume enclosed by the surface of revolution is equal to $2\pi rA$, where A is the area of the plane and r the radius of the circle described by its centroid.

(3) If A, B, and C are three points on one line, and D, E, and F are three points on another, and if the three lines AE, BF, and CD meet DB, EC, and FA, respectively, then the three points of intersection are collinear (*see* diagram). This theorem can

Pappus' theorems (3).

be regarded as a special case of *Pascal's theorem when the conic degenerates to two lines.

parabola A type of *conic that has an *eccentricity equal to 1. It is an open curve symmetrical about a line (its *axis*). The point at which the curve cuts the axis is the *vertex*. In a Cartesian coordinate system the parabola has a standard equation of the form

$$y^2 = 4ax$$

Here, the axis of the parabola is the x-axis, the directrix is the line $x = -a$, and the focus is the point $(a, 0)$. The length of the chord through the focus perpendcular to the axis, the *latus rectum*, is equal to $4a$. The parametric equations of the parabola are

$$x = at^2, \quad y = 2at$$

The *focal property* of the parabola is that for any point P on the curve, the tangent at P (APB) makes equal angles with a line from the focus F to P and with a line CP parallel to the x-axis, i.e. $\angle FPA = \angle CPB$. This is also called the *reflection property*, since for a parabolic reflector light from a source at the focus would be reflected in a beam parallel to the x-axis (the *optical property*), and sound would be similarly reflected (the *acoustical property*). *See also* projectile; cubical parabola.

parabolic Denoting or concerning a *parabola or *paraboloid.

parabolic spiral *See* spiral.

paraboloid A surface such that sections parallel to at least one plane are *parabolas. There are two types:

(a)

(b)

paraboloid (a) Elliptical and (b) hyperbolic paraboloids.

parabola Reflection property of the parabola.

(1) The *elliptical paraboloid* has an equation, in Cartesian coordinates, of the form

$$\frac{y^2}{b^2} + \frac{z^2}{c^2} = 2ax$$

In this case, the sections parallel to either the x–z or x–y coordinate planes are parabolas. Sections parallel to the y–z plane are ellipses. A *paraboloid of revolution*, formed by rotating a parabola about its axis, is a special case of an elliptical paraboloid in which the ellipses are circles. The shape is used in reflectors, radar antennae, etc. on account of the focal property of the *parabola.

(2) The *hyperbolic paraboloid* has an equation of the form

$$\frac{x^2}{a^2} - \frac{y^2}{b^2} = 2cz$$

Here, sections parallel to the x–z and y–z coordinate planes are parabolas. Those parallel to the x–y plane are hyperbolas.

paradox An *argument involving a set of apparently true premises $P_1, \ldots, P_n$ and a further premise Q such that a *contradiction is derivable from both

$$P_1 \& \cdots \& P_n \& Q$$

and

$$P_1 \& \cdots \& P_n \& \sim Q$$

A distinction is often made between *logical paradoxes* (e.g. *Russell's paradox and *Burali-Forti's paradox) and *semantic paradoxes* (e.g. the *liar paradox, and *Richard's, *Berry's, and the *Grelling–Nelson paradoxes). The former arise from purely formal considerations that are independent of any interpretation, while the latter involve semantic concepts, such as 'truth' and 'denotation'. *See also* implication (material).

parallactic angle Symbol: q. An angle at a point on the *celestial sphere between two segments of *great circles: one from the

point to the zenith and the other from the point to the north celestial pole. These two great circles, together with a third great circle joining the zenith to the pole, form a spherical triangle called the *astronomical triangle*. The parallactic angle is given by

$$\sin q = (\cos \phi \sin t) / \sin \zeta$$

where ϕ is the terrestrial latitude, t the hour angle, and ζ the zenith distance. It is also given by

$$\sin q = (\cos \phi \sin A) / \cos \delta$$

where A is the azimuth and δ the declination.

parallel Describing lines, curves, planes, or surfaces that are always equidistant, and that will never meet no matter how far they are produced. Parallel lines and curves must both lie in the same plane.

parallel axes theorem A result that relates moments of inertia of a body: if the moment of inertia of a body about an axis through its centre of mass is I, then the moment of inertia about any other axis parallel to it is $I + Md^2$, where M is the mass of the body and d is the distance between the axes.

parallelepiped A *prism all of whose faces are parallelograms. A *right parallelepiped* has lateral faces that are square or rectangular. If the bases are also square or rectangular, it is a *rectangular parallelepiped* or *cuboid*. A parallelepiped in which the lateral edges are not perpendicular to the base is an *oblique parallelepiped*. The

parallelepiped An oblique parallelepiped.

volume of a parallelepiped is the distance between the bases (the *altitude*) multiplied by the area of a base. The total surface area of a rectangular parallelepiped (lateral area + bases) is

$$2(ab + bc + ca)$$

where a, b, and c are the lengths of the sides, and the volume is abc.

A *parallelotope* is a parallelepiped whose sides a, b, and c are in the ratio $4:2:1$.

parallel of latitude A line of *latitude.

parallelogram A *quadrilateral that has both pairs of opposite sides equal. The area of a parallelogram is the length of any side multiplied by the perpendicular distance from that side to the opposite side.

parallelogram law The law stating that if the two shorter sides and the two longer sides of a *parallelogram represent the magnitude and direction of two *vectors **a** and **b**, then the sum of these vectors, i.e. their resultant **c**, is represented by the diagonal of the parallelogram (*see* diagram). It can be seen that

$$\mathbf{a} + \mathbf{b} = \mathbf{b} + \mathbf{a}$$

i.e. vector addition is commutative. Applied to velocities, the result is a *parallelogram of velocities*; to forces, a *parallelogram of forces*.

parallelogram law

parallelotope *See* parallelepiped.

parallel postulate The fifth postulate of *Euclidean geometry, often stated as: for

a given point outside a given line, only one line can be drawn through the point parallel to the given line. This statement of the postulate is sometimes called *Playfair's axiom* and is not the original version given by Euclid. *See* non-Euclidean geometry.

parallel transversal theorem *See* intercept theorem.

parameter 1. A *constant or *variable that distinguishes special cases of a general mathematical expression. For example, the general form of the equation for a line,

$$y = mx + c$$

contains parameters m and c, representing the gradient and y-intercept of any specific line. *See* parametric equations.
2. In statistics, the term usually refers to a constant occurring in the frequency function associated with a specific family of distributions, or to constants that determine the precise form of the deterministic component of a model that also contains a random element. Examples of the former are the parameters μ and σ^2 for the *normal distribution; λ for the *Poisson distribution; and n and p for the *binomial distribution. Examples of the latter are the parameters α, β_1, and β_2 in the linear *regression model

$$\mathrm{E}(Y|x_1, x_2) = \alpha + \beta_1 x_1 + \beta_2 x_2$$

parametric equations Equations that determine the *coordinates of points on a curve in terms of a single common *variable. In two-dimensional Cartesian coordinates, if the parameter is p the equations have the form $x = \mathrm{f}(p)$ and $y = \mathrm{g}(p)$. For instance, the circle

$$x^2 + y^2 = 16$$

has parametric equations

$$x = 4\cos\theta \quad \text{and} \quad y = 4\sin\theta$$

Each value of θ over the range $0°–360°$ determines a point on the circle. In this case the parameter θ is the angle the radius makes with the *x*-axis. The standard parametric equations of the *ellipse

$$\frac{x^2}{a^2} + \frac{y^2}{b^2} = 1$$

are

$$x = a\cos\phi \quad \text{and} \quad y = b\sin\phi$$

and depend on the eccentric angle ϕ of the point on the ellipse.

Parametric equations can also be used to give coordinates of points on a *surface in terms of two common variables. More generally, the term describes any system of equations that gives one set of equations explicitly in terms of a second set of independent variables (the parameters). *See* line; plane; hyperbola; parabola.

parametric form *See* line.

parametric methods Methods of *inference about *parameters based on the assumption that observed data constitute a *random sample from a specified family of distributions, as distinct from *nonparametric or *distribution-free methods. For data that are samples from a continuous distribution, parametric inference is often based on the *normal distribution or the *exponential distribution, while for discrete data it is often based on the *binomial distribution, *Poisson distribution, or *multinomial distribution. Because of the *central limit theorem, inference based on assumptions of normality are often not misleading, even when the data indicate some departure from normality, but in cases of doubt a *nonparametric method or use of the *bootstrap is often preferable.

parametrization The description of a curve, surface, etc. by *parametric equations.

Pareto distribution (V.F.D. Pareto, 1897) A *random variable X with *frequency function

$$\mathrm{f}(x) = \alpha k^{\alpha} x^{-(\alpha+1)}, \quad x \geqslant k$$

where k and α are positive parameters has a Pareto distribution. Pareto proposed the distribution because he believed that the cumulative *distribution function gave a good approximation to the proportion of incomes in a population that were less than x. Variation between regions or countries may be accounted for by changing the parameters k and α.

parity Two integers that are both odd or both even have *even parity*. If one is odd and the other even they have *odd parity*.

parity check matrix A *matrix $\mathbf{H}$ whose rows form a basis for the *dual code of a code C defined over a *Galois field F. A codeword $\mathbf{x}$ is in C if and only if the matrix product $\mathbf{H}\mathbf{x}^{\mathbf{T}}$ evaluated in the field F equals the zero vector.

parsec Symbol: pc. A unit of length used in astronomy, equal to the distance at which a baseline of 1 *astronomical unit subtends an angle of 1 second. 1 parsec = 3.085×10^{16} metre or approximately 3.26 light years. The name is a contraction of 'parallax second'.

partial correlation coefficient The *correlation coefficient between two *random variables in a conditional distribution when one or more other variables are held fixed. For example, if X_1 represents height, X_2 weight, and X_3 age, a high positive correlation between X_1 and X_2 may partly reflect the high positive correlation of each with X_3. The partial correlation coefficient eliminates the effect of age. The partial correlation coefficient between X_1 and X_2 with X_3 fixed is written as $r_{12.3}$.

partial derivative The rate of change of a *function of several variables with respect to one of the variables involved, the other variables being treated as constants. If $u = \mathrm{f}(x, y, z, \ldots)$, the partial derivative of u with respect to x is the rate of change of u when x increases, written as $\partial u/\partial x$, with y, z, $\ldots$ held constant. For example, if

$V = \pi r^2 h$, then

$$\frac{\partial V}{\partial r} = 2\pi r h \quad \text{and} \quad \frac{\partial V}{\partial h} = \pi r^2$$

See also total differential.

partial differential equation (PDE) *See* differential equation.

partial differentiation The *differentiation of a *function of more than one variable with respect to one variable, the others being treated as constants; i.e. the process of finding *partial derivatives.

partial fractions Fractions whose algebraic sum is a given fraction. For instance, $\frac{1}{2}$ and $\frac{1}{3}$ are partial fractions of $\frac{5}{6}$ since

$$\frac{5}{6} = \frac{1}{2} + \frac{1}{3}$$

The *decomposition* of a given fraction into partial fractions is achieved by first factorizing the denominator. For example, the fraction $(x+3)/(x^2 + 3x + 2)$ can be put in the form

$$\frac{A}{x+2} + \frac{B}{x+1}$$

A and B are found by putting this expression in the form

$$\frac{A(x+1) + B(x+2)}{x^2 + 3x + 2}$$

Then

$$x + 3 = (A + B)x + (A + 2B)$$

Coefficients of like powers are equated to give

$$A + B = 1 \quad \text{and} \quad A + 2B = 3$$

i.e. $B = 2$ and $A = -1$. The partial fractions are thus $-1/(x+2)$ and $2/(x+1)$. Decomposition into partial fractions is a method of simplifying certain expressions for integration (*see* integration by partial fractions).

partial order A relation $\leqslant$ between the elements of a *set S that satisfies the following three conditions:

(1) *Reflexive condition*: $a \leqslant a$ for each a in S.

(2) *Antisymmetric condition*: for a and b in S, $a \leqslant b$ and $b \leqslant a$ can both hold only if $a = b$.

(3) *Transitive condition*: if a, b, and c are in S, then $a \leqslant b$ and $b \leqslant c$ together imply $a \leqslant c$. If $b \leqslant a$, then also $a \geqslant b$; and if $a \leqslant b$ but $a \neq b$ then $a < b$.

An example of a set with a partial order is the set of natural numbers with $n \leqslant m$ if and only if n divides m.

If every pair of elements a and b in the set is *comparable* (i.e. either $a \leqslant b$ or $b \leqslant a$) then the *partially ordered set* (*poset*) is called *totally ordered* or a *chain*. The above partially ordered set of natural numbers is not totally ordered since, for example, 3 and 5 are not comparable. An example of a totally ordered set is the set of real numbers with the relation $\leqslant$ being the ordinary 'less than or equal to' relation.

Another standard way of classifying such sets is to use the concepts of *upper bound* and *lower bound* for some of the elements. An upper bound for a subset S' of the poset S is an element u of S such that $a \leqslant u$ for each a in S'. It is a *least upper bound* (l.u.b.), or *supremum* (sup), for S' if $u \leqslant v$ for every other upper bound v of S'. Similarly, an element l of S is a lower bound for S' if $l \leqslant a$ for each a in S'; and it is a *greatest lower bound* (g.l.b.), or *infimum* (inf), for S' if $k \leqslant l$ for every other lower

partial order

bound k of S'. The poset is a *lattice* if every pair of its elements has both a l.u.b. and a g.l.b., as in the diagram; here each element is represented by a dot, and a line segment joining element a to a higher element b indicates that $a < b$.

Every totally ordered set is a lattice. A poset which is a lattice is the example given above, of the set of the natural numbers with $n \leqslant m$, meaning that n divides m. In that case the l.u.b. of a pair of numbers is their least common multiple, and the g.l.b. is their highest common factor.

partial quotient 1. *See* division.
2. *See* continued fraction.

partial sum (of an infinite series) Any sum of a finite number of consecutive terms in an infinite *series, starting with the first. For the series

$$a_1 + a_2 + \cdots + a_n + \cdots$$

the partial sums are

$$s_1 = a_1$$

$$s_2 = a_1 + a_2$$

$$s_n = a_1 + a_2 + \cdots + a_n$$

If the sequence $s_1, s_2, \ldots, s_n$ of partial sums tends to a limit S as $n \to \infty$, then S is the sum of the infinite series. *See* convergent series.

particle A mathematical concept, used especially in mechanics, of an entity that possesses mass and an observable position in space and time but has negligible size. When a particle is subject to forces, these act at one point. Its kinematic behaviour is completely described by specifying its position vector at each instant. In mechanics, matter is considered to be made up of collections of particles (*see* rigid body). In practice, the theoretical results obtained for a particle are a good approximation when the size of a real body is small compared with the linear dimensions of the

system being studied, as for a planet moving around the sun.

particular integral A *particular solution used, together with the *complementary function, in solving linear *differential equations.

particular solution Any solution of a *differential equation that does not involve arbitrary constants.

partition 1. (of a set) A partition or *dissection* of a *set A is a collection of mutually *disjoint nonempty *subsets of A (i.e. the intersection of any pair of subsets is the empty set) whose *union equals A. For instance, the even numbers and the odd numbers constitute a partition of the set of natural numbers.
2. (of an interval) A partition or *dissection* of an *interval $[a, b]$ is a finite set of points $\{x_0, x_1, x_2, \ldots, x_n\}$ such that

$$a = x_0 < x_1 < x_2 < \cdots < x_n = b$$

3. (of an integer) A representation of a positive integer as a sum of positive integers. For example, the partitions of 4 are 4, $3 + 1, 2 + 2, 2 + 1 + 1,$ and $1 + 1 + 1 + 1$.
4. (of a matrix) A separation of all the elements of a *matrix into a number of matrices of lower order, called *submatrices* or *blocks*. For instance, a partition of the matrix

$$\begin{pmatrix} 1 & 2 & 3 \\ 4 & 5 & 6 \end{pmatrix} \quad \text{is} \quad \left(A \mid B \right)$$

where

$$A = \begin{pmatrix} 1 & 2 \\ 4 & 5 \end{pmatrix} \quad \text{and} \quad B = \begin{pmatrix} 3 \\ 6 \end{pmatrix}$$

are the submatrices.

partition function The *function $\mathrm{p}(n)$ that gives the number of *partitions of a *natural number n. Thus $\mathrm{p}(1) = 1$, $\mathrm{p}(2) = 2$, $\mathrm{p}(3) = 3$ and $\mathrm{p}(4) = 5$. For $|q| < 1$ the partition function satisfies

$$\sum_{n=0}^{\infty} \mathrm{p}(n)q^n = \prod_{i=1}^{\infty} \frac{1}{1-q^i}$$

where p(0) is interpreted as 1.

Pascal, Blaise (1623–62) French mathematician and physicist noted for his *Essai pour les coniques* (1640, Essay on Conic Sections), which contained *Pascal's theorem. Later, in 1653, he constructed his arithmetical triangle, and in his final years he described the cycloid and solved the problem of its quadrature. Other work of Pascal's was concerned with probability theory and with the invention of the first calculating machine (1642).

pascal Symbol: Pa. The *SI unit of pressure, equal to the pressure resulting from a force of 1 newton acting uniformly over an area of 1 square metre. [After B. Pascal]

Pascal's theorem The theorem that if a hexagon is inscribed in a *conic, the three points of intersection of opposite pairs of sides all lie on a straight line. The dual theorem (*see* duality) – that the opposite vertices of a hexagon circumscribed about a conic are connected by three lines that intersect in a point – is called *Brianchon's theorem*.

Pascal's triangle A triangular arrangement of numbers as shown above. The numbers give the coefficients for the expansion of $(x+y)^n$. The first row is for $n=0$, the second for $n=1$, etc. Each row has 1 as its first and last number. Other numbers are generated by adding the two numbers

```
            1
          1   1
        1   2   1
      1   3   3   1
    1   4   6   4   1
  1   5  10  10   5   1
1   6  15  20  15   6   1
          etc.
```

Pascal's triangle

immediately to the left and right in the row above.

See also binomial theorem; Zhu Shijie.

path 1. *See* walk.
2. A path (or *arc*) between two points *a* and *b* in a *topological space X is a *continuous map f: $[0, 1] \rightarrow X$ with $\mathrm{f}(0)=a$ and $\mathrm{f}(1)=b$. *See* connected space.

payoff matrix *See* game theory; decision theory.

PDE *Abbreviation for* *partial differential equation.

Peano, Giuseppe (1858–1932) Italian mathematician and logician who developed a clear notation for the new discipline of mathematical logic as well as proposing five simple axioms for number theory (*see* Peano's postulates). He is also remembered for his discovery in 1890 of the space-filling curve now known as *Peano's curve.

Peano arithmetic The form of *number theory based on *Peano's postulates.

Peano's curve A *space-filling curve* discovered by Peano in 1890. It may be

Peano's curve The first three stages of its generation.

developed by first drawing the diagonal of a square. The square is then divided into nine equal squares and certain diagonals are joined, as shown in the diagram. In the next stage each small square is subdivided into nine and, again, certain diagonals are joined. Continuing this process indefinitely gives a curve that passes through every point in the original square.

Peano's curve is a *fractal with *similarity dimension 2, and is an example of a fractal with integral dimension.

Peano's postulates A set of five *axioms, originally formulated by Dedekind, for *number theory:

(1) 0 is a natural number.
(2) Every natural number x has another natural number as its successor (often denoted by $S(x)$ or x').
(3) For all $x, 0 \neq S(x)$.
(4) If $S(x) = S(y)$ then $x = y$.
(5) If P is a property and 0 has P, and whenever a number x has P, then $S(x)$ also has P, and it follows that all numbers have P. This axiom is the principle of *induction.

Pearson, Karl (1857–1936) English mathematician who introduced into statistics such basic concepts as the standard deviation, the coefficient of variation, and the chi-squared test. As the founder of the journal *Biometrika* and its editor from 1901 until his death, Pearson exercised a considerable influence on the manner in which statistics came to be applied to biology.

Pearson distributions Karl Pearson showed that the *frequency function $f(x)$ of many distributions satisfies a differential equation of the form

$$f'(x) = \frac{(x - d)f(x)}{a + bx + cx^2}$$

Setting $a = -1$ and $b = c = d = 0$ gives the standard *normal distribution.

Pearson further divided these distributions into 12 types, based mainly on the nature of the roots of the equation $a + bx + cx^2 = 0$, which determine such features as the range and shape of the distributions.

Pearson's product moment correlation coefficient The product moment correlation coefficient. *See* correlation coefficient.

pedal curve A curve generated from a given curve: the *locus of the feet of perpendiculars from a fixed point to all the *tangents of the given curve.

pedal triangle 1. The triangle formed inside a given triangle by joining the feet of the three lines drawn from each vertex of the given triangle perpendicular to the opposite side. These three perpendiculars (the altitudes of the given triangle) bisect the interior angles of the pedal triangle. This is sometimes called the *orthic triangle*. 2. (of a point with respect to a triangle) The triangle formed by joining the feet of the three perpendiculars from the given point to the sides of the triangle. If the point lies on the *circumcircle of the triangle then the pedal triangle degenerates into a straight line: the *Simson line of the point.

Peirce, Charles Sanders (1839–1914) American mathematician, logician, and philosopher who in 1883, on the basis of earlier work by Boole and De Morgan, developed the first comprehensive formal theory of relations.

Pell's equation The *Diophantine equation

$$x^2 - Ay^2 = 1$$

where A is a positive integer which is not a perfect square. It can be solved by considering the *continued fraction expansion of $\sqrt{A}$. It is named after the English mathematician John Pell (1611–85).

pencil A set of geometrical objects sharing a common property. All the planes passing through a given line form a pencil of planes.

All the circles that lie in the same plane and intersect at two common points form a pencil of circles. All the spheres intersecting in a given circle form a pencil of spheres.

pendulum A body mounted so that it can swing freely about a fixed point under the influence of gravity. The *simple pendulum* is a mathematical model in which a particle of mass m is suspended by a weightless rod of length l and swings in a vertical plane. When the amplitude of the swing, i.e. the angular displacement θ is small, the motion is approximately simple *harmonic and the period of oscillation is

$$T = 2\pi \sqrt{\left(\frac{l}{g}\right)}$$

where g is the *acceleration of free fall.

In an actual pendulum, often called a *compound pendulum*, a *rigid body of convenient shape, such as a bar, swings about a horizontal axis through a point a distance h from the body's *centre of mass. When the amplitude of the swing, θ, is small, the motion is approximately simple harmonic and the period is

$$T = 2\pi \sqrt{\left(\frac{k^2 + h^2}{gh}\right)} = 2\pi \sqrt{\left(\frac{I}{Mgh}\right)}$$

where k is the *radius of gyration about an axis through the centre of mass and parallel to the axis of swing, I is the body's *moment of inertia about the axis of swing, and M is its mass.

pendulum property *See* cycloid.

Penrose tiles (R. Penrose, 1974) Two special tiles whose rules of combination make them *aperiodic, i.e. they can be used to tile the plane but cannot form a *periodic tiling (diagram (a)). The two tiles are a *kite* and a *dart*. Both have two sides of length 1 and two sides of length $\phi = \frac{1}{2}(1 + \sqrt{5})$, the *golden mean. Three angles of the kite are $72°$ and the other angle is $144°$. Two angles of the dart are $36°$, one is $72°$, and the other is $216°$.

Penrose tiles (a)

To form a *Penrose tiling*, the vertices must match: a vertex labelled a cannot be placed next to one labelled b. Diagram (b) shows part of a Penrose tiling.

Penrose tiles (b)

There are three-dimensional analogues that can be used to describe the structure of certain chemical substances.

pentagon A *polygon that has five interior angles (and five sides).

pentagram A symmetrical five-pointed star *polygon formed by drawing all the diagonals of a regular pentagon. See golden section.

pentahedron (*plural* **pentahedra**) A *polyhedron that has five faces. Particular examples are a triangular prism and a square pyramid.

percent Symbol %. Indicating hundredths. A fraction can be expressed as a *percentage* by multiplying it by 100, e.g. $\frac{1}{4}$ is 25 percent. A change in a quantity from a to b is a change of $100(b-a)/a$ percent.

percentage decrease *See* percentage increase.

percentage error The quantity

$$\frac{\text{absolute error}}{\text{true value}} \times 100$$

often applied to a *mean or other measure. If the true value of a population mean is 45 and an estimate of it based on a sample is 47.2, the percentage error of the estimate is

$$\frac{47.2 - 45}{45} \times 100 = 4.9\%$$

to two significant figures.

percentage increase The quantity

$$\frac{\text{increase in value}}{\text{original value}} \times 100$$

often applied to a price or other measure. For example, if the price of an item increases from £400 to £500, then the percentage increase is $\frac{500-400}{400} \times 100 = 25\%$.

Similarly, a *percentage decrease* is defined to be

$$\frac{\text{decrease in value}}{\text{original value}} \times 100$$

Thus if the value of an item is reduced from £500 to £400, then the percentage decrease is $\frac{500-400}{500} \times 100 = 20\%$.

percentage point A term used mainly in connection with statistical tables for significance testing. If a *statistic T has a continuous distribution such that $\Pr(T < t_L) = \theta/100$, then t_L is the lower θ percentage point. If $\Pr(T > t_U) = \theta/100$ then t_U is the upper θ percentage point. For example, if T has a standard normal distribution, the lower and upper 5 percentage points relevant to one-tail tests are -1.64 and 1.64. Values outside the interval $(-1.64, 1.64)$ indicate significance at the 5% level in an appropriate one-tail *hypothesis test.

The term is also used in association with two-tail tests, especially when a statistic T has a symmetric distribution with zero mean. In this case the value t such that $\Pr(|T| > t) = \theta/100$ is the θ percentage point. Thus, if T has a standard normal distribution, the 5 percentage point for a two-tail test is 1.96, implying significance at the 5% level if T lies outside the interval $(-1.96, 1.96)$. There is a close connection between percentage points and percentiles. *See* quantiles.

percentile *See* quantile.

perfect code A type of *error-correcting code that is particularly effective. A k-perfect code has the maximum possible number of codewords such that it is k-error correcting. A perfect code may well have Hamming distance less than $2k + 1$ between its codewords.

perfect number A natural number that is equal to the sum of its *proper divisors. Thus, the number 6 has proper divisors 1, 2, and 3, which add to give 6. The first four perfect numbers are 6, 28, 496, and 8128. It is known that if $2^n - 1$ is prime, then n is prime and $2^{n-1}(2^n - 1)$ is a perfect number. All even perfect numbers are of this type; it is not known whether there are any odd perfect numbers. Numbers for which the sum of their proper divisors is less than the number are called *deficient* or *defective numbers*; ones for which the sum exceeds the number are *abundant numbers*. *See also* amicable numbers; Mersenne numbers.

pericycloid An *epicycloid in which the rolling circle encloses the fixed circle.

perigon (round angle) An angle equal to one complete turn (360° or 2π radians).

perimeter The length of a *closed curve. The curve may be a smooth curve (e.g. an ellipse or circle) or a broken curve (e.g. a polygon).

period Symbol: T. The time taken to make one complete *oscillation or cycle. If a particular form of motion is represented by

$$x = a \cos(\omega t + \alpha)$$

the motion repeats itself after a time $2\pi/\omega$, where ω is the *angular frequency; this is the period T of the motion, the motion being described as *periodic*. The constants a and α are the *amplitude and initial *phase of the motion. *See also* periodic function.

period doubling The change in the periodic orbits of a parametrized family of transformations in which a periodic orbit of period n becomes a pair of periodic orbits of period $2n$. This occurs in the flip *bifurcations, and is related to the *Feigenbaum number.

periodic decimal *See* decimal.

periodic function A *function f of a real variable x for which there exists a number a (> 0) such that $f(x + a) = f(x)$ for all x; a is a period of f and the least possible period is called the *fundamental period* or simply the *period* of f. For example, sin x is a periodic function with period 2π since $\sin(x + 2\pi) = \sin x$ for all x.

periodic motion Any to-and-fro motion that is repeated in an identical manner at regular intervals. The duration of these intervals is the *period of the oscillation.

periodic point A point x which first returns to its initial position after n applications of an *iterated map T is a periodic point for T of period n. The set of points $\{x, T(x), T^2(x), \ldots, T^{n-1}(x)\}$ is called the *periodic orbit* of the point. For example, $z = i$ is a point of period 2 for the map $T: z \mapsto z^3$ of the complex plane, because $T(i) = -i$ and $T(-i) = i$; it has periodic orbit $\{i, -i\}$.

periodic tiling A tiling or tessellation of the plane that can be translated in two different nonparallel directions without essential change. A tiling that cannot be so translated without change is said to be a *nonperiodic tiling*. The three-dimensional analogue of periodic tiling is fundamental in *crystallography. *See* aperiodic tiling.

permanent For a square *array, the sum of all possible products of numbers (*elements) in which each product contains exactly one number from each row and each column. The permanent of the matrix

$$\begin{pmatrix} 1 & 2 & 3 \\ 4 & 5 & 6 \\ 7 & 8 & 9 \end{pmatrix}$$

is

$$1 \times 5 \times 9 + 1 \times 6 \times 8 + 2 \times 4 \times 9$$
$$+ 2 \times 6 \times 7 + 3 \times 4 \times 8 + 3 \times 5 \times 7$$
$$= 450$$

The permanent differs from the determinant in that signs are not associated with the products being summed. The permanent is much more expensive to compute than the determinant.

permutation 1. The number of ways of selecting $r \leqslant n$ objects from n distinguishable objects when order of selection is important; denoted by ^nP_r or $_nP_r$. Since the first may be chosen in n ways, the second in $n - 1$ ways, the third in $n - 2$ ways, and so on,

$$^nP_r = n(n-1)(n-2)\cdots(n-r+1)$$

When $r = n$, $^nP_n = n!$ The permutations of two objects from four objects A, B, C, and D are AB, AC, AD, BC, BD, CD, BA, CA, DA, CB, DB, and DC. Note that the second six are simply the first six reversed. *See* combination.

2. A one-to-one mapping (*see* one-to-one function) of a *set of elements onto itself. In this sense the permutation is regarded as an operation that may involve rearranging the members of the set. For a set of three items a_1, a_2, and a_3, the notation

$$\begin{pmatrix} 1 & 2 & 3 \\ 3 & 1 & 2 \end{pmatrix}$$

indicates a permutation in which a_1 is replaced by a_3, a_2 by a_1, and a_3 by a_2. This type of permutation, in which each member of the set replaces a successive member, is a *circular* (or *cyclic*) permutation (*see* diagram). Permutations can be regarded as a combination of *transpositions* of pairs of members of the set. Any permutation can always be effected either by an even number or an odd number of transpositions (but not both). In the former case the permutation is an *even permutation*, and in the latter case it is an *odd permutation*. The example above is even (two transpositions: (12) then (23)).

A *permutation* (or *substitution*) *group* is a *group whose elements are permutations, where combination of two permutations is interpreted as applying them successively. In particular, if there are n members of a set, the total number of permutations is $n!$, and these form a permutation group. For example, the six ($= 3!$) permutations of three members of a set are

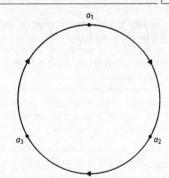

permutation A cyclic permutation.

$$P_1 = \begin{pmatrix} 1 & 2 & 3 \\ 1 & 2 & 3 \end{pmatrix}, \quad P_2 = \begin{pmatrix} 1 & 2 & 3 \\ 1 & 3 & 2 \end{pmatrix}$$

$$P_3 = \begin{pmatrix} 1 & 2 & 3 \\ 2 & 1 & 3 \end{pmatrix}, \quad P_4 = \begin{pmatrix} 1 & 2 & 3 \\ 2 & 3 & 1 \end{pmatrix}$$

$$P_5 = \begin{pmatrix} 1 & 2 & 3 \\ 3 & 1 & 2 \end{pmatrix}, \quad P_6 = \begin{pmatrix} 1 & 2 & 3 \\ 3 & 2 & 1 \end{pmatrix}$$

Here P_1 is the identity element of the group. The product of two members is itself a member: for example, $P_4 P_2 = P_6$. Each permutation has an inverse: for example, $P_4 P_5 = P_1$. The combination is associative.

A permutation group of all the permutations of a set (i.e. a group of order $n!$ when n is the number of members) is a *symmetric group*. A group of all the even permutations (of order $n!/2$) is an *alternating group*. A permutation group of order n (the same as the number of elements in the set) is a *regular group*. *See also* Cayley's theorem; group; permutation matrix.

permutation group *See* permutation.

permutation matrix A square *matrix having an element equal to 1 in each row (or column), the other elements in the row being zero, used to represent a given *permutation. For example, for permutation of three items, 1, 2, and 3:

$$\begin{pmatrix} 0 & 1 & 0 \\ 0 & 0 & 1 \\ 1 & 0 & 0 \end{pmatrix} \begin{pmatrix} 1 \\ 2 \\ 3 \end{pmatrix} = \begin{pmatrix} 2 \\ 3 \\ 1 \end{pmatrix}$$

Here

$$\begin{pmatrix} 0 & 1 & 0 \\ 0 & 0 & 1 \\ 1 & 0 & 0 \end{pmatrix}$$

is the permutation matrix mapping 123 into 231, i.e. the permutation matrix of the permutation

$$\begin{pmatrix} 1 & 2 & 3 \\ 2 & 3 & 1 \end{pmatrix}$$

For a given permutation group, the corresponding permutation matrices form an isomorphic group under matrix multiplication.

permutation test (E.J.G. Pitman, 1937) A type of *distribution-free *hypothesis test. For example, if the null hypothesis is that two independent samples of 5 and 7 observations are from identical populations and the alternative hypothesis is that the population distributions differ only in their means, a test may be based on a *statistic d, the absolute difference between the sample means. Large values of d support rejection of the null hypothesis. Suppose that $d = 3.5$. If the null hypothesis holds, the $5 + 7 = 12$ observations may be treated as a pooled sample from one population. The value of d is calculated for all $\binom{12}{5} = 792$ equally likely sample pairs of 5 and 7 obtainable by *random sampling without replacement from the pooled sample. If $d \geqslant 3.5$ for, say, 30 of the 792 pairs, then $p = 30/792 = 0.0379$ is the *p-value that gives the exact probability of making an error of the first kind if the null hypothesis is rejected when the observed $d = 3.5$. A p-value may be calculated for any observed d for any sample sizes m and n, and appropriate computer software enables exact p-

values to be quickly computed for small or medium values of m and n, and gives asymptotic approximations for large values of m and n.

The permutation test analogues of one- or two-sample *t-tests are sometimes called *Pitman tests*. The idea extends to location tests based on rank and other transformations, and to any number of samples and to tests about *correlation, *dispersion, and other properties. Permutation tests are sometimes called *randomization tests*, but some writers restrict the latter term to tests using the original data only.

perpendicular A line or plane that is at right angles to another line or plane; a normal.

perpendicular axes theorem A result that relates the *moments of inertia of a *lamina about three mutually perpendicular axes. If the moments of inertia of a lamina about two perpendicular axes Ox and Oy in its plane are I_x and I_y, respectively, then the moment of inertia about an axis through O perpendicular to the plane of the lamina is $I_x + I_y$.

perspective Two planar figures whose points can be put in a *one-to-one correspondence in such a way that the lines joining pairs of corresponding points pass through a common point P are in *perspective from a point*, and P is the *centre of perspective*. If the correspondence is such that pairs of corresponding lines meet in points lying on a common line, then the figures are in *perspective from a line* and that line is the *axis of perspective*.

A consequence of *Desargues's theorem is that two planar figures which are in perspective from a point are also in perspective from a line, and vice versa.

PERT *See* critical path analysis.

peta- *See* SI units.

Peurbach, Georg (1423–61) Austrian mathematician and astronomer who produced in his *Theoricae novae planetarum*

(1454, New Theory of the Planets) a popular description of the Ptolemaic system. He was also responsible for an influential table of sines and chords published posthumously in 1541.

p-group A *group in which the *order of each element is a power of the *prime number p. A finite group is a p-group if and only if its *order is a power of the prime p.

phase 1. (of a periodic phenomenon) For a particular value of the independent *variable, the part or fraction of the *period through which the variable has advanced, as measured from some arbitrary origin. **2.** *See* harmonic motion.

phase space Given a *flow

$$x(t) = (x_1(t), \ldots x_n(t))$$

describing a solution to a differential equation in n-dimensional Euclidean space, phase space is the space of all the vectors of the form

$$(x_1(t), \dot{x}_1(t), \ldots, x_n(t), \dot{x}_n(t))$$

in $2n$-dimensional Euclidean space. For example, a solution to the harmonic motion equation $\ddot{x} + \omega^2 x = 0$ is the flow $x = a \sin \omega t$, where a and ω are positive constants. Since $\dot{x} = a\omega \cos \omega t$, the phase space is the set of all points $(a \sin \omega t, a\omega \cos \omega t)$, i.e. an ellipse, as shown in the diagram.

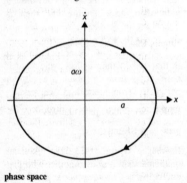

phase space

phi function *See* Euler's phi function.

physical quantity A characteristic of matter or energy, instances of which can be reproduced and quantified. The physical quantity itself is defined by specifying the method used to measure the ratio of two magnitudes of the quantity. If one of these magnitudes is taken as a standard, any other instance of that physical quantity can be expressed in terms of that standard. The standard so obtained is called a unit of measurement. For example, the physical quantity called 'mass' can be defined by specifying the way in which two masses are compared using a simple balance. If one of the masses is taken as a standard and given a name (such as 'kilogram' or 'pound') any other mass can be expressed in terms of this unit. In general, the magnitude of a physical quantity is the product of a number and a unit.

pi Symbol: π. The ratio of the length of the circumference of a circle to its diameter. The symbol π was first used in this sense by the Welsh writer William Jones in 1706. Ancient approximations to π include 3 (Old Testament), 25/8 (Babylonian), 256/81 (Egyptian), 22/7 (Greek), 355/113 (Chinese), and $\sqrt{10}$ (Indian). In 1429 the Arab mathematician Al-Kashi calculated a value of π correct to 16 decimal places. At present, over a trillion digits of π are known. The problem of *squaring the circle is equivalent to finding a construction for π that uses only unmarked straightedge and compasses.

Pi was proved to be *irrational by Lambert in 1767, and *transcendental by Lindemann in 1882.

π(x) Symbol for the *function that gives the number of *prime numbers less than or equal to the real number x. Thus $\pi(10) = 4$, and $\pi(1000) = 168$.

Picard iteration An iterative method for finding the numerical solution of ordinary differential equations. The proof that it is effective is usually based on the *contraction mapping principle. The method was first introduced by Charles Émile Picard (1856–1941).

Pick's theorem (G. Pick, 1899) If the vertices of a polygon are points of the integer *lattice, then the area A of the polygon is given by the formula $A = i + \frac{1}{2}b - 1$, where i is the number of points in the interior of the polygon and b is the number of points on the boundary, including the vertices.

Pick's theorem For this polygon, $i = 2$, $b = 6$, and $A = 2 + 3 - 1 = 4$.

pico- *See* SI units.

pictogram A visual representation of statistical information using drawings or pictures – *icons* – of a relevant nature to indicate data patterns. Suitable icons are used to indicate appropriate units, and the information is presented in what is essentially a pictorial *bar chart. The diagram shows a pictogram for numbers of cars produced by a factory in each of four months. Each 🚗 represents 100 cars. Part

pictogram of factory output of cars per month, each complete symbol representing 100 cars.

symbols give a crude representation of a proportion of 100 cars.

PID *Abbreviation for* *principal ideal domain.

piecewise continuous *See* continuous function.

pie chart A circle with sectors marked with areas representing the proportion of units in each of a set of given categories. For example, if 50 percent of the people on a beach were children, 25 percent adult males, and 25 percent adult females, a semicircle would represent the children and two quarter circles the adult males and females.

piercing point *See* trace.

pigeonhole principle (Dirichlet's principle) If n objects are put into p pigeonholes, where $1 \leqslant p < n$, then some pigeonhole must contain at least two objects. An alternative formulation is that if n objects are coloured with $p < n$ colours, then at least two of them have the same colour.

pint 1. An *imperial unit of capacity or volume equal to $\frac{1}{8}$ of a *gallon.
2. A unit of liquid measure in the *United States customary system equal to $\frac{1}{8}$ of a US *gallon.

pitch 1. *See* helix.
2. Angular movement of an aircraft, spacecraft, projectile, etc. about a horizontal axis at right angles to the direction of motion. *Compare* roll; yaw.

Pitman, Edwin James George (1897–1993) Australian statistician best known as a pioneer in distribution theory and inference, especially for the development of exact tests based on permutation or randomization theory. He also introduced the concept of asymptotic relative efficiency as a guide for indicating which of two or more statistical procedures might be preferable under specific distributional assumptions.

Pitman tests *See* permutation test.

pivotal condensation *See* Gaussian elimination.

plaintext In cryptology, the text to which the *cipher is to be applied; that is, the original message.

planar graph *See* graph.

plane 1. A surface such that a (straight) line that joins any two points of the surface lies in the surface. The *general form* of the equation of a plane in Cartesian coordinates is

$$Ax + By + Cz + D = 0$$

The *normal form* is

$$lx + my + nz = p$$

where l, m, and n are the *direction cosines of the normal from the origin and p is the length of this normal.

The equivalent *vector form* is $\mathbf{r}.\mathbf{n} = p$, where $\mathbf{r}$ is the position vector of a point on the plane and $\mathbf{n}$ is a unit vector normal to the plane.

The *intercept form* is

$$\frac{x}{a} + \frac{y}{b} + \frac{z}{c} = 1$$

a, b, and c being intercepts on the x-, y-, and z-axes, respectively.

The equation of a plane that passes through three points (x_1, y_1, z_1), (x_2, y_2, z_2), and (x_3, y_3, z_3) is (*see* determinant):

$$\begin{vmatrix} x & y & z & 1 \\ x_1 & y_1 & z_1 & 1 \\ x_2 & y_2 & z_2 & 1 \\ x_3 & y_3 & z_3 & 1 \end{vmatrix} = 0$$

The plane that passes through the three points with position vectors $\mathbf{a}$, $\mathbf{b}$, and $\mathbf{c}$ has the parametric equation

$$\mathbf{r} = \mathbf{a} + \lambda(\mathbf{b} - \mathbf{a}) + \mu(\mathbf{c} - \mathbf{a})$$

2. Lying entirely in one plane, as in *plane curve*.

plane of symmetry A plane about which a geometrical figure is symmetrical. A geometric figure has a plane of symmetry if every point in the figure has a corresponding point in the figure such that the plane bisects at right angles the line segment joining the points. *See also* reflection.

plane section *See* section.

planetary motion, laws of *See* Kepler's laws.

plane trigonometry *See* trigonometry.

plastic Describing a material that has been stretched beyond its range of *elasticity: when the *stress is removed the material cannot return to its original shape but assumes a permanent deformation. The ability to undergo such an irreversible deformation without fracturing is referred to as *plasticity*. *See* Hooke's law.

Plateau problem The problem of finding the minimum surface that is bounded by a given twisted curve. The problem is named after the Belgian physicist Joseph Antoine Ferdinand Plateau (1801–83), who experimented with soap films on wire formers.

Plato (*c*.428–348 BC) Greek philosopher whose name has become identified with the view that mathematical objects have a real existence independent of human thought. His name is also linked with the five regular polyhedra or Platonic solids – the tetrahedron, cube, octahedron, dodecahedron, and icosahedron (*see* polyhedron) – first described by him in *Timaeus*. Plato's insistence that mathematics be an essential part of the education of the guardians of his ideal Republic did much to establish the high reputation of mathematics in Western civilization.

platykurtic *See* kurtosis.

Playfair, John (1748–1819) Scottish mathematician noted for the proposal in his

Elements of Geometry (1795) of an alternative version of Euclid's *parallel postulate since known as *Playfair's axiom*.

Plimpton 322 An Old Babylonian cuneiform tablet in the Plimpton Collection at Columbia University. It was excavated at the ancient Iraqi city of Larsa and is dated to around 1800 BC. It is a table of numbers with 15 rows and 4 columns. Allowing for errors, the middle two numbers in each row are pairs such as 119 and 169 or 4601 and 6649, belonging to *Pythagorean triples. Several theories have been proposed to explain the composition and purpose of the tablet.

plot *See* experimental design.

Plücker, Julius (1801–68) German mathematician noted for his *Analytisch-geometrische Entwicklungen* (2 vols, 1828–31, Developments in Analytic Geometry). He proposed taking straight lines rather than points as the fundamental elements of the coordinate system, formulated the principle of duality, and introduced much of the modern notation. In 1835 Plücker published the first complete classification of plane cubic curves.

plug-in estimator A statistic calculated from a sample that is the sample-distribution analogue of the population-distribution parameter or characteristic it is estimating. For a random sample of n observations x_i, $i = 1, 2, \ldots, n$, the sample mean $\bar{x}$ is the plug-in estimator of the population mean μ, and the sample variance $\Sigma_i (x_i - \bar{x})^2 / n$ is the plug-in estimator of the population variance, σ^2. The latter is a biased estimator of σ^2; however, for a normal distribution it is the *maximum likelihood estimator. Plug-in estimators are widely used with the *jack-knife and the *bootstrap.

plus/minus sign The sign $\pm$, sometimes called the *plus or minus sign*. It is a shorthand way of writing the *plus and *minus signs together. For example, if $x^2 = 49$

then we can write $x = \pm 7$, meaning that the equation has the two solutions $x = +7$ and $x = -7$. We could also write $a = b \pm 2$ to mean that a and b are related by $a = b + 2$ or $a = b - 2$.
2. A way of indicating the precision of an observation. For instance, the value of an observed quantity might be given as 6.0 ± 0.3 units, meaning that the true value lies somewhere between $6.0 - 0.3 = 5.7$ units and $6.0 + 0.3 = 6.3$ units.

plus sign 1. The sign $+$ used to denote a *positive number such as $+7$. It was first used in this sense by Johannes Widmann in 1489.
2. The sign $+$ denoting *addition, as in $5 + 3$. It was first used in this sense by Henricus Grammateus in 1518. *See also* minus sign; plus/minus sign.

Poincaré, Jules Henri (1854–1912) French mathematician noted for his investigations in the 1880s of automorphic functions. Poincaré also made substantial contributions to the three- and n-body problems in his *Les Méthodes nouvelles de la mécanique céleste* (3 vols, 1892–9, New Methods in Celestial Mechanics), while other work of influence in astronomy was his later study of rotating fluid bodies. With over 500 published memoirs, Poincaré contributed to most branches of mathematics and physics, including thermodynamics, relativity, divergent series, probability theory, set theory, and topology, while in his less technical writings Poincaré sought to develop an intuitive view of mathematics and science.

Poincaré conjecture The conjecture that, if M is an n-manifold (*see* manifold) and M is homotopy-equivalent to the n-sphere S^n, then M is homeomorphic to S^n. The Poincaré conjecture has long been known to be true for $n = 1$ or 2, and was proved for $n \geqslant 5$ by S. Smale in 1960 and by M. Freedman for $n = 4$ in 1982. It was proved true for $n = 3$ by Grigori Perelman in 2004. He used the *Ricci flow*, a method that uses properties of solutions of certain *partial differential equations.

Poincaré duality theorem Let M be an n-manifold, with nth *homology group $H_n(M)$ infinite and cyclic (such a *manifold is said to be *orientable*). The Poincaré duality theorem states that, for such M, $H_r(M)$ is isomorphic to the $(n-r)$th *cohomology group $H^{n-r}(M)$, for all r. Its essence is that, in n-dimensional space, two subspaces of dimensions r and $n-r$ will usually meet at a single point.

Poincaré group *See* homotopy group.

Poinsot, Louis (1777–1859) French mathematician who in his *Éléments de statique* (1803) made a major contribution to mechanics by showing how anomalies in the application of the parallelogram of forces could be removed by introducing the notion of a *couple and that of a *wrench. The axis of a wrench is sometimes called the *Poinsot central axis*. He also developed a theory of *regular star polygons, and in 1809 added two regular star polyhedra to the two discovered in 1619 by Kepler. In 1812 Cauchy showed there could be only four, and they are now known as the *Kepler–Poinsot solids*.

point An element of geometry having position but no magnitude. A point in three-dimensional space is defined by its coordinates (x, y, z).

point estimate *See* estimation.

point group *See* crystallography.

point of contact A point of *tangency.

point of inflection *See* inflection.

point of osculation A point at which two branches of a curve have a common tangent so as to form a double *cusp of the first kind.

point–slope form *See* line.

Poisson, Siméon-Denis (1781–1840) French mathematician, a student of Laplace and Lagrange. He is well known for his work on probability theory and for discovering in 1837 the *Poisson distribution. He worked in this area mainly towards the end of his life; he had earlier established a reputation in celestial mechanics, and also in electricity and magnetism, where his work on integrals and Fourier series found many applications.

Poisson distribution A discrete *random variable X with *frequency function

$$\Pr(X = r) = \frac{1}{r!}\, e^{-\lambda} \lambda^r$$

where $r = 0, 1, 2, \ldots$. The mean and variance are both λ. The *binomial distribution tends to the Poisson distribution when $n \to \infty$, $p \to 0$, and $np = \lambda$.

Poisson process A *stochastic process in which events occur at random, in the sense that the distribution of the number of events occurring in any time interval depends only on the length of that interval and has a *Poisson distribution with mean λt, where t is the length of the interval and λ a constant. This is one of the simplest stochastic processes and is often used as a first approximation to describe, for example, traffic flow past an observation point on a motorway or the distribution of initiation time of calls in a telephone system over time periods when call density is reasonably constant. The distribution of the time elapsing between events such as cars passing a given point is the *exponential distribution if the process is a Poisson process. *See also* gamma distribution.

Poisson's ratio Symbol: σ. The ratio of lateral *strain to longitudinal strain in a body under tensile or compressive *stress, i.e. when a force of tension or compression is applied to its ends.

polar 1. A straight line associated with a *conic and a point P (the *pole*). Let a variable secant or chord through P cut the conic at L and M (*see* diagram (a)). The

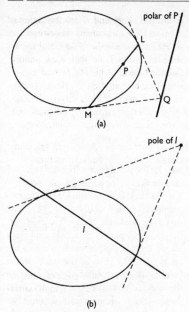

(a)

(b)

polar (a) Polar and (b) pole.

tangents to the conic at L and M meet at Q. Then Q always lies on a particular straight line – the polar of the point P.

2. A straight line joining the points of contact of the *tangents (real or imaginary) that can be drawn from a point to a conic is the *polar* of that point with respect to the conic.

The point of intersection of the tangents to the conic at the (real or imaginary) points of intersection of the conic and a straight line *l* is called the *pole* of that line with respect to the conic (*see* diagram (b)).

polar angle *See* polar coordinate system.

polar axis *See* polar coordinate system.

polar coordinate system A *coordinate system in which the position of a point is determined by the length of a line segment from a fixed origin together with the angle or angles that the line segment makes with a fixed line or lines. The origin is called the

pole and the line segment is the *radius vector* (*r*). In two dimensions, one reference axis is required (called the *polar axis*). The angle θ between the polar axis and the radius vector is called the *vectorial angle* (other terms are *polar angle*, *azimuth*, *amplitude*, and *anomaly*). By convention, positive values of θ are measured in an anticlockwise sense, negative values in a clockwise sense. The coordinates of the point are then specified as (r, θ). Polar coordinates in a plane are useful for dealing with systems that have central symmetry.

It is possible to change between polar and Cartesian coordinates. If the pole of the polar system coincides with the origin of the Cartesian system, and if the polar axis coincides with the *x*-axis, then a point (r, θ) has Cartesian coordinates given by

$$x = r \cos \theta, \quad y = r \sin \theta$$

For example, the point with polar coordinates $(3, 90°)$ has Cartesian coordinates $(0, 3)$. Similarly, a point (x, y) in a Cartesian coordinate system has polar coordinates given by

$$r = \sqrt{(x^2 + y^2)}, \quad \theta = \tan^{-1}(y/x)$$

where θ is such that

$$x : y : r = \cos \theta : \sin \theta : 1$$

For example, the point with Cartesian coordinates $(-1, -1)$ has polar coord-

polar coordinate system

inates $(\sqrt{2}, 225°)$. Polar coordinate systems are also used in three dimensions.

See spherical coordinate system; cylindrical coordinate system.

polar decomposition A generalization to matrices of the polar form of a *complex number. Any square complex matrix A has a polar decomposition $A = UH$, where U is a *unitary matrix and H is a Hermitian matrix with non-negative *eigenvalues.

polar equation An equation in *polar coordinates. For example,

$$r = 2\cos\theta$$

is the polar equation of the circle with Cartesian equation

$$(x-1)^2 + y^2 = 1$$

polar form *See* complex number.

polar normal *See* polar tangent.

polar tangent In *polar coordinates, the line segment on the *tangent to a curve lying between the point of contact and the intersection with a line through the pole perpendicular to the radius vector of the point of contact. The *polar normal* is the line segment on the normal between the point of contact and the intersection with the perpendicular through the pole. The projections of the polar tangent and polar normal on this perpendicular are the *polar subtangent* and *polar subnormal*, respectively.

polar triangle A triangle constructed from the *poles of a given *spherical triangle. For a given triangle ABC, the arc BC has two poles. The pole nearest A is taken (say A'). Similarly, B' is the pole of AC nearest B, and C' the pole of AB nearest C. The spherical triangle $A'B'C'$ is the polar triangle of ABC. The converse is also true: ABC is the polar triangle of $A'B'C'$.

A relationship holds between the angles (or sides) of a spherical triangle and the sides (or angles) of its polar triangle, as

follows. If A, B, and C are the angles of ABC and a, b, and c are its sides (a opposite A, etc.) and similarly A', B', and C' are the angles of $A'B'C'$ with a', b', and c' the sides, then:

$$A = 180° - a'$$

$$A' = 180° - a$$

$$B = 180° - b', \ldots$$

pole 1. The point from which distances are measured in a *polar coordinate system.
2. (of a circle on a sphere) One of the two points at which a diameter of the sphere perpendicular to the plane of the circle cuts the sphere. A pole of an arc on a sphere is one of the poles of the circle of which the arc is part. The poles of the earth are the poles of the geographical equator. Poles on the celestial sphere are poles of great circles on the sphere. *See* celestial equator; ecliptic; horizon; galactic equator.
3. (of a line) *See* polar.
4. (of an analytic function) *See* singular point.
5. (of a projection) *See* stereographic projection.

Polish notation *See* prefix notation.

polyalphabetic substitution cipher A *substitution cipher that uses two or more *alphabets; there is a rule that decides which alphabet is applied to which characters in the plaintext. *Compare* monoalphabetic substitution cipher.

polygon A figure formed by three or more points (vertices) joined by line segments (sides). The term is usually used to denote a closed plane figure in which no two sides intersect. In this case the number of sides is equal to the number of *interior angles. If all the interior angles are less than or equal to 180°, the figure is a *convex polygon*; if it has one or more interior angles greater than 180°, it is a *concave polygon*. A polygon that has all its sides equal is an *equilateral polygon*; one with all its interior

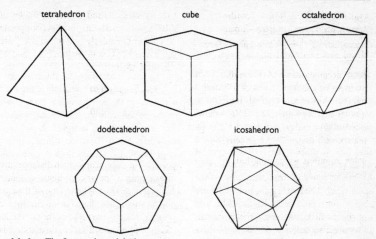

tetrahedron   cube   octahedron

dodecahedron   icosahedron

polyhedron The five regular polyhedra.

angles equal is an *equiangular polygon*. Note that an equilateral polygon need not be equiangular, or vice versa, except in the case of an equilateral triangle. A polygon that is both equilateral and equiangular is said to be *regular*. The *exterior angles of a regular polygon are each equal to $360°/n$, where n is the number of sides.

The distance from the centre of a regular polygon to one of its vertices is called the *long radius*, which is also the radius of the *circumcircle of the polygon. The perpendicular distance from the centre to one of the sides is called the *short radius* or *apothem*, which is also the radius of the *inscribed circle of the polygon.

A *regular star polygon* is a figure formed by joining every mth point, starting with a given point, of the n points that divide a circle's circumference into n equal parts, where m and n are *relatively prime, and $n \geqslant 3$. This star polygon is denoted by $\{n/m\}$. When $m = 1$, the resulting figure is a regular polygon. The star polygon $\{5/2\}$ is the *pentagram.

The term 'polygon' is also applied to figures in spherical and hyperbolic geometry.

polygon of forces *See* triangle of forces.

polyhedral angle A configuration in three dimensions of three or more *half-lines coming from a common point with the planes bounded by the lines. The point is the *vertex*, the half-lines are the *edges*, and the planes are the *faces*. A polyhedral angle is a *solid angle; the plane angles between adjacent edges are *face angles* of the polyhedron. Polyhedral angles are classified according to the number of faces as *trihedral* (three), *tetrahedral* (four), etc.

polyhedron (*plural* **polyhedra**) **1.** A solid with a surface composed of plane polygonal surfaces (*faces*). The sides of the polygons, joining two faces, are its *edges*. The corners, where three or more faces meet, are its *vertices*. Generally, the term is used for closed solid figures (*see* Euler's formula). A *convex polyhedron* is one for which no plane containing a face cuts any other face; otherwise the polyhedron is *concave*.

A *regular polyhedron* is one that has identical (congruent) regular polygons forming its faces and has all its polyhedral angles congruent. There are only five possible convex regular polyhedra (Euclid: Book XIII) (*see* diagram):

tetrahedron – four triangular faces

cube – six square faces
octahedron – eight triangular faces
dodecahedron – twelve pentagonal faces
icosahedron – twenty triangular faces.

The five regular solids played a significant part in Greek geometry. They were known to Plato and are often called the *Platonic solids*. Kepler used them in his model of the solar system.

A *uniform polyhedron* is a polyhedron that has identical *polyhedral angles at all its vertices, and has all its faces formed by regular polygons (not necessarily of the same type). The five regular polyhedra are also uniform polyhedra. Right prisms and antiprisms that have regular polygons as bases are also uniform. In addition, there are thirteen *semiregular polyhedra*, the so-called *Archimedean solids*. For example, the icosidodecahedron has 32 faces – 20 triangles and 12 pentagons. It has 60 edges and 30 vertices, each vertex being the meeting point of two triangles and two pentagons. Another example is the truncated cube, obtained by cutting the corners off a cube. If the corners are cut so that the new vertices lie at the centres of the edges of the original cube, a cuboctahedron results. Truncating the cuboctahedron and 'distorting' the rectangular faces into squares yields another Archimedean solid. Other uniform polyhedra can be generated by truncating the four other regular polyhedra or the icosidodecahedron.

A *regular star polyhedron* is a solid whose faces are all *regular star polygons. There are four regular star polyhedra, the so-called *Kepler–Poinsot solids*.

See also antiprism; polytope; prism; prismatoid; pyramid.

2. *See* combinatorial topology; equi-decomposable.

polynomial A mathematical expression that is a sum of terms, each term being a product of a constant and a non-negative (or zero) power of a variable or variables. For one variable, the general form is

$$a_0 + a_1x + a_2x^2 + \cdots + a_nx^n$$

The highest power (n) of the polynomial is its *degree* or *order*. Polynomials are described as linear, quadratic, cubic, quartic, quintic, etc., according to their degree (1, 2, 3, 4, 5, etc.). The constants a_i are the *coefficients* of the polynomial: a_0 is the *constant term* and a_n is the *leading coefficient*. They may be real or complex. A *polynomial function* is a *function whose values are given by a polynomial. A *polynomial equation* is an equation obtained by setting a polynomial equal to zero. A polynomial in several variables is a sum of terms which are multiples of products of non-negative (or zero) powers of the variables. For example, $5x^2y^2 + 2z - 1$ is a polynomial in x, y, and z of degree four.

polynomial ring A *ring, denoted by $R[x]$, which consists of all *polynomials in x with coefficients from the commutative ring R. Often the ring R will be a *field F, in which case the ring $F[x]$ is a *principal ideal domain.

polynomial time A measure of the *complexity of an algorithm. An algorithm is said to run in polynomial time if the number of elementary operations required to complete its computation can be expressed as a polynomial function of the size of the input.

More formally, an algorithm runs in polynomial time if integers A and k exist such that, for input data of length n, the computation is always completed in at most An^k steps. For example, if the basic computational step is to add or multiply a pair of digits, then squaring an n-digit decimal integer by the standard method runs in polynomial time, for it requires fewer than $4n^2$ steps, and the formal definition is thus satisfied with $A = 4$ and $k = 2$.

Algorithms which do not run in polynomial time are said to run in *exponential time*. For example, if the basic step is to print a letter, a program to print a word with n different letters (or symbols) and all its possible permutations would have to print a letter $n(n!)$ times. This will exceed An^k for any fixed A and k, once n is large enough, so the algorithm will run in exponential time.

In general, algorithms which run in exponential time can sometimes take so long to run, even on very powerful computers, that they are unusable. In contrast, unless n is very large, algorithms running in polynomial time are often more practical. *See also* NP problem.

polytope A term used to describe figures analogous to three-dimensional polyhedra in higher dimensions. While three-dimensional space allows five regular polyhedra, in four dimensions six regular polytopes can be constructed, including a remarkable figure with 600 tetrahedral faces known as the 600-cell. Thereafter, in all higher spaces from five-dimensional space onwards, there can be only three regular polytopes – analogues of the tetrahedron, cube, and octahedron of three-dimensional space. *See also* polyhedron.

Poncelet, Jean Victor (1788–1867) French mathematician who, in his *Traité des propriétés projectives des figures* (1822, Treatise on the Projective Properties of Figures), revived the study of projective geometry and formulated within it the important principle of duality.

pons asinorum The name given to the fifth proposition in Book I of Euclid, that the angles opposite the equal sides in an isosceles triangle are equal. The name is Latin for 'bridge of asses'; the proof is considered the first difficult one encountered by the student.

population In *sampling theory and more generally, a collection of items about which information is sought. A sample is taken and inferences are made about the characteristics of the population on the basis of sample evidence. For example, if, in a random sample of 300 adult Londoners, 75 (i.e. 25 percent) are smokers, then 25 percent is the appropriate estimate of the proportion of adult Londoners who smoke. *Confidence limits may be attached to this estimate for a random sample, but not for samples such as *quota samples.

Statistical inferences are made about *populations, sometimes hypothetical, for which it is believed the observations available could reasonably be regarded as a *random sample. For example, measurements may be made of the thickness of ten sheets of metal randomly selected from the daily production of a factory; while the actual population sampled is only that day's production, inferences are often taken to apply to a hypothetically infinite population of all such sheets the factory has ever produced or will ever produce under similar conditions. This concept is based on the assumption that production standards do not change measurably from day to day.

Loosely, the term is sometimes used in phrases such as 'a sample from a normal population' to imply that we are assuming that the values of the characteristic we are observing have, in the population we are sampling, a *normal distribution.

poset *Abbreviation for* partially ordered set. *See* partial order.

positional system A *number system in which the notation depends on the position of the digits in the number.

position vector Symbol: **r**. A *vector that gives the position of a point P relative to a fixed reference point O, generally the origin of a *coordinate system. If two points have the same position vector then they coincide. The position vector is an alternative to specifying a point by means of its coordinates relative to a chosen set of axes. If in time δt point P moves to Q, changing its position vector by $\delta \mathbf{r}$, the velocity of P is then given by Lim $\delta \mathbf{r}/\delta t$ as $\delta t \to 0$.

position vector

positive angle A rotation angle measured from an initial axis in an anticlockwise sense.

positive definite A symmetric matrix **A** is positive definite if the *quadratic form $x^T Ax$ is positive for all nonzero vectors **x**. This is equivalent to the condition that all the *eigenvalues of **A** are positive. *See also* scalar product.

positive number A real number that is greater than zero.

positive semidefinite A symmetric matrix **A** is positive semidefinite if the *quadratic form $x^T Ax$ is non-negative for all nonzero vectors **x**. This is equivalent to the condition that all the *eigenvalues of **A** are non-negative.

positive series A *series whose terms are all positive numbers.

Post, Emil Leon (1897–1954) American mathematical logician who, in his *Introduction to a General Theory of Elementary Propositions* (1921), proved the consistency and completeness of elementary logic, provided a decision procedure, and generalized the subject to include many-valued logics.

posterior distribution *See* Bayesian inference.

posterior probability *See* prior probability.

postfix notation (reverse Polish notation) A notation for *operators in which the operator is written after its arguments. Thus

23+ represents what would be written in *infix notation as 2+3, while 34×2+ represents 3×4+2, and 23+4× represents (2+3)×4. Postfix notation is convenient for implementation on computers and is the input format on some pocket calculators.

Similarly, there is a postfix logical notation, widely used by computer engineers, in which the alphabetic prefixes in J. Łucasiewicz's Polish notation are used as postfixes. Thus the simple formula K*pq* would become *pq*K, while the more complex formula CK*pq*ApN*q* in Polish notation becomes *pq*K*pq*NAC in reverse Polish notation. *Compare* prefix notation (Polish notation).

postulate An *axiom. The term is usually used in certain contexts, e.g. Euclid's postulates or Peano's postulates.

potency (of a set) *See* cardinal number.

potential At a point in a *conservative field, say a gravitational or electrostatic field, the *work done in bringing unit mass or unit charge to this point from a point infinitely distant from the cause of the field; this gives, say, the *gravitational potential* or *electrostatic potential*. Since these fields are conservative, potential is a function only of the position of a particular point. It varies in magnitude from point to point, and hence a *potential function* is a scalar function of position and is usually denoted by $\phi(\mathbf{r})$.

Field strength $\mathbf{g}(\mathbf{r})$ is a vector function of position given by either $\nabla\phi$ or $-\nabla\phi$ (where ∇ is the differential operator *del), depending on the sign convention adopted. Here $\nabla\phi$ is the *gradient of ϕ, or grad ϕ, and is given by

$$\nabla\phi = \frac{\partial\phi}{\partial x}\mathbf{i} + \frac{\partial\phi}{\partial y}\mathbf{j} + \frac{\partial\phi}{\partial z}\mathbf{k}$$

where $\mathbf{r} = x\mathbf{i} + y\mathbf{j} + z\mathbf{k}$.
See also field.

potential energy *Energy possessed by virtue of position. It is a scalar quantity, usually denoted by V, and can be defined only in a *conservative field of force. It is the negative value of the *work done by a conservative force in displacing a particle from its standard position to any other position. The zero of potential energy is usually the potential energy at a point infinitely distant from the source of the field. In the case of bodies situated above the earth's surface, the surface is usually taken as the zero of potential energy: for a small object of mass m at an altitude h, the potential energy is mgh, where g is the acceleration of free fall. In an isolated system, the total energy – potential plus *kinetic energy – is conserved: in moving from point A to point B potential energy might be acquired at the expense of kinetic energy; this potential energy is released on returning to A, with an equivalent gain in kinetic energy.

potential function *See* potential.

pound Symbol: lb. The *avoirdupois unit of mass, equal to 0.453 592 37 kilogram. Formerly defined in terms of a platinum standard of mass, it was redefined by the UK Weights and Measures Act (1963) in terms of the *kilogram.

poundal Symbol: pdl. An *f.p.s. unit of force, equal to the force required to impart to a mass of 1 pound an acceleration of 1 foot per second per second. 1 poundal = 0.138 255 newton.

pound-force Symbol: lbf. An *f.p.s. unit of force, equal to the force required to impart to a mass of 1 pound an acceleration equal to the standard acceleration of free fall. 1 pound-force = 32.1740 poundals = 4.448 newtons.

power 1. *See* exponent.
2. *See* residue.
3. *See* hypothesis testing.
4. Symbol: P. The rate at which *work is done. It is now usually measured in watts (joules per second). *See also* horsepower.

power series A *series of the form

$$c_0 + c_1 x + c_2 x^2 + \cdots + c_n x^n + \cdots$$

where x is a real variable, and $c_0, c_1, c_2, \ldots$ are constants that can be positive, negative, or zero; these constants are called the *coefficients* of the series. The variable can also be complex and is then usually denoted by z. The sine, cosine, logarithmic, and exponential functions can be represented as power series (*see* expansion).

A power series in x may converge for all values of x or for no value except $x = 0$ (*see* convergent series). Alternatively it may be absolutely convergent if $|x| < L$ and divergent if $|x| > L$. The constant L is known as the *limit of convergence*. The interval

$$-L < x < L$$

is the *interval of convergence* of the power series. When $x = \pm L$ the series may converge or diverge.

Likewise a power series in a complex variable z can converge for all values of z or for no value except $z = 0$. Alternatively it may converge absolutely for all values of z within a circle of radius R and diverge for any z outside this circle. The circle is the *circle of convergence* of the series and R is the *radius of convergence*.

Two power series can be added or multiplied together, term by term, to give a convergent series only for those values of x (or z) within the smaller of the two intervals (or radii) of convergence. A power series can be differentiated term by term for all x (or z) within its interval (or circle) of convergence, and integrated term by term between any limits within this region.

See also binomial series; cosine series; exponential series; logarithmic series; sine series; Taylor's theorem.

power set The power set of a given *set A consists of all sets included in A. It is denoted by PA or $\mathcal{P}(A)$:

$$PA = \{B : B \subseteq A\}$$

Thus, if a set has n elements, then its power set will have 2^n elements. For example, if A is $\{1, 2\}$ then PA is $\{\emptyset, \{1\}, \{2\}, \{1, 2\}\}$. The *cardinal number of PA is sometimes denoted by $2^{\bar{A}}$, where $\bar{A}$ is the cardinal number of A.

Pr *See* probability.

precession The slow change in the direction of orientation of the *axis of rotation of a spinning body that arises when the body is subjected to an *external force (*see* torque). It can be seen as the wobbling motion of a spinning top when its axis is not vertical. If the applied torque and rotational speed are constant, the extremities of the axis trace out circles in what is one complete period of precession; the earth's rotational axis precesses in a similar way, with a period of 25 800 years. The motion of the axis of rotation at any instant is perpendicular to the direction of the torque.

precision A quality associated with the spread of data obtained in repetitions of an experiment as measured by *variance; the lower the variance, the higher the precision. The precision of an estimator is measured by its standard error, and in general this may be decreased and precision increased by taking additional observations. *See* efficiency.

predicate In *logic, an expression that, when combined with one or more singular terms, forms a sentence. An n-place predicate is one that can form a sentence only when combined with n singular terms. For example, the sentence 'Tom is taller than Dick' contains the two-place predicate 'is taller than' and two singular terms, 'Tom' and 'Dick'. An n-place predicate of a formal language is often interpreted by having an n-place relation assigned to it as its semantic value. *See* predicate calculus.

predicate calculus A particular system of rules for manipulating symbols in *logic. Used without qualification, the term

means first-order predicate calculus, which consists of:

(1) Symbols of the following types:

(a) A (possibly empty) set of individual constants $a_1, a_2, \ldots$;

(b) an infinite set of variables $x_1, x_2, \ldots$;

(c) a (possibly empty) set of function letters $f_1, f_2, \ldots$;

(d) a set of predicate letters $p_1, p_2, \ldots$;

(e) a set of logical constants that will include truth-functional connectives and quantifiers;

(f) punctuation devices, such as '(' and ')'.

(2) Formation rules that recursively define the set of terms and *wffs.

(3) Rules of inference, typically *modus ponens* and generalization.

Although the predicate calculus may be approached from the standpoint of *natural deduction or regarded as a *logistic system, and although there are many alternative axiomatizations of the predicate calculus (*see* axiom), it is customary to use '*the* predicate calculus' to refer to one of the standard formulations that have been shown to be *complete, *sound, and *consistent. *See also* interpretation; logic; *compare* propositional calculus.

predicated variable *See* regression.

prediction error *See* cross-validation.

predictor variable *See* regression.

prefix notation (Polish notation) A notation for *operators in which the operator is written before its arguments. Thus $+23$ represents what would be written in *infix notation as $2+3$, while $\times 34+2$ represents $3\times4+2$, and $+23\times4$ represents $(2+3)\times4$.

It was originally introduced in the 1920s by the Polish logician J. Łucasiewicz as a bracket-free notation in which alphabetic prefixes replaced logical connectives as follows: Np for $\sim p$, Kpq for $p \& q$, Cpq for $p{\to}q$, Apq for $p \vee q$, and Epq for $p \leftrightarrow q$. Thus $(p \& \sim q){\to}(p \vee \sim r)$ becomes CKpNqApNr. *Compare* postfix notation (reverse Polish notation).

pre-image *See* function; measurable function.

premise *See* argument; syllogism.

present value *See* interest.

pressure Symbol: p. At a point in a liquid or gas, the *force exerted per unit area on an infinitesimally small plane situated at that point. Pressure can be regarded as a compressive *stress. If the fluid is at rest, the pressure at any point is the same in all directions. The SI unit of pressure is the pascal (newton per square metre); gas pressure is also measured in millibars or atmospheres. The pressure in a static liquid increases with depth h: $p_h = \rho g h$, where ρ is the liquid density, taken as constant, and g is the acceleration of free fall. In a gas under isothermal conditions, pressure decreases exponentially with height h. For an ideal gas,

$$p_h = p_0 \exp(-\rho_0 g h / p_0)$$

where ρ_0 and p_0 are the density and pressure at $h = 0$.

primality The state of being *prime. In principle, the simplest test for primality is by trial division. However, even with a computer performing a million divisions per second the method is impractical for large numbers; testing a 50-digit number would take 10^{11} years. It is possible to prove that a number is not prime by using Fermat's theorem. Thus, if for a number a, $a^n - a$ is not exactly divisible by n, then n must be composite. The converse is not true: exact division does not prove that n is prime (*see* pseudoprime). However, there are various tests that will give an unequivocal indication of primality, and these can be performed quickly on large computers (e.g. in less than a second for a 100-digit number). In 2002, M. Agrawal, N. Kayal, and N. Saxena showed that there is a *polynomial time algorithm that can decide whether or not a given integer is a prime.

prime A whole number larger than 1 that is divisible only by 1 and itself. So 2, 3, 5, 7, ..., 101, ..., 1093, ... are all primes. Each prime number has the property that if it divides a product then it must divide at least one of the factors (Euclid, *c*.300 BC). No other numbers bigger than 1 have this property. Thus 6, which is not a prime, divides the product of 3 and 4 (namely 12), but does not divide either 3 or 4. Every natural number bigger than 1 is either prime or can be written as a product of primes. For instance, $18 = 2 \times 3 \times 3$, 37 is prime, $91 = 7 \times 13$ (*see* fundamental theorem of arithmetic).

There is no largest prime, since if p is a prime it is always possible in theory to find another prime which is larger than p (*see* Euclid's proof of the infinity of primes). However, in practice fast computers and sophisticated tests are needed to find extremely large primes. For example, only as recently as 2006 was the *Mersenne number $2^{32\,582\,657} - 1$ shown to be prime.

The term can also be used analogously in some other situations where division is meaningful. For instance, in the context of all the integers, an integer n other than 0 and ± 1 is a *prime integer* if its only integer divisors are ± 1 and $\pm n$. The positive prime integers are just the ordinary natural number primes 2, 3, 5, ... and the negative prime integers are $-2, -3, -5, \ldots$. Every prime integer shares the important property that if it divides a product of two integers then it must divide at least one of the factors.

See also Gaussian integer.

prime factorization *See* factorization.

prime ideal An *ideal I in a *ring R is prime if whenever $x, y \in R$ are such that $xy \in I$, then at least one of x and y lies in I.

prime number theorem The statement that the number of *primes not exceeding a given natural number n is approximately $n/\ln n$, in the sense that the ratio of the number of such primes to $n/\ln n$ eventually approaches 1 as n becomes larger and larger. Here $\ln n$ is the natural logarithm (to the base e) of n. The result was first guessed by Legendre and Gauss, and eventually proved in 1896 by Hadamard and Vallée-Poussin using difficult methods of complex analysis. In 1949 A. Selberg and P. Erdős found a proof that avoids complex analysis (but it is still very difficult).

prime pair *See* twin primes.

prime symbol (accent) The mark $'$ placed above and to the right of a letter, as for example in x' (read as 'x prime'). Two or more such marks can be used, as in x'' (read as 'x double prime'), x''' ('x triple prime'), etc. Prime symbols are used in mathematics in a number of ways:

(1) To indicate feet and inches; for instance $6'3''$ (six feet three inches).

(2) To indicate minutes and seconds of arc in angular measure; for instance, an angle of $10°3'27''$ (ten degrees, three minutes, and twenty-seven seconds). For decimal fractions, in this use, the prime symbol is often printed before the decimal point, as in $3'.75$ (3.75 minutes). The symbol is sometimes called a *minute mark*.

(3) To represent a constant value of a variable. For example, (x, y) are the coordinates of a variable point and (x', y') the coordinates of a fixed point on the resulting curve.

(4) To represent related variables or constants. For example, a transformation of coordinates (x, y) to coordinates (x', y').

(5) To denote related points in geometry. For example, the triangle ABC compared with a similar triangle $A'B'C'$.

(6) To indicate first and higher derivatives. For example, for a function $f(x)$, the first derivative can be denoted by $f'(x)$, the second derivative by $f''(x)$, etc.

primitive An undefined expression of a *formal language.

primitive curve A curve from which some other curve is derived.

primitive element A *generator for the *multiplicative *cyclic group of all the nonzero *congruence *classes modulo a *prime number p. For many primes 2 is a primitive element, but not for all. For example, 2 has *order 3 modulo 7, but for $p = 7$ a primitive element must have order 6; for this group 3 and 5 are primitive elements. The proof that there is a primitive element for every prime is not elementary, and determining the smallest primitive element can be difficult.

primitive polynomial A *polynomial whose coefficients are a set of integers that have a highest common factor of 1.

primitive root A complex number z is a primitive nth root of unity if $z^n = 1$ but $z^k \neq 1$ for every $k < n$. The primitive nth roots of unity are of the form $\exp(2\pi ri/n)$, where r is *coprime to n; they are the zeroes of the *cyclotomic polynomial $\Phi_n(x)$.

Prim's algorithm *See* tree.

principal The money on which *interest is paid.

principal axes *See* moment of inertia.

principal component analysis A statistical technique for analysing data. The first step is to determine a linear function of two or more variables that accounts for as much as possible of the total variation (as measured by the sum of the variances of the individual variables). This linear function is called the *first principal component*. Successive principal components are orthogonal to the first and to one another, each accounting for as much as possible of the variation remaining at the stage at which it is formed. The analysis requires computation of *eigenvalues and eigenvectors of the *characteristic equation for the *covariance or *correlation matrix of the data set. Principal component analysis is sometimes regarded as a form of factor analysis, but the model for the latter is different.

principal diagonal *See* diagonal.

principal directions *See* curvature.

principal ideal *See* ideal.

principal ideal domain (PID) An *integral domain in which every *ideal is principal. For example, $\mathbb{Z}$, the set of all integers, is a PID.

principal moments of inertia *See* moment of inertia.

principal parts (of a triangle) The lengths of the three sides and the sizes of the interior angles, as distinguished from other properties, such as lengths of medians or sizes of exterior angles, which are *secondary parts*.

principal value *See* complex number.

principle of moments The principle that if a body is in *equilibrium under the action of a system of coplanar forces, then the sum of the *moments of the forces about any point in the plane is zero.

prior distribution *See* Bayesian inference.

prior probability The probability of an event or of some hypothesis assigned before data are collected. After data are collected the *posterior probability* is calculated using *Bayes' theorem or *Bayesian inference.

prism A solid figure formed from two congruent *polygons with their corresponding sides parallel (the *bases*) and the parallelograms (*lateral faces*) formed by joining the corresponding vertices of the polygons. The lines joining the vertices of the polygons are *lateral edges*. Prisms are named according to the base – for example, a *triangular prism* has two triangular bases (and three lateral faces); a quadrangular prism has bases that are quadrilaterals. Pentagonal, hexagonal, etc. prisms have bases that are pentagons, hexagons, etc.

A *right prism* is one in which the lateral edges are at right angles to the bases (i.e. the lateral faces are rectangles) – otherwise the prism is an *oblique prism* (i.e. one base

is displaced with respect to the other, but remains parallel to it). If the bases are regular polygons and the prism is also a right prism, then it is a *regular prism*.

See also antiprism.

prism (a) Right pentagonal and (b) oblique triangular prisms.

prismatic Denoting or concerning a *prism.

prismatic surface A surface generated by all the lines that are parallel to a given line and intersect a *broken line that is not in the same plane as the given line. The broken line is the *directrix* of the surface; the parallel lines are its *generators* (or *elements*). If the broken line is closed (i.e. a closed polygon), then the surface is a *closed prismatic surface*.

prismatoid A *polyhedron with vertices that all lie in one or other of two parallel planes. The two faces of the prismatoid lying in these planes are its *bases*. These need not necessarily both have the same number of sides (*see* below). The *lateral faces* are formed by lines drawn between vertices in the two planes (*lateral edges*). The lateral faces of a prismatoid are trapeziums, parallelograms, or triangles, or a mixture of these.

A *prismoid* is a prismatoid in which: (a) both bases have an equal number of sides; and (b) the lateral faces are quadrilaterals (either trapeziums or parallelograms). A prism is a special case of a prismoid in which the bases are identical. A *frustum of a pyramid is a prismoid in which the bases are (geometrically) similar.

prismoidal formula A formula for the volume (V) of a *prismatoid. It is usually given in one of two equivalent forms: either

$$V = \frac{1}{4}h(B + 3A)$$

where h is the altitude, B the area of a base, and A the area of a section parallel to the base at two-thirds of the distance to the other base; or

$$V = \frac{1}{6}h(B_1 + B_2 + 4A_m)$$

where B_1 and B_2 are the areas of the bases and A_m is the area of a section midway between the bases. The formula can also be applied to other solids, e.g. elliptical or circular cones.

prisoner's dilemma A situation in *game theory in which the players independently arrive at what appear to be optimum strategies but which give each of them payoffs inferior to those they could obtain if both opted for the same alternative strategy. It takes its name from illustrative examples such as the following.

Two criminal suspects (A and B) arrested by the police are separately each offered the following deal: if both confess they will serve 2 years in prison; if neither confesses, both will serve 1 year; and if only one confesses he will go free while the other serves 3 years. These options are summarized in the table.

Suspect A reasons as follows. B must either confess or not confess. If he confesses, then if I also confess I get 2 rather than 3 years; on the other hand, if B remains silent, then if I confess I go free rather than serving 1 year. Therefore, whatever B does – confess or remain silent – it is in my best interest to confess. By a similar line of reasoning B will conclude that it is in his best interest to confess. It follows that each prisoner will serve 2

		Prisoner B	
		Confess	Remain silent
Prisoner A	Confess	2, 2	0, 3
	Remain silent	3, 0	1, 1

Prisoner's dilemma The left-hand number in each cell is A's sentence, the right-hand number is B's sentence.

years, whereas if they had both remained silent they would have been imprisoned for only 1 year. Thus careful and plausible reasoning by both prisoners has led to a dominant strategy of confession which fails to yield the best outcome.

The dilemma was first posed by Merrill Flood and Melvin Dresher in 1950 while considering the possibilities of thermonuclear war, and seemed to suggest at the time that defence analysts could reason themselves logically into a nuclear war.

A more coherent strategy becomes possible in the *iterated prisoner's dilemma*. In this scenario A and B repeatedly interact by, for example, trading goods, rice for corn perhaps. Participant A might be tempted to receive the corn but hang on to his rice. What is B's best strategy? It has been shown that B can adopt nothing better than cooperating on the first deal and thereafter copying A's behaviour – a strategy known as *tit-for-tat*.

probability A measure associated with an *event A and denoted by $\Pr(A)$, which takes a value such that $0 \leqslant \Pr(A) \leqslant 1$. Operations on probabilities are governed by a set of *probability axioms*. In general, the higher the value of $\Pr(A)$, the more likely it is that an event will occur at any one performance of an experiment. If an event cannot happen, then $\Pr(A) = 0$, but the converse is not true. If an event is certain to happen, then $\Pr(A) = 1$; again, the converse is false. Numerical values can be assigned in simple cases by one of two methods:

(1) If the *sample space can be divided into subsets of n $(n \geqslant 2)$ equally likely outcomes and the event A is associated with r $(0 \leqslant r \leqslant n)$ of these, then $\Pr(A) = r/n$. Thus if a coin is tossed there are two equally likely outcomes, heads and tails. One of these is favourable to heads, so $\Pr(\text{heads}) = \frac{1}{2}$. If a die is cast there are six equally likely outcomes: $\{1, 2, 3, 4, 5, 6\}$. Two of these, $\{3, 6\}$, are favourable to the event 'score divisible by 3', thus $\Pr(\text{score divisible by 3}) = 2/6 = \frac{1}{3}$.

(2) If an experiment can be repeated a large number of times, n, and we record the number of experiments, r, say, in which the event A occurs, then r/n is called the *relative frequency* of A. If this tends to a limit as $n \to \infty$ this limit is $\Pr(A)$. *Mutually exclusive events* are events that cannot both occur in the one experiment. If two events are mutually exclusive, $A \cup B$ denotes the event 'either A or B occurs' and one axiom states that in this case

$$\Pr(A \cup B) = \Pr(A) + \Pr(B)$$

If A and B are not mutually exclusive we may deduce that

$$\Pr(A \cup B) = \Pr(A) + \Pr(B) - \Pr(A \cap B)$$

where $A \cap B$ means that both A and B occur. If we are interested in the probability that B occurs only in those experiments in which A is known to have occurred, the probability is called the *conditional probability* of B given A, and written as $\Pr(B \mid A)$. The multiplication rule of probability is

$$\Pr(A \cap B) = \Pr(A) \cdot \Pr(B \mid A)$$

If $\Pr(B \mid A) = \Pr(B)$ we say that A and B are *independent* and then

$$\Pr(A \cap B) = \Pr(A) \cdot \Pr(B)$$

See also subjective probability.

probability density function *See* frequency function.

probability function *See* frequency function.

probability generating function For a *random variable X, the *expectation of t^X, where t is a constant. It is denoted by $P(t) = E(t^X)$. It is useful for discrete variables taking non-negative integral values only. If $P(t)$ is convergent, the coefficient of t^r in its Taylor series expansion (*see* Taylor's theorem) is $\Pr(X = r)$. The rth derivative of $P(t)$ at $t = 1$ is $P^{(r)}(1) = E[X(X-1)\cdots(X-r+1)]$ and is called the rth factorial moment of X. In particular, $E(X) = P'(1)$ and $\text{Var}(X) = P''(1) + P'(1) - [P'(1)]^2$. For the binomial distribution,

$$P(t) = \sum_{r=0}^{n} t^r \binom{n}{r} p^r q^{n-r} = (pt + q)^n$$

from which, clearly,

$$\Pr(X = r) = \binom{n}{r} p^r q^{n-r}$$

is the coefficient of t^r in the binomial expansion of $P(t)$. Also, the rth derivative with respect to t is

$$P''(t) = n(n-1)\cdots(n-r+1)$$
$$\times p^r (pt + q)^{n-r}$$

from which $P'(1) = np$, $P''(1) = n(n-1)p^2$, and

$$\text{Var}(X) = n(n-1)p^2 + np - n^2p^2 = npq$$

probability mass function *See* frequency function.

probability paper Graph paper so scaled that the *distribution function for a specified distribution, most commonly the *normal distribution, becomes a straight line. If the sample distribution function (*see* random sample) for a sample believed to be from this distribution is plotted on the paper, then it should also lie near a straight line; departures indicate that the sample is probably not from that distribution.

probable error For a sample from a *normal distribution the probable error is

$$0.6745 \times \text{standard error}$$

It is so called because 50 percent of the normal distribution lies within the range $\mu \pm 0.6745\sigma$. *Confidence intervals are now usually quoted in preference.

probit analysis (C. Bliss, 1934) A method for analysing *quantal responses. It is used, for example, to compare insecticides in experiments where the proportion of insects showing a quantal response, such as death, are recorded. Probit analysis is a special case of a *generalized linear model.

problem of Apollonius In his lost treatise *On Contacts*, Apollonius posed this problem: given any three points, straight lines, or circles, or any combination, to draw a circle which passes through the point or points and is tangential to the lines or circles (as the case may be). This leads to ten distinct possibilities. Two cases – three points and three straight lines – were solved by Euclid (Book IV), the remainder apparently by Apollonius. The case of three circles proved to be of sufficient interest to attract the attention of several leading 17th-century mathematicians, including Viète and Newton.

problem of Pappus In Book VII of the *Collection*, Pappus discussed a problem known as the four-line locus derived from Apollonius: given four plane lines, to find the locus of a point P such that the product of the distances from P to any two of them is proportional to the product of the distances to the other two. The locus is in fact a conic section. Pappus generalized the problem to cover cases for n lines, where $n > 4$. It was from consideration of this problem in 1631 that Des-

cartes was led to his discovery of analytical geometry.

Proclus (*c*.410–485) Greek mathematician and author of a commentary on Book I of Euclid's *Elements* which has survived and contains much material unavailable elsewhere. It includes his attempt to prove Euclid's fifth postulate.

produce In geometry, to extend a line.

producer's risk *See* acceptance sampling.

product The result of multiplying two or more numbers, *vectors, *matrices, *sets, etc. *See also* multiplication; Cartesian product; continued product; infinite product; intersection.

product formulae Formulae in plane trigonometry for products of trigonometric functions:

$$\sin x \cos y = \tfrac{1}{2}[\sin(x+y) + \sin(x-y)]$$

$$\cos x \sin y = \tfrac{1}{2}[\sin(x+y) - \sin(x-y)]$$

$$\sin x \sin y = \tfrac{1}{2}[\cos(x-y) - \cos(x+y)]$$

$$\cos x \cos y = \tfrac{1}{2}[\cos(x+y) + \cos(x-y)]$$

See also factor formulae.

product moment *See* moment.

product moment correlation coefficient *See* correlation coefficient.

product of inertia *See* moment of inertia.

product rule (for differentiation) A method for differentiating a product using the formula

$$\frac{\mathrm{d}}{\mathrm{d}x}(uv) = u\frac{\mathrm{d}v}{\mathrm{d}x} + v\frac{\mathrm{d}u}{\mathrm{d}x}$$

For example, it is possible to differentiate the function $y = x \sin x$ using $u = x$ and $v = \sin x$, so that

$$\frac{\mathrm{d}u}{\mathrm{d}x} = 1 \quad \text{and} \quad \frac{\mathrm{d}v}{\mathrm{d}x} = \cos x$$

The formula then gives

$$\frac{dy}{dx} = x\cos x + \sin x$$

Compare quotient rule.

product topology A topology on the product of *topological spaces. A set $U \subset X \times Y$ is open if every point $(x, y) \in U$ is contained in a subset of U of the form $V \times W$, where V and W are open in X and Y, respectively. A product topology can be constructed on an arbitrary product (possibly with an infinite number of factors). For *metric spaces the *Cartesian metric defines the product topology.

programming The act of planning and producing a set of instructions to solve a problem by computer.

progression A simple *sequence of numbers in which there is a constant relation between two consecutive terms. The most common progressions are the *arithmetic, *geometric, and *harmonic sequences.

projectile A body thrown or projected with a particular initial speed and direction. Its subsequent motion depends only on external forces such as gravitational force and air resistance. The path of a projectile is called its *trajectory*. *See* ballistics.

projection A *mapping of a geometric figure onto a plane according to certain rules. *See* central projection; Mercator's projection; orthogonal projection; projective geometry; stereographic projection.

projective geometry A branch of geometry originated by Girard Desargues in the 17th century out of his work on *conics. Desargues was influenced by perspective in art and struck by the fact that a projection of a conic is also a conic. He assumed that parallel lines meet at an ideal point (infinity) and (like Kepler) he considered the parabola to have a second focus at infinity. His method was to consider properties of conics that are unchanged under projection. Desargues used in particular the complete *quadrangle because of its harmonic ratios. He showed, for instance, that if a quadrangle is inscribed in a conic, the line through two of the diagonal points is the *polar line of the third diagonal point.

Projective geometry can be defined as the study of those properties of plane figures that are unchanged under *central projection. It was misunderstood and neglected in Desargues's time, but it did inspire Pascal in his early work (*see* Pascal's theorem). The subject was revived by Poncelet in the early 19th century.

In projective geometry, there is a duality between points and lines; for every theorem whose statement involves points and lines there is a dual theorem whose statement can be derived from the original by interchanging the roles of points and lines.

Projective geometry can be studied with coordinates in any given field. When the field is finite, important combinatorial structures arise. *See* combinatorics.

projective plane The projective plane over a *field F is the set of points represented by nonzero triples (x, y, z) with $x, y, z \in F$, and if λ is a nonzero element of F, the triples (x, y, z) and $(\lambda x, \lambda y, \lambda z)$ represent the same point. If F is the field of real (or complex) numbers, the plane is called the *real* (or *complex*) *projective plane*. If F is a finite field with q elements, then the projective plane over F has $q^2 + q + 1$ points.

Topologically, the real projective plane is *homeomorphic to the union of a Möbius strip and a disc which are glued together along their boundaries,

prolate *See* ellipsoid.

proof A chain of reasoning using rules of *inference, ultimately based on a set of *axioms, that leads to a conclusion. More precisely, a proof is a sequence B_1, ..., B_n of *wffs of a *formal system S such that for each B_i, with $1 \leqslant i \leqslant n$, either B_i is an axiom or B_i is immediately inferred from some previous wffs of the sequence by a single application of a rule of inference of S. *See* consequence; deduction; indirect proof; induction; theorem.

proof by contradiction *See* indirect proof.

proof by contraposition The proof of a statement $p \rightarrow q$ by first proving its *contrapositive $\sim q \rightarrow \sim p$. Thus we can prove that 'If $2^n - 1$ is prime, then n is prime' by first proving its contrapositive 'If n is composite, then $2^n - 1$ is not prime'.

proof theory (metalogic, metamathematics, logical syntax) The study of *proofs and provability as they occur within *formal languages. As proofs are simply finite sequences of formulae, proof theory does not need to involve any interpretations that a formal language may have. The study of purely formal properties of formal languages, such as deducibility, independence, simple completeness, and, particularly, consistency, all fall within the scope of proof theory. *See also* logic.

proper class *See* von Neumann set theory.

proper divisor A divisor of an integer which is not equal to the integer itself. For example, 1, 3, and 7 are the proper divisors of 21. An older name for proper divisor is *aliquot part*. *See* amicable numbers; perfect number.

proper divisors of zero *See* integral domain.

proper fraction A fraction in which the numerator is less than the denominator. For example, $\frac{3}{4}$ is a proper fraction ($\frac{4}{3}$ is an *improper fraction*).

proper inclusion A *set A is properly included (*see* inclusion) in a set B, denoted by $A \subset B$, if and only if it is a *proper subset of B.

properly divergent *See* divergent series; divergent sequence.

proper subset (proper subclass) A *set A is a proper *subset of a set B, denoted by $A \subset B$, if and only if A is included (*see* inclusion) in but not equal to B:

$$(A \subset B) \leftrightarrow (A \subseteq B) \,\&\, (A \neq B)$$

For example, if A is $\{1, 2, 3\}$, B is $\{1, 2, 3, 4\}$, and C is $\{1, 2, 3\}$, then A is a proper subset of B but, although A is a subset of C, it is not a proper subset of C.

proportional *See* variation.

propositional calculus (sentential calculus) A *formal system that contains:

(1) Symbols of the following types:

(a) an infinite set of propositional variables $A_1, A_2, \ldots$;

(b) truth-functional connectives, such as '&' and '$\sim$'; and

(c) punctuation, such as '(' and ')'.

(2) Formation rules, such as: if B and C are *wffs then '$B \,\&\, C$' is a wff.

(3) Rules of inference, such as *modus ponens*.

Although the propositional calculus may be approached from the standpoint of *natural deduction or regarded as a *logistic system, and although there are many alternative axiomatizations of the propositional calculus (*see* axiom), it is customary to use '*the* propositional calculus' to refer to one of the standard formulations that have been shown to be *complete, *sound, and *consistent. *See also* interpretation; logic; *compare* predicate calculus.

p-series The *series

$$1 + \left(\frac{1}{2}\right)^p + \left(\frac{1}{3}\right)^p + \cdots + \left(\frac{1}{n}\right)^p + \cdots$$

If $p > 1$ the series converges; if $p \leqslant 1$ the series diverges. When $p = 1$ it becomes the *harmonic series. *See also* Riemann zeta function.

pseudo-inverse (E.H. Moore, 1920; R. Penrose, 1955) A generalization of the notion of *inverse that applies to rectangular matrices. If A is an $m \times n$ matrix then its pseudo-inverse (also called the *Moore–*

Penrose pseudo-inverse) is the unique $n \times m$ matrix X satisfying the four *Moore–Penrose conditions*:

(i) $AXA = A$
(ii) $XAX = X$
(iii) $AX = (AX)^*$
(iv) $XA = (XA)^*$

where * denotes the Hermitian conjugate. The pseudo-inverse is denoted by A^+. If $A = UDV^*$ is a *singular value decomposition then $A^+ = VD^+U^*$, where D^+ is obtained by inverting the nonzero diagonal elements of D. When A has full *rank, $A^+ = (A^*A)^{-1}A^*$ if $m > n$, and $A^+ = A^*(AA^*)^{-1}$ if $m < n$. For example,

$$\begin{pmatrix} 2/3 & 0 \\ 5/6 & 1/2 \\ 1/3 & 1 \end{pmatrix}^+ = \begin{pmatrix} 5/6 & 2/3 & -1/3 \\ -1/2 & 0 & 1 \end{pmatrix}$$

pseudoprime A *composite number that sometimes exhibits behaviour more typical of *prime numbers. For example, if p is a prime and a is any integer, then p will divide $a^p - a$. However, it does not necessarily follow that if $a^n - a$ is divisible by n, then n is prime. A composite number n that exactly divides $a^n - a$ is said to be a *Fermat pseudoprime* to the base a. An example is 341, which is divisible by 11 and 31 and which divides exactly into $2^{341} - 2$. Thus 341 is a Fermat pseudoprime to the base 2. Numbers exist that are Fermat pseudoprimes to any base (e.g. 561, 1105, and 1729); these are known as *Carmichael numbers* after the American mathematician R.D. Carmichael, who discovered them in 1909. In 1992 it was shown that there are infinitely many of them. *See also* Fermat's theorem.

pseudo-random numbers *See* random numbers.

pseudosphere *See* tractrix.

Ptolemy, Claudius (2nd century AD) Greek astronomer and mathematician, author of the *Syntaxis mathematica* (Mathematical Collection), more commonly known as the *Almagest*. It contains a corrected and extended version of Hipparchus' table of chords together with a clear description of just how the table was constructed. Much use was made of the principle, since known as *Ptolemy's theorem. He is also known to have made an attempt to prove Euclid's fifth postulate.

Ptolemy's formulae *See* addition formulae.

Ptolemy's theorem A convex quadrilateral can be *inscribed in a circle if and only if the product of the lengths of one pair of opposite sides added to the product of the lengths of the other pair is equal to the product of the lengths of the diagonals. Thus, in a cyclic quadrilateral ABCD,

$$AB \cdot DC + AD \cdot BC = AC \cdot BD$$

public key cryptography Any method of sending messages in which the *encryption method can be made public but the *decoding method is known only to the intended recipients. It depends on using a *trapdoor function which is easy to calculate but whose inverse is not easy to find without specific knowledge. The best-known example is the *RSA cipher.

pulley A simple *machine that consists of a wheel with a grooved or flat rim around which a rope, belt, etc. can run. When a force is applied to the rope, the direction or point of application of the force can be changed and a weight can be lifted. A number of such wheels can be pivoted in parallel, using a single rope. In a frictionless system the *mechanical advantage is the ratio of the weight W to be moved to the applied pull P in the rope. This is equivalent to the number n of forces (ropes) supporting the weight. The wheel can also be mounted on a shaft so that it is driven by or drives a belt passing around it.

pulley

pulsatance *See* angular frequency.

pure imaginary *See* complex number.

pure mathematics *See* mathematics.

pure strategy *See* game theory.

pure surd *See* surd.

p-value In *hypothesis testing, the probability under the null hypothesis that a *statistic takes a value as extreme as or more extreme than that observed in the relevant tail (one-tail test) or tails (two-tail test). In the case of a parametric test, this is called an *exact p-value* if all the conditions for the parametric test hold. If *nonparametric methods are used, an appropriate *permutation test leads to exact *p*-values. For sufficiently small *p* the outcome of a test may be classed as significant at level *p*. This concept of significance differs from that of a pre-fixed significance level α (commonly fixed at $\alpha = 0.05$, 0.01, or 0.001).

For example, if a *t*-test is performed on a sample of ten observations from a normal distribution with unknown mean μ to test the hypothesis $\mu = 0$ against the alternative $\mu \neq 0$, the statistic t has nine degrees of freedom, and if $t = 2.41$ then $p = \Pr(|t| \geqslant 2.41) = 0.0392$, indicating significance at the exact 0.0392 level. If and only if $p \leqslant \alpha$ is a result significant at a pre-

chosen level α. Modern computer software usually gives exact *p*-values in output.

pyramid A solid figure (a *polyhedron) formed by a *polygon (the *base*) and a number of triangles (*lateral faces*) with a common vertex that is not coplanar with the base. Line segments from the common vertex to the vertices of the base are *lateral edges* of the pyramid. Pyramids are named according to the base: a triangular pyramid (which is a tetrahedron), a square pyramid, a pentagonal pyramid, etc.

If the base has a centre, a line from the centre to the vertex is the *axis* of the pyramid. A pyramid that has its axis perpendicular to its base is a *right pyramid*; otherwise, it is an *oblique pyramid*. If the base is a regular polygon and the pyramid is a right pyramid, then it is also a *regular pyramid*.

The *altitude* (*h*) of a pyramid is the perpendicular distance from the base to the vertex. The volume of any pyramid is $\frac{1}{3}Ah$, where A is the area of the base. In a regular pyramid, all the lateral edges have the same length. The *slant height* (*s*) of the pyramid is the altitude of a face; the total surface area of the lateral faces is $\frac{1}{2}sp$, where p is the perimeter of the base polygon.

(a)   (b)

pyramid (a) Right square and (b) oblique triangular pyramids.

pyramidal Denoting or concerning a *pyramid.

pyramidal surface A surface generated by all the lines that pass through a given point and intersect a *broken line that is not in the same plane as the given point. The point is the *vertex* of the surface, the broken line is its *directrix*, and the lines forming the surface are *generators* (or *elements*). The surface has two parts (called *nappes*) on each side of the vertex. If the broken line is closed (i.e. if it forms a closed polygon) the surface is a *closed pyramidal surface*.

Pythagoras (6th century BC) Greek mathematician and founder of the Pythagorean school, which claimed to have found the principles of all things in numbers. What precisely Pythagoras contributed himself is no longer clear, but amongst the achievements of his school the most significant is undoubtedly the discovery of the irrational numbers. Other discoveries include the numerical ratios determining the intervals of the musical scale, perfect and amicable numbers, figurate numbers, and *Pythagoras' theorem. Although the theorem had been known to the Babylonians over a thousand years before, its first general demonstration is attributed to the Pythagoreans.

Pythagoras' theorem In a right-angled triangle, the sum of the squares of the lengths of the sides a and b containing the right angle is equal to the square of the hypotenuse:

$$a^2 + b^2 = c^2$$

where c is the length of the side opposite the right angle.

The *converse is also true, i.e. if the sides of a triangle are such that $a^2 + b^2 = c^2$, then the angle opposite the side of length c is a right angle.

Pythagorean triple Three positive integers a, b, and c such that $a^2 + b^2 = c^2$. For example, $(3, 4, 5)$, $(6, 8, 10)$, and $(9, 12, 15)$ are Pythagorean triples. The triple (a, b, c) is a *primitive Pythagorean triple* when a, b, and c have no common factor greater than 1. So $(6, 8, 10)$ and $(9, 12, 15)$ are not primitive Pythagorean triples but $(3, 4, 5)$, $(5, 12, 13)$ and $(8, 15, 17)$ are. Pythagorean triples that are not primitive are of the form (ka, kb, kc) where k is an integer greater than 1 and (a, b, c) is primitive.

Every primitive Pythagorean triple can be obtained by calculating the numbers

$$2uv, \quad u^2 - v^2, \quad u^2 + v^2$$

where u and v are positive integers having no common factor greater than 1, with one of them odd and the other even, and $u > v$. For example, putting $u = 4$ and $v = 3$ gives the primitive triple $(7, 24, 25)$, and substituting $u = 6$ and $v = 1$ gives the triple $(12, 35, 37)$.

The numbers in a Pythagorean triple are the lengths of the sides of a right-angled triangle with integer sides. *See* Pythagoras' theorem.

Pythagorean identities *See* trigonometric functions.

Q

ℚ Symbol for the set of all *rational numbers.

QED *Abbreviation for* quod erat demonstrandum [Latin: which was to be demonstrated], sometimes added at the end of a proof.

QEF *Abbreviation for* quod erat faciendum [Latin: which was to be done], sometimes added after the completion of a geometrical construction.

QEI *Abbreviation for* quod erat inveniendum [Latin: which was to be found], sometimes added after the completion of a calculation.

Qin Jiushao (Ch'in Chiu-shao) (*c.* AD 1200) Chinese mathematician and calendar-maker who, following work by Liu Yi (*c.*1200), described in his *Shushu jiuzhang* (1247, Mathematical Treatise in Nine Sections) the solution of polynomial equations of arbitrary degree by the method of 'iterated multiplication', equivalent to the iterative methods given by Ruffini and *Horner in the 19th century. This work also contains methods for solving simultaneous linear congruences in the computation of calendars which are sufficient to establish what is now known as the *Chinese remainder theorem.

QR factorization For a matrix A with at least as many rows as columns, a factorization into the product of a matrix Q with *orthonormal columns and an *upper triangular matrix R, thus $A = QR$. This fac-

torization is computed by the *Gram–Schmidt method.

quadrangle A plane figure consisting of four points joined by lines. No three of the points are collinear. In a *simple quadrangle* the points are joined by four lines, which may or may not intersect. In a *complete quadrangle* (*see* diagram) the four points are joined by six lines; these lines intersect in a further three *diagonal points*.

The *harmonic property* of the complete quadrangle is that the pair of lines through any diagonal point and the pair of lines joining that point to the other diagonal points form a *harmonic pencil. Thus, in the diagram, the pencil of lines GF, GE, GA, GD is harmonic, and if GE produced meets AD at X, then $\{F, X; A, D\} = -1$.

See cross-ratio; *compare* quadrilateral.

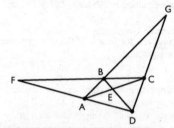

quadrangle Complete quadrangle ABCD: E, F, and G are diagonal points.

quadrangular prism A *prism that has bases that are quadrilaterals.

quadrant One of the four regions into which a plane is divided in a *Cartesian coordinate system.

quadrantal angle An angle that is a multiple of 90°, i.e. 90°, 180°, 270°, 360°, 450°, 540°, etc.

quadrantal spherical triangle *See* spherical triangle.

quadrants, law of *See* species.

quadratic Describing an expression, equation, etc. of the second *degree*. A *quadratic polynomial* is a polynomial of the second degree. A *quadratic equation* is an equation formed by putting a quadratic polynomial equal to zero. For one variable it has the form

$$ax^2 + bx + c = 0$$

A *quadratic* *form* is a homogeneous polynomial of the second degree; one in two variables is

$$ax^2 + 2hxy + by^2$$

which can also be written in matrix notation as

$$(x \quad y)\begin{pmatrix} a & h \\ h & b \end{pmatrix}\begin{pmatrix} x \\ y \end{pmatrix}$$

A *quadratic curve* is a curve with an algebraic equation of the second degree.

quadratic congruence A *congruence of the type

$$ax^2 + bx + c \equiv 0 \pmod{n}$$

where n is a given natural number, a, b, and c are given *integers, and x is an unknown integer. By using the method of completing the square, the congruence can be recast as

$$(2ax + b)^2 \equiv (b^2 - 4ac) \pmod{n}$$

Solving the original congruence is then equivalent to solving the simple congruence

$$y^2 \equiv (b^2 - 4ac) \pmod{n}$$

and the *linear congruence

$$2ax + b \equiv y \pmod{n}$$

quadratic convergence *See* order (**12**).

quadratic formula A formula giving the roots of a *quadratic equation. For the equation

$$ax^2 + bx + c = 0$$

the formula is

$$x = \frac{-b \pm \sqrt{(b^2 - 4ac)}}{2a}$$

for $a \neq 0$. *See also* completing the square.

quadratic reciprocity, law of A result that relates the solubility of the two *congruences

$$x^2 \equiv p \pmod{q} \quad \text{and} \quad y^2 \equiv q \pmod{p}$$

where p and q are different odd primes. The result is that if one of the primes is congruent to 1 modulo 4, then the congruences are either both soluble or both insoluble; and if both p and q are congruent to -1 modulo 4, then just one of the congruences is soluble. In terms of the *Legendre symbol, this can be expressed as

$$\left(\frac{p}{q}\right) = \begin{cases} \left(\frac{q}{p}\right) & \text{if either } p \equiv 1 \pmod{4} \\ & \text{or } q \equiv 1 \pmod{4} \\ -\left(\frac{q}{p}\right) & \text{if } p \equiv q \equiv -1 \pmod{4} \end{cases}$$

Two associated results are that

$$\left(\frac{-1}{p}\right) = \begin{cases} +1 & \text{if } p \equiv 1 \pmod{4} \\ -1 & \text{if } p \equiv -1 \pmod{4} \end{cases}$$

and

$$\left(\frac{2}{p}\right) = \begin{cases} +1 & \text{if } p \equiv \pm 1 \pmod{8} \\ -1 & \text{if } p \equiv \pm 3 \pmod{8} \end{cases}$$

These three results form an important part of the rules for evaluating arbitrary Legendre symbols.

quadratic residue A *residue of order 2. *See* Legendre symbol.

quadratrix A *transcendental curve, invented by *Hippias of Elis, containing the points of intersection of a uniformly rotating radius of a curve with a certain line moving uniformly parallel to itself. The quadratrix is constructed by allowing the radius OP of a circle (*see* diagram) to rotate uniformly about its centre O from OA to OC; simultaneously the line DE, parallel to AB, moves at the same rate from AB to OC. The locus of the intersection of the radius OP and the moving line DE forms the quadratrix AP'Q. The curve allowed an angle such as ∠ POQ to be trisected (*see* trisection) by constructing lines D'E' and D"E" parallel to DE such that DD' = D'D" = D"O.

quadratrix

quadrature Historically, the process of determining a square that has an area equal to the area enclosed by a closed curve. At one time a synonym for *integration, it is now usually confined to *numerical integration.

quadric A curve or surface that has an algebraic equation of the second *degree, i.e. a *conic (curve) or *conicoid (surface).

quadrilateral A plane figure formed by four intersecting lines. A *simple quadrilateral* is a polygon with four sides. A *complete quadrilateral* (*see* diagram) is the figure formed by four lines and their six points of intersection. These points are joined by a further three *diagonal lines*.

The *harmonic property* of the complete quadrilateral is that any diagonal line intersects the quadrilateral and the other two diagonal lines in a harmonic range (*see* cross-ratio). Thus, in the diagram, if DB produced meets GF at Y, then {Y, E; B, D} = −1. *Compare* quadrangle.

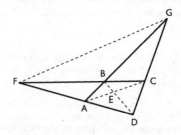

quadrilateral Complete quadrilateral: AC, BD, and FG are diagonal lines.

quality control The use of statistical methods, including *control charts, *cusum charts, and *acceptance sampling, to determine whether processes or goods produced are meeting certain specifications, and to indicate when corrective action should be taken if standards are not being met.

quantal response A situation in which an individual subjected to a stimulus shows only one possible response, if any, e.g. death. *See* probit analysis; logistic regression.

quantifier A logical constant used to indicate the *quantity* of a proposition. Thus, the *general* sentence 'All men are mortal' has as its logical form

$$(\forall x)(\mathrm{Man}(x) \supset \mathrm{Mortal}(x))$$

and the *particular* sentence 'Some men are mortal' has the logical form

$$(\exists x)\text{Mortal}(x)$$

The quantifier $(\forall x)$ is called the *universal quantifier* and is read as 'for all x'. It is common to write this simply as (x). The quantifier $(\exists x)$ is called the *existential quantifier*, and is read as 'for some x'. When constructing a formal system it is customary to define $(\exists x)A$ as $\sim(\forall x)\sim A$. *See also* predicate calculus.

quantiles If, for a *random variable X with *distribution function $F(x)$, we can, given a number p, find x_p such that $F(x_p) = p$, we say that x_p is the pth *quantile* of X. If $p = \frac{1}{2}$ then x_p is the median. If $p = r/4 \, (r = 1, 2, 3)$ we call x_p the rth *quartile*; if $p = r/10 \, (r = 1, 2, \ldots, 9)$ we call x_p the rth *decile*; and if $p = r/100 \, (r = 1, 2, \ldots, 99)$ we call x_p the rth *percentile*. For many discrete distributions no unique value of x_p can be found by using this definition, but the difficulty may be overcome by making suitable modifications. Quantiles may also be defined for data sets. The data must first be arranged in ascending order. If there are $2n + 1$ (an odd number of) observations, the median is the middle ordered value x_{n+1}. If there are $2n$ (an even number of) observations, the median is the mean of the two ordered observations x_n and x_{n+1}. *See* order statistics.

quantum mechanics A branch of mechanics developed in the early 20th century from results of experiments that could be explained only by assuming that certain physical quantities (e.g. energy, momentum) are *quantized* – i.e. they can take only certain discrete values. An aspect of quantum theory is wave–particle duality, the observation that particles can act as waves and vice versa. Erwin Schrödinger developed a form of quantum mechanics known as *wave mechanics*, based on solving wave equations of systems of particles.

Werner Heisenberg produced an equivalent operator formalism known as *matrix mechanics*. Modern quantum mechanics considers that all possible physical states of a system correspond to space vectors in a Hilbert space. Quantum mechanics differs from classical (or relativistic) mechanics in the way in which measurements on a system affect its state, and in the consequent fact that the information obtained is probabilistic rather than definite. Quantum effects become important for microscopic systems (elementary particles and atoms).

quart 1. An obsolescent *imperial unit of volume or capacity, equal to $\frac{1}{4}$ of a *gallon. **2**. A unit of liquid measure in common use in the USA; it equals $\frac{1}{4}$ of a US *gallon.

quartic (biquadratic) Describing a mathematical expression of the fourth *degree or order. Thus, a *quartic polynomial* is one of the form

$$ax^4 + bx^3 + cx^2 + dx + e$$

A *quartic function* is a function $f(x)$ whose value for a value of x is given by a quartic polynomial in x. A *quartic equation* is an equation of the form

$$ax^4 + bx^3 + cx^2 + dx + e = 0$$

To solve this equation, it can be recast in the form

$$y^4 + py^2 + qy + r = 0$$

by substituting $x = y - b/4a$ and dividing throughout by the coefficient of y^4. This, the *reduced quartic*, is set identically equal to

$$(y^2 + \lambda y + m)(y^2 - \lambda y + n) = 0$$

After equating coefficients and eliminating m and n from the three equations, the following *cubic in λ^2 is obtained:

$$\lambda^6 + 2p\lambda^4 + (p^2 - 4r)\lambda^2 - q^2 = 0$$

Since the cubic is solvable in terms of radicals, we can obtain λ, m, and n in terms of radicals and p, q, and r. The solution of the reduced quartic (and hence of the original quartic) is completed by solving each of the quadratic equations

$$y^2 + \lambda y + m = 0, \quad y^2 - \lambda y + n = 0$$

This method was given by Descartes in 1637. However, the first method of obtaining x in terms of the coefficients of a quartic equation was given by Ferrari, and was published in 1545 in the book by Cardano that also contained the first solution of the cubic.

A *quartic curve* is a curve with an algebraic equation of the fourth degree.

quartile *See* quantile.

quartile deviation The semi-interquartile range, $\frac{1}{2}(Q_3 - Q_1)$, where Q_1 and Q_3 are the first and third quartiles (*see* quantiles).

quasi-crystal *See* crystallography.

quaternion An entity of the form

$$x_0 + x_1 i + x_2 j + x_3 k$$

where x_0, x_1, x_2, and x_3 are real numbers. Quaternions were introduced in 1843 by Hamilton as a way of generalizing complex numbers in a plane to three dimensions. Quaternions combine by the normal laws of algebra with the exception of multiplication, which is not commutative. Multiplication is by the distributive law using

$$i^2 = j^2 = k^2 = ijk = -1$$

$$ij = k = -ji, \quad \text{etc.}$$

The set of quaternions (denoted by $\mathbb{H}$) can be regarded as a *vector space of dimension 4 over $\mathbb{R}$, the real field, with a basis 1, i, j, k. *See also* Frobenius's theorem.

quaternion group A non-Abelian multiplicative *group of 8 elements which are most commonly represented by the unit *quaternions ± 1, $\pm i$, $\pm j$, and $\pm k$. A *matrix representation of the same group is given by eight 4×4 matrices (*see* *generator (**2**)), with matrix multiplication as group operation.

Quetelet, Lambert-Adolphe-Jacques (1796–1874) Flemish astronomer and mathematician often referred to as the 'father of modern statistics'. He not only collected much basic statistical data on a wide range of phenomena but also attempted to analyse them and use them to test traditional views in such disciplines as medicine and criminology. It was Quetelet who introduced the notion of the 'average man'.

queuing theory A study of *stochastic processes involving customer-wait and service-time patterns where there is a random element in customer arrivals and/or times taken to serve by one or more servers. The theory applies not only to systems like banks and post offices, but also to, for example, berthing-and-unloading schedules for oil tankers at a refinery.

Quine, Willard Van Orman (1908–2000) American mathematical logician who, in his *Mathematical Logic* (1940), established an influential approach to set theory. In his later writings he also developed a consistently radical critique of a number of important issues, including the legitimacy of modal logic, the nature of logical truth, abstract entities, and meaning.

quintal system A *number system using the base five.

quintic Describing an expression of the fifth *degree.

quota sample A sample in which the units are not selected randomly but the interviewer is told to choose a certain number of units in each of a number of categories, e.g. 30 women, 16 men, half of each to be over 40, etc. The method is widely used in opinion polls and market research. While sampling error cannot be estimated, a well-designed quota sample often has low sampling error. There is a danger of bias being introduced by interviewer choice, but the prime difficulty in interpreting results of opinion polls, whatever method of sampling is used, often arises from 'don't know' responses and people changing their minds between the opinion poll and the event to which it relates, e.g. a forthcoming election. *See* sample survey; sampling theory.

quotient The result of dividing one number or *polynomial by another. In

$$q = a \div b$$

q is said to be 'the quotient of a by b'. *See* division.

quotient group *See* normal subgroup.

quotient ring *See* ideal.

quotient rule (for differentiation) A method for differentiating a quotient using the formula

$$\frac{d}{dx}\left(\frac{u}{v}\right) = \frac{1}{v^2}\left(v\frac{du}{dx} - u\frac{dv}{dx}\right)$$

For example, it is possible to differetiate the function $y = (\sin x)/x$ by using $u = \sin x$ and $v = x$, so that

$$\frac{du}{dx} = \cos x \quad \text{and} \quad \frac{dv}{dx} = 1$$

The formula then gives

$$\frac{dy}{dx} = \frac{x\cos x - \sin x}{x^2}$$

Compare product rule.

R

$\mathbb{R}$ Symbol for the set of all *real numbers.

$\mathbb{R}^n$ Symbol for *Euclidean space of n dimensions.

RA *Abbreviation for* *right ascension.

radial Directed along a radius.

radial component *See* velocity; acceleration.

radian Symbol: rad. The SI *supplementary unit of plane angle, equal to the angle subtended by an arc of unit length at the centre of a circle of unit radius. *See* angular measure.

radian measure *See* angular measure.

radical The *root of a quantity as indicated by the sign $\sqrt{\ }$ (the *radical sign*). A number (the *index*) placed to the left of the sign shows the type of root, e.g. $\sqrt[4]{\ }$ is a fourth root; if there is no number the root is a square root.

radicand The number or expression under a *radical sign; for instance, x in $\sqrt{x}$.

radius (*plural* **radii**) **1.** The distance from the centre of a circle to any point on its circumference, or the distance from the centre of a sphere to any point on its surface.
2. Any line segment joining the centre of a circle (or sphere) to a point on its circumference (or surface).
3. (of a conic) *See* focal radius.
4. *See* polygon.

radius of convergence *See* power series.

radius of curvature *See* curvature.

radius of gyration Symbol: k. A length representing the distance in a rotating system between the point or line about which rotation takes place and the point at which (or from which) a transfer of energy has the maximum effect. In a system with total mass m and *moment of inertia I, the radius of gyration about the rotational axis is given by $\sqrt{(I/m)}$; k can be considered as the radius of a thin ring, mass m, coaxial with the rotational axis and with moment of inertia equal to that of the body.

radius vector *See* polar coordinate system.

radix **1.** A *root.
2. A number that is the base of a number system or *logarithm.

Radon's theorem (J.K.A. Radon, 1921) Any set S of $n+2$ points in $\mathbb{R}^n$ can be *partitioned into two subsets whose *convex hulls meet; a point of intersection is called a *Radon point* of S.

Radon transform A *transform discovered by the Austrian mathematician Johann Karl August Radon (1887–1956) in 1917. In two dimensions, the transform reconstructs a function from the values of its *integral over a set of lines in the plane. The basic idea is similar to that of the *Cauchy–Crofton formula*, which calculates the length of a curve by counting the number of times the curve crosses lines; it was discovered in the 19th century and is named after A.-L. Cauchy and the Irish mathematician M.W. Crofton.

In three dimensions, integrals over planes are used in the Radon transform. It is the basis for the reconstruction of images from medical computed tomography scans. A numerical adaptation is the commonly used *Hough transform.

Ramanujan, Srinivasa Aaiyangar (1887–1920) Indian mathematician, largely self-taught, who while in Europe between 1914 and 1917 published 21 papers, some in collaboration with G.H. Hardy, mainly on number theory.

Ramsey's theorem (F.P. Ramsey, 1928) A theorem of *set theory describable in terms of *graphs whose edges are coloured. If one uses k colours to paint the various edges of a complete graph G (each edge coloured with a single colour) and G has a sufficiently large number of vertices, then there is a complete *subgraph of some given size entirely of one colour. For example, if one colours the edges of the complete graph with 6 vertices either black or white, then the graph must contain either a black or a white triangle.

More generally, given any set of positive integers $n_1, n_2, \ldots, n_k$, there is an R (the *Ramsey number* associated with this set of integers) such that if the edges of the complete graph with R or more vertices is edge coloured with k colours, then, for at least one of the i, there is a complete subgraph with n_i vertices with all its edges having the ith colour. The Ramsey number is often denoted $R(n_1, n_2, \ldots, n_k)$, so the example above says that $R(3, 3) = 6$. Ramsey's theorem can be regarded as a generalization of the *pigeonhole principle.

random In everyday use, 'random' is synonymous with 'haphazard', but in statistics it has a special meaning within a probabilistic framework. A *random sample of r items from n is a selection in which each item has an equal chance of selection. Thus, if we select four numbers at random

from 1 to 20 without replacement the selections 1, 2, 3, 4 and 3, 16, 7, 19 are equally likely. Most people would not regard the first selection as haphazard, but they would the second. See random numbers; random sample; random variable; random walk.

random error See error.

randomization test See permutation test.

randomized blocks A widely used *experimental design in which the experimental units are grouped into blocks so that all units within any one block are as similar as possible with regard to some chosen characteristic that might affect observations. The number of units in a block must be equal to (or be a multiple of) the number of treatments. Each treatment is applied to exactly one unit (or to an equal number of units) in each block, and is allocated to units within each block at random.

When the experiment is analysed by the *analysis of variance, a component representing variability between blocks can be removed from the residual mean square, often leading to an increase in *precision. For example, if five different growth hormones are to be tested on piglets, then using five piglets from the same litter to form each block would reduce genetic differences. In more sophisticated designs using blocks, it is not necessary to have the number of units per block equal to or a multiple of the number of treatments, but the analysis becomes more complicated.

See also Latin square.

random numbers A sequence of digits or numbers with the property that, in the long run, all digits or numbers in the sequence will occur equally often, and in which the occurrence of any one digit or number in a particular position in the sequence is no guide to the occurrence of earlier or later members of the sequence. The traditional method of generating random numbers is to draw numbered tickets or marbles from a container, but computer-generated *pseudo-random numbers* are now widely used. A number of tests are available to verify whether these have the essential properties of randomness. Random numbers are widely used in sample selection, and in allocating treatments to units in designed experiments (*see* experimental design), and also in *Monte Carlo and *simulation studies.

random sample 1. A *sample selected from a finite *population is called a *random sample* if it is chosen in such a way that every possible sample of the same size has an equal probability of selection. When obtaining a sample of n items from a population of N items, if each item may occur once and only once in the sample the procedure is referred to as sampling *without replacement*. If each selected item is returned to the population after selection for possible reselection, the procedure is called sampling *with replacement*. In sampling with replacement some items may thus occur more than once in a sample. When sampling with or without replacement, if each item eligible for selection at any stage has an equal probability of selection, the procedure is called *simple random sampling*. *See also* cluster sample; sample survey; sampling theory; stratified sample.
2. A sample of values of a *random variable X with a known *distribution is obtained by selecting values so that:

(1) for a continuous distribution the probability that the chosen value lies in the interval $(x, x + \delta x)$ is $f(x)\,\delta x$, where $f(x)$ is the *frequency function;
(2) for a discrete distribution the probability that the value x_i is obtained is $f(x_i)$, given by the *frequency function.

Tables of randomly selected numbers from certain distributions have been published, but in *simulation or *Monte Carlo studies it is usual to generate random samples by computer.

The distribution of a random sample x_1, $x_2, \ldots, x_n$ is discrete, and the frequency function is given by $\Pr(X = x_i) =$ $p_i = 1/n$, $i = 1, 2, \ldots, n$. The sample cumulative distribution is a step function with step $1/n$ at each ordered sample value x_i. Many functions of the sample distribution are used as estimates of the population equivalents; the sample mean $\bar{x} = \Sigma x_i/n$, for example, is an analogue of the population mean μ.

See also plug-in estimator.

random variable A *variable X that may take any one of a finite or countably infinite set of real values, each with an associated probability, is a *discrete random variable*. The probabilities associated with each value are the elements of the *frequency function. If X may take continuous values in a range (finite or infinite) with probability $f(x)\,\delta x$ associated with each infinitesimal interval $(x, x + \delta x)$, where $f(x)$ is the frequency function, then it is a *continuous random variable*. The convention is to use capital italic letters, e.g. X, Y, to denote a random variable and the corresponding lower-case letter, with suffix if needed, to denote an observed value of that variable. The distinction between a random variable and an ordinary mathematical variable is the association of a probability distribution with the former. An alternative name for a random variable is a *variate*. *See* distribution.

random walk A simple random walk is exemplified by a particle, at some integral point $x = k$ on the x-axis, which moves at time t_1 either to $x = k + 1$ (a step to the right) with *probability p or to $x = k - 1$ (a step to the left) with probability $1 - p$. A step to the left or right with these probabilities is repeated from this new position at time t_2 and then at times $t_3, t_4, \ldots$. The walk may cease if an *absorbing barrier* is reached. Many of the principles can be illustrated by a gambling game in which a player with initial capital k wins one unit with probability p or loses one unit with probability $1 - p$. In this case there is an absorbing barrier at $x = 0$ when the gambler becomes bankrupt. The concept may be generalized to allow several possible steps at each time t_i, or by the introduction of reflecting or elastic barriers (the latter allowing either reflection or absorption with specified probabilities) and to walks in two or more dimensions. A random walk is an example of a *Markov chain.

range 1. The set of values that can be assumed by the dependent variable for a given *function. For example, if for every number in the *domain $-1 \leqslant x \leqslant 1$ the function f is defined by $y = f(x) = 2x^3$, then the range of f is $[-2, 2]$. See also function.

2. The *set of values taken by a variable.

3. The difference between the largest and smallest values in a data set, or for a *random variable the length of the shortest interval which includes all nonzero values of the *frequency function. The range may be infinite.

rank 1. The ordinal associated with an ordered observation.

2. To arrange a set of objects in order, lowest to highest, on the basis of a characteristic. This may be a physical measurement such as height of individuals, or a subjective judgement as in the ranking of participants by judges in a talent contest or of preferences in a tasting test. In many cases it is possible to order observations according to some criterion without assigning exact measurements to individuals. For example, it is often possible to rank objects by height quite accurately without ever making precise measurements of height. Many nonparametric statistical tests are based on ranks and use these even if precise measurements of a characteristic are available. Ranked data are sometimes referred to as *ordinal data*. See nonparametric methods; order statistics.

3. (of a matrix) The maximum number of linearly independent columns of a *matrix, or (which is the same) the maximum number of linearly independent rows of the matrix. Equivalently, it is the order of the greatest nonzero *determinant that can be taken out of the matrix by selecting rows and columns. An $m \times n$ matrix is of *full rank* if its rank equals the minimum of m and n. The rank of a linear transformation (or matrix) is the dimension of its *image. See also augmented matrix.

rate of change See derivative.

rate of convergence See order (**12**).

ratio The quotient of two numbers or quantities indicating their relative sizes. The ratio of a to b is written as $a : b$ or a/b. The first term is the *antecedent* and the second the *consequent*.

The value of a ratio is unaltered if both terms are multiplied or divided by the same quantity. Thus $12 : 15$, $4 : 5$, and $24 : 30$ are equivalent ratios.

A *unitary ratio* has one of its terms equal to 1. For example, $1 : 8$ and $2 : 1$ are unitary ratios. The ratio notation can be extended to indicate the relative size of more than two quantities. For example, the ratio $a : b : c$ states that the ratios of the first to the second quantity, the second to the third, and the first to the third are equivalent to $a : b$, $b : c$, and $a : c$, respectively. Thus 25, 50, and 75 are in the ratio $1 : 2 : 3$.

See also inverse ratio; cross-ratio; division in a given ratio.

rational function (rational expression) The quotient of two *polynomial functions

$$f(x) = \frac{f_1(x)}{f_2(x)}$$

defined when $f_2(x) \neq 0$. An example is

$$f(x) = \frac{2x^2 + 3x + 4}{x^3 + 2}$$

When any factors common to f_1 and f_2 have been removed, the zeroes of the denominator are the poles of f (see singular point).

rationalize To remove *radicals from an equation, expression, etc. For example, the equation

$$\sqrt{(x+1)} = 2x$$

can be rationalized by squaring both sides to give

$$x + 1 = 4x^2$$

The denominator of $2/(\sqrt{5}+1)$ can be rationalized by multiplying both parts of the fraction by $\sqrt{5}-1$ and simplifying to $(\sqrt{5}-1)/2$.

rational number A number that is either an *integer or can be written as a quotient of two integers. For example, 1, 7, 540, 2/3, and 1/9 are rational numbers. Each rational number has a *decimal expansion that is either finite or periodic. The set of all rational numbers is denoted by $\mathbb{Q}$. *Compare* irrational number; *see also* Dedekind cut; real number.

rational operation Any of the operations addition, subtraction, multiplication, and division.

rational root theorem The theorem that if a *polynomial equation with integral *coefficients has a *root that is a rational number p/q (in its lowest terms), then the leading coefficient is divisible by q and the constant term is divisible by p.

ratio test A test for convergence or divergence of a given infinite *series, attributed usually to d'Alembert but also to Cauchy. In the series of positive terms

$$a_1 + a_2 + \cdots + a_n + a_{n+1} + \cdots$$

suppose that a_{n+1}/a_n tends to a *limit A as $n \to \infty$. Then when $A < 1$ the series converges (absolutely), when $A > 1$ the series diverges, and when $A = 1$ the test gives no information. *See* convergent series.

ray *See* half-line.

reaction The *force that results from the application of a force to a body in pushing,

pulling, lifting, or supporting the body. The reaction is exerted by the body itself, and acts in the opposite direction to the applied force. The force thus opposed is known as the *action*. By Newton's third law of motion, action and reaction are equal in magnitude but act in opposite directions.

real axis *See* Argand diagram.

real function (function of a real variable) A *function whose domain and codomain are sets of real numbers. An example is the function $f: x \mapsto x^3$, which maps the real numbers $\mathbb{R}$ to $\mathbb{R}$. *Compare* complex function.

real number A number that can be written in the form $\pm n.a_1 a_2 a_3 \ldots$, where n is an *integer and each a_i is one of the *digits 0 to 9, for example, 2, $\frac{1}{3} = 0.333\ldots$, -1.5, and $\pi = 3.14159\ldots$. Real numbers are either *rational or *irrational. The set of all real numbers is denoted by $\mathbb{R}$. The real numbers can be formally defined in terms of *Dedekind cuts of the rational numbers or Cauchy sequences (*see* metric space) of rational numbers.

There is a *one-to-one correspondence between the set of real numbers and the points of an infinite directed line containing a fixed origin. The positive real number $+a$ corresponds to the point whose distance from the origin is a units measured in the positive direction, and the negative number $-b$ corresponds to the point whose distance from the origin is b units measured in the negative direction. The number zero corresponds to the origin. In this context the line is called a *number line* or *real line*, and may be denoted by $\mathbb{R}^1$.

Compare complex number; *see* decimal; Euclidean space.

real part *See* complex number.

reciprocal **1.** The number or expression produced by dividing 1 by a given number or expression. Thus, the reciprocal of 2 is $\frac{1}{2}$ and the reciprocal of $1 + x$ is $1/(1 + x)$.

2. *See* inverse (of a matrix).

reciprocal curve The curve generated from a given curve by replacing each *ordinate by its reciprocal. Thus, the reciprocal curve of $y = 2x$ is $y = 1/(2x)$ (and vice versa).

reciprocal equation An equation that is unchanged (i.e. has the same *roots) if the variables are replaced by their reciprocals. Thus $x^2 + 1 = 0$ is a reciprocal equation since replacing x by $1/x$ gives $(1/x^2) + 1 = 0$, which simplifies to $1 + x^2 = 0$.

reciprocal matrix *See* inverse (of a matrix).

reciprocal ratio *See* inverse ratio.

reciprocal series (of a given series) The *series whose terms are each reciprocals of the terms of the given series. A *harmonic series is the reciprocal series of an arithmetic series.

reciprocal spiral *See* spiral.

Recorde, Robert (*c*.1510–58) Welsh mathematician noted for *The Whetstone of Witte* (1557), the first significant algebra textbook written in English, which introduced into mathematics the familiar sign = to represent equality. Recorde also produced comparable works on arithmetic, *The Grounde of Artes* (1543), and on geometry, *The Pathway to Knowledge* (1551).

rectangle A *quadrilateral with all four angles right angles. The pairs of opposite sides are equal. If all four sides are equal, the rectangle is a square.

rectangular coordinate system A *coordinate system in which the axes are perpendicular. *See* Cartesian coordinate system.

rectangular distribution *See* uniform distribution

rectangular hyperbola *See* hyperbola.

rectifiable Describing a curve that has a finite length.

rectify To find the length of (a curve).

rectilinear motion Motion along a straight line.

recurrence relation A relation between successive values of a *function or *sequence that allows the systematic calculation of values, given an initial value (or values) and the relation. For example, the *Fibonacci sequence may be generated by the recurrence relation

$$a_{n+1} = a_n + a_{n-1}$$

and the initial values $a_1 = a_2 = 1$.

Formulae of this type are sometimes called *recursive relations*, and the computation is then described as *recursive. Sophisticated recurrence relations are used in large-scale computational problems and may reduce round-off difficulties inherent in more direct types of calculation.

See difference equation; dynamic programming.

recurring decimal *See* decimal.

recursive A *function or *sequence is defined recursively if

(1) the value of f(0) and
(2) the value of f$(n + 1)$, given the value of f(n)

are both stated. For example, the *factorial function may be defined by

(1) f(0) = 1 and
(2) f$(n + 1) = (n + 1)$f(n) for $n = 0$, 1, 2, ...

Recursive definitions are also called *inductive* definitions or *recursions*. *See* recurrence relation.

reduced row echelon form An $m \times n$ matrix A is in reduced row echelon form if it is in *row echelon form and the first nonzero element in each row is 1 and is the only nonzero in its column. For example, the matrices

$$\begin{pmatrix} 1 & 0 & 2 & 0 \\ 0 & 1 & -3 & 0 \\ 0 & 0 & 0 & 1 \\ 0 & 0 & 0 & 0 \end{pmatrix}, \quad \begin{pmatrix} 1 & -1 & 0 & 1 \\ 0 & 0 & 1 & 2 \\ 0 & 0 & 0 & 0 \end{pmatrix}$$

are in reduced row echelon form. Reduced row echelon form is produced by *Gauss–Jordan elimination. *Reduced column echelon form* is defined by replacing 'row' by 'column' in the above definition.

reducible equation *See* reducible polynomial.

reducible fraction A common fraction such as $\frac{4}{6}$ in which the numerator and denominator have a *common factor greater than unity. *Compare* irreducible fraction.

reducible polynomial A *polynomial is reducible over a *field F if it can be factored (*see* factor) into two polynomials having coefficients in F. For instance, $x^2 - 1$ is reducible over $\mathbb{R}$ since it can be factored into $(x - 1)(x + 1)$, in which the coefficients are real numbers. The polynomial $x^2 + 1$ is an *irreducible polynomial* over $\mathbb{R}$ because its factors, $x + i$ and $x - i$, have coefficients in $\mathbb{C}$, the field of complex numbers.

A *reducible equation* over a field F is an equation of the form $P = 0$, where P is a reducible polynomial over F. An *irreducible equation* is similarly defined.

reducible radical A *radical that can be written in a rationalized form, i.e. a form not containing radicals. For example, $\sqrt{4}(= 2)$ and $\sqrt{16}(= 4)$ are reducible radicals. *Compare* irreducible radical.

reductio ad absurdum *See* indirect proof.

reduction formulae 1. Formulae in plane trigonometry that give trigonometric functions of an angle plus or minus a number of right angles in terms of functions of that angle. For example:

$$\tan(90^\circ \pm \theta) = -(\pm \cot \theta)$$
$$\sin(90^\circ \pm \theta) = \cos \theta$$
$$\cos(90^\circ \pm \theta) = -(\pm \sin \theta)$$
$$\tan(180^\circ \pm \theta) = \pm \tan \theta$$
$$\sin(180^\circ \pm \theta) = -(\pm \sin \theta)$$
$$\cos(180^\circ \pm \theta) = -\cos \theta$$
$$\tan(270^\circ \pm \theta) = -(\pm \cot \theta)$$
$$\sin(270^\circ \pm \theta) = -\cos \theta$$
$$\cos(270^\circ \pm \theta) = \pm \sin \theta$$

2. Formulae expressing an *integral in terms of a simpler integral, in particular one of reduced *power. Examples of reduction formulae are given in the Appendix.

Reed–Solomon code (I.S. Reed and G. Solomon, 1960) Over a *field $\mathbb{F}_q = \{a_1, a_2, \ldots, a_q\}$, with $a_q = 0$ and q elements, the kth Reed–Solomon code is the *linear code of length $q - 1$, defined as the set of all $(q - 1)$-tuples $(f(a_1), f(a_2), \ldots, f(a_{q-1}))$, where f ranges over all *polynomials of degree at most $k - 1$ with coefficients in $\mathbb{F}_q$.

These codes are in very common use, for example in storage devices (e.g. tape, CD, DVD, barcodes), mobile communications (e.g. cellular telephones, microwave links), satellite communications, modems and in digital television.

re-entrant angle An interior angle in a (concave) *polygon that is greater than 180°. *Compare* salient angle.

reference angles *See* related angles.

reference axis *See* axis.

reflection 1. (in a point) A *transformation, involving a fixed point C, such that the line segment joining a point to its image is bisected at C.

2. (in a line) A transformation, involving a *mirror line* or *axis l*, such that the line segment joining a point to its image is perpendicular to *l* and has its mid-point on *l*. Reflection of points of the plane in the line with equation $y = x \tan \theta$ maps the point with coordinates (x, y) onto the point

$$(x \cos 2\theta + y \sin 2\theta, \ x \sin 2\theta - y \cos 2\theta)$$

In particular, reflection in the *y*-axis maps (x, y) onto $(-x, y)$.

Reflection in a plane is defined similarly, i.e. the line segment joining a point to its image is perpendicular to the plane, and its mid-point lies in the plane.

See also symmetry.

reflection matrix *See* orthogonal matrix.

reflection property The *focal property of a conic. *See* ellipse; hyperbola; parabola.

reflex angle An angle between $180°$ and $360°$.

reflexive relation A *relation R on a *set *A* is reflexive if, for all $a \in A$, $a \, \mathrm{R} \, a$. The relation 'is divisible by', for example, is reflexive on the set of natural numbers as every number is divisible by itself. Relations such as 'greater than', which are not reflexive, are described as *irreflexive*.

Regiomontanus, also known as **Johann Müller** (1436–76) German astronomer and mathematician whose posthumously published *De triangulis omni modis* (1533, On All Classes of Triangles) is one of the first works on trigonometry as a discipline independent of astronomy. It was also one of the first works to substitute for the chords of antiquity the sine and cosine of the Arab mathematicians.

region A connected subset of a Euclidean space, sometimes called a *domain*. Special types, such as open, closed, and bounded regions, are often considered. For example, the set of points forming the interior of a circle is an open region, while a circle together with its interior represents a closed region. *See* connected set; open set.

region of convergence *See* functional series.

regressand *See* regression.

regression A *model that describes the dependence of the mean value of one random variable on one or more other variables; for example, a formula giving the average weight of an infant of a given height and age. More formally, if the mean value of a random variable *Y* for a fixed value *x* of another variable is written as $\mathrm{E}(Y|x)$ and called the *mean of Y conditional upon x*, then, when *x* varies, $\mathrm{E}(Y|x)$ is a function of *x* called the *regression of Y on x*.

The simplest case is that of linear, or straight-line, regression:

$$\mathrm{E}(Y|x) = \beta_0 + \beta_1 x$$

The random variable *Y* is called the *response, dependent* or *effect variable*, or the *regressand*, and *x* is called the *explanatory, fixed, independent, predicated, predictor*, or *cause variable*, or the *regressor*. The parameters β_0 and β_1 are *regression coefficients*, and are usually unknown. Regression analysis is concerned with estimating these parameters. If *Y* is assumed to be normally distributed with constant but usually unknown variance σ^2, which is independent of *x*, then, given a sample of *n* independent pairs of observations $(x_i, y_i), i = 1, 2, \ldots, n$, the *maximum likelihood estimators b_0 and b_1 of β_0 and β_1 are given by the method of *least squares, i.e. by minimizing $\Sigma_i (y_i - \beta_0 - \beta_1 x_i)^2$, and are

$$b_1 = \sum_i (x_i - \bar{x})(y_i - \bar{y}) \Big/ \sum_i (x_i - \bar{x})^2$$

$$b_0 = \bar{y} - b_1 \bar{x}$$

where $\bar{x} = \Sigma_i x_i / n$ and $\bar{y} = \Sigma_i y_i / n$.

An alternative equivalent model is $y_i = \beta_0 + \beta_1 x_i + \varepsilon_i$, where the ε_i are independently normally distributed with zero mean and variance σ^2.

Regression may be extended in many ways, including *multiple linear regression* where there are $p > 1$ explanatory variables $x_1, x_2, \ldots, x_p$ and

$$E(Y|x_1, x_2, \ldots, x_p)$$
$$= \beta_0 + \beta_1 x_1 + \beta_2 x_2 + \cdots + \beta_p x_p \quad (1)$$

The expression (1) considered as a function of the parameters β_i is a linear function of these parameters, and the term *linear regression model* is used in this sense for models that are not necessarily linear in the x_i. For example, if $x_i = x^i$, the regression function (1) is a polynomial of degree p.

More complicated methods of estimation are needed if the assumptions of the normality of Y and the independence or constancy of its variance are dropped.

Multiplicative models can often be reduced to linear models by taking logarithms, as in the *loglinear model. More generally, the right-hand side of (1) may be generalized to any function $f(\mathbf{x}, \boldsymbol{\theta})$, where $\mathbf{x}$ is a vector of p explanatory variables and $\boldsymbol{\theta}$ is a vector of q unknown parameters, giving

$$E(Y|\mathbf{x}) = f(\mathbf{x}, \boldsymbol{\theta})$$

If, in the linear regression model, the conditional mean $\mu = E(Y|\mathbf{x})$ is replaced by some function of the mean $g(\mu)$, the model is a *generalized linear model.

See also Gauss–Markov theorem; least squares; line of best fit.

regressor *See* regression.

regula falsi *See* false position, rule of.

regular function *See* analytic function.

regular graph *See* graph.

regular group A *permutation group that has an order equal to the number of members of the set of objects permuted.

regular polygon A *polygon that has all its sides equal and all its interior angles equal.

regular polyhedron A *polyhedron that has regular congruent faces and congruent *polyhedral angles.

regular prime A *prime p that does not divide the *class number of the *field obtained by adjoining to the rational numbers a primitive pth *root of unity. The distinction between regular and irregular primes was made by E. Kummer in 1850 in his important work on *Fermat's last theorem.

regular prism A right *prism that has regular polygons as bases.

regular pyramid A *right pyramid whose base is a regular polygon.

regular sequence *See* metric space.

regular star polygon *See* polygon.

regular star polyhedron *See* polyhedron.

Reinhold, Erasmus (1511–53) German mathematician and astronomer noted for his important *Tabulae prutenicae* (1551, Prussian Tables), the first tables of planetary motion to be based on the heliocentric theory of Copernicus.

related angles (reference angles) Angles that have the same absolute values for their *trigonometric functions. For example, $20°$, $160°$, $200°$, and $340°$ are related angles.

relation 1. An association between, or property of, two or more objects. Thus '$x = y$' and 'a lies between b and c' are relations, but 'N is prime' is not. A *binary relation* or *correspondence* (e.g. 'is equal to') involves two objects, a *ternary relation* (e.g. 'lies between') involves three, and an *n-ary relation* involves n objects.

A relation may be specified by listing all the instances for which it holds. More formally, a binary relation or correspondence R on sets X and Y is defined as the set of all ordered pairs (x, y) with $x \in X$ and $y \in Y$ for which the statement 'x has relation R to y' is true (written as $x R y$).

If $Y = X$ then R is a *relation on X*. For example, the relation 'is a factor of' on the set of positive integers is the set of ordered pairs of positive integers (a, b) for which a divides b, i.e. $(1, 1)$, $(2, 6)$, $(3, 12)$, etc.

The *inverse* of a binary relation R (on sets X and Y) is the relation S (on sets Y and X) such that $y S x$ if and only if $x R y$. For example, the inverse of the relation 'is a factor of' on the positive integers is the relation 'is a multiple of' on the positive integers and consists of pairs $(1, 1)$, $(6, 2)$, $(12, 3)$, etc.

A mapping or *function f with domain X and codomain Y may be regarded as a binary relation R, with $x R y$ equivalent in meaning to 'x is mapped by f to y' or '$y = f(x)$'.

A binary relation R on a set X is

(1) *reflexive* if $x R x$ for all $x \in X$;
(2) *symmetric* if $x R y$ always implies $y R x$;
(3) *transitive* if $x R y$ and $y R z$ together always imply $x R z$.

A relation satisfying properties (1), (2), and (3) is an *equivalence relation*. The relation of equality on a set is an example of an equivalence relation. The relation 'is a factor of' on the positive integers is reflexive and transitive, but not symmetric.

See also equivalence class; partial order.

2. (in a group) *See* generator (of a group).

relative *See* index.

relative acceleration *See* relative velocity.

relative complement *See* difference.

relative density (specific gravity) The ratio of the *mass of a solid or liquid to the mass of an equal volume of water at $4\,^{\circ}\mathrm{C}$ (or some other specified temperature). For gases, relative density is the ratio of the density of the gas to the density of air or hydrogen at the same temperature and pressure.

relative error If y is an approximation to a number x, then its relative error is $|x - y|/|x|$, which is defined only for $x \neq 0$.

relative frequency *See* frequency.

relatively prime (coprime) Describing two *integers that have no divisors in common other than $+1$ and -1. Thus 5 and 12, -18 and 35, and 72 and 91 are relatively prime pairs, but 6 and 9 are not relatively prime.

relative maximum *or* **minimum** *See* turning point.

relative velocity If two bodies P and Q have velocities $\mathbf{v}_P$ and $\mathbf{v}_Q$, then the velocity of P relative to Q is $\mathbf{v}_P - \mathbf{v}_Q$. The velocity of P is then the *vector sum of the velocity of P relative to Q and the velocity of Q itself. A similar relation holds for the relative acceleration of P and Q: the acceleration of P is the vector sum of the acceleration of P

(a) (b)

relative velocity (a) Velocities of P and Q: (b) velocities relative to Q.

relative to Q and the acceleration of Q itself.

These relations hold only for speeds very much smaller than the speed of light, c. For two bodies P and Q moving in the same direction with speeds v_P and v_Q, the relativistic expression for the magnitude of the velocity of P relative to Q is

$$\frac{|v_P - v_Q|}{1 - (v_P v_Q / c^2)}$$

See also relativity.

relativistic mass The mass of a body when it moves at speeds approaching the speed of light, c $(= 3 \times 10^8$ metres per second). According to the special theory of *relativity (and as experimentally verified), the mass m of a moving body exceeds the *rest mass, m_0, of the body and is a function of the body's speed v:

$$m = \frac{m_0}{\sqrt{(1 - v^2/c^2)}}$$

The increase in mass is negligible except at very high speeds. The total energy of the system is given by

$$E = mc^2 = \frac{m_0 c^2}{\sqrt{(1 - v^2/c^2)}}$$

and the relativistic momentum by

$$p = \frac{m_0 v}{\sqrt{(1 - v^2/c^2)}}$$

It can be shown that

$$E^2 = p^2 c^2 + m_0^2 c^4$$

These equations are the basis of *relativistic mechanics.

relativistic mechanics The study of the motion of particles or bodies that move at speeds comparable to the speed of light, c, i.e. at relativistic speeds. The equations must conform to the principles of the special and general theories of *relativity, and reduce to the equations used in *classical

mechanics (or nonrelativistic quantum mechanics) for speeds considerably less than c. There is conservation of mass–energy and of momentum in relativistic systems. *See also* relativistic mass.

relativity A theory of physics conceived by Albert Einstein and developed in two stages. The *special theory of relativity*, published in 1905, is concerned with the phenomena of physics as experienced by observers moving relative to one another at constant velocity. It is thus restricted to observers in inertial *frames of reference. The *general theory of relativity* was published in 1916 and extends the theory to observers in noninertial, i.e. accelerated, frames of reference. The two theories led to a re-analysis of the concepts of space and time and of the interrelationship between measurement and observer.

(1) The special theory has two fundamental postulates. The first is a generalization of work by Poincaré and Lorentz. It states that the laws of physics can be expressed in the same mathematical form in all inertial frames of reference: it is impossible to distinguish between two inertial frames by any physical experiment, be it mechanical, optical, or electrical. The second postulate, which follows from the first, states that in free space every inertial observer measures the same value of the speed of light relative to himself: the speed of light in free space must thus be a universal constant.

The equations for transforming the position and motion in one inertial frame to a different inertial frame must satisfy these two postulates. The equations used are those of the *Lorentz transformation*. For example:

$$x' = \beta(x - vt)$$

$$y' = y$$

$$z' = z$$

$$t' = \beta(t - vx/c^2)$$

where $\beta = 1/\sqrt{(1 - v^2/c^2)}$ and v is the magnitude of the relative velocity of the two frames. The Lorentz transformation replaces the *Galilean transformation* of classical mechanics and forms the basis of the mathematical treatment of special relativity. It does show, however, that the idea of the universality of time is invalid. It was Minkowski who realized that the two postulates of special relativity require space and time to be treated not as separate entities but as a unified four-dimensional concept, referred to as *spacetime. Spacetime subsequently became the frame of all later extensions of the theory of relativity.

Rigorous development of the two postulates enabled Einstein to invalidate many of the tacit assumptions of classical physics and to show that *Newton's laws of motion hold only for low speeds, i.e. speeds small in comparison with the speed of light, $c (= 3 \times 10^8$ metres per second). The equations of special relativity must (and do) reduce to those of classical mechanics for low speeds. The special theory was able to explain certain predictions, such as the *Lorentz–Fitzgerald contraction, and certain experimental observations that had already been made. There were also some startling predictions that followed from the theory. These include the relationship between mass and energy expressed in the *mass–energy equation and the concepts of *relativistic mass, *rest mass, and *time dilation. These have since been verified experimentally to considerable accuracy.

(2) The general theory of relativity is not restricted to inertial frames of reference and is consequently much more complex mathematically than the special theory. It is based on the *equivalence principle – that the physical effects of a gravitational field are indistinguishable from the physical effects experienced by an observer in an accelerated frame of reference. In addition the laws of physics in an accelerated frame cannot be distinguished from the laws in inertial frames: the laws are therefore invariant with respect to all possible frames of reference. The mathematical consequence is, essentially, a geometrical theory of gravitation. The geometry of spacetime is affected by the presence of matter: matter curves space in its vicinity. It is the curvature of space that controls the motions of bodies. The curvature is described in terms of four-dimensional *differential geometry, and for a particular collection of matter can be calculated from the *field equations* of general relativity. These are tensor equations for what is known as the *metric tensor*; the metric tensor completely describes the space.

General relativity reduces to the Newtonian theory of gravitation for small masses and low speeds. The theory has not, however, been conclusively proved, although there is substantial evidence. Experimental tests must verify the predictions of the general theory where they deviate from those of Newtonian theory and also where they deviate from those of variants of general relativity.

reliability A term used in several senses in specialized statistical applications, but chiefly as the *probability $R(t)$ that a device will not fail in the interval $(0, t)$. If the lifetime distribution function is $F(t)$, then $R(t) = 1 - F(t)$.

remainder 1. A number remaining after one number is divided into another an exact number of times.
2. (of a series) The infinite *series that starts after a specified term of a given series. For the series

$$a_1 + a_2 + \cdots + a_n + \cdots$$

the remainder after N terms is given by

$$R_N = a_{N+1} + a_{N+2} + \cdots$$

If the original series is convergent, then so is a remainder of the series. If S is the sum of the given series and s_N is the *partial sum of the first N terms, then

$$S = R_N + s_N$$

R_N can usually only be estimated, but can still be used to give a good approximation of S.

remainder theorem The theorem that a *polynomial P(x) divided by $x - a$ has a *remainder equal to P(a), i.e. the remainder is the value obtained by substituting a for x in the polynomial. If the remainder is zero the theorem reduces to the *factor theorem.

removable discontinuity *See* discontinuity.

removable singularity *See* singular point.

repeated decimal *See* decimal.

repeated root *See* multiple root.

repellor *See* chaos.

replication In a designed experiment, the number of experimental units to which each treatment is applied. Equal replication of all treatments is a common feature of many *experimental designs.

representation (of a group) A *homomorphism of a group of abstract symbols into a group of more familiar objects, such as a group of permutations or a group of matrices. In the former case it is a permutational representation, and in the latter case a matrix representation. A representation can be either one-to-one (injective), in which case it is called *faithful*, or not. For example, the group generated by two symbols J and K, which satisfy the relations $J^2 = K^2 = (JK)^2$ and $J^4 = I$, has eight elements that can be expressed as I (the identity element), J, K, JK, J^2, J^3, K^3, and KJ. It has a matrix representation since it is *isomorphic to a certain group of 4×4 matrices, and it also has a permutational representation as a certain group of permutations of eight symbols. *See* generators.

representative sample A *sample that in certain respects is typical of the *population from which it is chosen. *See* quota sample; sample survey; stratified sample.

repunit A natural number, such as 111, that has all its digits equal to 1. In base 10, the repunit with n digits each equal to 1 is known to be prime for $n = 2, 19, 23, 317$, and 1031.

residuals In statistics, the differences e_i between observed values y_i and values $\hat{y}_i$ predicted by a model, i.e. $e_i = y_i - \hat{y}_i$. Residuals are sometimes called *errors*, but care should be taken to distinguish between a residual and the realized value ε_i of a random variable which specifies the stochastic or random component in a model. For example, in simple linear *regression, a model

$$y_i = \alpha + \beta x_i + \varepsilon_i$$

is specified, where α, β, and ε_i are unknown, and ε_i is the realized value of a random variable ε, often assumed to be normally distributed with mean zero and (usually unknown) variance σ^2. If the *least-squares estimators of α and β are a and b, then $e_i = y_i - a - bx_i$, and the e_i are correlated, even though the ε_i are independent. *See also* residual variation.

residual sum of squares *See* analysis of variance.

residual variation The variation that is not accounted for by a *model fitted to data and which is determined by the *residuals. For example, from a set of data (x_i, y_i), $i = 1, 2, \ldots, n$, a linear regression $y = a + bx$ may be determined. If the estimate of y_i from this model is denoted by $\hat{y}_i$, the residual is given as $e_i = y_i - \hat{y}_i = y_i - a - bx_i$.

The residual variation is measured by the *error mean square in an *analysis of variance. Although it is determined by a different method in practice, the error mean square in the above example has the value

$$\frac{1}{n-2} \sum_{i=1}^{n} e_i^2$$

The divisor $n - 2$ represents the *degrees of freedom.

residue 1. The residue of a complex function $f(z)$ at a pole a is the coefficient of $(z - a)^{-1}$ in its *Laurent expansion about a. It is important because of *Cauchy's residue theorem*: that a *contour integral of a function $f(z)$ along a closed path Γ can often be evaluated in terms of the residues of f at the poles enclosed by Γ.

For example,

$$\int_{\Gamma} \frac{c}{z - a} \, dz = 2\pi i c$$

if Γ is a simple closed curve enclosing the point a.

2. If m and n are natural numbers, and a is an integer not divisible by m, then a is a residue of m of order n if there is a number x such that

$$x^n \equiv a \pmod{m}$$

If the congruence does not have a solution for x, then a is a *nonresidue* of m of order n. For example, the congruence

$$x^3 \equiv 1 \pmod{9}$$

has a solution $x = 4$, so 1 is a residue of 9 of order 3.

When $n = 1$, the congruence $x^n \equiv a \pmod{m}$ always has the solution $x = a$, so every a not divisible by m is a residue of order 1. When $n = 2$, a residue of order 2 is called a *quadratic residue*. For example, if $m = 7$ the numbers 1, 2, and 4 are quadratic residues since $1^2 \equiv 1 \pmod{7}$, $3^2 \equiv 2 \pmod{7}$, and $2^2 \equiv 4 \pmod{7}$; whereas the numbers 3, 5, and 6 are quadratic nonresidues.

A necessary condition for a to be a residue of m of order n, with a and m coprime, is that

$$a^{\phi(m)/d} \equiv 1 \pmod{m}$$

where $\phi(m)$ is *Euler's phi function of m, and d is the greatest common divisor of $\phi(m)$ and n. The condition is known as *Euler's criterion*.

See congruence modulo n.

residue class *See* congruence class.

resolution (of vectors) The process of determining two or more *vectors that have an equivalent effect to a given vector; the given vector is said to be *resolved* into *components.

resonance A phenomenon occurring in an oscillating system undergoing *forced oscillation whereby the system responds with maximum amplitude to the periodic driving force. This happens when the frequency of the driving force equals the frequency of the natural undamped oscillation of the system (*see* free oscillation), and can be a source of potential danger in mechanical structures.

response variable *See* regression.

restitution Restoration to some original state, especially of shape following an elastic deformation. *See also* Newton's law of restitution.

rest mass Symbol: m_0. A constant property of any material particle or body, equal to the mass of the particle or body when it is at rest. The *rest-mass energy*, E_0, of the particle or body is given by $m_0 c^2$, where c is the speed of light. The concept of rest mass is important in the theory of *relativity. Classical, Newtonian physics makes no distinction between rest mass and mass in general. When a particle or body is in motion, its mass m increases (*see* relativistic mass). For all velocities the total energy, $E = mc^2$, is equal to the sum of the rest-mass energy and the kinetic energy. This gives the relativistic expression for kinetic energy.

resultant 1. The *vector produced by adding two or more vectors.
2. The *vector quantity that has an equivalent effect to two or more given vectors. For a system of forces, say, acting at the same point, the resultant is a single force given by the vector sum of the forces (*see* parallelogram law). For a system of parallel or coplanar forces, the resultant can be a single force or a single *couple.
3. *See* eliminant.

retraction Given a *topological space X and a subspace Y, a continuous map f: $X \rightarrow Y$ is called a retraction if f keeps all points of Y fixed.

reverse Polish notation *See* postfix notation.

revolution *See* axis; solid of revolution; surface of revolution.

Rheticus, Georg Joachim (1514–76) Austrian mathematician and astronomer best known for his services as amanuensis to Copernicus. He was also responsible for the posthumously published *Opus palatinum de triangulis* (1596, The Palatine Work on Triangles), a table of trigonometric functions, which he was one of the first to define as ratios of the sides of a right triangle rather than by chords.

Rhind papyrus One of the prime sources for the history of Egyptian mathematics, and one of the earliest surviving mathematical texts. It is a scroll about 6 metres long, kept in the British Museum, and dating from 1650 BC. Despite the widespread belief that the ancient Egyptians were highly sophisticated mathematicians, the papyrus deals with very simple arithmetical problems, many of which are concerned with the division of 2 by the odd numbers 3–101. As they recognized only unit fractions, this led them to establish,

for example, that $2/9 = 1/6 + 1/18$. Typical problems considered are:

(1) A quantity and its half together become 16. What is the quantity?
(2) Find the volume of a cylindrical granary of diameter 9 and height 10.

In terms of symbolism, the Egyptians lacked a place-value notation, zero, decimal points, and signs for plus, minus, multiplication, and division. The papyrus is named after the Scottish antiquarian Alexander Henry Rhind (1833–63).

rhombohedron (*plural* **rhombohedra**) A hexagonal *prism.

rhomboid A *parallelogram that has adjacent sides unequal.

rhombus (rhomb) A *parallelogram that has all its sides equal.

rhumb line *See* loxodrome.

Ricci-Curbastro, Gregorio (1853–1925) Italian mathematician who in 1884 began to develop his absolute differential calculus, later called *tensor analysis.

Ricci flow *See* Poincaré conjecture.

Richardson extrapolation (deferred correction) Let f(h) be an approximation to an unknown quantity a, of the form f(h) $= a + c_1 h^2 + c_2 h^4 + \cdots$, where h is small and the c_i are unknown constants. Given the values f(h) and f($h/2$), Richardson extrapolation forms the new approximation $\hat{f} = \frac{1}{3}(f(h) - 4f(h/2))$ to a. From

$$f(h) = a + c_1 h^2 + c_2 h^4 + \cdots$$

$$f(h/2) = a + c_1 h^2/4 + c_2(h/2)^4 + \cdots$$

it can be seen that the new approximation is obtained by subtracting 4 times the second equation from the first to eliminate the h^2 terms, and so $\hat{f}$ can be expected to be a more accurate approximation to a than either f(h) or f($h/2$). The process is named after the English mathematician Lewis Fry Richardson (1881–1953). Given a sequence f(h), f($h/2$), f($h/4$), ... of approximate values, this process can be repeated to eliminate higher and higher powers of h, in a procedure known as *extrapolation to the limit*. *See* Romberg integration.

Richard's paradox A *paradox discovered by J.A. Richard in 1905. All the English words (or phrases) that denote real numbers can be enumerated as follows. Group together all English words of one letter and order them lexicographically, and then repeat the process for words of two letters, and then three letters, and so on. If we remove from this enumeration all those words that do not denote real numbers, then we are left with an enumeration E of English words denoting real numbers. Call the nth real number in E the nth *Richard number*. Consider the expression 'the real number whose nth decimal place (for each n) is 1 if the nth decimal place of the nth Richard number is not 1, and whose nth decimal place is 2 if the nth decimal place of the nth Richard number is 1'. This expression seems to denote a Richard number, say the kth, but by definition it differs from the kth Richard number in the kth decimal place.

Riemann, Georg Friedrich Bernhard (1826–66) German mathematician noted for his 1854 lecture *Über die Hypothesen welche der Geometrie zu Grunde liegen* (On the Hypotheses that Lie at the Foundations of Geometry) in which he developed his system of *non-Euclidean geometry. He further expressed for the first time the intimate connections between our understanding of space and our geometrical assumptions. In 1859, while searching for a better approximation to the number of primes than the prime number theorem, he introduced the *Riemann zeta function, and also formulated the Riemann hypothesis.

Riemann–Christoffel curvature tensor *See* Riemannian geometry.

Riemann hypothesis *See* Riemann zeta function.

Riemannian geometry A type of *non-Euclidean geometry developed by Riemann in 1854. In Euclidean geometry, the distance between two neighbouring points on a plane is given by a relationship of the form

$$ds^2 = dx^2 + dy^2$$

where rectangular Cartesian coordinates are used. More generally, the relationship can be written as

$$ds^2 = A\,dx^2 + B\,dx\,dy + C\,dy^2$$

where A, B, and C depend on x and y. Gauss considered this case and showed that it is possible to determine the *curvature at a point intrinsically in terms of A, B, and C. Riemann generalized this approach into the study of any type of *metric space in any number of dimensions. What is now called a *Riemannian space* is a space with n coordinates $(x_1, x_2, \ldots, x_n)$ in which the distance between neighbouring points is given by a quadratic form,

$$ds^2 = \sum g_{ij}(x)\,dx_i\,dx_j$$

where the $g_{ij}(x)$ are functions of $x_1, x_2, \ldots, x_n$. In the original form of Riemannian geometry, ds^2 was required to be always positive, although this is not the case in applications to general relativity theory. Usually, the coefficients $g_{ij}(x)$ are taken to have a nonvanishing determinant. The $g_{ij}(x)$ are the components of a symmetric covariant *tensor field (the *metric tensor*). In Riemannian geometry, the distance between two points can be determined by an integral of ds. *Riemannian curvature* is defined by an expression involving the metric tensor of the Riemannian space and a tensor known as the *Riemann–Christoffel curvature tensor* after Gauss and Elwin Bruno Christoffel (1829–1900).

Riemannian geometry had a profound effect on the way people thought about geometry and on the development of tensor analysis. It was also essential in the formulation of general *relativity and in later attempts to develop a unified field theory. The term is sometimes used in a more restricted sense to describe a particular type of non-Euclidean geometry in which the plane is interpreted as a sphere and a line as a great circle on the sphere. In this form of non-Euclidean geometry, Euclid's *parallel postulate is replaced by the postulate that no line can be drawn parallel to a given line through a point lying outside the line. Moreover, Euclid's second postulate (that a line can be extended indefinitely in both directions) is not applicable. This non-Euclidean geometry is also called *elliptic geometry*.

Riemannian metric A measurement of distance on a differential manifold M. It is usually described in terms of a scalar product $g_{ij}(x)$ on each tangent space $T_x(M)$ to M. Using the Riemannian metric, M can be defined as a *metric space. *See* Riemannian geometry.

Riemannian space (Riemann space) *See* Riemannian geometry.

Riemann integral *or* **sum** *See* integration.

Riemann–Roch theorem A theorem, stated and proved in the middle of the 19th century by Riemann and his student Gustav Roch (1839–66). It calculates the exact number of independent *holomorphic (i.e. analytic) functions that can be defined on any particular *Riemann surface. The formula for this number involves the numbers of independent differential forms of special kinds and some of these are equal to (easily calculated) topological invariants of the surface. The *Atiyah–Singer index theorem* is a far-reaching generalization which also applies to functions of several variables; it was proved in 1963 by M.F. Atiyah and I.M. Singer. The Riemann–

Roch theorem has also been used in the study of codes.

Riemann sphere *See* extended complex plane.

Riemann surface A surface on which a *holomorphic function is defined as a single-valued function without branches. The Riemann surface of the logarithmic function has a spiral form; that of the function $\sqrt{(z^3 + z + 1)}$ is homeomorphic to a torus. *See* genus.

Riemann zeta function The function

$$\zeta(s) = \sum_{n=1}^{\infty} n^{-s}$$

It is known as the Riemann zeta function (or *zeta function*), though it goes back to Euler, who in 1744 showed that

$$\sum_{n=1}^{\infty} n^{-s} = \prod_{p} (1 - p^{-s})^{-1}$$

where the product runs over all prime numbers p. The sum defining $\zeta(s)$ converges when the real part of the complex variable s is greater than 1 (for example, $\zeta(2) = \pi^2/6$ and $\zeta(4) = \pi^4/90$) and gives an *analytic function of s in the part of the complex plane Re(s) > 1.

The definition of this function of s can be extended across the rest of the plane, and results in a function $\zeta(s)$ that is analytic throughout the complex plane, apart from a singularity at $s = 1$. This extended function is zero when $s = -2k$ for each natural number k, but it is also zero for other values of s that all have their real part between 0 and 1. The *Riemann hypothesis*, put forward in 1859, is that these other values all have Re(s) = $\frac{1}{2}$. To date, this has not been proved, although Hardy showed in 1914 that $\zeta(s)$ has infinitely many zeroes with Re(s) = $\frac{1}{2}$. The hypothesis is important in work on the distribution of primes.

right angle An angle equal to one-quarter of a complete turn ($90°$ or $\frac{1}{2}\pi$ radians).

right-angled triangle A triangle that has one interior angle equal to 90°. *See* Pythagoras' theorem.

right ascension (RA) Symbol: α. The angular distance of a point on the *celestial sphere from the vernal *equinox. It is measured eastward along the celestial equator from the vernal equinox to the place at which an hour circle through the point intersects the celestial equator. Generally, right ascension is measured in units of time rather than degrees (24 hours, corresponding to 360°). Sometimes *hour angle is used instead. *See* equatorial coordinate system.

right coset *See* coset.

right-handed triad *See* Cartesian coordinate system.

right prism A *prism that has lateral edges that are perpendicular to its bases.

right pyramid A *pyramid that has its vertex directly above the centre of its base.

rigid body A collection of *particles – a body – in which the distance between any two particles does not change with time. A rigid body therefore suffers no perceptible distortion in shape or size when subject to forces. This concept of an ideal body is used in mechanics.

rigidity modulus A *modulus of elasticity that is used in relation to *shear in an elastic body. It is the ratio of the shear stress (tangential force per unit area) to the resulting angular deformation of the body.

ring A *set R, together with two *binary operations, that satisfies certain *axioms. The operations are referred to as 'addition' (+) and 'multiplication' (.), although these operations need not necessarily have the meanings they have in arithmetic. Given any three members of R, a, b, and c, the axioms are:

(1) The commutative law holds for addition, i.e.

$$a + b = b + a$$

(2) The associative law holds for both addition and multiplication, i.e.

$$(a + b) + c = a + (b + c)$$

$$a \cdot (b \cdot c) = (a \cdot b) \cdot c$$

(3) There is an element (the additive identity element) in R such that

$$a + 0 = 0 + a = a$$

(4) For every element a in R there is an inverse element $-a$ in R such that

$$a + (-a) = 0$$

(5) The distributive laws apply, i.e.

$$a \cdot (b + c) = a \cdot b + a \cdot c$$

$$(a + b) \cdot c = a \cdot c + b \cdot c$$

These axioms define a ring.

If multiplication is also commutative, i.e.

$$a \cdot b = b \cdot a$$

the ring is a *commutative ring*. If there is a multiplicative identity element 1, for which

$$a \cdot 1 = 1 \cdot a = a$$

the ring is called a *ring with unity* or a *ring with identity*. The set of all 2×2 *matrices with the operations of matrix addition and multiplication form a noncommutative ring with unity. A commutative ring with unity for which there are no *proper divisors of zero* is an *integral domain (i.e. there are no nonzero elements a and b with $a \cdot b = 0$).

The set of all integers with the operations of addition and multiplication form an integral domain. If every nonzero member a of R also has an associated multiplicative inverse (a^{-1}) such that

$$a \cdot a^{-1} = 1$$

then the integral domain is a *field.

A ring with unity in which every element has a multiplicative inverse is a *division ring* or *skew field*. If multiplication is commutative, then it is a field.

rise (y-step) The difference between the *ordinates of two points in a *Cartesian coordinate system. *Compare* run.

rising factorial *See* hypergeometric series.

Robert of Chester (*c*.1100) English scholar who translated numerous scientific texts from Arabic into Latin, including the *Algebra* of al-Khwarizmi.

Roberval, Gilles Personne de (1602–75) French mathematician who made important contributions to the early history of the calculus. He determined the area of the cycloid and of the parabola, as well as claiming for himself the discovery of the method of indivisibles. His most important work, however, was on the problem of tangents. Curves were taken by Roberval to be paths of moving points, and a tangent was therefore defined by determining the instantaneous direction of the moving point at any position on the curve.

Robinson, Abraham (1918–74) German-American mathematician who contributed to *model theory, a branch of mathematical logic, and to applied mathematics; but he is best known for founding the theory of *nonstandard analysis.

robustness A statistical test or *estimation procedure that is little affected by departures from assumptions on which it is based is said to be *robust*. For example, the *t-test for independent samples is little affected by departures from normality if the observations are from nearly symmetric distributions having approximately the same variance, but it may be unreliable if the distributions are skew or if the

variances are very different. Robustness to *outliers is important in practice. Nonparametric tests tend to be more robust than their parametric counterparts in these circumstances (*see* nonparametric methods).

rod An idealized material object having length and density, but no thickness. If its density is constant, the rod is said to be a *uniform rod*.

roll Angular movement of an aircraft, spacecraft, projectile, etc. about an axis coincident with the direction of motion. *Compare* pitch; yaw.

Rolle's theorem The theorem that if a *function $f(x)$ is continuous over a certain interval $a \leqslant x \leqslant b$, its first differential $f'(x)$ exists in $a < x < b$, and $f(a) = f(b)$, then there exists a point between a and b, say c, at which $f'(c) = 0$. It is named after the French mathematician Michel Rolle (1652–1719). *See also* mean-value theorem.

rolling friction The *friction encountered when a body rolls over a surface, as happens with ball bearings. In rolling motion there is a point or a line of contact between the rolling body and the surface that changes continuously, without the body sliding. Rolling motion between two materials generally produces much less friction than when they slide.

Romberg integration (W. Romberg, 1955) In *numerical integration over an interval $[a, b]$, the application of *Richardson extrapolation with repeated *trapezoidal rule approximations having subinterval widths $h_k = (b - a) / 2^{k-1}$, $k = 1, 2, \ldots$.

root 1. (of an equation) A number that, when substituted for the *variable in a given equation, satisfies the equation (i.e. makes both sides equal). Thus, the quadratic equation

$$x^2 - x - 6 = 0$$

has two real roots, $x = 3$ and $x = -2$. A *zero* of the function $f(x)$ is a root of the equation $f(x) = 0$. *See* solution of equations.

2. A number that produces a given number when raised to a given *power. Thus, 2 is the fourth root of 16 ($2^4 = 16$). Note that -2, 2i, and $-2i$ are also fourth roots of 16. *See also* radix.

3. (of a congruence) An *integer a such that the congruence

$$f(x) \equiv 0 \pmod{n}$$

is satisfied when $x = a$, i.e.

$$f(a) \equiv 0 \pmod{n}$$

See congruence modulo n.

4. *See* tree.

root mean square deviation *See* standard deviation.

root of unity An element a of a *field that satisfies $a^k = 1$ for some natural number k is called a *k-th root of unity*, and it is a *primitive k-th root of unity* if $a^k = 1$ and $a^r \neq 1$ for any natural number r smaller than k. When k is 2 or 3, the relevant roots are called *square roots* or *cube roots*, respectively. In the field of complex numbers, the *cube roots of unity* are 1, $\frac{1}{2}(-1 + i\sqrt{3})$, and $\frac{1}{2}(-1 - i\sqrt{3})$, the last two numbers being *primitive cube roots*.

rose A type of plane *curve given in *polar coordinates by an equation of the form

$$r = a \sin n\theta$$

where a is a constant and n is a positive integer. The curve consists of a number of loops arranged around the pole. If n is odd, the rose has n loops; if n is even it has $2n$ loops.

rotation 1. Motion of a body about a single fixed point or about two fixed points, i.e. about a fixed line. There is therefore motion about an axis – the *axis of rotation – that passes through either one fixed point or through two fixed points; this results in an angular displacement.

2. (in the plane) A *transformation involving a fixed point C of the plane – the *centre of rotation*. A rotation through angle θ about C maps a point P onto a point P' such that $CP = CP'$, $\angle PCP' = \theta$, and the turn from CP to CP' is anticlockwise if θ is positive, clockwise if θ is negative. A rotation of θ about the origin maps the point with Cartesian coordinates (x, y) onto the point $(x \cos\theta - y \sin\theta, x \sin\theta + y \cos\theta)$.

3. (in space) A transformation involving a fixed directional line l – the *axis of rotation*. A rotation through angle θ about l maps a point P onto a point P' such that $NP = NP'$, and $\angle PNP' = \theta$ where N is the foot of the perpendiculars from P and P' to l. In turning from NP to NP', a right-handed screw will move in the positive or negative direction of l according to whether θ is positive or negative.

See also symmetry.

rotation matrix *See* orthogonal matrix.

rotation of axes A *transformation from one *coordinate system to another in which the axes are rotated through a fixed angle. In a planar *Cartesian coordinate system if (x, y) are the coordinates of a point in one system of axes and (x', y')

rotation of axes

are the coordinates of the same point in the other system of axes, then

$$x = x' \cos \theta + y' \sin \theta$$

$$y = -x' \sin \theta + y' \cos \theta$$

where the angle θ is such that a positive (anticlockwise) rotation of θ will map the second set of axes onto the first.

Roth's theorem (K.F. Roth, 1955) If α is a real *algebraic *irrational number and v is a real number greater than 2, then there are only finitely many pairs of integers p and q such that

$$\left| \alpha - \frac{p}{q} \right| < \frac{1}{q^v}$$

rough Generating *friction. A rough surface can be contrasted with a smooth, i.e. frictionless, surface.

roulette A curve that is the *locus of a point on (or associated with) a curve that rolls without slipping on another curve or on a straight line. *See* astroid; cardioid; cycloid; deltoid; nephroid; hypocycloid; epicycloid; trochoid; hypotrochoid; epitrochoid.

round angle (perigon) An angle equal to one complete turn (360° or 2π radians).

rounding The process of replacing a number by the nearest number with a certain number of decimal places or *significant figures. With decimal numbers and rounding to two places of decimals, 1.576 and 1.5751 would both be rounded to 1.58, while 1.572 and 1.5749 would both be rounded to 1.57. A rule is needed for breaking ties in the ambiguous case where there are two nearest numbers to choose between. A commonly used rounding rule rounds up in the case of ties: if the first digit dropped is 5 or more the preceding digit is increased by 1, while if the first digit dropped is less than 5 the preceding digit is unchanged. Another way of resolving ties is to choose the number with an even

last figure; thus 1.575 would be rounded to 1.58, while 1.565 would be rounded to 1.56. The act of rounding produces a rounding error (*see* error). *Compare* truncation.

row A horizontal line of elements in an *array, as in a *determinant or *matrix.

row echelon form An $m \times n$ matrix A is in row echelon form if
(1) every row comprising only zero entries is below all rows containing nonzero entries; and
(2) the first nonzero entry in each nonzero row appears in a column to the right of the first nonzero entry in the row above it.

For example, the matrices

$$\begin{pmatrix} 1 & -1 & 2 & 3 \\ 0 & 2 & 1 & 0 \\ 0 & 0 & 0 & 3 \\ 0 & 0 & 0 & 0 \end{pmatrix}, \quad \begin{pmatrix} 2 & 1 \\ 0 & 2 \\ 0 & 0 \end{pmatrix}$$

are in row echelon form. Row echelon form is produced by *Gaussian elimination. *Column echelon form* is defined by replacing 'row' by 'column' in the above definition. *See also* reduced row echelon form.

row rank The dimension of the *row space of a matrix. It is equal to the *column rank and the *rank of the matrix.

row space The vector space of all *linear combinations of the rows of a matrix.

row vector (row matrix) A *matrix having a single row of elements.

RSA cipher (R.L. Rivest, A. Shamir, and L.M. Adelman, 1977) A commonly used cipher for *public key cryptography. It is based on the fact that it is relatively easy to check whether a number is *prime but difficult to factor a product of two large primes (each of approximate size 10^{100} or greater).

Let $N = pq$ be the product of two large primes; N is publicly known, but is difficult to factorize in a reasonable time if one doesn't know the factorization. The plaintext is first split into pieces and each converted to an integer x; the message is sent as x^e modulo N for a publicly known e. If f is such that ef equals 1 modulo $(p-1)(q-1)$, then $(x^e)^f = x^{ef}$ is congruent to x modulo N, and if the recipient either knows f (the private key) or the factorization $N = pq$, then the message can be decoded.

Ruffini–Horner method   *See* Horner's method.

ruled surface A surface that can be generated by a moving straight line. A conical surface is an example of a ruled surface.

rule of false position *See* false position, rule of.

rule of signs *See* Descartes's rule of signs.

run 1. A sequence that follows a specified pattern in a series of observations. For example, in the sequence $(2, 3, 4, 2, 7, 6, 3, 0, 3, 2, 1)$ composed of the digits $[0, 9]$, the numbers '2, 3, 4' constitute an *up run* (monotonically increasing), '7, 6, 3, 0' constitute a *down run* (monotonically decreasing), and '2, 3, 4, 2' constitute a run below the median of 4.5. If, in ten tosses of a coin, the sequence of heads (H) and tails (T) is HHHTTHHTHT, there are three runs of heads where a run is a sequence of one or more heads followed by a tail or no further observation. Many tests for randomness of computer generated pseudo-random numbers use properties of runs. *See* random numbers.
2. (*x*-step) The difference between the *abscissae of two points in a *Cartesian coordinate system. *Compare* rise.

Runge–Kutta method   (C.D.T. Runge, 1895; W.M. Kutta, 1901) A numerical method for solving differential equations of the form $dy/dx = f(x, y)$ given an *initial condition $y(a) = y_a$. A Runge–Kutta method generates a sequence of approxi-

mations $y_n \approx y(x_n)$ in which y_{n+1} is a linear combination of values of the function $f(x, y)$ evaluated with x in the *interval $[x_n, x_{n+1}]$ and various arguments y. There are many different Runge–Kutta methods, the best known being the four-stage method

$$y_{n+1} = y_n + \frac{1}{6}(k_1 + 2k_2 + 2k_3 + k_4)$$

where, with $h = x_{n+1} - x_n$,

$$k_1 = h\,f(x_n, y_n)$$
$$k_2 = h\,f(x_n + \tfrac{1}{2}h, y_n + \tfrac{1}{2}k_1)$$
$$k_3 = h\,f(x_n + \tfrac{1}{2}h, y_n + \tfrac{1}{2}k_2)$$
$$k_4 = h\,f(x_n + h, y_n + k_3)$$

Russell, Bertrand Arthur William (1872–1970) English mathematical logician and philosopher who in 1902, while working on the foundations of mathematics, discovered *Russell's paradox. To avoid this and other such antinomies, Russell developed his theory of types, which he included in *Principia mathematica* (3 vols, 1910–13). This work, written in collaboration with A.N. Whitehead, was an attempt to derive the whole of mathematics from purely logical assumptions.

Russell's paradox A *paradox of *set theory put forward by Bertrand Russell in 1902. Some sets are not members of themselves. An example is the set of all men. Other sets, such as the set of all things that are not men, are members of themselves (the set itself is not a man). Now consider the set S whose members are those sets that are not members of themselves. Is S a member of S? If it is, then it is not, and if it is not, then it is. This paradox can be derived from the *axiom of abstraction. It influenced the development of set theory in fostering the idea that sets are defined by their members rather than by general conditions.

S

S_n Symbol for the *symmetric group for a set of n elements.

Saccheri, Girolamo (1667–1733) Italian mathematician who in his *Euclides ab omni naevo vindicatus* (1733, Euclid Cleared from Every Stain) attempted to prove Euclid's parallel (fifth) postulate by the method of *reductio ad absurdum*. He failed however to find any obvious contradiction and narrowly missed becoming the first to discover a non-Euclidean geometry.

saddle point 1. For a surface $z = f(x, y)$ a saddle point occurs at a point where the *partial derivatives $\partial z/\partial x$ and $\partial z/\partial y$ are both zero but there is no local maximum or minimum. A *stationary point of a function f of two variables is a saddle point if the *Hessian of f at the point is negative. **2.** *See* game theory.

saddle point

St Petersburg paradox A paradox in probability, based on a coin-tossing game.

A player A proposes to a player B that they play a game in which A will toss a coin until the first appearance of a head, and if this is at the kth toss ($k = 1, 2, 3, \ldots$) he will pay £2^k to B. For the game to be fair, B must pay A in advance a stake equal to the *mean payment A can expect to make to B. This mean payment in £ is

$$\sum_{k=1}^{\infty} \frac{1}{2^k} \times 2^k = 1 + 1 + 1 + \cdots$$

which has no limit (i.e. the series is divergent). Thus it is not possible for B to pay his stake, and a *fair game is impossible. Note also that though his expectation is infinite, B can receive only a finite amount of money in a finite time. Indeed, in the long run this sum will be £8 or less in 87.5 percent of all games and will exceed £64 in less than 1.6 percent.

The paradox was first noted by Nicholas Bernoulli in a letter to de Montmort in 1713, and later investigated by his cousin Daniel Bernoulli. It is one of the earliest examples of a distribution with an infinite *expectation.

salient angle An interior angle in a *polygon that is less than 180°. *Compare* reentrant angle.

salient point A point at which two branches of a curve meet and have different *tangents. For example, $y = |x|$ has a salient point at the origin.

sample A finite *subset of a *population. A sample containing n items is called a *sample of size n*. *See* quota sample; random sample; representative sample; stratified sample.

sample correlation coefficient *See* correlation coefficient.

sample distribution function (empirical distribution function) If $x_{(i)}$ is the ith *order statistic in a sample $x_1, x_2, \ldots, x_n$ of size n, then the sample (or empirical) distribution function, $S_n(x)$, takes values

$$S_n(x) = \begin{cases} 0 & \text{for } x \leqslant x_{(1)} \\ i/n & \text{for } x_{(i)} \leqslant x < x_{(i+1)}, \\ & \quad i = 1, 2, \ldots, n-1 \\ 1 & \text{for } x \geqslant x_{(n)} \end{cases}$$

It is important because of its close relationship to the population *cumulative distribution function, $F(x)$. In particular, its mean value $E[S_n(x)] = F(x)$, its variance $\text{Var}[S_n(x)] = F(x)(1 - F(x))/n$, and $S_n(x)$ is a consistent estimator of $F(x)$ for any fixed x. This last property implies that $S_n(x) \to F(x)$ as $n \to \infty$.

sample space The *set, S, of all possible outcomes of an experiment. The possible scores when a single die is cast form the sample space $S = \{1, 2, 3, 4, 5, 6\}$. The sums of all possible scores when a pair of dice are cast form the sample space

$$S = \{2, 3, 4, 5, 6, 7, 8, 9, 10, 11, 12\}$$

See also event.

sample statistic *See* statistic.

sample survey A study to estimate *population characteristics in which those characteristics are observed for only a portion of that population, known as a sample. *See also* area sampling; census; cluster sample; quota sample; random sample; representative sample; sampling theory; stratified sample.

sampling distribution The *distribution of a *statistic. For example, for a random sample of size n from a distribution $N(\mu, \sigma^2)$ the sample mean $\bar{x}$ is an observed value of a random variable, $\bar{X}$, say, which has a distribution $N(\mu, \sigma^2/n)$. Any statistic is a *random variable; its value varies from sample to sample.

sampling error The difference between an estimate of a parameter based on a sample and the true parameter value. Because an *estimator is a *random variable it has a distribution (often called a *sampling distribution), so the estimate, in general, will not equal the true parameter value. For example, the mean m of a sample of n observations $x_1, x_2, \ldots, x_n$ from a normal distribution with mean λ and standard deviation ρ has itself a normal distribution with mean λ and standard deviation $\rho/\sqrt{n}$. If ρ is unknown, $\rho/\sqrt{n}$ is estimated by $s/\sqrt{n}$, where s is given by

$$s^2 = \sum_i \frac{(x - m)^2}{(n-1)}$$

and $s/\sqrt{n}$ is called the *standard error.

sampling frame *See* frame.

sampling theory The theory of methods of obtaining *samples and making inferences about *population characteristics on the basis of sample measurements. Simple random samples allow straightforward estimates with valid measurements of *sampling error; precision may be improved by using a *stratified sample or other modifications. A number of special methods including *area sampling, *cluster sampling, and multistage and multiphase sampling are in use. In practice, circumstances may preclude the use of strictly random samples, but some samples can reasonably be assumed to be almost equivalent to random samples. Techniques such as *quota sampling do not admit estimation of the sampling error.

satisfaction In *logic, an ordered *n-tuple is said to *satisfy* an open sentence (*see* variable) if and only if the *predicate of the open sentence is true of the ordered n-tuple. For example, 'x was the father of y' is satisfied by the ordered pair (Laertes, Odysseus) because Laertes stands in the relation 'was the father of' to Odysseus. *See also* interpretation.

scalar **1.** A number as distinguished from a *vector.
2. A *tensor of order zero.

scalar field *See* field.

scalar matrix A *diagonal matrix in which all the elements on the leading diagonal are equal.

scalar product For simple geometric *vectors in Euclidean space, the product of two vectors to give a *scalar. The scalar product is written as $\mathbf{A} \cdot \mathbf{B}$ and is equal to the products of the lengths of the vectors and the cosine of the angle between them, i.e. $|\mathbf{A}| \, |\mathbf{B}| \cos \eta$. It can be applied to various situations of physical interest. For example, the work done when a force $\mathbf{F}$ produces a displacement $\mathbf{s}$ is the scalar product $\mathbf{F} \cdot \mathbf{s} = |\mathbf{F}| \, |\mathbf{s}| \cos \eta$, where η is the angle that the force makes with the direction of motion.

More generally, if $\mathbf{A}$ is a vector defined by the n-tuple $(a_1, a_2, \ldots, a_n)$ and $\mathbf{B}$ is a vector defined by the n-tuple $(b_1, b_2, \ldots, b_n)$, the scalar product $\mathbf{A} \cdot \mathbf{B}$ is

$$a_1 b_1 + a_2 b_2 + \cdots + a_n b_n$$

In a *vector space, the scalar product (or *inner product*) associates a number $\mathbf{u} \cdot \mathbf{v}$ with all pairs of vectors $\mathbf{u}$ and $\mathbf{v}$, and has the following properties:

(1) $\mathbf{u} \cdot \mathbf{v} = \mathbf{v} \cdot \mathbf{u}$, i.e. scalar multiplication is commutative for all elements of the vector space;
(2) $\mathbf{u} \cdot (\mathbf{v} + \mathbf{w}) = \mathbf{u} \cdot \mathbf{v} + \mathbf{u} \cdot \mathbf{w}$, i.e. scalar multiplication is distributive over addition;
(3) for a number n, $n\mathbf{u} \cdot \mathbf{v} = n(\mathbf{u} \cdot \mathbf{v}) = \mathbf{u} \cdot n\mathbf{v}$.

If the scalar product $\mathbf{u} \cdot \mathbf{u}$ is greater than zero for all nonzero members $\mathbf{u}$ of the vector space, it is said to be *positive definite*. In this case the vector space is called an *inner product space*.

The scalar product is sometimes called the *dot product*.

scalar quantity A quantity, such as mass, length, time, density, or energy, that has size or magnitude but does not involve the concept of direction. It is thus treated mathematically as a *scalar.

scalar triple product *See* triple product.

scale factor *See* enlargement.

scalene triangle A triangle that has all three sides unequal.

scales of measurement The basic scales of measurement are (i) nominal, (ii) ordinal, (iii) linear interval, and (iv) linear ratio.

Nominal scales merely allocate items to nonordered categories, e.g. people to the nationalities English, French, German, Italian, etc. The only meaningful relationship between items so assigned is that of *equality* ($=$) (belonging to the same category) or *inequality* ($\neq$) (belonging to different categories).

Ordinal scales allocate items to ordered categories, e.g. different brands of television set may be graded as excellent, good, average, poor, or very poor. In addition to equality and inequality, the ordering relationships ($<$ and $>$) are meaningful.

Linear interval scales have an arbitrary zero, but measurements of an item as x and y on two such scales satisfy a relationship of the form

$$y = mx + c$$

where m and c are nonzero constants. An example is provided by the Fahrenheit (F) and Celsius (C) scales for measuring temperature. Corresponding measures of a temperature satisfy

$$F = \frac{9}{5}C + 32$$

In addition to complying with the equality and ordering relationships, measurements on these scales are such that for any two items the ratios of their differences for each scale is constant. For example, on the Celsius and Fahrenheit scales corresponding temperatures in degrees are

C	0	10	20	40
F	32	50	68	104

Here the difference ratios $(104 - 68)/(40 - 20)$ and $(50 - 32)/(10 - 0)$ are both 9/5. However, it is meaningless to describe an object with a temperature of $20°$ on the Celsius scale as being twice as hot as one with a temperature of $10°$ because on the Fahrenheit scale the corresponding temperatures are $50°$ and $68°$.

Linear ratio scales have a common fixed zero, and corresponding measurements of an item on two such scales satisfy a relationship $y = mx$. As well as satisfying meaningful relationships of the types described for the previous categories of measurement, a meaning is now attachable to the ratio of the measurements of two items on two such scales. For example, weight scales in grams or in ounces are linear ratio scales. If we record the weight of two items in grams and one weighs 10 grams and the other 20 grams, the ratio of their weights is 2. This ratio remains at 2 if we weigh each item in ounces, pounds, kilograms or on any other scale that has the fixed origin corresponding to zero weight.

There are also several nonlinear scales of measurement, a well-known one being the *logarithmic scale*, on which a value is measured by a number that is proportional to the logarithm of the value. A commonly used logarithmic scale has points marked at equal intervals corresponding to values of ..., 1, 10, 100, 1000, ... (whose logarithms to base 10 are ..., 0, 1, 2, 3, ...). Using logarithmic scales for x and y, the graph of $y = ax^n$ is a straight line. Using a linear scale for x and another scale, the *normal probability scale*, the graph of the cumulative *distribution function of the standard normal distribution is a straight line.

scatter diagram A two-dimensional plot of the n points for a set of n paired observations (x_i, y_i). The diagram may indicate some relationship between the variables such as a linear or quadratic trend.

schema (*plural* **schemata**) In *logic, a method of representing a possibly infinite number of *wffs of some object language by using metalinguistic expressions that take object language as substitution instances (*see* metalanguage). Thus, we might adopt $A \supset (B \supset A)$ as an axiom schema of some formal system S, and if p, q, and r are wffs of S then

$$p \supset (q \supset p)$$

is an axiom of S, as is

$$(p \vee r) \supset ((q \& r) \supset (p \vee r))$$

Similarly, it is possible to construct valid schemata, proof schemata, and theorem schemata.

Schooten, Frans van, the Younger (*c.*1615–*c.*1660) Dutch mathematician and author of an important Latin translation of the *Géometrie* of Descartes. The second edition, containing various related texts and commentaries, was published in two volumes in 1659–61 as *Geometria a Renato Des Cartes* and introduced the new Cartesian analytical methods to the mathematicians of Europe.

Schröder–Bernstein theorem (Cantor–Bernstein theorem) The theorem that if the *cardinal number of *set A is less than or equal to that of B, and the cardinal number of B is less than or equal to that of A, then the two sets have equal cardinal numbers. This was conjectured by Cantor in 1895, and proved independently by F.W.K.E. Schröder (1896) and F. Bernstein (1898).

Schur decomposition (I. Schur, 1909) Any square matrix A has a Schur decomposition $A = QTQ^*$, where Q is a *unitary matrix and T is an upper *triangular matrix, and where Q^* denotes the Hermitian conjugate of Q. The *eigenvalues of A appear on the diagonal of T.

Schwarz's inequality *See* Cauchy–Schwarz inequality.

scientific notation *See* exponential notation.

screw 1. A cylindrical or conical body with a helical groove cut in its surface, forming the thread. It can be considered as a wedge wound in the form of a *helix. When the end of the screw is placed in contact with a material, a rotation about its axis will cause a translation of the screw along this axis and into the material. A screw is a simple machine.
2. In three-dimensional geometry, the combination of a *rotation about a line and a *translation along this line. *See also* wrench.

s.d. *Abbreviation for* *standard deviation.

s.e. *Abbreviation for* *standard error.

sec Secant. *See* trigonometric functions.

secant 1. A line that cuts a given curve. If a secant line cuts a curve at two points, the segment of the line between the two points of intersection is a chord of the curve.
2. *See* trigonometric functions.

secant method An interative method for solving a nonlinear equation $f(x) = 0$ in one variable (*see* diagram). Given distinct initial guesses x_0 and x_1 for a root, the

secant method to solve $f(x) = 0$.

secant method consists of the *iteration

$$x_{n+1} = x_n - \frac{f(x_n)(x_n - x_{n-1})}{f(x_n) - f(x_{n-1})}$$

for $n = 1, 2, 3, \ldots$

The secant method can be derived by using the approximation

$$f'(x_n) \approx \frac{f(x_n) - f(x_{n-1})}{x_n - x_{n-1}}$$

in *Newton's method.

sech Hyperbolic secant. *See* hyperbolic functions.

second 1. Symbol: ″. A unit of angle equal to 1/60 of a minute. *See* angular measure.
2. Symbol: s. The *SI unit of time, equal to the duration of 9 192 631 770 periods of the radiation corresponding to the transition between the two hyperfine levels of the ground state of the caesium-133 atom. This definition came into force in 1964; before then the standard was the mean solar second, defined as 1/86 400 of the mean solar *day.

secondary diagonal An alternative name for the main antidiagonal of a square array. *See* diagonal.

secondary parts (of a triangle) Properties such as the lengths of medians or sizes of exterior angles, as distinguished from the lengths of the sides and sizes of interior angles, which are the *principal parts*.

second kind *See* first kind.

second of arc *See* degree of arc.

second-order convergence *See* order.

second-order differential equation A *differential equation that contains a second-order derivative (d^2y/dx^2, say) and no higher-order derivatives.

section (plane section) A plane geometric configuration formed by cutting a given figure with a plane. For instance, a section of a conical surface is a *conic. A *cross-section* is a section in which the plane is at right angles to an axis of the figure. For example, a cross-section of a right circular cylinder is a circle.

sectionally continuous *See* continuous function.

sector A part of a circle lying between two radii and either of the arcs that they cut off. The area of a sector is $\frac{1}{2}r^2\theta$, where r is the radius and θ the angle in radians subtended by the arc at the centre of the circle.

segment 1. A part of a line or curve between two points on the line or curve.
2. A region lying between a *chord of a circle and the corresponding arc cut off by the chord. A chord divides a circle into two segments: the *major segment* is the region between the chord and the longer (major) arc; the *minor segment* is the region between the chord and the shorter (minor) arc. The area of a segment is given by

$$A = \frac{1}{2}r^2(\theta - \sin\theta)$$

where r is the radius and θ the angle in radians that the arc subtends at the centre of the circle.
3. *See* spherical segment.

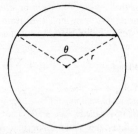

segment

selection function *See* choice.

self-similar *See* fractal.

semantics In *logic, the study of the relationships that hold between the expressions of a *formal language and a logical *domain. The study of interpretations falls within the scope of semantics. An interpretation assigns *semantic values* (entities in the domain) to the expressions of a formal language via *semantic rules*.

semiaxis A line segment that is one half of an axis of a conic. *See* ellipse; hyperbola.

semicircle Half a circle; either of the two parts of a circle cut off by a diameter.

semiconjugate axis *See* hyperbola.

semicubical parabola A plane *curve with the Cartesian equation

$$y^2 = kx^3$$

It has a *cusp at the origin. *See also* cubical parabola.

semigroup A *set S together with a *binary operation $\circ$ on it that satisfies one condition: that the operation is *associative*, i.e. for any three elements a, b, and c of S

$$a \circ (b \circ c) = (a \circ b) \circ c$$

A simple example of a semigroup is the set of all even integers with the operation of multiplication. *See* group.

semi-interquartile range *See* quartile deviation.

semilogarithmic graph *See* graph.

semimajor axis *See* ellipse; ellipsoid.

semimean axis *See* ellipsoid.

semiminor axis *See* ellipse; ellipsoid.

semitransverse axis *See* hyperbola.

sense The 'direction' of an *inequality, i.e. whether it signifies 'greater than' or 'less than'.

sentential calculus *See* propositional calculus.

separation (of a set) A *set X is separated into a pair of nonempty *subsets A and B if

$A \cup B = X$ and $A \cap B = \varnothing$. If the set is ordered the separation can be one of two possible types. In a separation of the first kind, each member of one set is less than every member of the second set with the separating number belonging arbitrarily to one set. In a separation of the second kind, each member of one set is smaller than every member of the second, as before, but in addition one set lacks a greatest member and the other set has no smallest member. *See* Dedekind cut.

sequence A succession of terms

$$a_1, a_2, a_3, a_4, \ldots$$

formed according to some rule or law. Examples are

$$1, 4, 9, 16, 25$$

$$1, -1, 1, -1, 1, \ldots$$

$$\frac{x}{1!}, \frac{x^2}{2!}, \frac{x^3}{3!}, \frac{x^4}{4!}, \ldots$$

It is not necessary for the terms to be unequal. The terms are ordered by matching them one by one with the positive integers, 1, 2, 3, ... The nth term is thus a_n, where n is a positive integer. Sometimes the terms are matched with the non-negative integers, 0, 1, 2, ... A *finite sequence* has a finite (i.e. limited) number of terms, as in the first example above. An *infinite sequence* has an unlimited number of terms, i.e. there is no last term, as in the second and third examples. An infinite sequence can however approach a limiting value as the number of terms, n, becomes very great. Such a sequence is described as a *convergent sequence* and is said to tend to a *limit as n tends to infinity.

With some sequences the nth term (or *general term*) expresses directly the rule by which the terms are formed. This is the case in the three examples above, where the nth terms are n^2, $(-1)^{n+1}$, and $x^n/n!$, respectively, $n \geqslant 1$. A sequence is then a function of n, the general term being given by

$$a_n = f(n)$$

and having as its domain the set of positive integers (or sometimes the set of non-negative integers). A sequence with general term a_n is written as $\{a_n\}$ or (a_n).

Other sequences are defined by a *recurrence relation: a rule is given by which the nth term can be determined when one or more preceding terms are known. This is the case with the *Fibonacci sequence. *See* series.

sequential analysis (A. Wald, 1947) A method of *inference where observations are taken one at a time, and after each observation a decision is made whether to accept or reject one of two hypotheses or to take further observations before reaching a decision. The technique is attractive, for example, in comparing two treatments for a disease, where observations are made as cases are presented for treatment and there are ethical reasons for stopping the experiment as soon as one treatment can confidently be regarded as superior. The procedural rules are based on the *likelihood ratio.

series The indicated sum of the terms of a *sequence. In the case of a finite sequence

$$a_1, a_2, a_3, \ldots, a_N$$

the corresponding series is

$$a_1 + a_2 + a_3 + \cdots + a_N = \sum_1^N a_n$$

This series has a finite or limited number of terms and is called a *finite series*. The Greek letter Σ is the summation sign, whose upper and lower limits indicate the values of the variable n over which the sum is calculated; in this case the set of positive integers 1, 2, ..., N.

In the case of an infinite sequence

$$a_1, a_2, \ldots, a_n, \ldots$$

the corresponding series is

$$a_1 + a_2 + \cdots + a_n \cdots = \sum_1^\infty a_n$$

This type of series has an unlimited number of terms and is called an *infinite series*.

The nth term, a_n, of a finite or infinite series is known as the *general term*. An infinite series can be either a *convergent series or a *divergent series depending on whether or not it converges to a finite sum. Convergence is an important characteristic of a series.

See also alternating series; arithmetic series; asymptotic series; binomial series; cosine series; exponential series; Fourier series; geometric series; Gregory's series; harmonic series; inverse sine series; logarithmic series; oscillating series; p-series; sine series; tangent series; Taylor's theorem.

serpentine A plane curve with the equation in Cartesian coordinates

$$x^2y + b^2y - a^2x = 0$$

where a and b are constants. It passes through the origin, about which it is symmetrical. The x-axis is an asymptote.

set (class) A collection of any kind of objects. The objects that make up a set are called its *elements or members*. The statement 'a is an element of the set A' can be written as $a \in A$, and a set containing elements a, b, and c is denoted by $\{a, b, c\}$. Also allowed as a set is the *empty* or *null set*, denoted by $\varnothing$, which is the set that contains no elements.

Sets are often specified by a condition for membership in the set; $\{x: x$ is a man$\}$ designates the set of men. The assumption that any condition can be used to specify a set leads to *Russell's paradox. The *axiom of extensionality states that two sets are identical if and only if they have exactly the same elements.

set theory The study of *sets was originally developed by Cantor as a means of investigating the theory of infinite series. In 1874 he published his famous proof that the *cardinal number of the set of real numbers is greater than that of the set of natural numbers. Set theory has been especially important in the foundations of mathematics, where it has been used to axiomatize the theory of numbers. Current axiomatizations of set theory have been influenced by the need to avoid *Russell's paradox.

sexagesimal Involving the number 60.

sexagesimal measure *See* angular measure.

sextic Having a *degree or order of six. For example, a *sextic equation* is an equation of the sixth degree.

sgn *See* signum function.

Shanks, William (1812–82) English mathematician noted for his calculation in 1873 of the first 707 places of π. It was shown in 1946 that he made a mistake and that the values from the 528th position were incorrect.

Shannon, Claude Elwood (1916–2001) American mathematician and author, with Warren Weaver, of the seminal *The Mathematical Theory of Communication* (1949), which founded the modern discipline of *information theory. Shannon showed how it was possible to measure the information content of a message. Earlier, in 1938, he had shown in his *A Symbolic Analysis of Relay and Switching Circuits* how Boolean algebra could be applied to computer design. Shannon also produced the first effective programs for chess-playing computers.

Shannon's theorem (C. E. Shannon, 1945) A theorem in *information theory that shows that for any transmission *channel subject to random errors, there is some effective *error-correcting code.

sheaf (bundle) A set of planes that all pass through a given point.

shear 1. Angular deformation of a body or part of a body without change in volume. It is a type of *strain in which some parallel planes in the body remain parallel but are relatively displaced in a direction parallel to themselves. The *stress associated with shear is the tangential shearing force per unit area. The shear is the angle, in radians, turned through by a line originally perpendicular to the direction of the stress. For example, if a pair of opposite faces of a rectangular block are deformed into parallelograms, and other faces remain rectangular (*see* diagram), the shear is equal to the angle θ. *See also* rigidity modulus.

shear

2. A *transformation of the points of the plane in which one line remains fixed, and all other points move parallel to the line by amounts proportional to their distance from the line. A shear preserves the areas of plane figures.

The shear of the plane with invariant line $x = 0$ and angle θ maps the point with Cartesian coordinates (x, y) to $(x + y \tan \theta, y)$.

3. A transformation of the points of space in which one plane remains fixed, and all other points move parallel to each other and to the plane by amounts proportional to their distance from the plane. A shear preserves the volumes of solid bodies.

sheet Any of the two or more separate parts that may form a given surface. *See* hyperboloid.

Sheppard's corrections (W.F. Sheppard, 1898) Adjustments to improve estimates of sample *moments when only *grouped data are available.

SHM *Abbreviation for* simple *harmonic motion.

short arc *See* arc.

short radius *See* polygon.

SI *Abbreviation for* Système International. *See* SI units.

side 1. **(arm)** One of the two lines forming an angle.
2. One of the lines joining the vertices of a *polygon.

Siegel–Tukey test *See* homogeneity of variance.

siemens Symbol: S. The *SI unit of electric conductance, equal to the conductance of a circuit or element that has a resistance of 1 ohm. [After E.W. von Siemens (1816–92)]

sieve of Eratosthenes (Eratosthenes, *c.*250 BC) A method of finding *prime numbers by writing down the numbers from 1 in increasing order, then striking out every second number after 2, every third number after 3 (in the original list), every fifth number after 5, and so on. The numbers remaining are primes. For a set of numbers from 1 to n it is necessary to sieve by prime numbers only up to the largest integer less than or equal to $\sqrt{n}$.

sievert Symbol: Sv. The *SI unit of dose of ionizing radiation, equal to the dose delivered by a point source of 1 milligram of radium, enclosed in a platinum container with walls 0.5 millimetre thick, to a sample 10 millimetres away over a period of 1 hour. It is equivalent to 1 joule per kilogram of irradiated material. [After R. Sievert (1896–1966)]

sigma function (sum function) The function $\sigma(n)$ that gives the sum of the positive *divisors of a *natural number n. Thus $\sigma(3) = 1 + 3 = 4$ and $\sigma(6) = 1 + 2 + 3 + 6 = 12$. If n has the prime factorization $p_1{}^{a_1} p_2{}^{a_2} \ldots p_r{}^{a_r}$, then

$$\sigma(n) = \prod_{i=1}^{r} \frac{p_i{}^{a_i - 1} - 1}{p_i - 1}$$

A *perfect number n is such that $\sigma(n) = 2n$.

sigma notation *See* summation sign.

sigmoid curve A *monotonically increasing curve between two horizontal *asymptotes and having a point of inflection. The normal distribution function and many other distributions have this form. It is sometimes called S-shaped because of its similarity to the integral sign, an old-fashioned form of S. Sigmoid curves also occur in growth studies when size variables are plotted against age. In this context the *logistic curve*

$$y = \frac{k}{1 + \exp(a - bx)}$$

where $b > 0$, is widely used.

sigmoid curve

sign *See* signum function.

signature 1. (of a permutation) A number defined to be $+1$ if the *permutation is even and -1 if the permutation is odd.
2. (of a quadratic form) The number of positive terms minus the number of negative terms.
3. (of a Hermitian matrix) The number of positive *eigenvalues minus the number of negative eigenvalues.

signed minor *See* cofactor.

signed number *See* directed number.

signed rank test *See* Wilcoxon signed rank test.

significance level *See* hypothesis testing.

significance test *See* hypothesis testing.

significant figures The run of figures (or digits) in a number that is relevant to its precision, as distinct from any additional zero digits that serve to indicate the number's magnitude.

For example, if populations are being quoted to the nearest thousand, the populations of three cities given as 1 702 000, 814 000, and 70 000 are correct to 4, 3, and 2 significant figures. Although they are not significant figures, the zeroes in the hundreds, tens, and units positions are essential in recording the magnitude of the populations.

In general, reading from left to right, the first nonzero digit of a number after *rounding is the first of the run of significant figures. For instance, rounded to three significant figures the numbers 1234.5 and 0.012 345 become 1230 and 0.0123.

sign test A nonparametric test of the hypothesis that a *population has a given *median, M. If the hypothesis is true, roughly half the n sample observations should have a value less than M, and half a value greater than M. Excessive numbers above or below M indicate rejection. The critical region for the test is the pair of tails of the *binomial distribution with parameters n, 0.5. An extension to a matched-pairs test of whether two populations have the same median is available. *See* nonparametric methods.

signum function The *function f defined by

$$f(x) = \begin{cases} 1 & \text{for } x > 0 \\ 0 & \text{for } x = 0 \\ -1 & \text{for } x < 0 \end{cases}$$

It is denoted by sgn x or sign(x).

similar Describing geometric figures or sets of points that are related by a *similarity. Two geometric figures are similar if one is directly or oppositely *congruent to an *enlargement of the other. *Similar polygons* have corresponding angles equal and corresponding sides proportional in length. Two triangles are similar if there is a correspondence between them satisfying one of the following conditions:

(1) The lengths of all three pairs of corresponding sides are in the same ratio.
(2) The lengths of two pairs of corresponding sides are in the same ratio, and the angles between them are equal.
(3) Two pairs of corresponding angles are equal.

similarity A *transformation that multiplies the distance between any two points by a constant. Thus, two points P and Q will have images P$'$, Q$'$ such that $P'Q' = cPQ$, where c is a positive constant. Two figures related by a similarity are *similar*. Any combination of translation, rotation, reflection, and enlargement is a similarity. A *spiral similarity* is composed of a rotation and an enlargement. A similarity with $c = 1$ is an *isometry. *See also* matrix.

similar matrices *See* matrix.

similitude *See* enlargement.

simple curve A curve that does not intersect itself.

simple discontinuity *See* discontinuity.

simple fraction *See* common fraction.

simple graph *See* graph.

simple group *See* normal subgroup.

simple harmonic motion *See* harmonic motion.

simple hypothesis *See* hypothesis testing.

simple interest *See* interest.

simple pole *See* singular point.

simple quadrangle A plane figure formed by four points, no three of which are *collinear, and four lines joining them. *See* quadrangle.

simple quadrilateral A *polygon with four sides. *See* quadrilateral.

simple root A number a that is a root of the *polynomial equation $f(x) = 0$, but is not a root of the equation $f(x)/(x - a) = 0$. If $a, b, \ldots$ are the different roots of $f(x) = 0$ then the polynomial can be factorized as $f(x) = (x - a)^k(x - b)^l \ldots$, and a is a simple root if and only if $k = 1$. *Compare* multiple root.

simplex *See* combinational topology.

simplex method A method or algorithm for solving *linear programming problems. Additional variables, called *slack variables*, are introduced to convert inequalities to equalities. The solution, obtained by an iterative process, may be set out in arrays called *tableaux*. Effectively, the conversion of inequalities to equalities enables us to define the boundaries of a simplex or region of feasible solutions satisfying the constraints, and the optimum solution then lies at a vertex of this simplex. The algorithm provides a systematic way of eliminating vertices until the optimizing vertex is located. In the diagram in the entry on *linear programming, the simplex is the stippled region and the optimizing vertex is at E. The method may be adapted to give information on the effects of altering constraints, to determine which constraints are critical, etc. *See also* Karmarkar's algorithm.

simplicial complex   *See* combinatorial topology.

Simpson, Thomas (1710–61) English mathematician noted for *Simpson's rule, which was published in 1743 in *Mathematical Dissertations on Physical and Analytical Subjects*.

Simpson's paradox (E.H. Simpson, 1951) A paradox in *contingency tables whereby an association that is significant in each of two tables may disappear or be reversed if the two tables are combined. Suppose, for example, that a new treatment N is being compared with a standard treatment S with 3000 patients, and a record is kept of the numbers of cures C and failures F; and the numbers in each category are as given in the first table.

	S	N
C	450	530
F.	850	1170

A *chi-squared test indicates a differential effect of treatments with the standard giving the greater proportion of cures. However, when the results are broken down into patients living in urban and rural areas, the data are as given in the second table.

| | Urban | | Rural | |
	S	N	S	N
C	100	350	350	180
F	500	1050	350	120

The chi-squared test indicates that in both urban and rural areas the new drug gives a better cure rate. The paradox arises because the overall data set combines two sets in which the numbers and proportions receiving each treatment are different, as also are the cure rates for each drug. The paradox reflects the danger of combining heterogeneous data.

Simpson's rule A rule for *numerical integration which approximates

$$\int_a^b f(x)\,dx$$

$$\approx \frac{b-a}{6}\left[f(a) + 4f(\tfrac{1}{2}(a+b)) + f(b)\right]$$

The area under the curve is thereby approximated by the area under a quadratic polynomial that passes through $(x, f(x))$ for the three values $x = a, \tfrac{1}{2}(a+b)$, and b.

The *repeated Simpson's rule* divides the interval $[a, b]$ into an even number n of subintervals of length $h = (b-a)/n$, based on equally spaced points $a = x_0$, $x_1, \ldots, x_n = b$, with corresponding ordinates $y_0, y_1, \ldots, y_n$, and applies Simpson's rule to groups of three successive points, giving

$$\int_a^b f(x)\,dx \approx \tfrac{1}{3}h(y_0 + 4y_1 + 2y_2 + 4y_3$$

$$+ \cdots + 2y_{n-2} + 4y_{n-1} + y_n)$$

Named after T. Simpson. *See also* Newton's rule; trapezoidal rule.

Simpson's rule with four subintervals.

Simson line (Wallace–Simson line) For any point on the *circumcircle of a triangle, the feet of the perpendiculars from the point to the sides of the triangle lie on a straight line. This line is often called the *Simson line* (or *simson*) of the point with respect to the triangle. Though it is named after the Scottish geometer Robert Simson (1687–1768), it does not appear in his work and was first discovered in 1797 by the Scottish mathematician William Wallace (1768–1843).

simulation A term applied to the study of a physical system in which there is a dynamic or probabilistic element (or perhaps both) by making use of a mathematical *model. For example, computer simulations may enable a manager to make a rapid assessment of the likely effects of different levels of investment, or of changing manufacturing procedures or the size of the workforce, on output and profit over a period of years. Government departments use simulation models to study the likely effects of tax changes, changes in interest rate, borrowing levels, etc. on public and private spending and demands for various resources and services. The usefulness of the method depends on how accurately the mathematical model reflects relevant aspects of physical reality. *See also* queuing theory.

simultaneous equations Two or more equations that apply simultaneously to given variables. The solution of simultaneous equations involves finding values of the variables that satisfy both equations. For instance, the equations

$$x + y = 6 \quad \text{and} \quad 2x + y = 4$$

can each be satisfied by an infinite set of pairs of values x and y. However, there is only one pair of values that satisfies both simultaneously, namely $x = -2$ and $y = 8$. The point $(-2, 8)$ is the point at which the two straight lines represented by the equations intersect on a graph. This is used in the *graphical solution of pairs of simultaneous equations – a technique that can be applied to pairs of simultaneous equations in two variables. Another simple method of solution is that of *elimination of the variables between the equations. *See* Cramer's rule; Gaussian elimination; Gauss–Seidel method.

simultaneous inequalities Two or more conditional *inequalities that hold simultaneously. The solution of a set of simultaneous inequalities is the set of values that satisfy all of them. For instance, the solution of the inequalities

$$x + y < 6, \quad x > 1, \quad y > 2$$

is the set of pairs (x, y) represented by the points enclosed by the three lines $x + y - 6 = 0$, $x = 1$, and $y = 2$.
    See linear programming; simplex method.

sine (sin) *See* trigonometric functions.

sine curve A graph of a sine function (*see* trigonometric functions). In rectangular Cartesian coordinates a graph of $y = \sin x$ is a regular undulating curve passing through the origin (*see* diagram). *See* cosine curve.

sine rule (law of sines) 1. A formula used for solving triangles in plane trigonometry:

$$\frac{a}{\sin A} = \frac{b}{\sin B} = \frac{c}{\sin C}$$

where a is the length of the side opposite angle A, b is opposite angle B, and c is opposite angle C.
2. A formula used in spherical trigonometry for solving *spherical triangles:

$$\frac{\sin a}{\sin A} = \frac{\sin b}{\sin B} = \frac{\sin c}{\sin C}$$

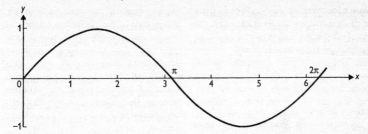

sine curve: $y = \sin x$.

sine series **1.** The *series expansion for a sine function:

$$\sin x = x - \frac{x^3}{3!} + \frac{x^5}{5!} - \frac{x^7}{7!} + \cdots$$

This is valid for all x. *See* trigonometric functions.
2. A *series in which the terms are *sine functions. *See* Fourier series.

single cusp *See* cusp.

singleton (unit set) A *set that contains only one element.

singularity *See* singular point.

singularity theory A theory that describes properties of differentiable functions in terms of their *singular points, for example in terms of the set of points where the derivative vanishes or the *Jacobian matrix is singular. *Catastrophe theory is part of singularity theory. *See also* Morse theory.

singular matrix A square *matrix whose *determinant is equal to zero; a square matrix that does not have an *inverse.

singular point (singularity) **1.** A point at which a function is not analytic (*see* analytic function). For instance, $f(z) = 1/(z-2)^2$ has a singular point at $z = 2$. If there is a neighbourhood of a singular point z_0 in which there is no other singular point, then there is said to be an *isolated singularity* at z_0. f has a *removable singularity* at z_0 if $f(z_0)$ can be redefined to make f analytic at z_0. For example,

$f(z) = \sin z/z$ has a removable singularity at $z = 0$.

A point a is a *branch point* of the function $f(z)$ if f has more than one value at points in a neighbourhood of a, but not at a itself. For example, the function $f(z) = (z-a)^{1/3}$ has three branches at $z = a$; that is, in every neighbourhood it has three possible values, corresponding to the three cube roots of $z - a$. To remove a branch point, it is usually necessary to redefine the function to be single valued on an associated *Riemann surface.

A function f has a *pole* of order k at z_0 if it can be written in the form

$$f(z) = \frac{\phi(z)}{(z - z_0)^k}$$

where ϕ is analytic at z_0 and $\phi(z_0) \neq 0$. When $k = 1$ the pole is called a *simple pole*. For instance,

$$f(z) = \frac{z}{(z-3)^2(z+1)}$$

has a simple pole at $z = -1$ and a pole of order 2 at $z = 3$. A pole is an isolated singularity. The *Laurent expansion of f about z_0 is

$$f(z) = \sum_{-k}^{\infty} a_n(z - z_0)^n$$

for z near z_0 since the coefficients a_n are zero for $n < -k$.

If the function has a singular point at z_0 that is neither a removable singularity nor a pole then it is said to have an *essential singularity* at z_0. If the essential singularity is isolated then a Laurent expansion can be found that has a principal part with infinitely many terms. For example:

$$f(z) = \exp\left(\frac{1}{z}\right) = 1 + \frac{1}{z} + \frac{1}{2! \, z^2} + \cdots$$

has an essential singularity at $z = 0$.

A *meromorphic function* is a function whose only singularities are poles.

See analytic continuation.

2. A point on a curve at which there is not a single smoothly turning tangent. Examples are *cusps, *isolated points, and *nodes.

singular value decomposition Any $m \times n$ matrix A has a singular value decomposition $A = UDV^*$, where U is an $m \times m$ *unitary matrix, V is an $n \times n$ unitary matrix, and D is an $m \times n$ matrix whose off-diagonal entries are zero and whose $\min(m, n)$ diagonal entries are nonnegative and arranged in non-increasing order down the diagonal. The diagonal entries of D are the *singular values* of A. The columns of V are the *right singular vectors* of A and the columns of U are the *left singular vectors* of A. The number of nonzero singular values equals the *rank of the matrix.

For example:

$$\begin{pmatrix} 2/3 & 0 \\ 5/6 & 1/2 \\ 1/3 & 1 \end{pmatrix} = \begin{pmatrix} 1/3 & 2/3 & 2/3 \\ 2/3 & 1/3 & -2/3 \\ 2/3 & -2/3 & 1/3 \end{pmatrix}$$

$$\times \begin{pmatrix} \sqrt{2} & 0 \\ 0 & 1/\sqrt{2} \\ 0 & 0 \end{pmatrix} \times \begin{pmatrix} 1/\sqrt{2} & 1/\sqrt{2} \\ 1/\sqrt{2} & -1\sqrt{2} \end{pmatrix}$$

sinh Hyperbolic sine. *See* hyperbolic functions.

sinusoidal Relating to a *sine curve.

SI units Système International d'Unités, a coherent system of units derived from

*m.k.s. units that is internationally used for scientific purposes. It consists of seven *base units and two *supplementary units (*see* table (a)), and a large number of *derived units, 18 of which have special names. Decimal multiples of SI units are expressed using a set of prefixes (*see* table (b)). Where possible a prefix representing 10 raised to a power that is a multiple of three should be used.

size 1. (of a graph) The number of edges of a *graph.

2. (of a sample) The number of items in a *sample.

skew curve *See* curve.

Skewes' number The function *Li(x) is an approximation to *$\pi(x)$, the number of *primes less than or equal to x, and Li$(x) > \pi(x)$ for all x within existing tables. In 1914, J.E. Littlewood proved that Li(x) must be less than $\pi(x)$ for infinitely many x. Then, in 1933, the South African mathematician Stanley Skewes (1899–1988) showed that the first x with Li$(x) < \pi(x)$ is less than $10^{10^{10^{34}}}$, known henceforth as Skewes' number.

skew field *See* division ring.

skew-Hermitian matrix *See* Hermitian conjugate.

skew lines Lines in space that are not parallel but do not intersect. Skew lines cannot lie in the same plane.

skewness The degree of asymmetry of a distribution. If l_i is the ith *moment about the mean, the *coefficient of skewness* is $c_1 = l_3/l_2^{3/2}$. It has the value 0 for a symmetric distribution. If c_1 is positive the skewness is called *positive skewness* and the distribution has a long tail to the right (*see* diagram); if c_1 is negative the skewness is called *negative skewness* and the distribution has a long tail to the left. Other measures of skewness include

(a) Base and supplementary SI units

Quantity	Name	Symbol
length	metre	m
mass	kilogram	kg
time	second	s
electric current	ampere	A
thermodynamic temperature	kelvin	K
luminous intensity	candela	cd
amount of substance	mole	mol
plane angle†	radian	rad
solid angle†	steradian	sr

† Supplementary units

(b) Prefixes for units

Prefix	Symbol	Factor	Prefix	Symbol	Factor
yotta	Y	10^{24}	deci	d	10^{-1}
zetta	Z	10^{21}	centi	c	10^{-2}
exa	E	10^{18}	milli	m	10^{-3}
peta	P	10^{15}	micro	μ	10^{-6}
tera	T	10^{12}	nano	n	10^{-9}
giga	G	10^{9}	pico	p	10^{-12}
mega	M	10^{6}	femto	f	10^{-15}
kilo	k	10^{3}	atto	a	10^{-18}
hecto	h	10^{2}	zepto	z	10^{-21}
deca	da	10^{1}	yocto	y	10^{-24}

$$\frac{mean - mode}{standard\ deviation} \quad \text{and} \quad \frac{Q_3 - 2M + Q_1}{Q_3 - Q_1}$$

where M is the median, and Q_1 and Q_3 are the first and third *quartiles. *See also* g-statistics.

skewness Frequency function with positive skewness.

skew-symmetric matrix *See* symmetric matrix.

slack variables *See* simplex method.

slant height 1. The length of a *generator of a right circular *cone.
2. The altitude of the lateral faces of a regular *pyramid.

slope 1. The angle that a line makes with the x-axis.
2. The *gradient of a curve at a given point.

slope–intercept form *See* line.

small circle A circle on a sphere that does not have its centre at the centre of the sphere; thus the radius of a small circle is less than the radius of the sphere. Each circle on a sphere is the intersection of a plane with the sphere. If the plane passes through the centre of the sphere, then the intersection is a *great circle; otherwise it is a small circle.

Smith, Henry John (1826–83) Irish mathematician noted for his work in number theory and his theorems on the possibility of expressing positive integers as the sums of five and seven squares. He also worked on the theory of elliptic functions.

smooth Generating no *friction. A smooth surface can be contrasted with a rough surface, which does generate friction.

smooth curve A curve for which the first *differential is continuous over all points.

smoothing Removal of erratic fluctuations in a time series by using a *moving average or fitting a trend curve (*see* time series).

smooth manifold *See* manifold.

Snedecor's *F*-distribution *See* *F*-distribution.

Snell, Willebrord van Roijen (1591–1626) Dutch mathematician and physicist best known for his formulation in 1621 of *Snell's laws* of refraction. He also worked on problems of geodesy. In 1621 he published an improvement on the classical method for calculating π.

snowflake curve A *fractal curve in the plane constructed by an iterated procedure. Starting from an equilateral triangle, each side is trisected and the middle third of each replaced by two sides of an equilateral triangle pointing outwards, to produce a six-pointed star (*see* diagram). The same process is applied to each side of the star, and repeated applications generate snowflake-like shapes. Continuing the process indefinitely gives the snowflake curve: a curve of infinite length containing a finite area 60 percent greater than that of the original triangle. It is a fractal with similarity dimension $\ln 4 / \ln 3$, and is closely related to the *Koch curve.

solid A three-dimensional geometric figure, e.g. a polyhedron or cone.

solid angle A configuration in three dimensions formed by all the *half-lines originating at a common point and passing through a closed plane *curve. There are two types of solid angle. In one the closed curve is a smooth curve, so that the solid angle is a *nappe of a conical surface. In the other type the closed curve is a polygon, so that the solid angle is a *polyhedral angle.

The idea of solid angle is an extension of plane angle to three dimensions, and it is possible to give a measure to a solid angle by an extension of radian measure. If a sphere, radius r, is considered with its centre at the vertex of the solid angle, then the solid angle in *steradians* is equal to A/r^2, where A is the area of the sphere intercepted by the solid angle. (Alternatively, the solid angle is the area intercepted on a unit sphere.) The total solid angle around a point is 4π steradians (i.e. $4\pi r^2 / r^2$). The *trihedral angle* formed by three mutually perpendicular half-lines is one-eighth of this, i.e. $\frac{1}{2}\pi$ steradians.

snowflake curve The first three stages of its generation.

solid geometry *See* geometry.

solid of revolution A solid generated by revolving a plane figure about an axis. For example, rotating a circle about a diameter generates a sphere. The volume of such a solid can be found by *integration. In a Cartesian coordinate system, if the axis lies along the x-axis the element of volume is a disc A dx, where A, the cross-sectional area of the disc, is πy^2. Thus, for a curve $y = f(x)$, the volume between $x = a$ and $x = b$ is given by the definite integral

$$\int_a^b \pi(f(x))^2 \, dx$$

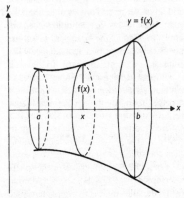

solid of revolution

solidus *See* division sign.

soliton *See* integrable system.

solstices (solstitial points) *See* equinoxes.

soluble group (*US*: **solvable group**) A group G that can be constructed successively (using *normal subgroups and factor groups) from *cyclic groups. G is soluble if it has a *composition series $H_0, H_1, \ldots, H_n$ such that all the *factor groups H_1/H_0, H_2/H_1, $\ldots$, H_n/H_{n-1} are *cyclic. The non-Abelian *symmetric group S_3 of all the symmetries of an equilateral triangle is soluble. Its elements can be written as I, σ_1, σ_2, σ_3, ρ, and ρ^2, where I is the *identity map that leaves the triangle unchanged, σ_1, σ_2, and σ_3 are reflections in the *medians, and ρ and ρ^2 are rotations about the centre of the triangle through $120°$ and $240°$, respectively. The sequence $H_0 = \{I\}$, $H_1 = \{I, \rho, \rho^2\}$ and $H_2 = S_3$ is a composition series, and the quotient groups H_1/H_0 and H_2/H_1 are 3- and 2-element cyclic groups, respectively. The symmetric groups S_n are not soluble if $n \geqslant 5$. *See* Galois theory.

solution of equations The process of finding the *roots of equations. In the case of polynomial equations, i.e. equations of the form

$$a_n x^n + \cdots + a_2 x^2 + a_1 x + a_0 = 0$$

the process is essentially one of finding the factors of the polynomial. In the simple case

$$x^2 - x - 6 = 0$$

the factors are $x - 3$ and $x + 2$, so that
$$(x - 3)(x + 2) = 0$$

and the roots are 3 and -2. (This follows because one factor or the other must be equal to zero (*see* factor theorem).) Here the factors are over the rational numbers. The equation
$$x^3 + x^2 - 3x - 3 = 0$$

can be factorized over the real numbers:
$$(x + \sqrt{3})(x - \sqrt{3})(x + 1) = 0$$

i.e. it has three real roots, but only one rational root (-1). The equation
$$x^2 + 49 = 0$$

has factors over the field of complex numbers:
$$(x + 7\mathrm{i})(x - 7\mathrm{i}) = 0$$

i.e. it has two complex roots ($\pm 7\mathrm{i}$).

Polynomial equations of degree 2 (i.e. quadratic equations) can be solved by inspection (in simple cases), by *completing the square, or by the *quadratic formula. Procedures can also be found for finding the roots of the *cubic and *quartic equations in terms of the coefficients. These involve rational operations and the extraction of roots. It can be shown that for polynomial equations of degree greater than four no such general method exists (*see* Galois theory).

Various methods exist for finding the number and nature of the roots of polynomial equations. Methods of approximate solution include *Newton's method. *See also* Descartes's rule of signs; simultaneous equations.

solution of triangles The process of calculating all the sides and angles of a triangle when sufficient data are available to specify the triangle. The method of solution depends on the type of triangle and the known parameters, as follows:

PLANE RIGHT-ANGLED TRIANGLES are determined by:

(1) *Two sides*. The third side is found by Pythagoras' theorem. One of the two acute angles is found by using a trigonometric ratio of two of the sides. The third angle is found by using the fact that the angles add to 180°.

(2) *One side and one additional angle*. In this case the third angle is found by subtraction from 180°. A second side can be found by a trigonometric function involving the known side. The third side is found by Pythagoras' theorem.

OBLIQUE PLANE TRIANGLES are determined by:

(1) *Two sides and the included angle*. The *cosine rule is used to find the third side and the other angles can then be determined by the *sine rule.

(2) *Two angles and the side between them*. The third angle is found by subtraction from 180° and the two other sides are found by using the *sine rule.

(3) *Three sides*. The unknown angles are found by using the *cosine rule or the *half-angle formulae of plane trigonometry.

In addition there is the *ambiguous case:

(4) *Two sides and the angle opposite one side*. The *sine rule is used to find a second angle (the third angle being obtained by subtraction from 180°). Ambiguity arises because if the sine of an angle is known, there are two possible angles that may have this (positive) sine – one acute and the other obtuse (the angles are supplementary).

RIGHT SPHERICAL TRIANGLES *Spherical triangles containing right angles are determined if two sides are known, or one side and an angle, or two angles other than the right angle. (This last condition does not apply to right plane triangles, which are not determined by two acute angles.) The solution of right spherical triangles is accomplished by using *Napier's rules of circular parts, together with the law of *species to select the appropriate quadrant.

OBLIQUE SPHERICAL TRIANGLES These are determined by:

(1) *Two sides and the included angle*. This can be solved using the *cosine rule of spherical trigonometry with the *half-angle formulae.

(2) *Two angles and the side between them*. Here the solution is obtained by using the *cosine rule with the *half-side formulae.

(3) *Three sides*. The solution is obtained by using the *half-angle formulae.

(4) *Three angles*. The solution is obtained by using the *half-side formulae.

The last case above (three angles) is peculiar to spherical triangles – plane triangles are not determined by three angles. In addition to the four cases above, there are two ambiguous cases in spherical trigonometry: two sides and the angle opposite one of them, and two angles and the side opposite one of them. These can be treated by the *sine rule of spherical trigonometry followed by *Napier's analogies. The solution of spherical triangles is sometimes helped by finding and solving the *polar triangle.

solution set The set of all possible solutions of an equation, inequality, or set of equations or inequalities.

solvable group *See* soluble group.

sorites In traditional logic, a series of *syllogisms in which all but the last conclusion is omitted: for example, 'All A is B, All B is C, All C is D, All D is E, therefore All A is E'. The term 'sorites' derives from the Greek word *soros*, meaning 'heap', and refers to the class of paradoxes first proposed by Eubulides of Miletus (4th century BC). One of these takes the form:

X is a heap of sand
X will remain a heap if one grain of sand is removed
X will remain a heap if a further grain of sand is removed

And so on, until eventually we have the absurd position that a single grain of sand is a heap. *See* Wang's paradox.

sound Describing a *logistic system in which every *theorem is a *valid *wff. Soundness is thus the converse of *completeness. If a system is sound and the axioms of the system are valid wffs, then all the theorems will also be valid; that is, the rules of inference preserve truth. In general, if A is a formal *consequence of $B_1, \ldots, B_n$, then, in a sound system, A will be a logical consequence of $B_1, \ldots, B_n$. Examples of sound systems include the propositional calculus and the predicate calculus. *See also* logic.

space *See* abstract space.

space coordinates Coordinates that determine the location of a point in three-dimensional space. *See* Cartesian coordinate system; cylindrical coordinate system; spherical coordinate system.

space curve A *curve in three-dimensional space.

space-filling curve *See* Peano's curve.

space group *See* crystallography; symmetry.

spacetime (spacetime continuum) The single concept into which space and time can be unified in order to describe the geometry of the universe. It replaces the idea of space and time as separate entities: spacetime has four dimensions compared with the three dimensions of ordinary (Euclidean) space. It is used in both the special and the general theory of *relativity, and was defined precisely by Hermann Minkowski.

The appropriate geometrical model of spacetime for special relativity is known as the *Minkowski universe*, which is described by means of Minkowski geometry. In the Minkowski universe spacetime is 'flat', much as space is 'flat' in Euclidean geometry. This is acceptable in the case of special relativity. General relativity, however, is concerned with the gravitational effects of matter, which cause spacetime to curve: massive objects produce distortions and ripples in local spacetime, and the motions of bodies are then dictated by the curvature. The geometry of curved spacetime is described by means of *differential geometry.

span *See* vector space.

sparse matrix A matrix that has a relatively large proportion of zero entries. The nonzeroes of a sparse matrix may, for example, be clustered around the leading diagonal, as in a *tridiagonal matrix:

or they may be scattered throughout the matrix with no obvious pattern.

Spearman's rank correlation coefficient *See* correlation coefficient.

special linear group *See* general linear group.

species In spherical trigonometry, two angles, two sides, or an angle and a side are of the same species if both are between $0°$ and $90°$ or if both are between $90°$ and $180°$. If one is between $0°$ and $90°$ and the other between $90°$ and $180°$, then they are of opposite species. The *law of species* (or *law of quadrants*) is applied to a right *spherical triangle. If A, B, and C are the angles and a, b, and c the respective sides opposite these angles, and C is the right angle, then:

(1) A and a are the same species and B and b are the same species;

(2) if $c < 90°$, a and b are the same species (i.e. a, b, A, and B are all the same species);

(3) if $c > 90°$, a and b are different species (as are A and B).

The rule is used in solving right spherical triangles. For example, for a right-angled triangle with side $c = 30°$ and angle $B = 30°$, the other sides can be found by using *Napier's rules of circular parts, which give relationships of the type

$$\sin b = \sin c \sin B$$

In the example, $\sin b = \frac{1}{2} \cdot \frac{1}{2} = \frac{1}{4}$, so side b is $\sin^{-1}\frac{1}{4}$; i.e. $14°\,29'$ or $165°\,31'$. The law of species can distinguish between these two: both b and B are of the same species so since B is an acute angle, b must also be less than $90°$ – i.e. it must be $14°\,29'$.

specific gravity *See* relative density.

speed The rate of change (i.e. the *derivative) of distance with respect to time. The direction of motion is not specified. Speed is thus the magnitude of the vector quantity *velocity. *See also* angular speed.

speed of light Symbol: c. The speed at which light and other electromagnetic waves travel in a vacuum. It is a universal constant and equals $299\,792\,458$ m s^{-1}. *See* wave; relativity.

sphere A closed surface that is the *locus of all points that are a fixed distance (the radius) from a given point (the centre). The surface area is $4\pi r^2$ and the enclosed volume is $\frac{4}{3}\pi r^3$. In rectangular Cartesian coordinates, the equation of a sphere is

$$(x - a)^2 + (y - b)^2 + (z - c)^2 = r^2$$

where (a, b, c) are the coordinates of the centre. The sphere is the closed surface that encloses the maximum volume for a given surface area.

More generally, the *n-sphere* S^n $(n \geqslant 0)$ is the *subspace of the $(n + 1)$-dimensional Euclidean space $\mathbb{R}^{n+1}$ of points $(x_1, \ldots, x_{n+1})$ such that $\sqrt{(x_1^2 + \cdots + x_{n+1}^2)} = 1$.

spherical angle An angle formed by two arcs of *great circles meeting on the surface of a sphere. The vertex of the angle forms the *pole of a (third) great circle and the two arcs forming the angle cut off another arc on this great circle. The length of this third arc (in degrees) gives the degree measure of the spherical angle.

spherical cone *See* spherical sector.

spherical coordinate system A *polar coordinate system in three dimensions. The location of a point P is made with reference to two axes at right angles taken from an origin (or *pole*) O. One coordinate is the *radius vector*, which is the distance OP from the pole to the point. The radius vector is given the symbol r (sometimes ρ). The other two coordinates are angles

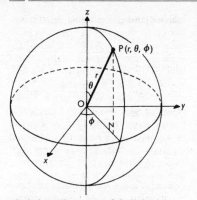

spherical coordinate system Spherical polar coordinate system.

measured with respect to two axes: the horizontal axis (corresponding to the x-axis of Cartesian coordinates) and the vertical axis (corresponding to the z-axis and called the *polar axis*). The plane of the two axes is called the *initial meridian plane*. The angle between the polar axis and the radius vector is the *colatitude* θ: the angle between the horizontal axis and the projection of the radius vector on the horizontal plane is the *longitude* ϕ. The point P is specified by three coordinates, written as (r, θ, ϕ).

The colatitude θ may vary between 0 and π radians; the longitude may have any value but is usually taken between 0 and 2π radians. Spherical coordinates are used in studying systems that possess spherical symmetry; examples occur in field theory, spherical harmonics, celestial mechanics (*see* astronomical coordinate system), and atomic structure. The method of locating a point is similar (but not identical) to the system of *geographical co-ordinates. Spherical coordinates are also called *spherical polar coordinates*.

It is possible to transform from a spherical coordinate system to a rectangular Cartesian coordinate system. If the pole of the spherical system coincides with the origin of the Cartesian system, the polar axis coincides with the z-axis, and the initial meridian plane coincides with the x–z

plane, then a point (r, θ, ϕ) in spherical coordinates has Cartesian coordinates given by

$$x = r \sin \theta \cos \phi$$

$$y = r \sin \theta \sin \phi$$

$$z = r \cos \theta$$

Similarly, a point (x, y, z) in Cartesian coordinates has spherical coordinates given by

$$r = \sqrt{(x^2 + y^2 + z^2)}$$

$$\theta = \tan^{-1}\left[\frac{\sqrt{(x^2 + y^2)}}{z}\right]$$

$$\phi = \tan^{-1}\left(\frac{y}{x}\right)$$

where θ is such that $0 \leqslant \theta < \pi$ and the value of ϕ is such that

$$x : y : r \sin \theta = \cos \phi : \sin \phi : 1$$

spherical degree A unit of area on the surface of a sphere equal to the area of a birectangular triangle (*see* spherical triangle) that has a third angle of one degree. A hemisphere has an area of 360 spherical degrees and a sphere has 720 spherical degrees. *See also* steradian.

spherical distance The distance between two points on a sphere, equal to the length of the minor *arc of a *great circle cut off by the points.

spherical excess *See* spherical polygon; spherical triangle.

spherical harmonic *See* harmonic.

spherical polar coordinate system *See* spherical coordinate system.

spherical polygon A figure formed on the surface of a sphere by three or more arcs of *great circles. The sum of the angles of a spherical polygon lies between $180(n - 2)^\circ$ and $180n^\circ$, where n is the number of sides. The difference between the sum and $180(n - 2)^\circ$ is the *spherical excess* of the polygon. The area of a spherical polygon is given by $\pi r^2 E/180$, where r is the radius of the sphere and E is the spherical excess.

spherical pyramid A closed surface formed by a *spherical polygon and lines from the vertices of the polygon to the centre of the sphere. The spherical pyramid includes the curved polygon surface together with the plane lateral faces. Its volume is $\pi r^3 E/540$, where r is the radius of the sphere and E the spherical excess of the polygon.

spherical sector A closed surface that is the *surface of revolution of a sector of a circle revolved (through 360°) about a diameter of the circle. The spherical sector is bounded by a *zone on the surface of the sphere (formed by the arc of the sector) and by one or two conical surfaces formed by the radius or radii of the sector. If the axis of revolution lies outside the sector, the figure has a zone of two bases and has two conical surfaces. If the axis passes through the sector, the spherical sector has a zone of one base and has one conical surface. In this case it is a *spherical cone*. The volume of a spherical sector (or cone) is $\frac{2}{3}\pi r^2 h$ where r is the radius of the sphere and h, is the altitude of the zone.

spherical segment A closed surface formed by two parallel planes cutting a sphere. The spherical segment, in general, has two circular bases with a *zone of the sphere between them. If one of the planes is a tangent plane, a segment of one base is formed. The volume of a spherical segment is

$$\frac{1}{6}\pi h(3r_1^2 + 3r_2^2 + h^2)$$

where h is the altitude of the zone and r_1 and r_2 are the radii of the bases. For a segment of one base, the formula becomes

$$\frac{1}{6}\pi h(3r^2 + h^2)$$

spherical triangle A figure on a sphere formed by three arcs of *great circles of the sphere. The angles of a spherical triangle are the *spherical angles formed between the arcs; the lengths of the sides are often specified by the angles they subtend at the centre of the sphere. Unlike plane triangles, the angles of a spherical triangle do not add to 180°: the sum can be any value in the range 180°–540°, and may contain one, two, or three right angles. A *right* spherical triangle has at least one right angle; a *birectangular spherical triangle* contains two right angles; a *trirectangular spherical triangle* has three right angles. A *quadrantal spherical triangle* is one for which one side is equal to 90° (i.e. subtends an angle of 90° at the centre of the sphere). The difference between the sum of the angles and 180° is the *spherical excess* of the triangle. Spherical triangles differ in other ways from plane triangles (*see* solution of triangles). The area of a spherical triangle is given by $\pi r^2 E/180$, where r is the radius of the sphere and E the spherical excess. *See also* polar triangle.

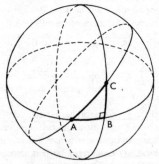

spherical triangle A right spherical triangle.

spherical trigonometry *See* trigonometry; solution of triangles.

spherical wedge A closed surface formed by two planes meeting along the axis of a sphere. The wedge includes the parts of the planes within the sphere together with the region of the surface that they cut off (the *lune). The volume enclosed by a spherical wedge is $\pi r^3 \theta / 270$, where θ is the angle between the planes and r is the radius of the sphere.

spheroid *See* ellipsoid.

spinode *See* cusp.

spiral A plane *curve, or part of a plane curve, whose equation in polar coordinates is

$$r = f(\theta)$$

for which r always increases (or always decreases) as θ increases.

The *Archimedean spiral* is defined by the equation

$$r = a\theta$$

The *hyperbolic* (or *reciprocal*) *spiral* has the equation

$$r\theta = a$$

The *parabolic* (or *Fermat's*) *spiral* has the equation

$$r^2 = a^2\theta$$

The *logarithmic* (or *logistic*) *spiral* has the equation

$$\ln(r/a) = \theta \cot b \quad \text{or} \quad r = a\exp(\theta \cot b)$$

This curve has the property that the tangent at any point makes an angle b with the radius vector, hence the alternative name *equiangular spiral*.

The *klothoid* or *Cornu spiral* is a curve having the *intrinsic equation

$$a^2 \kappa = s$$

the *curvature κ at any point being proportional to the *arc length s from the pole to the point. The curve may be defined parametrically by

$$x = bC(s/b), \quad y = bS(s/b)$$

where $b = a\sqrt{2}$, and C and S are *Fresnel integrals. This curve is named after the French physicist Marie Alfred Cornu (1841–1902). It is used in analysing intensities of diffraction patterns.

The *lituus* (plural *litui*; the name means 'trumpet') is a curve given by

$$r^2\theta = a^2$$

It is asymptotic to the polar axis.

(a)

(b)

spiral (a) Equiangular and (b) klothoid or Cornu spirals.

spline *See* approximation theory.

spur *See* trace.

square 1. A simple *quadrilateral with four equal sides and all four angles right angles.
2. A number or expression obtained by multiplying a given number or expression by itself. Thus, the square of 6 is 6×6 (written as 6^2).

square-free Describing an *integer that is not divisible by any square integer (apart from 1^2). For instance, 15 is square-free but 28 is not.

square matrix A *matrix having the same number of rows as columns.

square number A number that is the square of an integer: 1, 4, 9, 16, etc.

square root A number that when multiplied by itself gives a given number. For instance, 3 is a square root of 9, written as $\sqrt{9}$, since $3^2 = 9$. *See also* Hero's method.

squaring the circle The problem of constructing, using only unmarked straightedge and compasses, a square equal in area to a given circle. It dates from the time of Anaxagoras (5th century BC) and is one of the three classical problems of antiquity (the others being the *duplication of the cube and *trisection of an angle). In 1882 Lindemann demonstrated the impossibility of the construction by his proof that π is a transcendental number (*see* algebraic number).

standard deviation (s.d.) The positive square root of the *variance.

standard error (s.e.) The *standard deviation of the sampling distribution of a *statistic. For example, if the mean $\bar{x}$ of a sample of n is the statistic used to estimate the unknown mean μ of a normal distribution with variance σ^2, the s.e. is $\sigma/\sqrt{n}$. The term 'standard error' is also sometimes used for the estimate of that quantity, i.e. $s/\sqrt{n}$, where

$$s^2 = \frac{1}{n-1} \sum_i (x_i - \bar{x})^2$$

Confusion is avoided if $s/\sqrt{n}$ is called the *estimated standard error*.

standard form 1. (of a number) *See* exponential notation.
2. (of an equation) A simple form of an equation; a form in which the equation is usually written. For example, the standard form of the equation for a circle in a Cartesian coordinate system is that for a circle

with its centre at the origin:

$$x^2 + y^2 = r^2$$

standardized random variable If X is a *random variable (not necessarily normal) with mean μ and *standard deviation σ,

$$Z = (X - \mu)/\sigma$$

is called a standardized variable. Values of this variable are often called *Z-scores*, but this name is sometimes reserved for when μ and σ are replaced by sample estimates $\bar{x}$ and s. Only if X has a *normal distribution and μ and σ are known does Z have a standard normal distribution.

Standardization to a Z- or a *T-score is widely used in an educational context to introduce comparability in marks scored for different subjects; in a subject such as mathematics, if papers are marked out of 100 it is not unusual to find candidates obtaining unstandardized marks throughout the interval [0, 95] while in French marks may all lie in the interval [25, 75].

standard normal variable *See* normal distribution.

star polygon *See* polygon.

statics A branch of mechanics concerned with the forces and torques under which a body is in *equilibrium, i.e. at rest relative to its surroundings.

stationary point (critical point) 1. A point on a graph of a curve at which the *tangent is horizontal, i.e. a point at which there is a maximum, a minimum, or a horizontal point of *inflection. A stationary point occurs when the first derivative $f'(x)$ of the curve $y = f(x)$ is zero. *See* turning point.
2. A point on a surface at which there is a horizontal *tangent plane. A stationary point on a surface is either a maximum, a minimum, or a *saddle point. It occurs when the two partial derivatives $\partial z/\partial x$ and $\partial z/\partial y$ of the surface $z = f(x, y)$ are both zero.

3. For a function $f(x_1, x_2, \ldots, x_n)$ of two or more variables, a stationary (or critical) point is one at which each of the first-order partial derivatives $\partial f / \partial x_i$ vanishes, i.e. the *gradient of f, grad f, vanishes.

statistic A term originally used to describe any single figure derived from or contained in a set of data (statistics). For example, the *mean, median, smallest value, and percentage of the data with values exceeding 7 would each be a statistic in this sense. In formal statistical theory a statistic is described as any function of the sample values. An example of a statistic in this sense is the quantity t used in the *t-test, or any function of sample values used to estimate a population parameter, such as a sample mean, as an estimator of a population mean.

statistical control A term used in *quality control to indicate that a process is operating within statistically determined limits, indicating acceptable performance.

statistical inference *See* Bayesian inference; confidence interval; decision theory; estimation; fiducial inference; hypothesis testing; inference; sequential analysis.

statistical mechanics *See* mechanics.

statistical significance *See* hypothesis testing.

statistics **1.** A collection of numerical data; for example, official statistics on employment, or on imports and exports, or monthly meteorological records for the Isle of Tiree.
2. The science of collecting, studying, and analysing numerical data. The subject divides broadly into two branches. *Descriptive statistics* is concerned mainly with collecting, summarizing, and interpreting data. *Inferential statistics* is concerned with methods for obtaining and analysing data to make inferences applicable in a wider context (e.g. from sample to population). It is concerned also with the precision and reliability of such inferences

insofar as this involves probabilistic considerations. In this context statistics may be described as that branch of applied mathematics based on probability theory.
3. *Plural of* *statistic.

statute mile *See* mile.

Steiner, Jakob (1796–1863) Swiss mathematician best known for his *Systematische Entwickelung* (1832, Systematic Development) and his attempt to establish a comprehensive theory of geometry using stereographic projection.

Steiner's problem *See* Fermat point.

stem-and-leaf display (J.W. Tukey, 1977) A semi-graphical presentation of data. For example, for the data set 10, 27, 19, 11, 14, 41, 38, 59, 7, 21 we may consider the tens digits as *stems* and the units digits as *leaves* and arrange the data in order in a table:

Stem	Leaf
0	7
1	0 1 4 9
2	1 7
3	8
4	1
5	9

It is then easy to write a *five-number summary, and the leaf distribution has the pattern of a *histogram turned on its side. A complete ordering of the data is thus achieved with little more work than that required for grouping in classes with an interval of width 10.

step function A *function whose graph consists solely of disconnected horizontal line segments. An example is the graph of $y = [x]$, the *integer part of x.

steradian Symbol: sr. The SI *supplementary unit of solid angle, equal to the solid angle subtended by unit area at the centre of a sphere with unit radius.

stereographic projection A conformal projection (*see* conformal transformation) of a sphere onto a plane. A point P (the *pole*) is taken on the sphere and the plane is perpendicular to the diameter through P. Points on the sphere, A, are mapped by straight lines from P onto the plane to give points A'.

stereographic projection

Stevin, Simon (1548–1620) Flemish mathematician and engineer noted for his work in statics and hydrostatics. He is best known, however, for a work on arithmetic published in 1585 in both Flemish and French which contained the first comprehensive discussion of decimal fractions.

Stirling, James (1692–1770) Scottish mathematician and protégé of Newton. In 1717 he added four new cubic curves to the 72 already described by Newton in 1704. His book *Methodus differentialis* (The Differential Method, 1730) contained work on infinite series and their summation, and interpolation using finite differences. He is best known today, however, for *Stirling's formula.

Stirling's formula (J. Stirling, 1730) The formula

$$n! \approx \sqrt{(2\pi n)}(n/e)^n$$

It gives an approximation, for large values of the positive integer n, to *factorial n.

stochastic Implying random variation, generally used to describe systems which are not deterministic rather than systems which are deterministic apart from a random error. The stochastic nature of a system is often associated with time: for example, in a queuing system in a bank or post office, distributions may be specified for the intervals between customer arrivals and for the service times at one or more service points. The number of people in the queue at any given time is an example of a stochastic variable. *See also* stochastic matrix; stochastic process.

stochastic analysis The study of the basic mathematics describing stochastic processes, such as the differential equations associated with *Brownian motion.

stochastic matrix A term widely used for the matrix of transition probabilities for a *Markov chain. In this context it has the property that all elements are non-negative and all its row sums are unity. Such a matrix is said to be *doubly stochastic* if all column sums are also unity. The term is sometimes used more generally for any matrix comprised of *stochastic elements.

stochastic model *See* model.

stochastic process A random process. Common usage excludes essentially deterministic processes, which are subject only to random errors. *See* birth–death process; branching process; Markov chain; Poisson process; queuing theory; random walk; reliability; time series.

Stokes, Sir George Gabriel (1819–1903) Anglo-Irish mathematician and physicist known for his formulation in 1845 of *Stokes's law* of fluid resistance; in his honour the unit of kinematic viscosity was named the *stokes*. Other important work by Stokes was concerned with the propagation of sound, fluorescence, spectroscopy, the wave theory of light, and the nature of the aether.

Stokes's theorem (G.G. Stokes, 1854) The theorem that for a *vector function **F**

$$\int_S \nabla \times \mathbf{F} \cdot \mathbf{n} \, dA = \int_C \mathbf{F} \cdot d\mathbf{s}$$

where **n** is a *unit vector normal to S, i.e. the integral of curl **F** over a surface S is equal to the integral of **F** around the boundary C of the surface. There is a very general variant of the theorem, applicable in all dimensions, that uses integration of *differential forms. It includes the original Stokes's theorem and *Green's theorem as special cases. It can be regarded as the appropriate generalization to higher dimensions of the *fundamental theorem of calculus.

straight angle (flat angle) An angle equal to one-half of a complete turn ($180°$ or π radians).

straight line See line.

strain A measure of the *deformation of a body subjected to an *external force. The deformation can be a change in shape or size, and the strain can be expressed as the change in length per unit length, change in area per unit area, or change in volume per unit volume; these quantities are dimensionless. *Shear, another form of strain, is an angular deformation without change in volume, and is measured in radians. *Stresses are set up within a body under strain. See also Hooke's law; elasticity.

strange attractor See chaos.

Strassen's method (V. Strassen, 1969) A method for multiplying two $n \times n$ matrices with a number of operations proportional to $n^{\log_2 7}$, as opposed to the standard way of multiplying matrices that requires a number of operations proportional to n^3 ($\log_2 7 \approx 2.81$). The method is based on recursive application of the following formula for multiplying two 2×2 matrices A and B in only 7 (instead of the usual 8) multiplications. If $C = AB$, then

$$C = \begin{pmatrix} p_1 + p_4 - p_5 + p_7 & p_3 + p_5 \\ p_2 + p_4 & p_1 + p_3 - p_2 + p_6 \end{pmatrix}$$

where

$$p_1 = (a_{11} + a_{22})(b_{11} + b_{22})$$
$$p_2 = (a_{21} + a_{22})b_{11}$$
$$p_3 = a_{11}(b_{12} - b_{22})$$
$$p_4 = a_{22}(b_{21} - b_{11})$$
$$p_5 = (a_{11} + a_{12})b_{22}$$
$$p_6 = (a_{21} - a_{11})(b_{11} + b_{12})$$
$$p_7 = (a_{12} - a_{22})(b_{21} + b_{22})$$

stratified sample A *population may be divided into strata so that there is greater uniformity with respect to characteristics being measured within each stratum than there is between strata. Separate *random samples are taken within each stratum. The appropriate analysis enables more precise estimation of population characteristics. For example, a survey carried out in a school to determine attitudes to banning nuclear weapons might show that different proportions of boys and girls favour a ban. A two-strata sample might be used, each sex forming a stratum. A common practice, often giving optimum precision, is to take samples for each stratum of size proportional to population stratum size. For example, if there are 400 boys and 200 girls in the school and a sample of 60 were to be taken, this would contain 40 randomly selected boys and 20 randomly selected girls.

stress A measure of the internal reactions of a body subjected to an *external force. A system of internal forces is set up, in equilibrium, and the stress is expressed as the force per unit area. Stress is always associated with an accompanying deformation of the body, measured in terms of *strain. Stress can be *tensile, compressive, or *shear (see diagram).

Any real body under stress undergoes deformation to a greater or lesser degree. For small stresses most materials are elastic, i.e. they return to their original shape once the stress disappears. Again, for small stresses, strain is proportional to stress (*see* Hooke's law). At greater stresses some materials will crack while others become plastic and eventually fracture. *See also* elasticity; elongation; modulus of elasticity.

— original shape
---- shape under stress
stress Types of stress.

strict equivalence *See* equivalence.

strict implication *See* implication.

strict inequality *See* inequality.

string *See* alphabet.

strophoid A plane *curve that has the equation in Cartesian coordinates

$$y^2 = x^2 \frac{x+a}{a-x}$$

It can be generated by taking a fixed point Q on the x-axis at $(-a, 0)$ and drawing a line to cut the y-axis at R. There are points P and P' on QR such that

$$PR = RP' = OR$$

where O is the origin. As the position of R varies, P and P' trace a strophoid. The line $x = a$ is an asymptote.

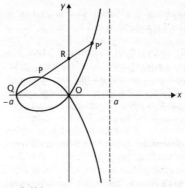

strophoid

Student's *t*-distribution *See* *t*-distribution.

Student's *t*-test *See* *t*-test.

subclass *See* subset.

subcomplex *See* combinatorial topology.

subdiagonal *See* matrix.

subfactorial For an *integer n, the expression

$$n!\left(\frac{1}{2!} - \frac{1}{3!} + \frac{1}{4!} - \cdots + \frac{(-1)^n}{n!}\right)$$

subfield For a *field F, a *subset of the members of the field is a subfield of F if the subset is a field with respect to the operations on F. This means that the subset has to form a *subgroup of the field's *additive group, and the nonzero elements of the subset have to form a subgroup of the field's *multiplicative group. Every subfield of F contains the zero and identity elements of F. *See* extension field.

subgraph A *graph whose vertices and edges form *subsets of those of another graph.

subgroup For a *group G, a subset S of the members of the group is a subgroup of G if the subset forms a group with respect to the *binary operation ∘ of G when it is applied just to members of S. If a and b are any elements in S, this means that $a \circ b$

has to be in S, or, in other words, when restricted in this way $\circ$ is a binary operation on S. Also, the *inverse of each member of S has again to be in S. For example, the set of all even integers is a subgroup of the group of integers $\mathbb{Z}$ under addition. Every subgroup of G contains the identity element of G. *See also* coset; Lagrange's theorem.

subharmonic function A continuous real-valued *function f with *domain D that is an open *subset of the complex *field, such that for every closed disc with centre a and radius r contained in D

$$f(a) \leqslant \frac{1}{2\pi} \int_0^{2\pi} f(a + re^{i\theta})\, d\theta$$

The integral is the average of f over the circumference of the disc and is called the *circumferential mean* of f. The function f is said to be *subharmonic* in D; if the inequality is reversed it is said to be *superharmonic* in D.

subjective probability *Probability based on a degree of belief. An axiomatic theory of subjective probability exists, but one difficulty is that not all rational people may assign the same subjective probability value to a given event. *See* Bayesian inference.

submatrix *See* partition (of a matrix).

subnormal In *Cartesian coordinates, the line segment on the x-axis lying between the intercept of a *normal with the x-axis and the foot of a perpendicular from the point at which the normal meets the curve to the x-axis. *See also* polar tangent.

subring For a *ring R, a *subset of the members of R is a subring of R if the subset is a ring with respect to the two operations of R.

subscript A number written to the lower right of a symbol, usually to identify an

element of a *sequence. *See also* superscript.

sub-sequence A *sequence within a sequence. Hence

$$a_2, a_4, a_6, \ldots, a_{2n}, \ldots$$

is a sub-sequence of

$$a_1, a_2, a_3, \ldots, a_n, \ldots$$

A sub-sequence is a function whose domain is a subset of the positive integers.

subset (subclass) A *set A is a subset of a set B, denoted by $A \subseteq B$, if and only if whatever is a member of A is also a member of B:

$$A \subseteq B \leftrightarrow (\forall x)(x \in A \rightarrow x \in B)$$

For example, if $A = \{1, 2, 3\}$, $B = \{1, 2, 3, 4\}$, and $C = \{1, 2, 3\}$ then

$$A \subseteq B, \quad A \subseteq A, \quad A \subseteq C$$

but B is not a subset of A or C. *See also* proper subset; inclusion; proper inclusion.

subspace 1. (of a vector space) A *subset of the elements of a *vector space that is itself a vector space. The subspace has to contain the zero element of the space.
2. (of a topological space) *See* topological space.

substitution The replacement of all occurrences of a variable or expression by another variable, expression, or quantity. For example, to evaluate $x^2 + 2x + 6$ when $x = 2$, all occurrences of x are replaced by 2, and the value of the expression is thus calculated to be 14. This is called 'making the substitution $x = 2$'. If, instead, the substitution $x = y - 1$ is made, the expression becomes $(y - 1)^2 + 2(y - 1) + 6$, which simplifies to $y^2 + 5$. It is now easily seen that the expression can never be less than 5 in value, and takes the minimum value when $y = 0$, i.e. when $x = -1$. *See* change of variable; elimination.

substitution cipher A *cipher in which each letter is replaced by a letter or a word. *See also* mono-alphabetic substitution cipher; polyalphabetic substitution cipher.

substitution group A *group whose members are *permutations.

subtangent In *Cartesian coordinates, the line segment on the x-axis lying between the intercept of a *tangent with the x-axis and the foot of a perpendicular from the point of tangency to the x-axis. *See also* polar tangent.

subtend An arc or line segment joining two points A and B is said to subtend the angle ACB at a third point C.

subtraction The inverse operation to addition; the process, for two given numbers, of finding a third which, added to one of the numbers, gives the other, written as

$$d = a - b$$

where d is the *difference*, a is the *minuend*, and b is the *subtrahend*. The subtrahend plus the difference gives the minuend. Analogous operations are defined for other entities, e.g. *matrices and *vectors.

subtraction formulae Formulae in plane trigonometry. *See* addition formulae.

subtrahend The quantity subtracted from another in finding a difference. *See* subtraction.

successive over-relaxation *See* Gauss–Seidel method.

sufficient condition *See* necessary condition.

sufficient statistic (R.A. Fisher, 1921) A *statistic that contains all the information in a sample that is relevant to the point *estimation of a specific parameter. For example, in estimating the mean μ of a normal distribution, the sample mean $\bar{x}$ is a sufficient statistic. Knowledge of the individual sample values provides no further information about μ, since the distri-

bution of the sample, conditional upon $\bar{x}$, is independent of the population mean μ. If a sufficient statistic exists, the *maximum likelihood estimator is a function of that sufficient statistic.

sum 1. The result of an *addition.
2. (of sets) *See* union.
3. (of an infinite series) The *limit of the *sequence of *partial sums of an infinite *series, i.e. the limit of the sum of the first n terms of the series, as $n \to \infty$. *See* convergent series.

sum function *See* sigma function.

summand *See* summation sign.

summation (of an infinite series) The process of finding the sum of a *convergent series or of attributing a sum to a *divergent series.

summation sign The sign Σ (Greek capital sigma) used to indicate summation of a set or sequence of numbers or variables (the *summands*). When the 1st to the Nth terms of a sequence

$$a_1, a_2, \ldots, a_n, \ldots$$

are to be summed, this is written as

$$\sum_{n=1}^{N} a_n$$

The summation of an infinite number of terms is written as

$$\sum_{n=1}^{\infty} a_n \quad \text{or simply} \quad \Sigma a_n$$

sup Supremum. *See* least upper bound.

superdiagonal *See* matrix.

superharmonic function *See* subharmonic function.

superscript A number written to the upper right of a symbol, usually indicating an *exponent. When in brackets, it also indicates the order of a derivative. *See also* subscript.

supplement *See* supplementary angles.

supplemental chords A pair of *chords joining a point on a circle to the two ends of a diameter. Supplemental chords are perpendicular (the angle between them is an angle in a semicircle).

supplementary angles Two angles that have a sum of 180°. Each angle is said to be the *supplement* of the other.

supplementary units The *SI units for plane angle (radian) and solid angle (steradian). They may be regarded as dimensionally independent physical quantities, and therefore included with the *base units, or as dimensionless *derived units. The decision to treat them as supplementary units which may be used to form derived units (e.g. angular velocity is measured in radians per second) has been universally accepted.

supremum (sup) *See* least upper bound.

surd An expression containing *irrational *roots; for example, $\sqrt{7}$, or $6 + \sqrt{5}$, or $\sqrt{3} + 2\sqrt{11}$. A *pure surd* contains only irrational terms; a *mixed surd* also contains rational terms.

surface In general, a surface is a set of points (x, y, z) in space whose coordinates satisfy an equation such as $z = F(x, y)$ or $G(x, y, z) = 0$, or are given in terms of two parameters. For example, $z = x + y$ is the equation of a *plane surface; $x^2 + y^2 + z^2 - 4 = 0$ is the equation of a spherical surface; and $x = r\cos\theta$, $y = r\sin\theta$, $z = \lambda$ are parametric equations of a circular cylindrical surface of radius r.

Alternatively, a surface may be defined as the *image of a continuous mapping of a region of the *Euclidean plane $\mathbb{R}^2$ (*see* manifold).

surface of revolution A surface generated by rotating a curve about an axis. For example, rotating a parabola about its axis of symmetry gives a paraboloid of revolution. The area of such a surface can be obtained by integration. In Cartesian coordinates, if the axis lies along the x-axis the element of area is $2\pi y \, ds$, where ds is an element of length of the curve. Since $ds^2 = dy^2 + dx^2$, the area of the surface between $x = a$ and $x = b$ is given by the definite integral

$$\int_a^b 2\pi y \sqrt{\left[1 + \left(\frac{dy}{dx}\right)^2\right]} \, dx$$

surgery (J.W. Milnor, 1961) A technique for making geometrical modifications to a *manifold, so as to produce a cobordant manifold (*see* cobordism) with simpler homotopy groups. Surgery has proved to be a very powerful tool in the study of the topology of manifolds.

surjection (onto, surjective function) A surjection from a *set A to a set B is a *function whose *domain is A and whose *range is the whole of B. For example, if $A = \{2, -2, 3\}$ and $B = \{4, 9\}$ then the function $f: x \to x^2$ is a surjection. *See also* bijection; injection.

survival analysis The analysis of data on times to occurrence of a defined *quantal response such as death or recovery from a disease, or of times to failure of a machine or some item of equipment. Typically the time distribution tends to be positively skewed (*see* skewness), or skewed to the right. The response of interest tends to occur over a short, or moderate timescale for some items, but being scattered over a longer timescale for the remaining items.

Data are often incomplete in that some items are lost to the study or *censored before the response of interest occurs. Although censored items provide less information than items that show the required response, they do contain some useful information. For instance, a person who withdraws from a medical study 100 days after treatment but before showing the response of interest certainly has a longer survival time than one who survives only 90 days, but one cannot say whether a person who withdraws without responding after 100 days might or might not have a longer survival time than one who survives, say, 120 days.

syllogism Traditional logicians distinguished between the following four forms of *categorical proposition: All S is P, No S is P, Some S is P, and Some S is not P, referred to respectively as A, E, I, and O propositions. A syllogism was then defined as an *argument consisting of three propositions such that the first two, the *premises*, entail the third, the *conclusion*, as in the famous example: All men are mortal, Socrates is a man, therefore Socrates is mortal. The forms of the propositions give the *mood* of the syllogism – in this case AII.

A syllogism must employ exactly three terms, each of which appears twice. The terms S and P appear respectively as the *subject* and *predicate* of the conclusion, and in each premise along with M, the middle term. The position of the terms in the premises allows the following four *figures* to be distinguished according to the location of the middle term:

$$
\begin{array}{llll}
(1) & M - P & (2) & P - M \\
    & S - M &   & S - M \\
    & S - P &   & S - P \\
\end{array}
$$

$$
\begin{array}{llll}
(3) & M - P & (4) & P - M \\
    & M - S &   & M - S \\
    & S - P &   & S - P \\
\end{array}
$$

As any of the four categorical forms can appear as premise and conclusion, 256 distinct *syllogistic forms* can be identified, of which 19 yield valid arguments. The example given earlier is a valid first-figure

syllogistic form of mood AII. Under certain other assumptions, however, the number of valid forms can either increase to 24 or reduce to 15:

(1) Five weakened forms can be recognized but are not normally counted. For example, the first-figure syllogism

    All men are mortal
    All Greeks are men
∴ Some Greeks are mortal

though technically valid, is more naturally recognized as a weaker form of

    All men are mortal
    All Greeks are men
∴ All Greeks are mortal

(2) If only I and O propositions are allowed to have *existential import, then a form such as

    All ghosts are men
    All men are mortals
∴ Some mortals are ghosts

will no longer be valid. Four such forms can be distinguished, and therefore reduce the number of valid forms under this assumption to 15.

Sylvester, James Joseph (1814–97) English mathematician best known for the work on invariants on which he collaborated with Cayley. In 1878 he became the founding editor of the first American mathematics research periodical, the *American Journal of Mathematics*.

Sylvester's law of inertia (J.J. Sylvester, 1852) The theorem that the inertia of a *symmetric (Hermitian) matrix is unchanged under congruence (conjunctive) transformations.

symbolic manipulation *See* computer algebra.

symmetric difference The symmetric difference of two sets A and B is the set of elements belonging to one but not both of A and B. It can be defined as the union of their *differences:

$$\{x : x \in (A \backslash B) \vee x \in (B \backslash A)\}$$

Thus, if $A = \{1, 2, 3, 4\}$ and $B = \{3, 4, 5, 6\}$, their symmetric difference will be the set $C = \{1, 2, 5, 6\}$. The symmetric difference of A and B is symbolized variously as $A \ominus B$, $A \nabla B$, or $A + B$.

symmetric form *See* line.

symmetric function A *function of several variables such that

$$f(x_1, \ldots, x_i, \ldots, x_j, \ldots, x_n)$$
$$= f(x_1, \ldots, x_j, \ldots, x_i, \ldots, x_n)$$

for every pair (x_i, x_j). f is sometimes called *totally symmetric*. For example,

$$f(x, y, z) = x^2 + y^2 + z^2 + 2xyz$$

is totally symmetric since the function is unchanged if any two of the three variables are interchanged. f is *totally skew symmetric* if

$$f(x_1, \ldots, x_i, \ldots, x_j, \ldots, x_n)$$
$$= -f(x_1, \ldots, x_j, \ldots, x_i, \ldots, x_n)$$

for every pair (x_i, x_j). For example,

$$f(x, y, z) = (x - y)(y - z)(z - x)$$

is totally skew symmetric.

symmetric group A *permutation group formed by all the permutations of a set. The symmetric group for a set of n elements is denoted by S_n.

symmetric matrix A square *matrix that is symmetrical about its leading diagonal. A symmetric matrix is equal to its *transpose. A *skew-symmetric (antisymmetric)* matrix is one that is equal to the negative of its transpose. *See also* Hermitian conjugate.

symmetric positive definite *See* positive definite.

symmetric relation A binary *relation R on a *set A is symmetric if, for all $x, y \in A$, $x \, \mathrm{R} \, y \rightarrow y \, \mathrm{R} \, x$. The relation 'cousin', for

example, is symmetric on the set of people. If, however, as with the relation 'inclusion',

$$x \, \mathrm{R} \, y \, \& \, y \, \mathrm{R} \, x \rightarrow x = y$$

then the relation is said to be *antisymmetric*. If, as with relations like 'greater than',

$$x \, \mathrm{R} \, y \rightarrow \sim(y \, \mathrm{R} \, x)$$

the relation is described as *asymmetric*.

symmetry In general, a figure or expression is said to be symmetric if parts of it may be interchanged without changing the whole. For example, $x^2 + 2xy + y^2$ is symmetric in x and y. A *symmetry operation (symmetry)* is any operation on a figure or expression that produces an identical figure or expression. A figure or expression which is not symmetric is called *asymmetric*.

A geometric figure has *reflectional symmetry* if points in the figure have corresponding points reflected in some point (*centre of symmetry*), line (*axis of symmetry*), or plane (*plane of symmetry*) (*see* reflection). Thus a circle has a centre of symmetry, and any diameter is an axis of symmetry. A right circular cone has an axis of symmetry. A right square pyramid has four planes of symmetry.

symmetry Three of the axes of rotational symmetry of the cube.

A geometric figure has *n*-fold *rotational symmetry* about an axis if a rotation of $360°/n$ is a symmetry of the figure. Such a symmetry is called a symmetry of *order n*. Thus an axis of symmetry can be a rotational axis. For example, an axis through the centre of a square perpendicular to the plane is a 4-fold axis since a rotation of $360°/4$ produces an identical figure; a diagonal is a 2-fold axis. The symmetry operations on a figure form a group (its *symmetry group*) if the product of two operations is defined to be one operation followed by the other. This is why group theory is useful in interpreting the molecular spectra of chemical compounds, and is also important in the study of crystal structure (*see* crystallography).

See also frieze group; wallpaper group.

symplectic geometry   A geometry on *manifolds based on the structure of *Hamiltonian mechanics. It can be used to study general systems since it allows the study of motion on configuration spaces with minimum use of explicit coordinates.

syntax 1. (of a formal language) A list of expressions of a *formal language together with a set of *formation rules.
2. (logical syntax) *See* proof theory.

synthetic division   *See* Horner's method.

T

tableaux *See* simplex method.

tacnode *See* cusp.

tac-point A point where two members of a family of curves intersect and have a common *tangent (i.e. a tacnode). For instance, the equation

$$(x - m)^2 + (y - m)^2 = 4$$

represents a *family of circles, given by different values of m. The circle with $m = 0$ (centre at the origin) and the circle with $m = 2\sqrt{2}$ have a tac-point at $(\sqrt{2}, \sqrt{2})$. A locus of tac-points is a *tac-locus*. In the case above the tac-locus is the line $y = x$.

tan Tangent. *See* trigonometric functions.

tangency *See* tangent.

tangent **1.** A line (*tangent line*) that touches a curve at only one point (the *point of contact* or *point of tangency*). For a given curve at a given point, the slope of the tangent can be found by taking the derivative at that point, and this enables the equation of the tangent line to be found. For example, to find the equation of the tangent line of the curve $y = 2x^2$ at the point $(3, 18)$, it is assumed that the tangent line has an equation of the form $y = mx + c$, where m is the slope of the tangent and c is its intercept on the y-axis. The slope m can be found by taking the derivative of the curve at the given point, i.e. $dy/dx = 4x$, so that at the point where $x = 3$ the slope is 12. Thus the tangent line is

$$y = 12x + c$$

The value of c is found by substituting the values $x = 3$ and $y = 18$ in this equation (since the point $(3, 18)$ must lie on the tangent line). This gives $c = -18$, so the equation of the tangent line is

$$y = 12x - 18$$

A tangent line to any space curve at a given point P can be defined by considering the secant line through P and another point on the curve Q. The tangent through P is the limiting position of this secant as Q approaches P.

A line can also be *tangent to a surface* at a certain point if the line is a tangent of another line or curve on the surface passing through the point.

In Cartesian coordinates the length of a tangent is taken to be the length of the line segment from the point of contact to the x-axis. The length of the *normal is similarly the length of the line segment of the normal

tangent Tangent and normal.

from the point to the x-axis. The *subtangent* is the projection of the length of the tangent on the x-axis and the *subnormal* is the projection of the length of the normal on the x-axis. *See also* polar tangent.
2. *See* trigonometric functions.

tangent–chord theorem *See* circle.

tangent curve 1. A graph of a tangent function (*see* trigonometric functions). In Cartesian coordinates the graph of $y = \tan x$ is a periodic curve with separate branches, each with a point of inflection at the x-axis and asymptotic to lines $x = \pm\pi/2, \pm3\pi/2, \pm5\pi/2$, etc.
2. Any of a set of curves that have a common point at which they have a common tangent. Two closed curves are *externally tangent* if each lies outside the other and *internally tangent* if one is inside the other. A curve can be tangent to a plane or other surface at a point if it is tangent to a line or curve on the surface passing through the point.

tangent curve

tangent formulae The *half-angle and *half-side formulae used in trigonometry.

tangential Being a *tangent or directed along a tangent.

tangential component *See* acceleration.

tangent plane A plane that touches a given surface at a particular point. Specifically, it

is a plane in which all the lines that pass through the point are tangents to the surface at the point. If the surface is a conical or cylindrical surface then the tangent plane will touch it along a line.

It is possible to find the equation of a tangent plane at a given point by finding the direction cosines (*see* direction angles) of the normal to the surface at that point. The normal to the surface is also the normal to the tangent plane, so if it has direction cosines l, m, and n, then the equation of the tangent plane at that point (x_1, y_1, z_1) is

$$l(x - x_1) + m(y - y_1) + n(z - z_1) = 0$$

The direction cosines of the normal are found by evaluating partial derivatives of the function representing the surface. *See* normal.

tangent-secant theorem *See* circle.

tangent series The *series expansion for the tangent function:

$$\tan x = x + \frac{x^3}{3} + \frac{2x^5}{15} + \frac{17x^7}{315} + \cdots$$

This is valid for $-\frac{1}{2}\pi < x < \frac{1}{2}\pi$. *See also* Gregory's series.

tangent space The set $T_x(M)$ of all vectors that are tangent to a differential *manifold at a point $x \in M$. For a surface, it is the tangent plane at a point.

tanh Hyperbolic tangent. *See* hyperbolic functions.

Tarski, Alfred (1902–85) Polish-American mathematical logician who in his *The Concept of Truth in Formalized Languages* (1935) introduced the distinction between language and metalanguage, which he considered to be necessary to avoid the paradoxes detected within the foundations of set theory. Tarski also made numerous contributions to decision theory and model theory, and pioneered the application of algebra to the study of formal systems.

Tartaglia, Niccolò (*c*.1500–1557) Italian mathematician noted for his discovery in 1535 of the long-sought solution to the general cubic equation. He revealed his method in confidence to Cardano who promptly published it, without consent but with acknowledgement, in his *Ars magna* (1545). Tartaglia also wrote the influential three-volume *General trattato di numeri et misure* (1556–60, Treatise on Numbers and Measures) as well as being the first, in 1543, to translate Euclid into a modern Western language, namely his native Italian.

tautochrone *See* isochrone; cycloid.

tautology A statement that is true under all assignments (*see* interpretation) of *truth values to its *atomic sentences. *Truth tables provide an effective procedure for determining whether or not a given *wff is a tautology. Tautologies are considered to be unhelpful propositions since they are true regardless of the truth values of their components, and thus give no extra information about real circumstances.

Taylor, Brook (1685–1731) English mathematician who made important contributions to Newton's newly developed calculus. In his book *Methodus incrementorum directa et inversa* (1715, Direct and Indirect Methods of Incrementation) he first formulated the expansion since known as *Taylor's theorem. In the same year he published a work on perspective, *Linear Perspective*.

Taylor's theorem A theorem which expresses a *function $f(x)$ as the sum of a *polynomial and a remainder:

$$f(x) = f(a) + f'(a)(x - a)$$

$$+ f''(a)\frac{(x - a)^2}{2!}$$

$$+ f'''(a)\frac{(x - a)^3}{3!} + \cdots + R_n$$

where R_n is the remainder after n terms:

$$R_n = \frac{1}{n!} h^n f^{(n)}(a + h\theta)$$

where $h = x - a$ and θ lies between 0 and 1.

If $n \to \infty$, the expansion is a *Taylor series*, which represents the function if $R_n \to 0$ as $n \to \infty$. If $a = 0$, the series is called a *Maclaurin series*.

Tchebyshev, Pafnuty Livovich (1821–94) Russian mathematician noted for his foundation in the mid 19th century of the St Petersburg mathematical school. Tchebyshev himself worked on number theory, proving in 1850 Bertrand's postulate that if $n > 3$ then there is at least one prime between n and $2n - 2$. He is best remembered, however, for his work in probability theory.

Tchebyshev polynomials A particular family of polynomials of degree 0, 1, 2, ... The first four Tchebyshev polynomials are

$$T_0(x) = 1$$

$$T_1(x) = x$$

$$T_2(x) = 2x^2 - 1$$

$$T_3(x) = 4x^3 - 3x$$

The general formula is

$$T_k(x) = \cos(k \cos^{-1} x)$$

The polynomials satisfy the recurrence relation

$$T_{k+1}(x) = 2xT_k(x) - T_{k-1}(x), \quad k \geqslant 1$$

The Tchebyshev polynomials form a set of *orthogonal polynomials with respect to the interval $[-1, 1]$ and the weight function $1/\sqrt{(1 - x^2)}$. The Tchebyshev polynomials have important applications in *interpolation.

Tchebyshev's inequality (P.L. Tchebyshev, 1874) If X is a *random variable and $g(x)$ a non-negative function of x, then the *probability that $g(X) \geqslant k$ $(k > 0)$ is less than or equal to the *expectation of $g(X)$ divided by k:

$$\Pr(g(X) \geqslant k) \leqslant E(g(X))/k$$

The particular case where

$$g(x) = (x - \mu)^2$$

where μ is the mean of X and $k = t^2\sigma^2$, σ^2 being the variance of X, implies that

$$\Pr(|X - \mu| \geqslant t\sigma) \leqslant 1/t^2$$

This is sometimes called the *Bienaymé–Tchebyshev inequality* (I.-J. Bienaymé, 1853; P.L. Tchebyshev, 1867).

t-distribution (W.S. Gosset, 1908) Also known as *Student's t-distribution*, 'Student' being Gosset's pseudonym. Essentially it is a *distribution of a variable which is proportional to the ratio of a standard normal variable (*see* normal distribution) to the square root of a chisquared variable (*see* chi-squared distribution) with k degrees of freedom; also identical to the square root of an *F*-distribution variable with 1, k degrees of freedom, $k \geqslant 1$. Given a sample of n observations, x_i, from a normal distribution with mean μ, the statistic

$$t = \frac{\bar{x} - \mu}{s/\sqrt{n}}$$

where

$$s^2 = \sum_{i=1}^{n} \frac{(x_i - \bar{x})^2}{n-1}$$

has a *t*-distribution with $n - 1$ degrees of freedom. *See also* t-test.

temporal logic A general term covering attempts to incorporate statements which contain temporal information within a logical framework. More specifically it is used to refer to the *modal logic of tensed propositions known as *tense logic.

tend to *See* limit.

tense logic (A.N. Prior, 1957) A form of *temporal logic in which it is accepted that propositions such as 'Socrates is sleeping' can be true at one time and false at another.

The simplest approach is to add to the *propositional calculus a set of operators, G, F, and P, defined by: $Gp = $ (It will always be the case that p), $Fp = $ (It will at some time be the case that p) and $Pp = $ (It has always been the case that p), together with a number of axioms, including $Gp \rightarrow Fp$ (what will always be, will be) and $p \rightarrow GPp$ (what is the case, will always have been the case). The operator F can be defined in terms of G through the definition 'Fp is equivalent and replaceable by $\sim G \sim p$', i.e. 'what will at some time be the case, will not always not be the case'.

With further additions more sophisticated tense logics have been developed in which, among other things, time can have no beginning or end, or can be circular or linear, dense or branching.

tension A *force that stretches or tends to stretch a body or structure. For example, a taut wire under tension might be elongated. *Tensile stress* is set up within the body or structure in reaction to such a force. *See also* stress.

tensor An abstract entity having a set of components that are functions of position in *n*-dimensional space.

Suppose that points have n coordinates $x^i \ (= x^1, x^2, \ldots, x^n)$ in some coordinate system, and corresponding coordinates $\bar{x}^i (= \bar{x}^1, \bar{x}^2, \ldots, \bar{x}^n)$ in a second coordinate system. (Note that suffixes are not exponents in tensor notation.) A set of n components, denoted by A^i, that are functions of the n coordinates x^i will become a set of n components $\bar{A}^i$ that are functions of the n coordinates $\bar{x}^i$ on a change of coordinates from the first to the second system. Similarly, A^{ij}, A^{ijk}, ... denote sets of n^2, n^3, ... components.

A tensor is a set of components that obeys some transformation law. The number of suffixes indicates the *order* of the tensor; their position indicates the type of tensor. A *contravariant tensor* of order 1 is a set A^i satisfying, for each i,

$$\bar{A}^i = \sum_{r=1}^{n} \frac{\partial \bar{x}^i}{\partial x^r} A^r$$

A contravariant tensor of order 2 is a set A^{ij} satisfying, for all i and j,

$$\bar{A}^{ij} = \sum_r \sum_s \frac{\partial \bar{x}^i}{\partial x^r} \frac{\partial \bar{x}^j}{\partial x^s} A^{rs}$$

A *covariant tensor* of order 1 is a set A_i satisfying, for each i,

$$\bar{A}_i = \sum_r \frac{\partial x^r}{\partial \bar{x}^i} A_r$$

A covariant tensor of order 2 is a set A_{ij} satisfying, for all i and j,

$$\bar{A}_{ij} = \sum_r \sum_s \frac{\partial x^r}{\partial \bar{x}^i} \frac{\partial x^s}{\partial \bar{x}^j} A_{rs}$$

Whereas contravariant tensors have superscripts and covariant tensors have subscripts, a *mixed tensor* has both. For instance, a mixed tensor of order 2 is a set A_j^i satisfying, for all i, j,

$$\bar{A}_j^i = \sum_r \sum_s \frac{\partial \bar{x}^i}{\partial x^r} \frac{\partial x^s}{\partial \bar{x}^j} A_s^r$$

Tensors of higher order are similarly defined.

A tensor of order zero is a scalar; a tensor of order 1 is a vector. The *Kronecker delta is an example of a mixed tensor. Strictly, a tensor applies to a point in each coordinate system; one applied to a region is a *tensor field*.

A *Cartesian tensor* is a tensor on $\mathbb{R}^3$ that transforms appropriately under rotations. The concept is commonly used in the study of continuum mechanics.

Tensor analysis was developed by Ricci-Curbastro as a generalization of vector analysis. It was used by Einstein in his formulation of the general theory of relativity, and is important in differential geometry and in physics.

tera- *See* SI units.

term 1. In general, a part of an equation or mathematical expression. In a polynomial, the terms are the expressions that are added together. For instance, x^2, $-xy$, and y^2 are the three terms of the trinomial

$$x^2 - xy + y^2$$

In a fraction, the terms are the numerator and the denominator.
2. In *logic, an expression that stands in the subject position of a sentence, and is thus in contrast to a *predicate. Terms are always interpreted as standing for an object, although sometimes only with respect to a sequence, as in the case of logical variables (*see* interpretation). *See also* predicate calculus.

terminal speed The limiting speed approached by a body as it moves through air or some other fluid that resists its motion; this resistive force varies as some power of the body's speed. The body reaches terminal speed when the resultant force on it is zero so that it has no acceleration. For a body falling freely through air, air resistance increases with speed until it balances the force of gravity. The body will then fall at constant terminal speed.

terminating decimal *See* decimal.

terminating fraction A finite *continued fraction.

ternary relation *See* relation.

tesla Symbol: T. The *SI unit of magnetic flux density, equal to a density of 1 weber of magnetic flux per square metre. [After N. Tesla (1870–1943)]

tessellation A covering or *tiling* of the entire plane with nonoverlapping shapes (*see* diagram (a)). A *regular tessellation* uses congruent regular polygons, all of one kind. There are three regular tessella-

tessellation (a)

tions: those using squares, equilateral triangles, and hexagons. A *semiregular tessellation* uses congruent regular polygons of more than one kind, arranged so that every vertex of the tessellation is congruent to every other vertex, e.g. one composed of triangles and squares (*see* diagram (b)). There are eight semiregular plane tessellations, of which two are mirror images of each other.

tessellation (b)

When a tessellation of the plane is possible using congruent copies of a shape (or set of shapes), the shape (or set) is said to *tessellate* the plane. For example, any scalene triangle or plane quadrilateral and its mirror image will tessellate the plane.

tesseract A regular *polytope in four-dimensional space which is the analogue of the cube in three-dimensional space.

tests of homogeneity The class of tests for equality of means, variances, proportions, etc, applied to samples from different populations. *See* homogeneity of variance; *F*-test; *t*-test.

tetrahedral angle 1. A *polyhedral angle with four faces.
2. If lines are drawn from the centre of symmetry of a regular *tetrahedron to the vertices, the plane angle between any two of these lines (109° 28′) is called the tetrahedral angle.

tetrahedron (*plural* **tetrahedra**) A solid figure that has four triangular faces (i.e. a triangular pyramid). A regular tetrahedron, in which the faces are equilateral triangles, is one of the five regular polyhedra. *See* polyhedron.

Thales of Miletus (*c.*625–*c.*547 BC) Greek mathematician and philosopher who is generally considered to be the first Western scientist and philosopher. His fame as a mathematician rests upon his supposed discovery of seven geometrical propositions, including the familiar Euclidean theorems; the angles at the base of an isosceles triangle are equal, and an angle inscribed in a semicircle is a right angle. According to one tradition, Thales acquired his mathematical learning from Egyptian scholars. He is reported to have predicted the solar eclipse of 585 BC.

Theaetetus (*c.*414–*c.*369 BC) In Plato's dialogue the *Theaetetus*, the character Theaetetus engages in a discussion on the nature of incommensurable magnitudes. It is thought that the material in Euclid's Book X dealing with the irrational numbers comes in fact from lost work by the historical Theaetetus. He is also credited with the discovery of two of the regular solids, the octahedron and the icosahedron.

Theodorus of Cyrene (*c.*425 BC) Greek mathematician who, according to Plato, demonstrated that not only was $\sqrt{2}$ irrational, but that so too were $\sqrt{3}$ and

$\sqrt{5}$, and the roots of all other non-square numbers up to 17.

theorem A statement derived from *premises rather than assumed. In logic, a theorem is a *wff A of a *formal system S such that $A = B_n$ for some *proof B_1, B_2, ..., B_n in S. 'A is a theorem of S' is denoted by '$\vdash_S A$' (the subscript is omitted if it is clear which formal system is intended). A *lemma* is a theorem which is proved and then used in the proof of another theorem. A theorem easily deduced from another theorem is a *corollary* of that theorem. *See also* converse; deduction; duality; metatheorem.

theory of equations The study of polynomials and their roots. *See* Galois theory; solution of equations.

theta functions Certain special functions of a complex variable that are more general than trigonometric and elliptic functions. The roots of every polynomial can be written in terms of theta functions.

third kind *See* first kind.

Thom, René Frédéric (1923–2002) French mathematician who, in his *Stabilité structurelle et morphogenèse* (1972, Structural Stability and Morphogenesis), created the discipline of catastrophe theory. In earlier work, he made fundamental contributions to the study of topology, and introduced the idea of cobordism.

Thomson, William *See* Kelvin.

thou *See* mil.

three-body problem Given *Newton's laws of motion and his law of *gravitation, is it possible to calculate accurately the future positions and velocities of n mutually attractive material bodies? Newton solved the problem for $n = 2$ (as if, for example, the sun were the sole gravitational influence on an orbiting planet). For $n = 3$, however, for something like the sun–earth–moon system, the problem still awaits a general solution. The problem was tackled repeat-edly by Euler, Lagrange, and Laplace in the 18th century, and by Poincaré in more recent times. While a number of limited solutions have been worked out, mathematicians must still rely largely on methods of approximation when called upon to work out from first principles the future positions and velocities of the moon.

three-valued logic Traditional logic was based upon the assumption that every proposition is either true or false, and was consequently committed to the law of the *excluded middle, $p \vee \sim p$. It was thus described as two-valued logic. In 1920 the Polish logician Jan *Łukasiewicz proposed for consideration a three-valued logic designed to accommodate future contingents. The proposition 'It will rain tomorrow' is neither true (T) nor false (F), he argued, but *undetermined* (U). Under this interpretation the excluded middle principle, $p \vee \sim p$, can sometimes take the value U and therefore can no longer be accepted as a logical law. Several other three-valued systems have been developed in which the T and F (or 1 and 0) of traditional two-valued logic are replaced by the three truth values 1, $\frac{1}{2}$, and 0.

Tikhonov's theorem (A.N. Tikhonov, 1930; E. Čech, 1937) The *product topology on any product of *compact spaces is compact. It is also spelt *Tychonoff's theorem*.

tiling *See* tessellation.

time Symbol: t. The continuous, irreversible passage of existence, or a part of this *continuum. The SI unit of time is the *second. *See also* day; year; spacetime.

time dilation One of the effects predicted by the special theory of *relativity and since verified experimentally. When two observers move at constant relative velocity, each will observe that the other's clock is operating more slowly, i.e. time is different for two observers moving relative to each other. If a clock at rest ticks n

times per second, then according to someone moving at speed v this clock will appear to tick $n\sqrt{(1 - v^2/c^2)}$ times per second, where c is the speed of light. The effect is significant only at very high speeds.

time series A set of observations, usually measurements or counts, ordered in time. Time series are widely used in economics to predict future trends in output, sales, inflation, etc. To get meaningful predictions over the long term, allowance should be made for seasonal fluctuations or other periodic features by suitable adjustments; smoothing (e.g. by using a *moving average) may also be needed to reduce the influence of short-term irregularities. An example of a time series is monthly sales of cars in a country over a ten-year period. These could be expected to show a seasonal trend and perhaps an annual peak (e.g. in the UK in August, when, until 1999, sales were boosted by a new letter in car registrations each year). There are also likely to be departures from any long-term trend due to factors such as economic recession. Time series of maximum or mean summer temperatures at a given meteorological station over periods of fifty years or more may provide evidence about a phenomenon such as global warming, but careful analysis is needed because meteorological phenomena are well known to exhibit trends which may persist for periods from two or three to fifty years but which are not part of longer-term trends. *See also* Box–Jenkins model; Durbin–Watson statistic.

tit-for-tat *See* prisoner's dilemma.

Toeplitz matrix A square *matrix with constant diagonals, illustrated for $n = 4$ by

$$\begin{pmatrix} a & b & c & d \\ e & a & b & c \\ f & e & a & b \\ g & f & e & a \end{pmatrix}$$

Named after the German mathematician Otto Toeplitz (1881–1940).

ton 1. (long ton) A UK unit of mass equivalent to 2240 lb. *See also* avoirdupois; troy system.
2. (short ton) A US unit of mass equivalent to 2000 lb.

tonne A unit of mass in the *metric system equal to 1000 kilograms. It is sometimes known as the *metric ton*. 1 tonne = 2204.62 lb = 0.9842 ton.

topological group A *group G which is also a *topological space, and where the functions m: $G \times G \to G$ and u: $G \to G$ are continuous maps. Here m is the multiplication in G and u(g) = g^{-1} for all $g \in G$. For example, the circle S^1, regarded as the multiplicative group of complex numbers of unit modulus, is a topological group.

topological space A *set, together with sufficient extra structure to make sense of the notion of continuity, when applied to functions between sets. More precisely, a set X is called a topological space if a collection T of subsets of X is specified, satisfying the following three axioms:

(1) the empty set and X itself belong to T;
(2) the intersection of two sets in T is again in T;
(3) the union of any collection of sets in T is again in T.

The sets in T are called *open sets*, and T is sometimes referred to as a *topology* on X. For example, the real line $\mathbb{R}^1$ becomes a topological space if we take as open sets those subsets U for which, given any $x \in U$, there exists $\varepsilon > 0$ such that $\{y \in \mathbb{R}^1 : |x - y| < \varepsilon\}$ is contained in U. (It is easily seen that the collection of such subsets satisfies axioms 1–3.) A similar definition is valid in any *metric space, but it is not in general required of every topological space that it should be metric. However, some of the most common examples do arise as metric spaces.

A subset A of a topological space X is called a *subspace* if it is given the structure of a topological space by specifying that the open sets of A consist of the intersection with A of all the open sets of X (this is the *subspace topology* on A).

Given topological spaces X and Y, a function $f: X \to Y$ is a *continuous map* (otherwise known as a *continuous mapping* or *continuous function*) if, for each open set U in Y, $f^{-1}(U)$, defined to be $\{x \in X: f(x) \in U\}$, is an open set in X. (If $X = Y = \mathbb{R}^1$, this definition reduces to the usual definition of a continuous real-valued function of a real variable.)

See also homeomorphism.

topology The study of those properties of geometrical figures that are invariant under continuous deformation (sometimes known as 'rubber-sheet geometry'). Unlike the geometer, who is typically concerned with questions of congruence or similarity of triangles, the topologist is not at all interested in distances and angles, and will for example regard a circle and a square (of whatever size) as equivalent, since either can be continuously deformed into the other. Thus such topics as *knot theory belong to topology rather than to geometry; for the distinction between, say, a granny knot and a reef knot cannot be measured in terms of angles and lengths, yet no amount of stretching or bending will transform one knot into the other.

More formally, topology is the study of those properties of *topological spaces that are invariant under *homeomorphism.

The subject has two main branches: *point-set topology* (sometimes known as analytic topology or general topology), which is concerned with the intrinsic properties of the various types of topological spaces; and *algebraic topology*, which seeks to classify topological spaces by using algebraic methods. It was originally called *analysis situs. See also* combinatorial topology.

torque *See* moment of a force.

torr Symbol: Torr. A unit of pressure, equal to 1/760 atmosphere or 1 millimetre of mercury, 1 torr $= 133.322$ pascals. [After E. Torricelli (1608–47)]

torus (anchor ring) A surface formed by revolving a circle about a line which is in the plane of the circle but does not intersect the circle. If r is the radius of the circle and R the distance of its centre from the line, then the area of the torus is $4^2 r R$, and the enclosed volume is $2^2 r^2 R$.

In topology, a torus can be described as the 2-manifold obtained from the square

$$\{(x_1, x_2) \in \mathbb{R}^2 : |x_1|, |x_2| \leqslant 1\}$$

by identifying opposite edges 'without twists', i.e. by identifying $(-1, x_2)$ with $(1, x_2)$ for all x_2 and $(x_1, -1)$ with $(x_1, 1)$ for all x_1.

The torus is homeomorphic to the topological product of two circles (*see* Cartesian product).

total differential The *differential of a function of more than one variable. If $z = f(x, y)$, the total differential of z is given by

$$dz = \frac{\partial f}{\partial x}\, dx + \frac{\partial f}{\partial y}\, dy$$

totient function The *function that counts the number of *totitives of a natural number. *See* Euler's phi function.

totitive A natural number not exceeding another natural number n and *relatively prime to n. For example, the totitives of 10 are 1, 3, 7, and 9. The number of totitives of n is denoted by $\phi(n)$, *Euler's phi function. Thus $\phi(10) = 4$.

tower *See* nested sets.

trace 1. (spur) The algebraic sum of the elements in the leading diagonal of a *matrix.
2. (piercing point) A point at which a space *curve intersects a coordinate plane.

tractrix A plane *curve that is the *involute of a *catenary. A tractrix is the locus of a point P that moves so that the length PP′ of a tangent at P cutting the x-axis at P′ is constant. If a is the length PP′, the parametric equations of the tractrix are

$$x = a(u - \tanh u), \quad y = a \operatorname{sech} u$$

where u is a variable parameter. The curve is symmetrical about the y-axis with a cusp on the y-axis. The x-axis is an asymptote. The surface of revolution formed by rotating a tractrix about the x-axis has constant negative curvature, and is known as a *pseudosphere*. It provides a model for the non-Euclidean hyperbolic geometry of Lobachevsky.

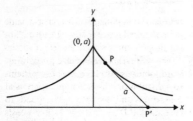

tractrix

trail See walk.

trajectory 1. The path of a moving particle or body.
2. The ordered subset of a space X consisting of the points x, $T(x)$, $T^2(x)$, ... associated with a point x and an *iterated map $T : X \to X$. See dynamical system.

transcendental curve A curve that has an equation involving *transcendental functions.

transcendental function See function.

transcendental number See algebraic number.

transfinite number or **set** See Cantor's theory of sets.

transform 1. A relationship between members of a *group A, B, and Y such that $A = Y^{-1}BY$. A is said to be the transform of B by Y, and A and B are said to be *conjugates*. The *conjugate set* of an element is the set of all its conjugates. If A, B, and Y are matrices, A is the transform of B by Y, provided Y is nonsingular; A is said to be *similar* to B.
2. See integral transform.

transformation 1. A change in the form of a mathematical expression, as by rearranging the terms.
2. A mapping (or *function). The term is essentially synonymous with 'function' but is commonly used for changes in coordinate systems. *Matrix notation is often used for transformations.

A point in two-dimensional space can be represented by a column vector $\begin{pmatrix} x \\ y \end{pmatrix}$. Transformation to a point $\begin{pmatrix} x' \\ y' \end{pmatrix}$ can occur as the result of matrix multiplication,

$$\begin{pmatrix} x' \\ y' \end{pmatrix} = \mathbf{T} \begin{pmatrix} x \\ y \end{pmatrix}$$

where $\mathbf{T}$ is the *transformation matrix*. Examples of transformation matrices are:
(a) reflection in the x-axis:

$$\begin{pmatrix} 1 & 0 \\ 0 & -1 \end{pmatrix}$$

(b) extension in the x-direction, by a factor k:

$$\begin{pmatrix} k & 0 \\ 0 & 1 \end{pmatrix}$$

See also linear transformation.
3. In *statistics, data are often transformed by taking logarithms or square roots (or arc sines for proportions) to obtain data closer to a normal distribution or to allow fitting of *linear models, etc. See also logarithmic transformation; normalizing transformation.

transformation of axes A change in the axes of a *coordinate system: either
(1) a change from one system to another, as in transformation from a *Cartesian

coordinate system to a *polar coordinate system; or
(2) a change in position of axes, as in *rotation or *translation of axes.

transition matrix *See* Markov chain.

transitive law *See* order properties.

transitive relation A *binary relation R on a *set A is transitive if for all $x, y, z \in A$

$$x R y \ \& \ y R z \rightarrow x R z$$

Thus the relation 'greater than' is transitive. Relations like 'greater by 1 than', however, for which

$$x R y \ \& \ y R z \rightarrow \sim (x R z)$$

are said to be *intransitive*.

translation 1. Motion in a straight line.
2. A *transformation such that the directed line segments joining points to their images all have the same magnitude and direction. A translation maps a point with position vector **r** onto a point with position vector **r** + **a**, where **a** is a constant vector. In the plane, a translation maps the point with Cartesian coordinates (x, y) onto the point $(x + a, \ y + b)$, where a and b are constants. The image of a figure under a translation is called a *translate* of the figure.

translation of axes A *transformation from one set of axes to another set parallel to the original axes. In a plane *Cartesian coordinate system, if (x, y) are the coordinates of a point P in one set of axes and (x', y') the coordinates in the second set of axes, then

$$x = x' + h, \quad y = y' + k$$

where (h, k) are the coordinates of the origin of the second system with respect to the first system. Translation of axes is used to simplify equations of curves. For example, the circle

$$(x - 3)^2 + (y - 5)^2 = 7$$

has its centre at the point $(3, 5)$. Translation of the x- and y-axes to new axes with their origin at $(3, 5)$ gives

$$x'^2 + y'^2 = 7$$

translation of axes

transportation problem The generic name in operational research for a group of *linear programming problems with a special structure that permits more straightforward solutions. The basic problem pertains to a distributor who has storage depots at S locations and needs to deliver goods to customers at D destinations. Delivery costs depend on distances and times taken, and mode of transport used. These costs are assumed to be known, as are the available stocks in each depot and the amount each customer requires. The problem is to determine how much should be sent from each depot to each customer so as to minimize total transport costs. Relevant information is set out in a tableau; a commonly used algorithm for solution is called the *north-west-corner rule*.

transpose A *matrix formed from a given matrix by interchanging the rows and columns. The transpose of a row vector is a column vector (and vice versa). The transpose of a matrix A is commonly denoted by A^{T}.

transposition A *permutation of a set that merely interchanges two elements.

transposition cryptography A method of constructing *ciphertext in which a *permutation, known to the intended recipient, is applied to the characters of the plaintext. For example, rugby → bgryu and wales → elwsa.

transversal A line cutting two or more other lines. If the transversal cuts two separate lines, then eight angles are formed. The four angles lying between the two lines are *interior angles*; the four lying outside the two lines are *exterior angles*. An interior (or exterior) angle formed by the transversal's cutting of one line and an interior (or exterior) angle formed at the other line together constitute a pair of *alternate angles* if they lie on opposite sides of the transversal. An interior angle at one line with an exterior angle at the other constitute a pair of *corresponding angles* if they lie on the same side of the transversal. If the two lines cut by the transversal are parallel, then alternate and corresponding angles are equal. *See also* parallel transversal theorem.

transverse axis *See* hyperbola.

transverse component *See* velocity; acceleration.

transverse wave A form of *wave motion in which the vibration or displacement of the transmitting medium occurs in a plane perpendicular to the direction of propagation of the wave. Surface ripples on water and electromagnetic waves (such as light or radio waves) are transverse. *Compare* longitudinal wave.

trapdoor function A *one-to-one function whose values are easy to calculate but whose *inverse is difficult to evaluate. It often takes the form of a straightforward *algorithm producing $y = f(x)$ where, however, it is known to be difficult to calculate x from the value y. Such functions are widely used to produce *ciphertext since they are easy to calculate, but the plaintext is difficult to find.

interior angles	d, c, A, B
exterior angles	a, b, D, C
alternate pairs	dB, cA, aC, bD
corresponding pairs	aA, bB, cC, dD

transversal

trapezium (*US*: **trapezoid**) A *quadrilateral that has one pair of opposite sides parallel, the other pair being nonparallel. The area of a trapezium is $\frac{1}{2}h(a + b)$, where a and b are the lengths of the parallel sides and h is the distance between them.

trapezoidal rule (trapezium rule, trapezoid rule) A rule for *numerical integration which approximates

$$\int_a^b f(x)\,dx \approx \tfrac{1}{2}(b - a)(f(a) + f(b))$$

The area under the curve is thus approximated by the area under a trapezium. The repeated trapezoidal rule breaks the interval $[a, b]$ into n subintervals of length $h = (b - a)/n$, based on equally spaced points $a = x_0, x_1, \ldots, x_n = b$ with corres-

trapezoidal rule

ponding ordinates y_0, y_1, ..., y_n, and applies the rule on each subinterval, giving

$$\int_a^b f(x) \, dx \approx \frac{1}{2}h(y_0 + 2y_1 + 2y_2 + \cdots + 2y_{n-1} + y_n)$$

See also Newton's rule; Simpson's rule.

travelling salesman problem Any problem equivalent to that of a travelling salesman who wishes to visit several cities B, C, D, ... and return to his starting point A in such a way that he covers the least possible total distance (*see* diagram). In terms of *weighted graphs, the problem is to find a closed *walk which includes all vertices and has the least total weight. In the diagram, a solution is the walk A to C to B to D to A. *See* Chinese postman problem; NP complete.

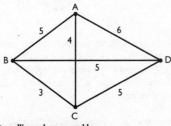

travelling salesman problem

treatment *See* experimental design.

tree A connected *graph with no *circuits. A disconnected graph with no circuits is called a *forest*. A theorem of Cayley (1889) states that the number of distinct *labelled trees* which can be drawn using n labelled points is n^{n-2}. Thus 4 points A, B, C, and D give rise to 16 different labelled trees (*see* diagram (a)).

A *rooted tree* (*see* diagram (b)) has one vertex designated as an origin and called the *root*. A rooted tree in which the root has *degree 2 and every other vertex has degree 1 or 3 is a *binary tree* (*see* diagram (c)).

tree (a) A labelled tree; (b) a rooted tree with root at the vertex labelled R; (c) a binary tree with root at the vertex labelled B.

Some *networks are trees, and the solution of some *network analyses gives rise to a tree. The vertices are often labelled and referred to as *nodes*. Information may be recorded in them (e.g. costs in the nodes of a *decision tree) and a cost, penalty, or probability may be associated with each edge. For example, the problem of joining all nodes in a graph by the minimum length of cable leads to a tree known as a *minimum spanning tree*. Several algorithms are available for determining a minimum spanning tree; two are *Kruskal's algorithm* and *Prim's algorithm* (J.B. Kruskal Jr, 1956; R.C. Prim, 1957).

tree diagram A diagram in the form of a *tree which is often useful for determining probabilities associated with sequences of experiments. If two balls are drawn (without replacement) from a jar containing five green and three red balls, we may want to know the probability that there are then four green balls and two red balls left in the jar. All possible outcomes may be represented on a tree diagram. The root node A in the diagram shown here corresponds to the original jar, and the contents are indicated at that node. The first draw leads to the vertex B or C, and on the edge joining those vertices the colour of the ball drawn and the probability of drawing that ball are given, and the contents of the jar after the first draw is indicated at B and C. The second draw is made from

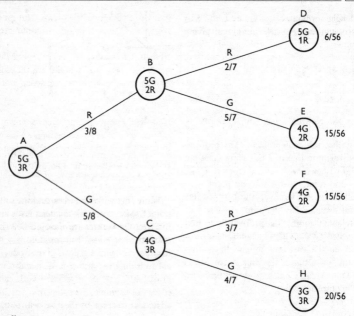

tree diagram

either B or C; the possible outcomes are indicated at nodes D, E, F, and H, and the relevant colour and probability information is shown on each edge originating from B or C. The probability of terminating at each of the nodes D, E, F, and H is the product of the probabilities on the edges leading to that node, and these are marked alongside each node. The condition that there are four green and two red balls is satisfied at nodes E and F, and the total probability associated with these nodes is the sum of the probabilities associated with each, i.e. $15/56 + 15/56 = 15/28$.

trend *See* time series.

trend line (trend curve) A term used, often rather loosely, in statistics for a line or curve, often a *polynomial of low degree or some simple piece-wise curve, to distinguish the general pattern of a relationship between variables that may appear to hold, or may be thought to hold, were it not for random or nuisance variation. Such curves range from curves fitted by eye to curves fitted by using *moving averages, least-squares *regression, or more sophisticated fitting procedures.

trend test 1. A hypothesis test involving k ($\geqslant 3$) samples of the null hypothesis $H_0: \mu_1 = \mu_2 = \cdots = \mu_k$ against the alternative $H_1: \mu_1 < \mu_2 < \cdots < \mu_k$ or $H_1: \mu_1 > \mu_2 > \cdots > \mu_k$, where μ_i is the ith population *mean (or sometimes the ith population *median). *See* Jonckheere–Terpstra test; Page test.
2. Any test for a monotonic (increasing or decreasing) overall trend in a successive sequence of observations. *See* time series.

trial A single performance of an experiment (e.g. tossing a coin) when the outcome is uncertain. Some writers distinguish between a trial and a series of trials by reserving the term *experiment* for the latter. *See also* Bernoulli trial.

triangle A plane closed figure formed by three line segments (the sides) joining three points (the vertices). Triangles are classified according to the relative lengths of their sides:

A *scalene triangle* has all three sides unequal.

An *isosceles triangle* has two sides equal, and unequal to the third.

An *equilateral triangle* has all three sides equal. Equilateral triangles are also equiangular – the angles are all equal to 60°.

Triangles are alternatively classified according to their angles:

An *acute triangle* is one in which all three interior angles are acute angles.

An *obtuse triangle* is one in which one interior angle is an obtuse angle.

A *right-angled triangle* is one in which one interior angle is a right angle.

An *oblique triangle* is one that does not contain a right angle.

Some theorems on triangles are:

(1) The angles of a triangle sum to 180°.

(2) In an isosceles triangle, the angles opposite the equal sides are also equal.

(3) The external angle of a triangle is equal to the sum of the two opposite interior angles.

(4) A line drawn between the mid-points of two sides of a triangle is parallel to the third side and equal to half of it.

See also Pythagoras' theorem; solution of triangles; spherical triangle; trigonometry.

triangle inequality The inequality $a + b > c$, where a, b, and c are the sides of a triangle. *See also* metric; space; norm (of a vector space).

triangle of forces If three forces acting at a point can be represented in magnitude and direction by the sides of a triangle taken in order, then their *resultant is zero. Such a triangle is called a *triangle of forces*. The *converse is also true.

More generally, if $n \geqslant 3$ forces acting at a point can be represented in magnitude and direction by the sides of a closed polygon taken in order, then their resultant is zero. The converse is also true. Such a polygon is called a *polygon of forces*. *See also* force polygon.

triangular distribution 1. The distribution of the sum of two discrete independent random variables each having the same *uniform distribution. The graph of the frequency function is triangular in shape. A simple example is provided by the sum of the scores on two fair dice thrown independently.

2. The distribution of the sum of two independently distributed random variables each having a continuous uniform (or rectangular) distribution over an interval $[a, b]$.

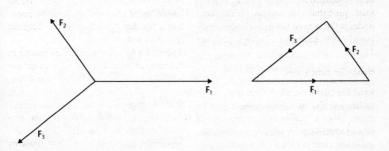

triangle of forces for three forces F_1, F_2, and F_3, whose resultant is zero.

In this case the *frequency function is

$$f(x) = \begin{cases} 0 & \text{for } x \leqslant 2a \\ \dfrac{x-2a}{(b-a)^2} & \text{for } 2a < x \leqslant a + b \\ \dfrac{2b-x}{(b-a)^2} & \text{for } a + b < x \leqslant 2b \\ 0 & \text{for } x > 2b \end{cases}$$

triangular matrix A *matrix in which all elements on one side of the leading *diagonal are zero. The matrix is *upper triangular* if all elements below the leading diagonal are zero, i.e. $a_{ij} = 0$ whenever $i > j$; it is *lower triangular* if all elements above the leading diagonal are zero, i.e. $a_{ij} = 0$ whenever $i < j$. The matrix

$$\begin{pmatrix} a & b & c \\ 0 & e & f \\ 0 & 0 & i \end{pmatrix}$$

is an upper triangular matrix.

triangular number An integer that can be represented by a triangular *array of dots: 1, 3, 6, 10, etc.

triangular prism A *prism that has triangular bases.

triangulation 1. A method of surveying or mapping an area using triangles with known base length and base angles.
2. A decomposition of a *topological space into subsets *homeomorphic to *simplexes of various dimensions which abut each other along their faces. In the case of a triangulation of a *surface, the simplexes are of dimension at most 2, so are triangles, edges, and vertices.

trichotomy law *See* order properties.

trident of Newton A plane curve with the equation

$$xy = ax^3 + bx^2 + cx + d$$

tridiagonal matrix A square *matrix whose elements are zero except on the principal diagonal and the first superdiagonal and first subdiagonal. The matrix

$$\begin{pmatrix} a & b & 0 \\ c & d & e \\ 0 & f & g \end{pmatrix}$$

is tridiagonal. *See also* sparse matrix.

trigonometric functions Functions of angles defined, for an acute angle, as ratios of sides in a right-angled triangle containing the angle. They are sometimes called *trigonometric ratios*. If ABC is a right-angled triangle with C as the right angle, and the sides of lengths a, b, and c are opposite the angles A, B, and C respectively (*see* diagram (a)), then the trigonometric functions (with their abbreviations) are as follows:

trigonometric functions (a)

Tangent

$$\tan A = a/b$$

Sine

$$\sin A = a/c$$

Cosine

$$\cos A = b/c$$

Cotangent

$$\cot A = b/a \quad \text{(also written as } \operatorname{ctn} A\text{)}$$

Cosecant

$$\csc A = c/a \quad \text{(also written as } \operatorname{cosec} A\text{)}$$

Secant

$$\sec A = c/b$$

As defined, three of these functions are reciprocals of the other three:

$$\cot A = \frac{1}{\tan A}$$

$$\csc A = \frac{1}{\sin A}$$

$$\sec A = \frac{1}{\cos A}$$

From these definitions it also follows that

$$\tan A = \frac{\sin A}{\cos A}, \quad \cot A = \frac{\csc A}{\sec A}$$

Other fundamental relationships (the *Pythagorean identities*) are based on Pythagoras' theorem:

$$\sin^2 A + \cos^2 A = 1$$

$$1 + \tan^2 A = \sec^2 A$$

$$1 + \cot^2 A = \csc^2 A$$

Trigonometric functions of certain angles can be obtained from simple right-angled and equilateral triangles (*see* diagram (b)):

trigonometric functions (b)

$$\tan 45^\circ = \cot 45^\circ = 1$$

$$\sin 45^\circ = \cos 45^\circ = 1/\sqrt{2}$$

$$\sin 30^\circ = \cos 60^\circ = 1/2$$

$$\cos 30^\circ = \sin 60^\circ = \sqrt{3}/2$$

$$\tan 30^\circ = \cot 60^\circ = 1/\sqrt{3}$$

Various other relationships between trigonometric functions can be used. *See* addition formulae; double-angle formulae; half-angle formulae; product formulae; reduction formulae.

By using rectangular coordinates the definitions of trigonometric functions can be extended to angles of any size in the following way (*see* diagram (c)). A point P is taken with coordinates (x, y). The radius vector OP has length r and the angle θ is taken as the directed angle measured anticlockwise from the x-axis. The three main trigonometric functions are then defined in terms of r and the coordinates x and y:

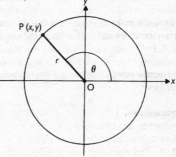

trigonometric functions (c)

$$\tan \theta = y/x$$

$$\sin \theta = y/r$$

$$\cos \theta = x/r$$

(The other functions are reciprocals of these.)

This can give negative values of the trigonometric functions. For example, an obtuse angle (between 90° and 180°) has a positive value of y and a negative value of x. Consequently, the sine of an obtuse angle is positive and the cosine and tangent are negative. The general definition also allows meaning to be given to trigonometric functions of negative angles by taking a negative angle as one measured clockwise from the x-axis, giving

$$\tan(-\theta) = -\tan\theta$$

$$\sin(-\theta) = -\sin\theta$$

$$\cos(-\theta) = \cos\theta$$

In addition, meaning can be given to trigonometric functions of angles that are multiples of right angles. Thus, $\sin 90° = 1$, $\cos 90° = 0$, $\tan 180° = 0$, etc. In defining trigonometric functions in this way the point P is taken to move around a circle, so the functions are known as *circular functions*.

The trigonometric functions are defined above for angles, but are extensively used for numbers. In this case $\sin x$, where x is a number, is defined as the sine of the angle equal to x radians; $\cos x$ and the other functions are defined similarly. These functions can also be expressed as infinite series:

Sine series

$$\sin x = x - \frac{x^3}{3!} + \frac{x^5}{5!} - \frac{x^7}{7!} + \cdots$$

Cosine series

$$\cos x = 1 - \frac{x^2}{2!} + \frac{x^4}{4!} - \frac{x^6}{6!} + \cdots$$

In addition, the series can be used to define the trigonometric functions of a complex number z.

See also cofunctions; Euler's identities; inverse trigonometric functions; hyperbolic functions.

trigonometric ratios A name sometimes used for trigonometric functions.

trigonometry The branch of mathematics concerned with solving triangles by using *trigonometric functions. It is of immense practical value in such fields as engineering, architecture, surveying, navigation, and astronomy. The subject is divided into *plane trigonometry* (concerned with plane triangles) and *spherical trigonometry* (concerned with *spherical triangles). Trigonometric functions also play an important role in analysis and are used to represent waves and other periodic phenomena.

The earliest rudiments of trigonometry are found in records from Egypt and Mesopotamia. There is a Babylonian stone tablet (*c.*1900–1600 BC) on which are listed ratios equivalent to the modern $\sec^2$. The Egyptian *Rhind papyrus (*c.*1650 BC) contains problems in which the ratios of the sides of a triangle are applied to pyramids. Neither the Egyptians nor the Babylonians had our present concept of angular measure, and ratios of the type described above were regarded as properties of triangles rather than of angles.

The important advances were made by the Greeks from the time of Hippocrates of Chios (*Elements*, *c.*430 BC), who studied the relationships between the arc of a circle (a measure of the central angle) and the chord of the arc. In 140 BC Hipparchus produced a table of chords (the first forerunner of our modern tables of sines). Menelaus of Alexandria (*Spherics*, *c.* AD 100) first used spherical triangles and introduced spherical trigonometry. Ptolemy (*Almagest*, *c.* AD 140) tabulated chords of angles between $\frac{1}{2}°$ and 180° at $\frac{1}{2}°$ intervals. He also investigated trigonometric identities.

Greek trigonometry was further developed by Hindu mathematicians who made the advance of replacing the chords used by the Greeks by half-chords of circles with given radii – i.e. the equivalent of our sine functions. The earliest such tables are in the *Siddhantas* (Systems of Astronomy) of the 4th and 5th centuries AD. Like numbers, modern trigonometry came to us from Hindu mathematicians via Arab mathematicians. Translations from Arabic into Latin in the 12th century introduced trigonometry into Europe.

The person responsible for 'modern' trigonometry was the Renaissance mathematician Regiomontanus. From the time of Hipparchus, trigonometry had been regarded simply as a tool for astronomical calculation. Regiomontanus (*De triangulis omni modis*, 1464, published 1533) was the first to treat trigonometry as a subject in its own right. Further advances were made by Nicolaus Copernicus in *De revolutionibus orbium coelestium* (1543) and by his student Rheticus. In *Opus palatinum de triangulis* (completed by his student in 1596), Rheticus established the use of the six main trigonometric functions, tabulated values for them, and concentrated on the idea that the functions represented ratios in a right-angled triangle (rather than the traditional half chords of circles).

Modern analytical geometry dates from the time of François Viète, who prepared tables of the six functions to the nearest minute (1579). Viète also derived the product formulae, tangent formulae, and multiple-angle formulae. It was towards the end of the 15th century that the name 'trigonometry' first came into use.

See also solution of triangles.

trihedral angle A *polyhedral angle with three faces.

trillion One thousand thousand million (10^{12}). The term has long been established in this sense in the USA. In the UK the term originally meant one million million million (10^{18}), being a contraction formed from *tri* = three and *million*, but since the 1970s it has commonly been used to mean 10^{12}.

trimmed mean An arithmetic *mean formed by discarding a proportion of the most extreme observations in a sample. The object is to reduce the influence of extreme observations, or *outliers, on the value of the mean. With severe trimming, the trimmed mean approaches the median.

trinomial A *polynomial that has three terms; for example,

$$ax^2 + bx + c$$

triple *See* ordered pair.

triple integral A *multiple integral involving three successive integrations. *See* volume.

triple product A product of three *vectors **A**, **B**, and **C**.

The *triple vector product* (or *vector triple product*) is the product

$$\mathbf{A} \times (\mathbf{B} \times \mathbf{C})$$

which is a vector. It is equal to

$$(\mathbf{A} . \mathbf{C})\mathbf{B} - (\mathbf{A} . \mathbf{B})\ \mathbf{C}$$

The *triple scalar product* (or *scalar triple product*) is defined as

$$\mathbf{A} . (\mathbf{B} \times \mathbf{C})$$

which is a scalar equal to

$$|\mathbf{A}|\ |\mathbf{B}|\ |\mathbf{C}|\ \sin \theta \cos \alpha$$

triple product The volume of the parallelepiped is given by the triple scalar product $\mathbf{A} . (\mathbf{B} \times \mathbf{C})$.

where θ is the angle between **B** and **C**, and α is the angle between $\mathbf{B} \times \mathbf{C}$ and **A**. Geometrically, a triple scalar product gives the volume of a parallelepiped of which **A**, **B**, and **C** are *coterminal edges (see diagram).

trirectangular Having three *right angles. *See* spherical triangle.

trisection The process of dividing anything into three equal parts. The points, lines, planes, etc. that trisect something are its *trisectors*. The problem of trisecting an angle using only unmarked straightedge and compasses is one of the three classical problems of Greek geometry (along with *squaring the circle and *duplication of the cube). It is now known that the construction is impossible. *See also* trisectrix; quadratrix.

trisector *See* trisection.

trisectrix Any of various curves that can be used to trisect an angle. The *trisectrix of Maclaurin* has the equation

$$x^3 + xy^2 + ay^2 - 3ax^2 = 0$$

It is symmetrical about the x-axis, with a loop and an asymptote $x = -a$. If a line is drawn from a point P with coordinates $(2a, 0)$ to cut the curve at P$'$, the angle that PP$'$ makes with the x-axis is three times the angle that OP$'$ makes with this axis.

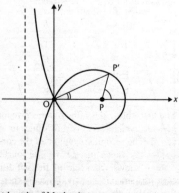

trisectrix of Maclaurin.

trivial group A *group consisting of one element.

trivial solution A solution of an equation or set of equations in which the values of all the variables are zero.

trochoid A plane *curve that is a generalization of a *cycloid in that the generating point lies anywhere on the radius (or radius produced) of the generating circle (see generator). If r is the radius of this circle and a the distance of the point from the centre, then the curve has parametric equations

$$x = r\theta - a\sin\theta, \quad y = r - a\cos\theta$$

troy system A British system of units of mass used for precious metals and gemstones. It is named after the city of Troyes in France, where it was first used, and is based on the grain (originally the mass of a grain of wheat). The grain in this system has the same mass as the grain in the *avoirdupois and *apothecaries' systems. The other troy units are:

4 grains = 1 carat
6 carats = 1 pennyweight
20 pennyweights = 1 troy ounce
            (of 480 grains)
12 troy ounces = 1 troy pound
            (of 5760 grains)
25 troy pounds = 1 troy quarter
4 troy quarters = 1 troy hundredweight
            (of 100 pounds)
20 troy hundredweights = 1 troy ton (of 2000 troy pounds)

truncated Describing the part of a solid figure cut off by one or more planes that do not intersect within the figure. *See also* frustum; polyhedron.

truncated distribution A distribution formed from another specified distribution by ignoring the part lying to the left or to the right of a fixed value of the random variable. For example, the length X of items produced by a manufacturing process may be normally distributed, but items below some fixed length x_0 may be scrapped. If the remainder are sold, the distribution of lengths of those sold will follow a truncated normal distribution. The mean and variance of the truncated distribution are related to the mean and variance of the original normal distribution in a way that depends on the chosen truncation value x_0.

truncation The process of dropping trailing digits from a number. For example, 1.576 is truncated to 1.57 to three digits. The act of truncation causes a *truncation error* (*see* error). *Compare* rounding.

truth *See* interpretation.

truth function In *logic, a *function whose arguments and values are *truth values. A compound sentence is said to be *truth-functional* if its truth value is wholly determined by the truth values of its parts. All the compound sentences of the propositional calculus are truth-functional. A *truth-functional connective* is a connective that stands for a truth function (*compare* implication (strict)). A set of truth functions is said to be (functionally) *complete* when every truth function of any number of arguments can be expressed by use of the members of the set. The set is *independent* if the truth functions it can express cannot be expressed by one of its proper subsets. A complete independent set of connectives is used when setting up the most economical versions of the propositional calculus. An example (there are many) of a complete independent set of connectives is that containing & (*see* and) and ~ (*see* not); the set containing & and ∨ (*see* or) is one which is not functionally complete. *See also* truth table; logic.

truth table A table for evaluating the truth value of a truth-functional (*see* truth function) *compound sentence on the basis of the *truth values of its parts. Truth tables are used both to define the truth-functional connectives and to test for validity. Each row of a truth table indicates the truth

value of a compound sentence, given a particular assignment of truth values to its components, and if there are sufficiently many rows then the truth value of a compound sentence under any assignment of truth values to its parts will be apparent.

For example, to define the connective ~ (not), we construct the following truth table in which the truth values true and false are represented by T and F, respectively:

A	$\sim A$
T	F
F	T

Alternatively, we could represent the truth values true and false by 1 and 0, and construct an equivalent truth table for the connective ~ :

A	$\sim A$
1	0
0	1

A valid *wff is true under all interpretations. In the *propositional calculus this amounts to a wff being true under all assignments of truth values to its atomic wffs; that is, when a wff is evaluated by a truth table it will take the value T (or 1) for every assignment of truth values to its atomic components.

For example, the following truth table shows that $(A \ \& \ B) \supset B$ is valid:

$(A \ \& \ B)$	$\supset$	B
T T T	T	T
T F F	T	F
F F T	T	T
F F F	T	F

The first, third, and fifth columns give all possible assignments of truth values to A and B. The second column gives the corresponding values of $A \ \& \ B$. The fourth column shows the value of the wff for each assignment; as only T occurs there the wff is valid. Truth tables provide an effective means for determining whether or not a wff is valid (*see* decidable).

truth value An object assigned as the semantic value (denotation) of a sentence when interpreting a *formal language. Usually, two truth values are used, represented variously by T or 1 for true and by F or 0 for false. *See* semantics; truth table.

Tschirnhaus, Ehrenfried Walther von (1651–1708) German mathematician who in 1682 began the study of caustic curves. He also worked on problems of maxima and minima and on the theory of equations, and became a minor participant in the priority dispute between Leibniz and Newton over the discovery of the calculus.

***T*-score** A *statistic occasionally used in preference to a *standardized random variable. If X is a random variable with mean μ and standard deviation σ, the T-score is given by the transformation

$$T = 50 + 10(X - \mu)/\sigma$$

If Z is a standardized variable, $T = 50 + 10Z$. T is mainly used to standardize educational data about a mean score of 50.

Tsu Chung Chi *See* Zu Chongzhi.

***t*-test** (W.S. Gosset, 1908) Also known as *Student's t-test*, 'Student' being Gosset's pseudonym. A test of whether a sample of n observations with mean $\bar{x}$ comes from a *normal distribution with mean μ_0. Under the null hypothesis, the statistic

$$t = \frac{\bar{x} - \mu_0}{s/\sqrt{n}}$$

where

$$s^2 = \frac{1}{n-1} \sum_i (x_i - \bar{x})^2$$

has the *t-distribution with $n - 1$ degrees of freedom. The test extends to hypotheses about differences between means for matched pairs, and for two independent samples from normal distributions with equal but unknown variances. Tables giving selected *quantiles of the distribution for different degrees of freedom are useful for hypothesis testing and to form *confidence intervals. Most standard statistical software packages include the *t*-test, and give *p-values for the observed value of t, and will also provide confidence intervals at any pre-specified level, making tables more or less obsolete.

Tukey, John Wilder (1915–2000) Often regarded as the most innovative and versatile American statistician of the 20th century, he is perhaps most widely known for developing the approach known as *exploratory data analysis, featuring especially *box-and-whisker diagrams and *stem-and-leaf displays. He also made major contributions in nonparametric methods, robustness studies, and the interpretation of sophisticated computer graphics.

Turing, Alan Mathison (1912–54) English mathematician and logician who in 1936 introduced the important idea of the *Turing machine to make precise the notion of computability. He also wrote a pioneering paper on mathematical biology.

Turing machine An abstract computer described by Alan Turing in 1936. It consists of an infinite tape divided into cells, each of which can contain the number 1 or 0 or be blank. The machine can read the tape one cell at a time and, depending on its internal state and the content of the cell, it can print or erase the numbers 1 or 0, move to the right or left, or halt.

The machine can be in any one of a finite number of active states 1, 2, 3, 4, ..., or in an inactive halt state H in which operations stop. While operating, the machine is controlled by the input from the tape according to a finite *instruction set*: a typical instruction being $\{5, b, 1, R, 3\}$ which stands for 'if in state 5, reading a blank cell, print 1, move to the cell one place to the right, and enter state 3'.

As an example, we can construct a Turing machine to add 2 and 3. For ease, numbers are written in unary, in which 2 is represented by 11, 3 by 111, and in general a number n by n 1's. We begin with the following tape:

A   0 0 0 1 1 0 1 1 1 0 0

and the instruction set presented in tabular form as

{1, 0, 0, R, 1}

{1, 1, 0, R, 2}

{2, 0, 1, R, H}

{2, 1, 1, R, 2}

If the machine starts in state 1, reading the first symbol on the left, and the instruction set above is applied to A, it will produce the following sequences on the tape:

B   0 0 0 0 1 0 1 1 1 0 0

C   0 0 0 0 1 1 1 1 1 0 0

The Turing machine will thus have succeeded in adding 2 and 3. Also, because the machine is now in state H, it will halt.

Turing demonstrated that any serial calculation done on any computer, however complex, could also be carried out on a Turing machine. He went on to establish the important result that not all functions are *computable*, a result clearly related to *Gödel's theorem and *Church's thesis. *See also* halting problem.

turning point A maximum or minimum point on a smooth curve; i.e. a point where the y-coordinate changes from increasing to decreasing, or vice versa, and the tangent is horizontal. A change from increasing to decreasing is a *maximum point*. If the value at this point is the largest value of the function, the point is an *absolute maximum*; otherwise it is a *relative maximum* (i.e. the maximum relative to other points

in the neighbourhood). *Minimum points* are similarly defined.

The positions of maxima and minima are usually found by taking the first derivative of the function and equating it to zero. This gives stationary points at which the tangent to the curve is horizontal, i.e. maxima, minima, and horizontal points of *inflection. To distinguish between the three, the second derivative of the function is evaluated at the point. If the second derivative is negative at the point, then the point is a maximum (the *slope* of the tangent changes from positive to negative). Conversely, if the second derivative is positive, the point is a minimum. If the second derivative is zero the position is more complicated: the point may be a maximum, minimum, or horizontal point of inflection. To distinguish between these, it is necessary to find the signs of the derivatives at two points, one each side of the point in question. At a maximum the first derivative changes sign from positive to negative; at a minimum it changes sign from negative to positive (for increasing values of the variable x). Alternatively, it may be simpler to find the actual values of the function itself on each side of the point and compare these with the value at the point. At a point of inflection, the *second* derivative changes sign at the point (detected by taking the second derivative at points on each side of the given point).

twin primes (prime pair) A pair of *prime numbers that differ by 2. Examples are 3 and 5, 5 and 7, 11 and 13, and 17 and 19. The problem of whether there are an infinite number of such pairs is still unsolved.

twisted curve *See* curve.

twisted product *See* bundle.

two-person game *See* game theory.

two-point form *See* line.

two-tail test *See* hypothesis testing.

two-way classification Classification of a set of observations in rows and columns, each representing one of two criteria. *See* contingency table; randomized blocks.

Tychonoff's theorem *See* Tikhonov's theorem.

Type I *or* **II error** *See* hypothesis testing.

U

unary operation An operation applying to one element of a *set. For example, taking the positive square root of a number is a unary operation. If the set is S, a unary operation u on S can be regarded as a *function whose *domain is S and whose *codomain is also S, and we can write u: $S \rightarrow S$. *Compare* binary operation.

unbiased estimator An *estimator T is said to be an unbiased estimator of a parameter θ if $E(T) = \theta$. If

$$E(T) - \theta = b \neq 0$$

then b is called the *bias* in T.

unbiased hypothesis test A test for which the *probability of observing a value of the *statistic in the critical region of size α is greater than α whenever the alternative hypothesis is true. Broadly, this implies that a result in the critical region (causing H_0 to be rejected) is more likely when H_1 is true than when H_0 is true. *See also* hypothesis testing.

unbounded function A *function that does not have both a lower and an upper *bound. A function f is unbounded if for any positive real number M there is a value of x, x_M, that depends on M such that $|f(x_M)| > M$. For example, the function $f(x) = 1/x$ defined on domain $0 < x < \infty$ is unbounded because by choosing x sufficiently small $1/x$ can be made as large as required. This function is bounded below but unbounded above. $f(x) = x \sin x$ defined on domain $0 < x < \infty$ takes positive and negative values and is unbounded below and above, because by choosing sufficiently large values of x, f(x) can be made sufficiently large and positive or large and negative. *Compare* bound.

unbounded set *See* bounded set.

undecagon A *polygon with eleven interior angles (and eleven sides).

undetermined multipliers *See* Lagrange multipliers.

uniform convergence A possible property of a *series whose terms are *continuous functions of a variable x in an interval. The sum of the series is a continuous function of the variable in the given interval. The series

$$u_0(x) + u_1(x) + u_2(x) + \cdots u_n(x) + \cdots$$

is said to be uniformly convergent in the interval (a, b) if it converges for every value of x between a and b, and if a positive integer N (independent of x) can be found such that the absolute value of the *remainder R_n of the given series, where

$$R_n = u_{n+1}(x) + u_{n+2}(x) + \cdots$$

is less than some arbitrary positive number ε (on which N is dependent) for every value of $n \geqslant N$ and for every value of x lying in the interval (a, b). There are several tests to determine whether a series is uniformly convergent in a given interval.

uniform distribution 1. A discrete distribution over a range $[0, n]$ having *frequency function

$$p(r) = \Pr(X = r) = \frac{1}{n + 1}$$

for all integral values r between 0 and n inclusive. Random digits have a uniform distribution over the interval $[0, 9]$

2. (rectangular distribution) A continuous distribution over the interval $[a, b]$ with *frequency function

$$f(x) = \frac{1}{b - a}$$

The graph of $f(x)$ has the shape of a rectangle of height $1/(b - a)$.

uniform gravitational field A gravitational field (*see* gravitation) in which identical particles experience identical forces independent of their position. The earth's gravitational field can be taken to be uniform for small bodies on or near its surface.

uniformly continuous function A *function with *domain X and *codomain Y that are both *metric spaces, for which, given any $\varepsilon > 0$, there exists a $\delta > 0$ depending only on ε, such that if the distance between any two points x_1 and x_2 in X is less than δ, the distance between $f(x_1)$ and $f(x_2)$ in Y is less than ε. If f is continuous on X and X is *compact, then f is uniformly continuous on X.

In particular, if X is the real interval (a, b) and Y a set of real numbers, then f is uniformly continuous on (a, b) if whenever

$$|x_1 - x_2| < \delta$$

then

$$|f(x_1) - f(x_2)| < \varepsilon$$

for any x_1 and x_2 in (a, b). For example, if X is $(0, 1)$ and $f(x) = x^2$, and given $\varepsilon > 0$, the value of δ is taken to be $\frac{1}{2}\varepsilon$, then when-

ever $|x_1 - x_2| < \frac{1}{2}\varepsilon$,

$$\begin{aligned}
|f(x_1) - f(x_2)| &= |x_1^2 - x_2^2| \\
&= |(x_1 - x_2)(x_1 + x_2)| \\
&< |\tfrac{1}{2}\varepsilon \times 2| = \varepsilon
\end{aligned}$$

Therefore f is uniformly continuous.

If a function is continuous on a closed interval then it is uniformly continuous on that interval. *See also* continuous function.

uniform motion Motion with constant velocity, speed, or acceleration.

uniform polyhedron A *polyhedron that has *regular polygons for all its faces and identical vertices.

unilateral surface A surface that has only one side, as in a *Möbius strip or *Klein bottle.

unimodal distribution A *distribution having only one *mode. For discrete distributions, if two adjacent values of X both have the same probability or frequency, and this is the modal value, it is usual to regard this also as a unimodal distribution. For example, the binomial distribution with $n = 3$ and $p = \frac{1}{2}$ gives

$$\Pr(X = 0) = \Pr(X = 3) = \frac{1}{8}$$

$$\Pr(X = 1) = \Pr(X = 2) = \frac{3}{8}$$

The modal values $X = 1$ and $X = 2$ are taken to constitute one mode. *Compare* bimodal distribution.

unimodular matrix A square *matrix that has a *determinant equal to unity.

union (join, sum) The union of two *sets A and B, denoted by $A \cup B$, consists of those elements that belong either to A or to B:

$$A \cup B = \{x \colon (x \in A) \vee (x \in B)\}$$

For example, if A is $\{1, 2, 3, 4\}$ and B is $\{1, 4, 5, 6\}$ then $A \cup B$ is $\{1, 2, 3, 4, 5, 6\}$. *Compare* intersection.

unique factorization theorem *See* fundamental theorem of arithmetic.

uniqueness theorem A theorem asserting that only one particular type of entity can exist. An example is the theorem that, given a plane and a point P outside the plane, only one plane can pass through P parallel to the given plane.

unit 1. A standard used in the measurement of a *physical quantity. *See* SI units; apothecaries' system; avoirdupois; British units of length; c.g.s. units; coherent units; derived units; f.p.s. units; imperial units; metric system; m.k.s. units; troy system; US customary system.
2. The number 1.
3. An *invertible element in a *ring with identity.
4. *See* experimental design.

unitary group The group $U(n)$ of all unitary $n \times n$ complex matrices.

unitary matrix A *matrix whose inverse is its *Hermitian conjugate. For example, the matrix

$$\begin{pmatrix} 1 & i \\ 0 & 1 \end{pmatrix}$$

is unitary.

unitary method A procedure for solving problems in *variation in which one of the variables is reduced to unity. For example, suppose that y is proportional to x, and it is required to find the value of y when $x = 7$, given that $y = 9$ when $x = 5$. The solution proceeds as follows:

when $x = 5$, $y = 9$

when $x = 1$, $y = \dfrac{9}{5}$

when $x = 7$, $y = \dfrac{9}{5} \times 7 = 12\dfrac{3}{5}$

unitary ratio *See* ratio.

unitary transformation *See* matrix.

unit circle *or* **sphere** A circle (or sphere) that has a radius 1 unit in length.

United States customary system The nonmetric system of weights and measures used in the USA. It is based on various units of measure in use in Britain in the 1700s, and differs from *imperial units principally in its units of volume (*see* gallon).

unit matrix *See* identity matrix.

unit set *See* singleton.

unit square *or* **cube** A square (or cube) with a side 1 unit in length.

unit vector A *vector of unit magnitude.

unity The number 1.

universal instantiation *See* logic.

universal quantifier *See* quantifier.

universal set Relative to a particular *domain, the universal set, denoted by $\mathscr{E}$ or I, is the *set of all objects of that domain:

$$\mathscr{E} = \{x : x = x\}$$

Compare null set.

universe of discourse *See* domain.

unknown A value or function that is to be found: a member of the *solution set of a given problem.

upper bound *See* bound.

upper limit (of integration) *See* integration.

upper triangular matrix *See* triangular matrix.

U-shaped distribution A *distribution over a finite *range for which the *frequency function has approximately equal maxima at or near the ends of the range. In many parts of the world the daily distribution of the proportion of the sky covered by cloud has a U-shaped distribution, completely clear or completely cloudy days being more common than partly cloudy days.

utility theory A form of *decision theory in which decisions are based on the concept of utility or benefit. The concept of utility as a benefit or degree of happiness, particularly in economic studies, predates an axiomatic theory of utility formulated by J. von Neumann and O. Morgenstern (1944). Interest often focuses on determining policies that maximize expected utility. A subjective element often enters into assessment of utilities.

A simple example illustrating the calculation of an expected utility is where organizers of a function must decide whether to hold it outdoors or indoors. On past experience they may be able to estimate likely attendances. Numbers attending would be a relevant utility measure as an income generator if each person paid the same admission charge.

Suppose the probability of rain is 0.1 and the probability of no rain 0.9, and the likely attendances are those in the following *contingency table:

	Rain $(p = 0.1)$	No rain $(p = 0.9)$
Hold outdoors	100	650
Hold indoors	250	400

If held outdoors the expected utility is $E(U) = 100 \times 0.1 + 650 \times 0.9 = 595$, while if held indoors the expected utility is $E(U) = 250 \times 0.1 + 400 \times 0.9 = 385$. To maximize expected utility the function should be held outdoors.

In practice, utilities may reflect factors such as attitudes to risk. This is often the case when associated with undesirable events of small probability. Most people insure their homes against fire damage, attaching a higher utility to protection against the small risk of their home being destroyed by fire than they do to the risk of forfeiting a relatively modest sum each year paid as a premium, although they may find it hard to quantify these utilities.

V

valid Describing a logical *argument in which if the premises are true then the conclusion must also be true. Otherwise, an argument is said to be *invalid*. More precisely, an argument is valid if and only if, in all *interpretations where the premises are true, the conclusion is also true. A *wff A is said to be valid (symbolically: $\models A$) if and only if it is true under all interpretations. *See also* consequence; logic.

Vallée-Poussin, Charles-Jean de la (1866–1962) Belgian mathematician who, in 1896, and independently of Hadamard, proved the *prime number theorem.

value 1. *See* absolute value.
2. *See* game theory.

Vandermonde, Alexandre-Théophile (1735–96) French musician and chemist regarded as the founder in 1772 of a notation and calculus of determinants.

Vandermonde determinant *See* Vandermonde matrix.

Vandermonde matrix An $n \times n$ *matrix defined in terms of given numbers x_1, x_2, ..., x_n, of the form illustrated for $n = 4$ by

$$\begin{pmatrix} 1 & 1 & 1 & 1 \\ x_1 & x_2 & x_3 & x_4 \\ x_1^2 & x_2^2 & x_3^2 & x_4^2 \\ x_1^3 & x_2^3 & x_3^3 & x_4^3 \end{pmatrix}$$

The determinant of the Vandermonde matrix, the *Vandermonde determinant*, is equal to

$$\prod_{1 \leqslant j < i \leqslant n} (x_i - x_j)$$

Thus for $n = 4$ the determinant is

$$(x_2 - x_1)(x_3 - x_1)(x_3 - x_2)(x_4 - x_1)$$
$$(x_4 - x_2)(x_4 - x_3)$$

The Vandermonde matrix is a *nonsingular matrix if and only if the x_i are distinct.

Vandermonde's theorem The theorem that if n is a positive integer and x and y have any values whatever, then

$$\binom{x+y}{n} = \binom{x}{n} + \binom{x}{n-1}\binom{y}{1}$$

$$+ \binom{x}{n-2}\binom{y}{2} + \cdots + \binom{y}{n}$$

where $\binom{x}{r}$ is a *binomial coefficient.

van der Pol's equation (B. van der Pol, 1927) A *differential equation first used to describe the behaviour of electronic circuits in early radios, but now used to model a number of other phenomena in physics and biology. The equation is a wave equation with a damping factor μ:

$$\frac{d^2 x}{dt^2} - \mu(1 - x^2)\frac{dx}{dt} + x = 0$$

vanish To become zero.

Var *See* variance.

variable 1. A mathematical entity that can stand for any of the members of a given *set. The members of the set constitute *val-

ues of the variable, and the set itself defines the variable's *range* (i.e. the possible values that it may take). In considering a function $f(x)$ of a variable x the function's value itself is also a variable. It is common to refer to the value of the function as the *dependent variable* and to x as the *independent variable*. Thus, in $y = 3x + 5$, y is regarded as the dependent variable and x as the independent variable. *See also* function; random variable.

2. An expression in *logic that can stand for any element of a set (called the *domain) over which it is said to *range*. Logical variables are in contrast to *constants, which can stand only for single fixed elements. A variable is said to be *free* in a *wff A if it is not preceded in A by a *quantifier. Wffs with free variables are called *open sentences*, and are neither true nor false. Variables that are not free are called *bound*, and if all the variables in a wff are bound, then the wff is said to be *closed*, and is either true or false (*see* interpretation). For example, as the variable y in

$$(\exists x)(x \text{ is the son of } y)$$

is free, the wff is neither true nor false; but as the variables x and y are bound in

$$(\exists x)(\exists y)(x \text{ is the son of } y)$$

then the wff is either true or false.

variables separable Describing a type of ordinary *differential equation in which the terms in y can be separated from the terms in x. The equation can then be solved by integration.

variance For a *random variable X the variance is the second *moment about the mean, denoted by $E(X - \mu)^2$ or Var (X). This is equivalent to $E(X^2) - (E(X))^2$, i.e. the second moment about the origin minus the square of the mean. For a sample, the variance is the second sample moment about the sample mean, i.e.

$$s_x^2 = \frac{1}{n} \sum_i (x_i - \bar{x})^2$$

or equivalently

$$s_x^2 = \frac{1}{n} \sum_i (x_i^2) - \left(\frac{1}{n} \sum_i x_i\right)^2$$

The unbiased sample estimator of a population variance is $s^2 = n s_x^2 / (n - 1)$. The positive square root of the variance is the *standard deviation.

variance, analysis of *See* analysis of variance.

variance ratio The ratio of two estimates of *variance with *degrees of freedom f_1 and f_2. If they both estimate the same variance the ratio will have an *F-distribution with f_1 and f_2 degrees of freedom. *See* analysis of variance.

variate *See* random variable.

variation **1. (mutual variation)** If two variables x and y are such that their ratio is always constant, then y is said to *vary directly* as x, or to be *directly proportional* to x. This is written as

$$y \propto x \quad \text{or} \quad y = kx$$

where k is the *constant of proportionality*. The shorter forms 'y *varies as* x' and 'y is *proportional to* x' are also used.

If y is proportional to the reciprocal of x, then y is said to *vary inversely* as x, or to be *inversely proportional* to x. This is written as

$$y \propto 1/x \quad \text{or} \quad y = k/x$$

where k is a constant.

If y varies as the product of two variables x and z, then y is said to *vary* or to *vary jointly as x and z*. This is written as

$$y \propto xz \quad \text{or} \quad y = kxz$$

where k is a constant. For example, the volume of a right circular cylinder varies jointly as the square of the radius and the vertical height. *See* unitary method.

2. (of a function) The least upper *bound of

$$\sum_{i=1}^{n} |f(x_i) - f(x_{i-1})|$$

where f is a real-valued *function with a *domain that is a real interval [a, b] and the bound is taken over all possible *partitions $a = x_0 < x_1 < \cdots < x_n = b$ of the interval. If the bound is finite, f is said to have *bounded* or *finite variation*.

3. *See* calculus of variations.

variation, coefficient of A relative measure of *dispersion for sets of data defined as

$$\frac{100 \times \text{standard deviation}}{\text{mean}}$$

It was proposed by Pearson as a means of comparing variability in different distributions, but it is sensitive to errors in the mean. It is usually looked upon as a somewhat crude but nevertheless useful yardstick.

variations in sign (of a polynomial) *See* Descartes's rule of signs.

variety *See* algebraic variety.

Vassiliev invariant *See* knot polynomial.

Veblen, Oswald (1880–1960) American mathematician who worked on differential geometry and mathematical physics. He was also influential in the early development of topology through his book *Analysis situs* (1922).

vector An entity in Euclidean space that has both magnitude and direction. A vector can be represented geometrically by a directed segment of a line. A *located vector* is one that can be described by an ordered pair of points in space (**AB** or $\overrightarrow{AB}$), interpreted as a line segment from point A to point B (*see* diagram (a)). Two vectors are equivalent if they have the same magnitude and the same direction, so any located vec-

vector (a) $\overrightarrow{AB}$ is the vector located at A, and $\overrightarrow{OP}$ is the position vector of P. The absolute value of $\overrightarrow{OP}$ is $\sqrt{(x^2 + y^2)}$.

tor is equivalent to a vector from some standard point O (the origin) to a point P, where AB is parallel to OP and the lengths AB and OP are equal. In two dimensions, a vector located at the origin is specified by two numbers (x, y) giving the coordinates of the end point. Such a vector is called the *position vector* of the point (x, y).

The length of a vector, without regard to direction, is called its *absolute value* (or *numerical value*). For the position vector of the point (x, y), the absolute value is $\sqrt{(x^2 + y^2)}$. A *unit vector* is a vector that has an absolute value of unity.

Two or more vectors can be added by placing the line segments end to end. The sum of the vectors (called the *resultant*) is the line segment from the initial point of the first vector to the final point of the last. In the case of two vectors, this is equivalent to the *parallelogram law for adding vector quantities. For a given vector **v**, the negative vector −**v** is one having the same absolute value as **v** and parallel to it, but having the opposite direction, so subtraction of vectors can be defined in terms of addition: $\mathbf{u} - \mathbf{v} = \mathbf{u} + (-\mathbf{v})$ (*see* diagram (b)). Vector addition is both commutative and associative. A vector **u** can also be multiplied by a scalar (i.e. by a number) *n*. If *n* is positive, the product

vector (b) Vector addition and subtraction.

$n\mathbf{u}$ is a vector with the same direction as $\mathbf{u}$ and with n times the absolute value.

Any two or more vectors that have a given vector as their resultant are *components* of the given vector. The component of a vector in a given direction is the projection of the vector along that direction. In particular it is often convenient to represent a vector as a sum of components that are multiples of unit vectors. For instance, in three dimensions the vector $\mathbf{u}$ from the origin to the point (x, y, z) can be written as $x\mathbf{i} + y\mathbf{j} + z\mathbf{k}$, where $\mathbf{i}$, $\mathbf{j}$, and $\mathbf{k}$ are unit vectors along the x-, y-, and z-axes, respectively (*see* diagram (c)). In multiplying a vector by a scalar, the individual components are multiplied: for example,

$$n\mathbf{u} = n(x\mathbf{i} + y\mathbf{j} + z\mathbf{k})$$

$$= nx\mathbf{i} + ny\mathbf{j} + nz\mathbf{k}$$

vector (c) The position vector $\mathbf{u}$ expressed as the sum of its components: $\mathbf{u} = x\mathbf{i} + y\mathbf{j} + z\mathbf{k}$.

In adding vectors, the corresponding components are added: for example, if

$$\mathbf{u} = a\mathbf{i} + b\mathbf{j} + c\mathbf{k} \quad \text{and} \quad \mathbf{v} = d\mathbf{i} + e\mathbf{j} + f\mathbf{k}$$

then $\mathbf{u} + \mathbf{v}$ is given by

$$(a + d)\mathbf{i} + (b + e)\mathbf{j} + (c + f)\mathbf{k}$$

It is also possible to define multiplication of two vectors (*see* scalar product; vector product) and three vectors (*see* triple product), as well as derivatives of vector functions (*see* curl; divergence; gradient). The idea of vectors in three-dimensional space can be extended to higher dimensions. In this case, a vector can be represented by an n-tuple $(x_1, x_2, \ldots, x_n)$. More generally, vectors can be regarded as mathematical objects that can be added and can be multiplied by numbers (say), but cannot necessarily be multiplied together to give other vectors. In this sense, a vector is an element of a *vector space.

See division in a given ratio.

vector bundle A collection of vector spaces V_x parametrized by the points x of a topological space X. Examples are the *tangent spaces $T_x(M)$ of a manifold M, in which case $X = M$, and, more generally, the spaces of various *tensors.

vector field *See* field.

vectorial angle *See* polar coordinate system.

vector product (cross product) A product of two *vectors to give a third vector

$$\mathbf{C} = \mathbf{A} \times \mathbf{B}$$

The notation $\mathbf{A} \wedge \mathbf{B}$ is also commonly used. The length of $\mathbf{C}$ is the product of the lengths of $\mathbf{A}$ and $\mathbf{B}$ multiplied by the sine of the angle between them:

$$|\mathbf{C}| = |\mathbf{A}|\ |\mathbf{B}| \sin \theta$$

The direction of C is perpendicular to the plane of $\mathbf{A}$ and $\mathbf{B}$. When the vectors are written in the order $\mathbf{A} \times \mathbf{B}$, then $\mathbf{C}$ points in the direction in which a right-handed screw would move in turning from $\mathbf{A}$ to $\mathbf{B}$. Note that vector multiplication is noncommutative since

$$\mathbf{A} \times \mathbf{B} = -\mathbf{B} \times \mathbf{A}$$

If

$$\mathbf{A} = a\mathbf{i} + b\mathbf{j} + c\mathbf{k} \quad \text{and} \quad \mathbf{B} = d\mathbf{i} + e\mathbf{j} + f\mathbf{k}$$

then

$$\mathbf{C} = (bf - ce)\mathbf{i} + (cd - af)\mathbf{j} + (ae - bd)\mathbf{k}$$

$$= \begin{vmatrix} \mathbf{i} & \mathbf{j} & \mathbf{k} \\ a & b & c \\ d & e & f \end{vmatrix}$$

The vector product is defined only in three-dimensional space. It can be applied in certain physical situations: for example, the force $\mathbf{F}$ on a charge q moving with velocity $\mathbf{v}$ in a magnetic field $\mathbf{B}$ is given by $\mathbf{F} = q\mathbf{v} \times \mathbf{B}$. *See also* scalar product; triple product; angular velocity; moment of a force.

vector product

vector quantity Any quantity, such as velocity, momentum, or force, that has both magnitude and direction and for which *vector addition is defined and meaningful; for a complete specification both the direction and magnitude must be stated. It is thus treated mathematically as a *vector.

vector space (linear space) A set V of mathematical objects (called *vectors) that is associated with a *field F of objects (called *scalars), with the following properties:

(1) There is an operation of addition, and the addition of any two vectors in the set produces another vector in the set.

(2) Multiplication of a vector by a scalar gives another vector in the set.

(3) Addition of vectors is associative, i.e.

$$\mathbf{u} + (\mathbf{v} + \mathbf{w}) = (\mathbf{u} + \mathbf{v}) + \mathbf{w}$$

(4) Addition is commutative, i.e.

$$\mathbf{u} + \mathbf{v} = \mathbf{v} + \mathbf{u}$$

(5) There is a *zero vector* $\mathbf{0}$, such that

$$\mathbf{u} + \mathbf{0} = \mathbf{0} + \mathbf{u} = \mathbf{u}$$

(6) Every vector $\mathbf{u}$ has a negative $-\mathbf{u}$, such that

$$\mathbf{u} + (-\mathbf{u}) = \mathbf{0}$$

(7) If n is a scalar and $\mathbf{u}$ and $\mathbf{v}$ are vectors, then

$$n(\mathbf{u} + \mathbf{v}) = n\mathbf{u} + n\mathbf{v}$$

(8) If n and m are scalars and $\mathbf{u}$ is a vector, then

$$(n + m)\mathbf{u} = n\mathbf{u} + m\mathbf{u}$$

(9) If n and m are scalars and $\mathbf{u}$ is a vector, then

$$(nm)\mathbf{u} = n(m\mathbf{u})$$

(10) $1\mathbf{u} = \mathbf{u}$, where 1 is the unit element in F.

The set V is said to be a vector space over the field F. Note that the elements of a vector space form an *Abelian group. This axiomatic definition of a vector space includes the geometrical vectors represented by directed line segments in three-dimensional Euclidean space. It also covers other mathematical objects such as matrices, polynomials, and functions. The study of vector spaces gives insight into the nature of fields. For instance, the field of complex numbers is a vector space over the field of real numbers.

A *linear combination* is an expression of the form

$$n_1 \mathbf{v}_1 + n_2 \mathbf{v}_2 + \cdots$$

where $\mathbf{v}_1, \mathbf{v}_2, \ldots$ are vectors and $n_1, n_2, \ldots$ are scalars. If m vectors $\mathbf{v}_1, \mathbf{v}_2, \ldots, \mathbf{v}_m$ can be taken, and all the elements in the vector space can be produced by linear combinations of these m vectors, then the m vectors are said to *span* or *generate* the vector space.

In addition, if for the set $\mathbf{v}_1, \mathbf{v}_2, \ldots, \mathbf{v}_m$ it is possible to choose scalars $c_1, c_2, \ldots, c_m$, which are not all zero and are such that

$$c_1 \mathbf{v}_1 + c_2 \mathbf{v}_2 + \cdots + c_m \mathbf{v}_m = \mathbf{0}$$

then the set of elements is said to be *linearly dependent*. In this case, one of the m vectors in the set is a linear combination of some or all of the others. Otherwise the set of m vectors is *linearly independent*. A *basis* of a vector space is a linearly independent set of vectors that span the space. The *dimension* of a vector space V is the number of elements in a basis, denoted by dim V.

See also module; norm (of a vector space); scalar product.

vector subspace A subset of a *vector space which is itself a vector space. Examples are the set of real numbers in the space of *complex numbers, and the set of complex numbers in the *quaternions.

vector triple product *See* triple product.

velocity Symbol: $\mathbf{v}$. The rate of change of position with time when the direction of motion is specified. Velocity $\mathbf{v}$ is thus a *vector quantity; its magnitude v is referred to as *speed*. It is expressed in metres per second $(\mathrm{m\,s^{-1}})$ or similar units. The *average velocity* during some interval is the difference in position vector at the beginning and end of the interval divided by the elapsed time. As this time interval

approaches zero, the average velocity approaches the *instantaneous velocity*. Thus when a point or particle moves in space its velocity is the first derivative of the position vector $\mathbf{r}$:

$$\mathbf{v} = \frac{d\mathbf{r}}{dt} = \frac{dx}{dt}\mathbf{i} + \frac{dy}{dt}\mathbf{j} + \frac{dz}{dt}\mathbf{k}$$

where $\mathbf{i}$, $\mathbf{j}$, and $\mathbf{k}$ are unit vectors. In one dimension

$$\mathbf{v} = \frac{ds}{dt}\mathbf{i}$$

where s is the distance from an origin.

The velocity of a point with *polar coordinates (r, θ) which moves in a plane curved path is conveniently described by two perpendicular components: a *radial component* of dr/dt (directed away from the origin) and a *transverse component* of $r\,d\theta/dt$ (anticlockwise). The resultant is given by the vector sum of the components, and is directed along the tangent to the curve. Its magnitude is $|ds/dt|$, where s is the arc distance of the point from a fixed point on the curve.

See also angular velocity.

velocity ratio *See* machine.

Venn, John (1834–1923) English mathematician who introduced in his *Symbolic Logic* (1881) diagrams of overlapping circles to represent relations between sets. They have since been known as *Venn diagrams. He had earlier, in his *Logic of Chance* (1866), formulated one of the first versions of the frequency theory of probability.

Venn diagram A diagram used to illustrate relationships between *sets. Commonly, a rectangle represents the *universal set and a circle within it represents a given set (all members of the given set are represented by points within the circle). A subset is represented by a circle within a circle, and *union and *intersection are indicated

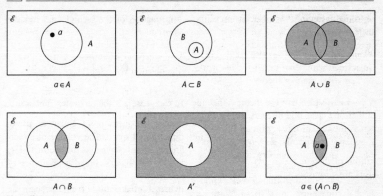

Venn diagram

by overlapping circles. *See also* complement; inclusion; member.

vernal equinox *See* equinoxes.

versiera *See* witch of Agnesi.

vertex (*plural* **vertices**) **1.** A point at which two or more lines or line segments meet on the boundary of a geometric figure (the edges of a polygon or polyhedron, the generators of a cone or pyramid, etc.).
2. *See* graph.

vertex matrix *See* adjacency matrix.

vertically opposite *See* opposite.

vibration *See* oscillation.

Viète, François (Franciscus Vieta) (1540–1603) French mathematician noted for his *In artem analyticam isagoge* (1591, Introduction to the Analytical Arts), one of the earliest Western works on algebra. In it he denoted unknowns by vowels and known quantities by consonants, and also introduced an improved notation for squares, cubes, and other powers. With his new algebraic techniques Viète succeeded in solving a number of problems classical authors had found unyielding to geometrical attacks. He developed new methods of solving equations in his *De aequationum recognitione et emendatione* (1615, On the Recognition and Emend-

ation of Equations). He was the first, in his *Canon mathematicus seu ad triangula* (1579, The Mathematical Canon Applied to Triangles), to tackle the problem of solving plane and spherical triangles with the help of the six main trigonometric functions.

Viète's product The *infinite product of nested square roots in the formula

$$\frac{2}{\pi} = \left(\frac{1}{2}\sqrt{2}\right)\left(\frac{1}{2}\sqrt{2 + \sqrt{2}}\right)$$

$$\times \left(\frac{1}{2}\sqrt{2 + \sqrt{2 + \sqrt{2}}}\right)\cdots$$

due to Viète in 1593. *See also* Wallis's product.

Vigenère cipher A *polyalphabetic substitution cipher which uses several *Caesar ciphers in conjunction. The use of a cipher of this kind can be traced back to Italian cryptographers of the 15th century who developed them because *monoalphabetic substitution ciphers were susceptible to attack by *frequency analysis. The French diplomat Blaise de Vigenère (1523–96) improved these ciphers by using 26 Caesar ciphers and including a word in the message which told the recipient in which order to use the ciphers.

vigesimal system A *number system using the base twenty.

Vinogradov, Ivan Matveyevich (1891–1983) Soviet mathematician noted for his work on analytical number theory.

Vinogradov's theorem The theorem that all sufficiently large odd integers can be written as the sum of three *primes. From the theorem it can also be shown that all sufficiently large even numbers can be written as the sum of three primes plus 3. The conjecture was published (along with *Goldbach's conjecture) in 1770 in Waring's book *Meditationes algebraicae*, and was proved by Vinogradov in 1937.

virtual-work principle A principle used in *statics: if a system in static *equilibrium undergoes an infinitesimal displacement consistent with the constraints on the system, then the total *work done on the system is zero.

Viviani, Vincenzo (1622–1703) Italian mathematician and physicist who in his *De maximis et minimis* (1659) attempted to reconstruct the fifth book of the *Conics* of Apollonius. He also published in 1674 an edition of Euclid. Viviani was one of the mathematicians who succeeded in determining the tangent to the cycloid.

volt Symbol: V. The *SI unit of electric potential, potential difference, and electromotive force, equal to the difference in potential between two points on a conductor carrying a constant current of 1 ampere when the power dissipated between these points is 1 watt. [After A. Volta (1745–1827)]

Volterra's integral equations Types of *integral equation named after the Italian mathematician Vito Volterra (1860–1940). An equation of the first kind has the form

$$f(x) = \lambda \int_a^x K(x, y) \, g(y) \, dy$$

An equation of the second kind has the form

$$g(x) = f(x) + \lambda \int_a^x K(x, y) \, g(y) \, dy$$

In each case g is the unknown function.

volume A measure of extent in three-dimensional space. The volume of a rectangular parallelepiped is the product of its length, width, and breadth. Volumes of polyhedra can also be calculated. For solid figures bounded by curved surfaces, the volume is found by *integration. For a *solid of revolution the element of volume can be taken as the volume of an elementary disc $A \, dx$, where A is the area. Volume can also be obtained by a triple integral of the form

$$\iiint dx \, dy \, dz$$

von Koch curve *See* Koch curve.

von Neumann, John (1903–57) Hungarian-American mathematician best known for his work with Oskar Morgenstern which led to their *The Theory of Games and Economic Behavior* (1944). He also made a major contribution to the development of the modern computer both as a practical design problem and as a means of investigating general theoretical questions on the nature of, and constraints upon, logical automata. In more traditional fields von Neumann worked on problems in set theory (*von Neumann set theory), the theory of operators, Lie groups, and shock waves. He also worked on quantum mechanics and succeeded in axiomatizing the subject, a result published in his definitive work *Mathematische Grundlagen der Quantenmechanik* (1932, Mathematical Foundations of Quantum Mechanics). *See* game theory.

von Neumann set theory In 1925 von Neumann proposed an alternative to the orthodox axiomatization of set theory introduced by Zermelo (*see* Zermelo–Fraenkel set theory). To avoid the paradoxes identified by Russell and others he introduced a radical distinction between sets and classes.

Every set in his system is a class, but not all classes are sets. Those that are not are termed *proper classes*.

vulgar fraction *See* common fraction.

W

waiting time *See* exponential distribution; gamma distribution.

walk In a *graph, an alternating sequence of vertices and edges from vertex v_0 to vertex v_k: $v_0, e_1, v_1, e_2, \ldots, v_k$, in which edge e_i joins vertices v_{i-1} and v_i. The number of edges in the walk is its *length*. Thus, in the diagram, P, f, Q, b, R, b, Q is a walk from P to Q of length 3. A *closed walk* has $v_0 = v_k$, e.g. P, f, Q, b, R, b, Q, f, P.

A walk in which all vertices are distinct (except possibly v_0 and v_k) is a *path*. A *closed path* or *cycle* has $v_0 = v_k$, e.g. P, f, Q, b, R, d, S, e, P.

A walk in which all edges are distinct is a *trail*. A *closed trail* or *circuit* has $v_0 = v_k$, e.g. Q, b, R, c, S, d, R, a, Q. (Note that there is no standard terminology. For example, the concepts described above as 'walk', 'path', 'cycle', and 'circuit' may elsewhere be termed 'path', 'simple path', 'circuit', 'cycle'.)

A graph is *connected* if, for every pair of vertices u, v there is a path from u to v; otherwise, it is *disconnected*.

Wallace–Simson line *See* Simson line.

Wallis, John (1616–1703) English mathematician noted for his pioneering work on the infinitesimal calculus. In his *Arithmetica infinitorum* (1655, The Arithmetic of Infinitesimals) he sought to determine π by expressing $\pi/2$ as an infinite product. He was also the first to explain the meaning of such exponential forms as x^0, x^{-n}, and $x^{n/m}$, and to introduce ∞ as the symbol for infinity.

Wallis's product The *infinite product in the formula

$$\frac{\pi}{2} = \lim_{n \to \infty} \frac{2.2}{1.3} \times \frac{4.4}{3.5} \times \cdots$$

$$\times \frac{(2n).(2n)}{(2n-1).(2n+1)}$$

It is equivalent to a result in Wallis's 1656 book *Arithmetica Infinitorum*. *See also* Viète's product.

wallpaper group A *symmetry group* of a two-dimensional pattern. It is the two-dimensional analogue of the symmetry group of a crystal. It can be shown that there are exactly 17 different wallpaper groups. *See* crystallography; *compare* frieze group.

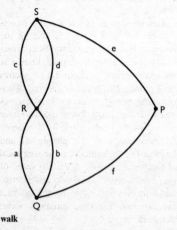

walk

Wang's paradox A *paradox proposed by the Chinese-American logician Hao Wang (1921–95), with features similar to the various *sorites paradoxes:

0 is a small number

If *n* is a small number then *n* + 1 is a small number

Therefore all numbers are small

Some have argued that there is no paradox, as all numbers relative to larger numbers are indeed small; others, that, as 'small' is a vague predicate, no coherent logical conclusions can be drawn.

Waring, Edward (1734–98) English mathematician noted for his *Meditationes algebraicae* (1770) which contained the first statement of *Wilson's theorem and his own conjecture, *Waring's problem.

Waring's problem The problem of proving the conjecture that any positive integer can be written as a sum of not more than 9 cubes or not more than 19 fourth powers of integers. The first part, on cubes, was mainly proved by A. Wieferich in 1909 (with a gap in the argument being filled by A.J. Kempner in 1912), and the second part, on fourth powers, was proved by R. Balasubramanian, J.-M. Deshouliers and F. Dress in 1986.

The term 'Waring's problem' is sometimes used for the more general problem of showing that there is a maximum number of terms to a given power necessary for expressing any positive integer. In other words, given a positive integer k, there is an integer $g(k)$ such that any positive integer can be expressed as the sum of not more than $g(k)$ positive integers each raised to the power k. Thus, if $k = 2$ then $g(2) = 4$; i.e. any positive integer can be expressed as a sum of up to four squares – a result proved by Lagrange (*see* Lagrange's theorem). Waring's original conjecture is that $g(3) = 9$ and $g(4) = 19$. Hilbert solved the general problem in 1909. It is now known that for every value of k up to 471 600 000, the value of $g(k)$ is given by the formula

$$g(k) = [(3/2)^k] + 2^k - 2$$

where $[x]$ denotes the *integer part of x. For example, when $k = 5$, the integer part of $(3/2)^5$ is 7, so the formula gives $g(5) = 7 + 32 - 2 = 37$ (a result proved earlier by J-j. Chen in 1964). It is suspected that the formula holds for all values of k.

watt Symbol: W. The *SI unit of power, equal to 1 joule of energy per second. [After J. Watt (1736–1819)]

wave Any disturbance that can be propagated from one point to another through a gaseous, solid, or liquid medium without any permanent displacement of the medium. Sound and light waves are two forms of wave motion. A sound wave is a type of *elastic wave*: a particle of the medium is displaced in such a way that it can transfer its momentum to an adjacent particle and then return to its original position; the adjacent particle then disturbs another particle, and so on. Light waves are a type of *electromagnetic wave*: the wave consists of oscillating electric and magnetic fields that do not disturb the particles of the medium, and so can travel through a vacuum; the fields oscillate at right angles to the direction of propagation.

The velocity at which a wave travels depends on what type it is and on the medium. Electromagnetic waves in a vacuum travel at a constant speed, known as the *speed of light*, c (about 3×10^8 metres per second); the speed is reduced when travelling through a medium. Elastic waves travel at very much lower speeds. Other properties of a wave include its *frequency v, *wavelength λ, *amplitude a, and *period T. The frequency is the reciprocal of the period, i.e. $v = 1/T$. The speed of propagation for both elastic and electromagnetic waves is given by the product $v\lambda$. See also beats; longitudinal wave; transverse wave.

wave equation The partial *differential equation

$$\nabla^2 \phi = \frac{1}{c^2} \frac{\partial^2 \phi}{\partial t^2}$$

where ∇ is the differential operator *del, ϕ is a scalar function of position and time, and c is a constant. *See* Bessel functions; Laplace's equation.

wavelength Symbol: λ. A property of a *wave, expressed as the distance travelled in the direction of propagation between two points at the same phase of disturbance in consecutive cycles of the wave.

wavelets Special functions through which one can express other functions, but which allow greater flexibility than the traditional *Fourier series. First introduced in the 1980s for purely mathematical reasons, wavelets have since been used extensively in many disciplines. *See also* harmonic analysis.

wave mechanics *See* quantum mechanics.

wavenumber The reciprocal of the *wavelength of a wave.

weber Symbol: Wb. The *SI unit of magnetic flux, equal to the flux that, when linking a circuit of one turn, produces in it an electromotive force of 1 volt as it is reduced to zero at a uniform rate in 1 second. [After W.E. Weber (1804–91)]

Wedderburn's theorem (J.H. Wedderburn, 1905) The theorem that a *division ring which is finite must be a *field. Wedderburn's theorem says in effect that the commutativity of multiplication follows from the other field axioms if the underlying set D is finite. There are examples of division rings that are not fields but, according to Wedderburn, they must be infinite. The best-known example is the division ring of *quaternions.

wedge *See* spherical wedge.

Weibull distribution (W. Weibull, 1939) A distribution of a positive continuous *random variable X with frequency function of the form

$$f(x) = \frac{k}{\lambda} \left(\frac{x}{\lambda}\right)^{k-1} \exp\left[-\left(\frac{x}{k}\right)\right]^k$$

where k and λ are positive *parameters. The distribution is used in reliability studies and in meteorology. It is one of the most commonly used *extreme value distributions.

Weierstrass, Karl Theodor Wilhelm (1815–97) German mathematician noted for his work on real and complex functions. In 1871 Weierstrass discovered a continuous curve with no tangent at any point. Throughout his career he emphasized the need to introduce into analysis much greater rigour and, in the tradition of Cauchy, he sought to define with greater precision such terms as continuity, limit, differential, and irrationals, which he saw as infinite sequences of rationals.

Weierstrass's theorem *See* dense set.

weight Symbol: W. The *force exerted on matter by the *gravity of the earth (or of whatever celestial body on which the matter is located). The weight of an object of *mass m is equal to mg, where g is the *acceleration of free fall. Since g varies with position (and with celestial body), weight is not a constant property of matter. *See also* mass.

weighted graph A *graph (or *network) in which numbers are attached to edges (or arcs). The numbers may represent distances associated with the edges, or times taken or costs incurred in travelling along them. The weight of a *walk, path, or trail is the sum of the weights of all its edges.

weighted least squares *See* lowess.

weighted mean *See* mean.

well-formed formula (wff) In *logic, a sequence of symbols from a *formal language constructed according to the *formation rules of the language.

well-ordered set If a *set A is an *ordered set and if every *subset of A has a first element, then A is a well-ordered set. Using the *axiom of choice, Zermelo was the first to prove the important theorem that every ordered set can be well ordered.

Weyl, Hermann Klaus Hugo (1885–1955) German mathematician noted for his work in mathematical physics, the foundations of mathematics, and pure mathematics, in which he contributed to group theory and the theory of Hilbert space. His work in mathematical physics provided some of the formalism necessary for the development of both relativity and quantum theory.

wff *Abbreviation for* *well-formed formula.

Whitehead, Alfred North (1861–1947) English mathematician and philosopher. After the publication of his *A Treatise on Universal Algebra* (1898), Whitehead began the collaboration with Bertrand Russell that led to *Principia mathematica* (3 vols, 1910–13), their attempt to derive the whole of mathematics from purely logical principles.

Whittaker, Sir Edmund Taylor (1873–1956) English mathematician who, in 1904, published a very influential text on mathematical analysis; later editions were co-written with G.N. Watson. He also made contributions to astronomy, mechanics, numerical analysis, and the historical and philosophical aspects of physical theories.

Wiener, Norbert (1894–1964) American mathematician well known for his work in mathematical logic, stochastic processes, and Fourier transforms. In 1948, however, he became an internationally known figure with the publication of his *Cybernetics, or Control and Communication in the Animal and the Machine*, the work which founded the modern discipline of *cybernetics.

Wilcoxon rank sum test (F. Wilcoxon, 1945) A *distribution-free test of the hypothesis that two independent samples come from the same *population, against alternatives that specify either a difference in median only or, more generally, that one population distribution is stochastically larger – i.e. that the *cumulative distribution functions $F(x)$ and $G(y)$ satisfy a relationship $F(u) \geqslant G(u)$ with strict inequality for at least some u. The observations in the combined samples are arranged in ascending order and replaced by their *ranks. The test statistic is the sum of the ranks of the observations in one of the samples. Tables of critical values are published for various sample sizes; alternatively, exact *p-values may be computed with some statistical software. For large samples a normal approximation is often used. An alternative formulation of the problem gives rise to the equivalent *Mann–Whitney test* (H.B. Mann and D.R. Whitney, 1947). The method may be extended to obtain *confidence intervals for median differences.

Wilcoxon signed rank test (F. Wilcoxon, 1945) A *distribution-free test for hypotheses about the mean or median of a symmetric distribution. Deviations of sample values from the hypothesized mean or median M are ranked in order of magnitude, and these deviations are replaced by their ranks together with a sign to indicate whether the deviation is positive or negative. The test statistic is usually taken to be the sum of the positive ranks, and critical values are tabulated for various sample sizes. The exact *p-value of the statistic may be computed with the aid of statistical software packages. For large samples, an approximation based on the *normal distribution is often used. The method may be extended to *matched pair samples to provide a test of whether the sample values in the pairs are identically distributed, because if they are then the within-pairs differences will have median zero. It is also possible to obtain a distribution-free *confidence interval for a population median or, in the case of paired samples, for a median difference.

Wilkins, John (1614–72) English mathematician and scientist who in his *Mathematical Magick* (1648) demonstrated, among other things, the use of mathematics in the design of machines.

Wilson's theorem The theorem that if p is a *prime then it divides $(p-1)! + 1$. Thus 5 divides $4! + 1 = 25$. The statement was first published by Waring in 1770 in his book *Meditationes algebraicae* and ascribed to the English mathematician John Wilson (1741–93). It was first proved by Lagrange in 1771. The converse of the theorem is also true and implies that, if n is a *composite number it cannot divide the corresponding expression $(n-1)! + 1$. So Wilson's theorem and its converse theoretically provide a test to determine whether any natural number greater than 1 is prime or composite. Unfortunately, the technique is impractical for even moderately large numbers.

winding number Given a point in a plane, the winding number of a *closed curve is the number of times the curve goes round the point in an anticlockwise sense. The number depends on the curve and the chosen point.

winsorization A method of estimation analogous to forming a *trimmed mean, except that instead of discarding the trimmed values, each is replaced by the closest value that is not trimmed. Many generalizations are possible. W.J. Dixon (1960) proposed the term winsorization to indicate that the procedure was first suggested by the American statistician Charles P. Winsor (1895–1951).

witch of Agnesi (versiera) A plane *curve obtained by first taking a point P on a given circle and the *tangent line through Q, where PQ is a diameter. A line from P is drawn to cut the circle at any point R and

to cut the tangent line at S. From R, a line is drawn parallel to QS and from S a line is drawn parallel to QP, the two lines intersecting at P'. The witch is the locus of all such points P' (i.e. for all points R on the generating circle). If the circle is drawn with P at the origin and Q at $(2a, 0)$, its equation is

$$xy^2 = 4a^2(2a - x)$$

a being the radius of the circle.

The curve was studied by Maria Agnesi in the 18th century. The Italian mathematician Guido Grandi (1671–1742) had previously named it the *versorio*, from the Latin *vertere* (to turn). Agnesi confused this with *versiera*, which has 'witch' as one of its colloquial meanings.

witch of Agnesi

with *or* without replacement *See* random sample.

word *See* alphabet.

work Symbol: *W*. A transfer of *energy that occurs when a *force is applied to a body so that the point of application is moved. Strictly the body should be moving, and then the force has a component in the direction of motion. Energy is transferred from the agent exerting the force to the body so that the body's kinetic energy is increased. If work is done on the agent by the moving body then *negative work* is done: an agent doing negative work is gaining energy from the body (*see* potential energy). Work, like energy, is measured in joules.

Work is a *scalar quantity. If the force is constant and acts in the direction of motion, the amount of work done on the body is given by the product of the magnitude F of the force and the distance s moved by the point of application. In general the work done during motion from a position s_1 to a position s_2 is given by

$$W = \int_{s_1}^{s_2} F \cos\theta \, ds$$

where θ is the angle between the direction of the force and the infinitesimal displacement ds. By using the *scalar product, this can be written as

$$W = \int_{s_1}^{s_2} \mathbf{F} . d\mathbf{s}$$

See power.

wrapped Cauchy distribution *See* directional data.

Wren, Sir Christopher (1632–1723) English astronomer, architect, and mathematician, noted in mathematics for his work on the hyperboloid and for his rectification of the cycloid in 1658.

wrench A force acting together with a couple whose plane is perpendicular to the line of action of the force. The line of action of the force is the *axis* of the wrench. Any system of forces is equivalent to a wrench. *See* Poinsot.

Wronskian The *determinant $W(x) \equiv$

$$\begin{vmatrix} f_1(x) & f_2(x) & \cdots & f_n(x) \\ f_1'(x) & f_2'(x) & \cdots & f_n'(x) \\ \vdots & \vdots & & \vdots \\ f_1^{(n-1)}(x) & f_2^{(n-1)}(x) & \cdots & f_n^{(n-1)}(x) \end{vmatrix}$$

defined on an *interval (a,b), where $\{f_1(x), f_2(x), \ldots, f_n(x)\}$ is a *set of n *functions each having continuous *derivatives up to the $(n-1)$th order in (a,b).

If the functions are linearly dependent on that interval, i.e. if there exist constants $c_1, c_2, \ldots, c_n$ that are not all zero and such that

$$c_1 f_1(x) + c_2 f_2(x) + \cdots + c_n f_n(x) = 0$$

for all x in (a,b), then $W(x) = 0$.

If $W(x) = 0$ for all x in (a,b), then there exists a subinterval of $(a\ b)$ on which the functions are linearly dependent. It is named after the Polish mathematician Józef Maria Hoene-Wroński (1776–1853). *See also* Hessian; Jacobian.

XYZ

x-axis *See* Cartesian coordinate system.

x-coordinate *See* abscissa.

x-step *See* run.

yard A *British unit of length, originally defined in terms of a bronze standard but redefined in the UK Weights and Measures Act (1963) as 0.9144 metre exactly.

Yates, Frank (1902–94) English statistician who made important contributions to the analysis of experiments, especially those involving factorial treatment structures, to sampling theory and practice, and pioneered the use of computers in statistics.

Yates's correction (F. Yates, 1934) A special case of a *continuity correction used in the *chi-squared test when applied to 2×2 *contingency tables.

yaw Angular movement of an aircraft, spacecraft, projectile, etc. about its vertical axis. *Compare* pitch; roll.

y-axis *See* Cartesian coordinate system.

y-coordinate *See* ordinate.

year A unit of time based on the period of revolution of the earth round the sun. It can be defined in various ways. The *civil year* (*calendar year* or *Julian year*) has an average value of 365.25 mean solar days; three successive years of 365 days are followed by a leap year of 366 days. The *tropical year* (or *solar year*) is the interval between two consecutive passages, in the same direction, of the sun through the earth's equatorial plane; its value is 365.242 199 mean solar days. The *anomalistic year*, the average interval between two consecutive passages of the earth through perihelion, is 365.259 641 mean solar days. The *sidereal year*, the interval in which the sun appears to perform a complete revolution with reference to the fixed stars, is 365.256 366 mean solar days. *See* day.

yield The *interest paid on an investment.

yocto- *See* SI units.

yotta- *See* SI units.

Youden square (W.J. Youden, 1940) An *experimental design that might more appropriately be called an incomplete *Latin square, since it is composed of some, but not all, rows of a Latin square. An example is

$$A \quad B \quad C \quad D \quad E \quad F \quad G$$
$$B \quad E \quad F \quad G \quad D \quad A \quad C$$
$$C \quad F \quad D \quad B \quad A \quad G \quad E$$

in which the rows may be looked upon as *randomized blocks and the columns as *balanced incomplete blocks. The analysis is fairly straightforward, and the efficiency reasonably high. The Youden square incorporates the desirable feature of a Latin square design of allowing the elimination of unwanted variability in more than one direction without the irksome restriction of having both the number of rows and the number of columns equal to the number of treatments.

Young's inequality If $a > 1$ and $b > 1$ are such that $1/a + 1/b = 1$, then

$$xy \leqslant \frac{x^a}{a} + \frac{y^b}{b}, \quad \text{for all} \quad x, y > 0$$

The case where $a = b = 2$ yields the *arithmetic–geometric mean inequality. Named after the English mathematician William Henry Young (1863–1942).

Young's modulus Symbol: E. A *modulus of elasticity that is used when an elastic body is under tension (or compression). It is the ratio of the applied force per unit area of cross-section to the resulting increase (or decrease) in length per unit length of the body. This is equivalent to the ratio of tensile (or compressive) *stress to the associated longitudinal *strain. It is named after the English physicist, physician, and Egyptologist Thomas Young (1773–1829).

y-step *See* rise.

$\mathbb{Z}$ Symbol for the set of all *integers.

$\mathbb{Z}^+$ Symbol for the set of all positive *integers, i.e. the natural numbers.

z-axis *See* Cartesian coordinate system.

z-coordinate *See* Cartesian coordinate system.

z-distribution *See* Fisher's z-distribution.

zenith A point on the *celestial sphere directly above an observer. The zenith is one of the poles of the horizon. *Compare* nadir.

zenith distance (coaltitude) Symbol: ζ. The angular distance of a point on the *celestial sphere from the zenith taken along a *great circle passing through the zenith, the point, and the *nadir. It is the complement of the altitude (i.e. $\zeta = 90° - h$) and is sometimes used instead of altitude. *See* horizontal coordinate system.

Zeno's paradoxes Four *paradoxes proposed by the Greek philosopher Zeno of Elea (5th century BC), demonstrating the difficulties in supposing that anything can be infinitely subdivided. The best known – that of Achilles and the tortoise – proposes a race in which the tortoise is given a start. To overtake the tortoise Achilles must first reach the tortoise's starting position. By then the tortoise will have moved ahead to a new position. By the time Achilles reaches this, the tortoise will have moved again. The paradox is that Achilles will never catch the tortoise, no matter how swiftly he runs.

zepto- *See* SI units.

Zermelo, Ernst (1871–1953) German mathematician who in his *Untersuchungen über die Grundlagen der Mengenlehre* (1908, Investigations on the Foundations of Set Theory) founded the modern discipline of axiomatic set theory. *See* Zermelo–Fraenkel set theory.

Zermelo–Fraenkel set theory In order to avoid the *paradoxes Bertrand *Russell and others had found in the foundations of set theory, Zermelo proposed a supposedly rigorous axiomatic basis for the new discipline in 1900. As modified by A.A.H. Fraenkel in 1922, Zermelo's system has formed the basis of most later axiomatizations. In addition to such familiar assumptions as the axiom of extensionality and the union axiom, Zermelo also found it necessary to assume the more controversial and less intuitively acceptable axiom of *choice and *axiom of infinity.

zero 1. *See* number system.
2. (of a function) If, for a *function $f(x)$, the value $x = a$ is such that $f(a) = 0$, then a is a zero of the function. *See* root.

zero angle (null angle) An angle of $0°$.

zero divisor A nonzero element in a *ring that can be combined with another nonzero element in the ring (using the ring's multiplication operation) to give the ring's zero element. For example, in the ring of all 2×2 *matrices with the operations of matrix addition and multiplication, the elements $A = \begin{pmatrix} 1 & 1 \\ 1 & 1 \end{pmatrix}$ and $B = \begin{pmatrix} 1 & -1 \\ -1 & 1 \end{pmatrix}$ are zero divisors since $AB = \begin{pmatrix} 0 & 0 \\ 0 & 0 \end{pmatrix}$, which is the ring's zero element.

zero matrix (null matrix) A *matrix in which all the elements are zero.

zero-sum game A term used in *game theory to describe a game where the sum of the gains and losses made by all players is zero. If the sum is not zero, the game is a *nonzero-sum game*. A simple example of a zero-sum game is a game of chess where the two players mutually agree that the loser shall pay the winner a fixed sum. Most games played at casinos are non-zero-sum games as there is a leakage of funds to (and more rarely from) the banker. Any nonzero-sum game involving n players can be modified to a zero-sum game involving $n+1$ players. For example, most casino games involving n players become zero-sum games if the banker is introduced as the $(n+1)$th player. *See* game theory; prisoner's dilemma.

zero vector (null vector) 1. A *vector whose *absolute value is zero.
2. *See* vector space.

zeta function *See* Riemann zeta function.

zetta- *See* SI units.

Zhu Shijie (Chu Shih-chieh) (c.1300) Chinese mathematician noted for his *Siyuan yujian* (1303, The Precious Mirror of the Four Elements). It contains what in the West became known as Pascal's triangle, and its use in extracting roots – both traceable in China back to Jia Xian (Chia Hsien) (c.1100). It also describes the 'method of four unknowns' – a system of notation for polynomials in four variables (the 'celestial, earthly, human, and material elements') and techniques for manipu-

lating them and solving problems. The section dealing with the summation of finite series includes a method using finite differences equivalent to the use of the forward difference formula in *Gregory–Newton interpolation.

zonal harmonic *See* harmonic.

zone A surface formed by two parallel planes cutting a sphere. If neither plane is a tangent plane the surface is a *zone of two bases*. If one of the planes is a tangent it is a *zone of one base*. The area of the zone is $2\pi Rh$, where h is the perpendicular distance between the planes, and R is the radius of the sphere.

zone

Z-score *See* standardized random variable.

z-transformation *See* Fisher's z-transformation.

Zu Chongzhi (Tsu Chung Chi) (AD 429–500) Chinese mathematician, astronomer, and calendar-maker. All his 51 recorded works are lost, including the *Zhui shu* (Method of Interpolation). With his son Zu Geng, he found π to lie between 3.141 592 6 and 3.141 592 7, and gave 355/113 as a 'close ratio'. They are also credited with a proof of the formula for the volume of a sphere by transforming known volumes using a principle identical to that of *Cavalieri.

APPENDIX

Table 1 Derivatives

y	$\dfrac{dy}{dx}$
x^n	nx^{n-1}
$u + v$	$\dfrac{du}{dx} + \dfrac{dv}{dx}$
uv	$u\dfrac{dv}{dx} + v\dfrac{du}{dx}$
$\dfrac{u}{v}$	$\dfrac{1}{v^2}\left(v\dfrac{du}{dx} - u\dfrac{dv}{dx}\right)$
ax	a
$\sin x$	$\cos x$
$\cos x$	$-\sin x$
$\tan x$	$\sec^2 x$
$\operatorname{cosec} x$	$-\operatorname{cosec} x\ \cot x$
$\sec x$	$\sec x \tan x$
$\cot x$	$-\operatorname{cosec}^2 x$
$\sin^{-1} x$	$\dfrac{1}{\sqrt{(1-x^2)}}$
$\cos^{-1} x$	$-\dfrac{1}{\sqrt{(1-x^2)}}$
$\tan^{-1} x$	$\dfrac{1}{1+x^2}$
e^x	e^x
e^{-x}	$-e^{-x}$
a^x	$a^x \times \ln a$

y	$\dfrac{dy}{dx}$
$\sinh x$	$\cosh x$
$\cosh x$	$\sinh x$
$\tanh x$	$\operatorname{sech}^2 x$
$\operatorname{cosech} x$	$-\operatorname{cosech} x \coth x$
$\operatorname{sech} x$	$-\operatorname{sech} x \tanh x$
$\coth x$	$-\operatorname{cosech}^2 x$
$\sinh^{-1} x$	$\dfrac{1}{\sqrt{(x^2+1)}}$
$\cosh^{-1} x$	$\dfrac{1}{\sqrt{(x^2-1)}}$
$\tanh^{-1} x$	$\dfrac{1}{1-x^2}$
$\operatorname{cosech}^{-1} x$	$-\dfrac{1}{x\sqrt{(1+x^2)}}$
$\operatorname{sech}^{-1} x$	$-\dfrac{1}{x\sqrt{(1-x^2)}}$
$\coth^{-1} x$	$-\dfrac{1}{x^2-1}$
$\cot^{-1} x$	$-\dfrac{1}{1+x^2}$
$\sec^{-1} x$	$\dfrac{1}{x\sqrt{(x^2-1)}}$
$\operatorname{cosec}^{-1} x$	$-\dfrac{1}{x\sqrt{(x^2-1)}}$
$\ln x$	$\dfrac{1}{x}$

Table 2 Integrals | 476

Table 2 Integrals

y	$\int y \, \mathrm{d}x$
x^n	$\dfrac{x^{n+1}}{n+1}$ for $n \neq -1$
$u \dfrac{\mathrm{d}v}{\mathrm{d}x}$	$uv - \int v \dfrac{\mathrm{d}u}{\mathrm{d}x} \, \mathrm{d}x$
$\dfrac{1}{x}$	$\ln |x|$
$\dfrac{\mathrm{f}'(x)}{\mathrm{f}(x)}$	$\ln |\mathrm{f}(x)|$
$\dfrac{1}{x \ln x}$	$\ln |\ln x|$
$\ln x$	$x \ln x - x$
$\dfrac{\ln x}{x}$	$\frac{1}{2}(\ln x)^2$
e^x	e^x
e^{ax}	$\dfrac{1}{a} \mathrm{e}^{ax}$
a^x	$\dfrac{a^x}{\ln a}$
$x^a \ln x$	$\dfrac{x^{a+1}}{a+1}\left(\ln x - \dfrac{1}{a+1}\right)$ for $a \neq -1$
$\sin x$	$-\cos x$
$\cos x$	$\sin x$
$\tan x$	$-\ln |\cos x|$
$\cot x$	$\ln |\sin x|$
$\sec x$	$\ln(\sec x + \tan x)$
$\operatorname{cosec} x$	$\ln(\operatorname{cosec} x - \cot x)$
$\sin^2 x$	$\frac{1}{2}x - \frac{1}{4}\sin 2x$
$\cos^2 x$	$\frac{1}{2}x + \frac{1}{4}\sin 2x$
$\sec x \tan x$	$\sec x$

y	$\int y \, \mathrm{d}x$
$\operatorname{cosec} x \cot x$	$-\operatorname{cosec} x$
$\dfrac{1}{1 + \cos x}$	$\tan \frac{1}{2}x$
$\dfrac{1}{1 - \cos x}$	$-\cot \frac{1}{2}x$
$\dfrac{1}{\sin x \cos x}$	$\ln |\tan x|$
$\operatorname{cosec}^2 x$	$-\cot x$
$\sec^2 x$	$\tan x$
$\sin^{-1} x$	$x \sin^{-1} x + \sqrt{(1 - x^2)}$
$\cos^{-1} x$	$x \cos^{-1} x - \sqrt{(1 - x^2)}$
$\tan^{-1} x$	$x \tan^{-1} x - \frac{1}{2} \ln(1 + x^2)$
$\cot^{-1} x$	$x \cot^{-1} x + \frac{1}{2} \ln(1 + x^2)$
$\sinh x$	$\cosh x$
$\cosh x$	$\sinh x$
$\tanh x$	$\ln \cosh x$
$\coth x$	$\ln |\sinh x|$
$\operatorname{cosech} x$	$\ln |\tanh \frac{1}{2}x|$
$\operatorname{sech} x$	$2 \tan^{-1} \mathrm{e}^x$
$\sinh^{-1} x$	$x \sinh^{-1} x - \sqrt{(1 + x^2)}$
$\cosh^{-1} x$	$x \cosh^{-1} x - \sqrt{(x^2 - 1)}$
$\tanh^{-1} x$	$x \tanh^{-1} x + \frac{1}{2} \ln(1 - x^2)$
$\coth^{-1} x$	$x \coth^{-1} x + \frac{1}{2} \ln(x^2 - 1)$
$\dfrac{1}{x^2 + a^2}$	$\dfrac{1}{a} \tan^{-1} \dfrac{x}{a}$

Note: the constant of integration has been omitted.

Table 2 Integrals

y	$\int y\,dx$	y	$\int y\,dx$
$\dfrac{1}{\sqrt{(a^2+x^2)}}$	$\sinh^{-1}\dfrac{x}{a}$	$\dfrac{x}{\sqrt{(a^2-x^2)}}$	$-\sqrt{(a^2-x^2)}$
$\dfrac{1}{x\sqrt{(x^2-a^2)}}$	$-\dfrac{1}{a}\sin^{-1}\dfrac{a}{x}$	$\dfrac{1}{\sqrt{(x^2-a^2)}}$	$\cosh^{-1}\dfrac{x}{a}$
$\dfrac{1}{\sqrt{(a^2-x^2)}}$	$\sin^{-1}\dfrac{x}{a}$	$\dfrac{1}{x\sqrt{(x^2+a^2)}}$	$-\dfrac{1}{a}\sinh^{-1}\dfrac{a}{x}$
$\dfrac{1}{x^2-a^2}$	$\dfrac{1}{2a}\ln\dfrac{a-x}{a+x}$ for $\lvert x\rvert < a,\,a>0$ $\dfrac{1}{2a}\ln\dfrac{x-a}{x+a}$ for $\lvert x\rvert > a,\,a>0$	$\sqrt{(a^2-x^2)}$	$-\tfrac{1}{2}a^2\cos^{-1}\dfrac{x}{a}+\tfrac{1}{2}x\sqrt{(a^2-x^2)}$
		$\sqrt{(x^2-a^2)}$	$\tfrac{1}{2}a^2\cosh^{-1}\dfrac{x}{a}+\tfrac{1}{2}x\sqrt{(x^2-a^2)}$
$\dfrac{x}{\sqrt{(a^2+x^2)}}$	$\sqrt{(a^2+x^2)}$	$\sqrt{(x^2+a^2)}$	$\tfrac{1}{2}a^2\sinh^{-1}\dfrac{x}{a}+\tfrac{1}{2}x\sqrt{(x^2+a^2)}$

Note: the constant of integration has been omitted.

Table 3 Reduction formulae | 478

Table 3 Reduction formulae

For a positive integer n

$$\int \sin^n ax \, dx = -\frac{\sin^{n-1} ax \cos ax}{na} + \frac{n-1}{n} \int \sin^{n-2} ax \, dx$$

For a positive integer n

$$\int \cos^n ax \, dx = \frac{\cos^{n-1} ax \sin ax}{na} + \frac{n-1}{n} \int \cos^{n-2} ax \, dx$$

For an integer $n > 1$

$$\int \tan^n ax \, dx = \frac{1}{a(n-1)} \tan^{n-1} ax - \int \tan^{n-2} ax \, dx$$

For a positive integer n

$$\int x^n e^{ax} \, dx = \frac{1}{a} x^n e^{ax} - \frac{n}{a} \int x^{n-1} e^{ax} \, dx$$

For a positive integer n

$$\int (\ln x)^n \, dx = x(\ln x)^n - n \int (\ln x)^{n-1} \, dx$$

Table 4 Centres of mass The position of the centre of mass of certain uniform bodies.

Body	Position of centre of mass
Triangular lamina	$\frac{2}{3}$ of the way along any median from its vertex
Solid hemisphere of radius r	$\frac{3}{8} r$ from the centre
Hemispherical shell of radius r	$\frac{1}{2} r$ from the centre
Cone or pyramid	$\frac{3}{4}$ of the way from the vertex to the mass centre of the base
Circular arc of radius r and angle 2α	$\dfrac{r \sin \alpha}{\alpha}$ from the centre of the circle
Circular sector of radius r and angle 2α	$\dfrac{2r \sin \alpha}{3\alpha}$ from the centre of the circle

Table 5 Moments of inertia Values for certain uniform bodies of mass M about certain axes.

Body	Axis	Moment of inertia
Rod of length $2a$	perpendicular to rod, through mid-point	$\frac{1}{3}Ma^2$
Rectangular lamina of sides $2a$, $2b$	perpendicular to lamina, through centre	$\frac{1}{3}M(a^2+b^2)$
Circular disc of radius a	perpendicular to disc, through centre	$\frac{1}{2}Ma^2$
Solid sphere of radius a	any diameter	$\frac{2}{5}Ma^2$
Spherical shell of radius a	any diameter	$\frac{2}{3}Ma^2$
Elliptical lamina of axes $2a$, $2b$	axis of length $2a$	$\frac{1}{4}Mb^2$
	perpendicular to lamina, through centre	$\frac{1}{4}M(a^2+b^2)$
Solid ellipsoid with axes $2a$, $2b$, $2c$	axis of length $2c$	$\frac{1}{5}M(a^2+b^2)$
Ellipsoidal shell with axes $2a$, $2b$, $2c$	axis of length $2c$	$\frac{1}{3}M(a^2+b^2)$
Right circular solid cone with base radius a	axis of cone	$\frac{3}{10}Ma^2$

Table 6 The Greek alphabet

Capital	Lower-case		Capital	Lower-case	
A	α	alpha	N	ν	nu
B	β	beta	Ξ	ξ	xi
Γ	γ	gamma	O	o	omicron
Δ	δ	delta	Π	π	pi
E	ε	epsilon	P	ρ	rho
Z	ζ	zeta	Σ	σ	sigma
H	η	eta	T	τ	tau
Θ	θ	theta	Y	υ	upsilon
I	ι	iota	Φ	ϕ	phi
K	κ	kappa	X	χ	chi
Λ	λ	lambda	Ψ	ψ	psi
M	μ	mu	Ω	ω	omega

Table 7 Common signs and symbols | 480

Table 7 Common signs and symbols

Symbol	Reference	Symbol	Reference
$=$	equals sign	$\sim, \neg$	not
$\neq, <, \leqslant, >, \geqslant$	inequality	$\supset, \rightarrow, \Rightarrow$	implication
$\approx, \cong$	approximation	$\equiv, \leftrightarrow, \Leftrightarrow$	equivalence
$\equiv$	identity, congruence modulo n	$\forall, \exists$	quantifier
		$f(x), f: x \mapsto y,$	function (map,
$+$	plus sign	$f: X \rightarrow Y$	mapping)
$-$	minus sign	$f^{-1}(x)$	inverse function
$\pm$	plus/minus sign	$x \rightarrow 0, x \rightarrow a, x \rightarrow \infty$	limit
$\times, .$	multiplication sign	$f(x) \sim g(x)$	asymptotic
$\div, /$	division sign	$f', f'', \ldots \dot{y}, \ddot{y}, \ldots$	derivative
$|$	divisor	$\dfrac{dy}{dx}, \dfrac{d^2 y}{dx^2}, \cdots$	derivative
$\sqrt{a}, \sqrt[n]{a}$	radical		
$a^m, a^{m/n}$	exponent	$\dfrac{\partial y}{\partial x}$	partial derivative
$a:b$	ratio		
$y \propto x$	variation	$\int$	integral
$|a|$	absolute value	$\iint, \iiint, \cdots$	multiple integral
$[a, b], (a, b)$	interval	$\overrightarrow{AB}, \mathbf{AB}, \mathbf{i}, \mathbf{j}, \mathbf{k}$	vector
$n!$	factorial	$\mathbf{r}$	position vector
$\dbinom{n}{r}$	binomial coefficient	$|\mathbf{A}|$	absolute value
$\sum$	summation sign	$\mathbf{A} . \mathbf{B}$	scalar product
$\prod$	continued product	$\mathbf{A} \times \mathbf{B}, \mathbf{A} \wedge \mathbf{B}$	vector product
$[x]$	integer part	$\|\mathbf{v}\|$	norm
$(x_1, x_2, \ldots, x_n)$	n-tuple	$\begin{vmatrix} a & b \\ c & d \end{vmatrix}, \det(A), |A|$	determinant
$a \in A, \{a, b, c\},$ $\{x: \ldots\}$	set	$\begin{pmatrix} a & b \\ c & d \end{pmatrix}, A$	matrix
$\subseteq$	subset	A^{-1}	inverse (of a matrix)
$\subset$	proper subset	A^{T}	transpose
$\cup$	union	$\bar{A}$	complex conjugate
$\cap$	intersection	A^*	Hermitian conjugate
$\backslash$	difference	$\mathrm{Pr}(A), \mathrm{Pr}(B|A),$	probability
A'	complement	$\mathrm{E}(X)$	expectation
$\varnothing$	empty set	$\mathrm{Var}(X)$	variance
$\mathscr{E}$	universal set	$\mathrm{Cov}(X, Y)$	covariance
$\bar{A}, n(A), |A|$	cardinal number	$\mathrm{Re}\,z, \mathrm{Im}\,z$	complex number
$\mathrm{P}A, \mathscr{P}(A)$	power set	$|z|$	modulus
$A \times B$	Cartesian product	$\arg z$	argument
$\&, \wedge$	and	$z^*, \bar{z}$	complex conjugate
$\vee$	or		

Note: symbols such as $\mathbb{R}$ and $\mathbb{Z}$ are to be found in the alphabetic section of the Dictionary.

PENGUIN REFERENCE LIBRARY

THE PENGUIN DICTIONARY OF ECONOMICS

EDITED BY GRAHAM BANNOCK, R. E. BAXTER & EVAN DAVIS

'Another winner from the Penguin Reference shelf' John David Charles Hilton

An undoubted classic, *The Penguin Dictionary of Economics* has enlightened over half a million economics students over the last thirty years. From *Hotelling's Law* to *hyperinflation*, this now fully updated dictionary explains a host of economic terms in accessible yet detailed entries for both local and international markets. Wide-ranging and illuminating, this comprehensive practical guide is an absolute must for students of economics and professionals (in business, finance or the public sector), and for anyone wishing to follow economic discussions in the media today.

- Has sold more than 600,000 copies worldwide

- Provides entries on major individual economists, from *Joseph Stiglitz* to *John Maynard Keynes* to *Adam Smith*

- Discusses economic theory, including development economics, industrial organisation, finance and game theory, as well as international monetary and welfare economics

- Surveys applied economics, major financial institutions and the history of economics

ONLY PENGUIN GIVES YOU MORE

PENGUIN REFERENCE LIBRARY

THE PENGUIN DICTIONARY OF ACCOUNTING

EDITED BY CHRISTOPHER NOBES

You're an accountant needing *golden handcuffs*, you're faced with a jargon-laden business exam, you're the proud owner of an incomprehensible tax form. Where do you turn? *The Penguin Dictionary of Accounting* is your answer. Demystifying the most complex terms in simple, easy-to-understand language, this is the ideal reference book for anyone who reads, speaks or writes accountancy – whether student, businessperson, investor, preparer of financial statements, auditor or accounts manager. Compiled by Christopher Nobes, PricewaterhouseCoopers Professor of Accounting at Reading University, this fully up-to-date edition is your first stop in unravelling the ambiguities of accounting.

Contains information in the fields of *tax*, *auditing* and *management accounting*

Includes relevant terms relating to finance and law, from *depreciation* to the *ratchet effect* via *hyperinflation*

Analyses all the terms used in International Financial Reporting Standards

Caters for accountants worldwide, making clear where terminology differs across countries

ONLY PENGUIN GIVES YOU MORE

PENGUIN REFERENCE LIBRARY

THE PENGUIN DICTIONARY OF BIOLOGY
EDITED BY M. THAIN & M. HICKMAN

'A marvellous compendium: accurate, clear and complete'
Matt Ridley, author of *Genome*

Worried about your *maternal effect* or *biological clock*? Need to know a *rhizoid* from a *rhizome*? Think you're going to fail your *zoology* or *botany* exam? *The Penguin Dictionary of Biology* is your saviour, defining some 6000 terms relating to this rich, complex and constantly expanding subject – from *amino acids*, *bacteria* and the *cell cycle* to *X-ray diffraction*, *Y chromosome* and *zygotes*. Long established as the definitive single-volume source, this dictionary has sold over 200,000 copies and is extensively updated for its eleventh edition.

- Contains over 400 new entries to take account of the latest thinking on *genetics*, *human physiology*, *disease* and *cell biology*
- Superbly complemented by *The Penguin Dictionary of Human Biology*, due in Autumn 2007
- Ideal for students, teachers, professionals and amateur biologists
- Extensively illustrated throughout, with charts demonstrating *vertebrate spermatogenesis*, *primate evolution* and much more

ONLY PENGUIN GIVES YOU MORE

PENGUIN REFERENCE LIBRARY

THE PENGUIN DICTIONARY OF ARCHITECTURE & LANDSCAPE ARCHITECTURE

EDITED BY JOHN FLEMING, HUGH HONOUR & NIKOLAUS PEVSNER

'Immensely useful, succinct and judicious … rich in accurate fact and accumulated wisdom' *The Times Literary Supplement*

Need a handbook to your architect's business or studies? Struggling with your latest DIY project? Interested in the art of architecture? This classic work, now in its fifth edition, covers every aspect of architecture and landscape architecture. Ranging from ancient times to contemporary trends, it adopts a truly international perspective, focussing on countries and cultures such as *Coptic*, *Tibet* and *De Stijl*.

- Provides biographies of architects and landscape architects, ancient and modern, from *Le Corbusier* to *Frank Lloyd Wright* to *Sir Christopher Wren*

- Surveys architecture and landscape architecture by country

- Discusses the history of urban design, including parks and motorways

- Covers structural and technical innovations and new terminology

- Includes clear and detailed drawings

ONLY PENGUIN GIVES YOU MORE

PENGUIN POCKET REFERENCE

PORTABLE   **DESIRABLE**   **INDISPENSABLE**

Penguin Pocket Babies' Names

Penguin Pocket Book of Facts

Penguin Pocket Crossword Finisher

Penguin Pocket Dictionary of Quotations

Penguin Pocket English Dictionary

Penguin Pocket Famous People

Penguin Pocket French Dictionary

Penguin Pocket German Dictionary

Penguin Pocket Italian Dictionary

Penguin Pocket Jokes

Penguin Pocket Kings and Queens

Penguin Pocket On This Day

Penguin Pocket Rhyming Dictionary

Penguin Pocket Roget's Thesaurus

Penguin Pocket School Dictionary

Penguin Pocket Spanish Dictionary

Penguin Pocket Spelling Dictionary

Penguin Pocket Thesaurus

Penguin Pocket Writer's Handbook

Our Penguin Pockets are part of the extensive Penguin Reference
Library – a resource that draws on over 70 years of experience bringing
reliable, useful and clear information to millions of readers around the
world. We want to make knowledge everybody's property.